Introduction to

Physical Gas Dynamics

Introduction to
Physical Gas Dynamics

Walter G. Vincenti

Department of Aeronautics and Astronautics
Stanford University

Charles H. Kruger, Jr.

Department of Mechanical Engineering
Stanford University

ROBERT E. KRIEGER PUBLISHING COMPANY
HUNTINGTON, NEW YORK
1975

Original Edition 1965
Second Printing, August, 1967
Reprint 1975
with corrections

Printed and Published by
ROBERT E. KRIEGER PUBLISHING CO., INC.
645 NEW YORK AVENUE
HUNTINGTON, NEW YORK 11743

Printed in the United States of America

Library of Congress Cataloging in Publication Data

Vincenti, Walter Guido, 1917-
Introduction to physical gas dynamics.

Reprint of the 1967 ed. published by Wiley, New York.
Includes bibliographies.
1. Gas dynamics. I. Kruger, Charles H., joint author. II. Title.
QC168.V55 1975 533'.2 75-5806
ISBN 0-88275-309-6

To Joyce and Margaret

Preface

This book is the outgrowth of a series of courses developed over the past five years in the departments of Aeronautics and Astronautics and of Mechanical Engineering at Stanford University. These courses were introduced to instruct our students in the general features of high-temperature and nonequilibrium gas flows. Requests for the lecture notes that were prepared have encouraged us to believe that a book on these subjects might be useful for similar purposes in other schools of engineering. Since a knowledge of kinetic theory of gases, statistical mechanics, chemical thermodynamics, and chemical kinetics is basic to our subject, material on these disciplines is included. In the book, as in the courses, emphasis is placed on the ideas and structure of the subject and on certain basic results, rather than on technological problems of current concern. The aim is to bring the student to the point where he can understand more advanced treatises in the relevant sciences as well as the pertinent research literature in gas dynamics. In this way we attempt to do for high-temperature and nonequilibrium flows what existing books do for the classical field of the dynamics of perfect gases.

In developing the material we have attempted to maintain a balance, appropriate for engineering purposes, between microscopic physics and macroscopic gas dynamics. The term "physical gas dynamics" is intended to convey this idea. The treatment itself thus alternates between the discussion of physical and chemical processes and the illustration of how these processes influence the behavior of fluid flow. The material of the book falls, in a still different sense, into two major parts. Except for a brief discussion of transport phenomena in Chapter I, the first six chapters deal with questions of equilibrium. Here the physics and chemistry predominate, and gas dynamics appears only in Chapter VI. The second six chapters are concerned with nonequilibrium phenomena—molecular vibration and chemical reactions (which are essentially similar), molecular transport, and radiative transport. These are taken up separately and in order, with a chapter on the process itself followed by a chapter on the related gas dynamics. We do not mean to suggest that the various phenomena arise so neatly separated in practical problems. Our concern is rather to illustrate the essential effects of each process with a minimum of complexity.

The assumed background for the book is that common to senior or first-year-graduate students in most engineering schools in the United States. For deliberate reasons, there is a considerable difference in the level at which the physics and chemistry and the gas dynamics begin. In the latter we assume that the student has had or is taking concurrently a course in classical gas dynamics based on material such as that found in the books by Liepmann and Roshko (*Elements of Gas Dynamics*, Wiley, 1957) or Shapiro (*The Dynamics and Thermodynamics of Compressible Fluid Flow*, Ronald, 1953). In the molecular physics and physical chemistry, on the other hand, we suppose only that the student has had the usual first undergraduate courses in physics and chemistry for engineers. At Stanford, and perhaps at other schools, these circumstances are in fact the case. This situation is now changing in engineering schools generally and eventually will cease to be true. In the meantime, the elementary material in the early chapters is available for first study, reference, or review. With regard to thermodynamics, we suppose that the student has had a general course in classical (i.e., equilibrium) thermodynamics as taught to most engineering students but has no acquaintance with the special topic of chemical thermodynamics. As to mathematics, Chapters I through VII presuppose a knowledge of elementary calculus plus a few topics from ordinary differential equations and advanced calculus. The later chapters assume also a familiarity with elementary partial differential equations to the extent required in gas dynamics at the level treated by Liepmann and Roshko or Shapiro. In general, the degree of sophistication of the book—and we trust that of the student—increases as the book progresses.

At Stanford, which operates on the quarter system, the material has been used in five ten-week courses, each meeting for three hours per week. A primary series of three courses extends over one academic year at the first-year graduate level and covers Chapters I through X. Only the first of these courses, covering the basic material of Chapters I through IV, is required as prerequisite for the third, which deals with the kinetic theory of Chapters IX and X. Most students, however, follow the complete series. A fourth graduate course, on radiative gas dynamics, is based on Chapters XI and XII. Again, only the first course in the primary series is a prerequisite. Chapters I, II, and IV also form the basis for a senior-level undergraduate course in kinetic theory and statistical mechanics.

A reader wanting to do independent, specialized study may be assisted by the groupings of topics and background material listed below, which can be pursued separately from a study of the remainder of the book.

Here Roman numerals indicate chapters, and Arabic numerals indicate sections. Items in parentheses are not essential for an understanding of the later material.

Equilibrium gas properties: III, IV, V
Equilibrium flow: III, IV, V, VI
Kinetic theory: I, II, (IV-9), IX, X
Transport properties: I, II, IX, X-1 to 8
Chemically reacting flow: III, IV, V-2 and 3, VII, VIII
Flow with vibrational nonequilibrium: IV-1 to 12, VII-1 to 3, 10, and 11, VIII
Radiative gas dynamics: (II-2 and 3), (IV-1 to 5), (IX-2), XI, XII

References throughout the book are listed alphabetically according to author at the end of each chapter. Citing of references in the text is by the author's name and the year of publication. The references have been chosen for the most part to provide fundamental derivations, alternative treatments, more advanced reading, or practical numerical data. Where we have relied heavily on some particular book, we have tried to acknowledge this fact. We have made no attempt in general to always cite original papers or to assess historical priority. This seemed hardly practical in a textbook covering such a diversity of subject matter. We have avoided using industrial or university reports, since such reports are often not available to the student even with considerable effort. A few exceptions have been made where important material was not available otherwise. It may be that our policies of referencing have led us to slight some of our colleagues. If so, we hope they will accept our apologies.

An index of mathematical symbols, akin to the usual subject index, is provided at the rear of the book.

WALTER G. VINCENTI
CHARLES H. KRUGER, JR.

Stanford, California
July, 1965

Acknowledgments

We are especially grateful to the John Simon Guggenheim Memorial Foundation, which helped to support one of us (W. G. V.) during a sabbatical year devoted to writing. Without the understanding aid of this exemplary organization the book could not have been completed in a reasonable time. The same can be said of the cooperation of certain of our associates at Stanford, in particular, Dean Joseph M. Pettit of the School of Engineering and our department heads, Nicholas J. Hoff and William M. Kays.

Many people, faculty and students, have contributed to the book over the past five years. We do not mention all of them only because they are so numerous. Special thanks are due, however, to the following, who have read and criticized various chapters at length:

Barrett S. Baldwin, Jr., Ames Research Center, NASA
George Emanuel, Aerospace Corporation
Robert H. Eustis, Stanford University
Robert A. Gross, Columbia University
Morton Mitchner, Stanford University
Frederick S. Sherman, University of California, Berkeley
Milton D. Van Dyke, Stanford University
Pierre Van Rysselberghe, Stanford University

Our appreciation goes as well to Krishnamurty Karamcheti, our colleague at Stanford, who participated in the inception and teaching of the courses from which the book evolved. Recognition is also properly given to several of our students: Harris McKee, who checked several chapters in detail, and Glenn Hohnstreiter and Frederick Morse, who aided in the preparation of the figures. The bulk of the typing was capably handled by Mrs. Katherine Bradley and Miss Christine Najera. Valuable assistance on the proofs and index was provided by Miss Margaret Vincenti and Marc Vincenti.

Certain of the material, particularly in Chapters VII, VIII, and XII, draws on our own research at Stanford, which was supported by the National Science Foundation and the Air Force Office of Scientific Research.

Finally, we want to acknowledge our great debt to our own teachers, whose high standards of excellence we hope we have in some measure maintained: Stephen P. Timoshenko and Elliott G. Reid (for W. G. V.) and Ascher H. Shapiro (for C. H. K.).

W. G. V.
C. H. K.

Contents

Chapter I

Introductory Kinetic Theory

1 INTRODUCTION

We begin our study of physical gas dynamics with a discussion of some of the pertinent problems from the kinetic theory of gases. In kinetic theory a gas is considered as made up at the microscopic level of very small, individual molecules in a state of constant motion. If there is no movement of the gas at the macroscopic level, this motion is regarded as purely random and is accompanied by continual collisions of the molecules with each other and with any surfaces that may be present. The movement of a given molecule is thus to be thought of as a kind of random banging about, with its velocity undergoing frequent and more or less discontinuous changes in both magnitude and direction. Indeed, it is this freedom of motion of the molecules, limited only by collisions, that differentiates a gas from the more ordered situation that exists in a liquid or a solid.

When there is a general macroscopic movement of the gas, as is the usual situation in gas dynamics, the motion of the molecules is not completely random. This statement is, in fact, merely two ways of saying the same thing—the general motion is only a macroscopic reflection of the nonrandomness of the molecular motion. In particular, the familiar flow velocity of continuum gas dynamics is, from the molecular point of view, merely the average velocity of the molecules taken over a volume large enough to contain many molecules but small relative to the dimensions of the flow field. Zero flow velocity, that is, an average molecular velocity of zero, corresponds to random absolute motion of the molecules (i.e., molecules of a given speed have no preferred direction). In a flowing gas the molecular motion, although not random from an absolute point of view, will appear almost so to an observer moving at the local flow velocity. These ideas will be made precise in Chapter IX when we consider the flow of a nonuniform gas in detail. For the time being we are interested primarily in the properties of the gas as such, and for this a consideration of the random motion is for the most part sufficient.

To make the foregoing ideas quantitative, we must introduce a precise assumption regarding the molecular model, that is, regarding the nature of the molecules and of the forces acting between them. Given such a model and the usual laws of mechanics, it would then be possible, in principle, to trace out the path of each molecule with time, assuming that the initial conditions are known (which, of course, they are not). Such a calculation, though of interest theoretically, would be enormously difficult. Fortunately it is also of little necessity. What we are really concerned with in most practical applications is the gross, bulk behavior of the gas as represented by certain observable quantities such as pressure, temperature, and viscosity, which are manifestations of the molecular motions averaged in space or time. It is the primary task of kinetic theory to relate and "explain" these macroscopic properties in terms of the microscopic characteristics of the molecular model.

The approach to this problem in the present chapter will be deliberately rough and nonrigorous—refinement will come later in Chapter II. The aim here is a qualitative understanding of certain ideas and some feeling for the numerical magnitudes that are involved. The results, however, are essentially correct.

The reader who wants to study kinetic theory more extensively than we shall be able to here will probably enjoy the book by Jeans (1940). His mechanical description (pp. 11–15) of the theory in terms of balls on a billiard table is particularly helpful in fixing the basic ideas. Among the other useful books in the field are those of Kennard (1938), Present (1958), and Loeb (1961). Other references will be given later in Chapter IX.

2 MOLECULAR MODEL

Further word on the molecular model is necessary. Aside from the fundamental attribute of mass, a molecule possesses an external force field and an internal structure. The external field is, as a matter of fact, a consequence of the internal structure, that is, of the fact that the molecule consists of one or more atomic nuclei surrounded by orbiting electrons. The somewhat arbitrary concentration of our attention separately on the "outside" and "inside" of the molecules, however, is a useful procedure. (For a discussion of the origin of interatomic and intermolecular forces at a beginning level, see Slater, 1939, Chapter XXII.)

The external force field is generally assumed to be spherically symmetric. This is very nearly true in many cases and to assume otherwise is almost prohibitively difficult. On this basis, the force field can be represented as in Fig. 1, which shows the force F between two molecules as a function of the distance r between them. At large distances the true representation (solid curve) shows a weak attractive force that tends to zero as the distance increases. At short distances there is a strong repulsive force that increases rapidly as the orbiting electrons of the two molecules intermingle. At

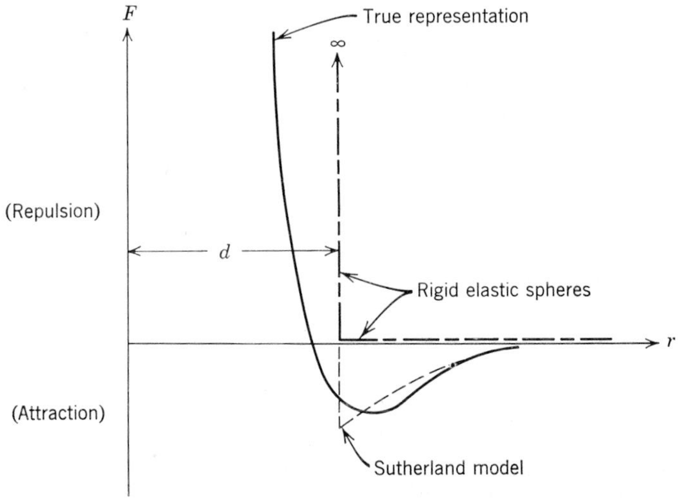

Fig. 1. Intermolecular force F as a function of the distance r between two molecules.

some intermediate distance, F passes through zero. The molecules, if they had no kinetic energy, would remain in equilibrium indefinitely at this distance.

Even with the simplest possible equation for the curve just described, the analysis is complicated. For this reason further approximation is often made. The simplest (Fig. 1) is to regard the molecules as rigid elastic spheres with zero attractive force when apart and an infinite repulsive force at the instant of contact. For like molecules, contact will take place when the distance between the center of the spheres is d, where d is the assumed diameter of the spherical molecule. A considerable amount of surprisingly accurate information can be obtained on the basis of this crude model, provided we chose d to give agreement between the resulting theory and the experimental data for some basic quantity such as viscous stress. Some improvement can be obtained at the expense of further complication by going to the so-called "Sutherland model." This model supplements the rigid sphere by adding a weak attractive force when the spheres are not in contact. Other models are also possible. One that we shall find useful in Chapter IX dispenses entirely with the idea of solid spheres and assumes a pure repulsive force varying with some inverse power of the intermolecular distance.

The internal structure of the molecules is important primarily for its effect on the energy content of the gas. Being composed of nuclei and electrons that have motion relative to the center of mass of the molecule, the molecules can possess forms of energy (rotation, vibration, etc.) above and beyond that associated with their translational motion. In addition, the internal structure may be expected to affect the collisional interaction of the molecules at short range. Inclusion of the effects of internal structure considerably complicates the kinetic theory.

In the present book the treatment of intermolecular forces in any but the simplest way will be reserved for the discussion of nonequilibrium kinetic theory in Chapters IX and X. The effects of internal structure, at least insofar as the internal energy is concerned, will be taken up in detail in the discussion of equilibrium statistical mechanics in Chapter IV. For the present the molecular model that will be assumed, except for certain digressions, is that of the structureless, perfectly elastic sphere with no attractive forces and with repulsive forces existing only during contact. The characteristics of this "billiard-ball" model are completely specified by giving the mass and diameter of the various types of molecules, the number of each type of molecule per unit volume, and some measure of the speed of the random motion. Our task is to relate the macroscopic properties of the gas to these assumedly given microscopic quantities.

3 PRESSURE, TEMPERATURE, AND INTERNAL ENERGY

Consider a gas mixture in a state of equilibrium inside a cubical box. Suppose that the box is at rest so that the molecular motion is purely random. We wish to study the pressure exerted on the walls of the box by this random motion.

For simplicity we assume that the molecules do not collide with each other but only with the walls. We also assume that upon collision with a wall a molecule is reflected specularly (i.e., angle of reflection equals angle of incidence) and with its speed unchanged. It must be emphasized that neither of these assumptions is true; the fate of a given molecule upon collision with the wall is not known, and molecules do in reality collide with each other with great frequency. Intermolecular collisions are, in fact, the chief mechanism for bringing the gas to its final state of equilibrium. Nevertheless, *once the equilibrium state has been established*, the gas behaves, at least insofar as the pressure is concerned, *as if* the assumptions were correct. This is because the condition of equilibrium requires that in any region the number of molecules of a given species and having velocity closely equal, in both magnitude and direction, to any given value must not vary with time. Thus, in a small region in the vicinity of a given point, for every molecule deflected away from its original velocity by a collision, there will at the same time be another of the same species deflected into this velocity. This molecule can be regarded as carrying on the motion of the original molecule just as if no collision had occurred. Somewhat similar ideas can be applied at the wall. Here the essential fact is that within a gas in equilibrium there can be no preferred direction. That is to say, in any region of the gas the number of molecules of a given species and having velocity closely equal in magnitude to a given value must be the same irrespective of the direction of the velocity. If this requirement is applied in a small region immediately adjacent to the wall, it implies that for every molecule arriving at a small element of the wall, another of the same species will at the same time leave that element with a specularly oriented velocity of equal magnitude. Thus the gas behaves as if a given molecule were reflected specularly and without change of speed.

Under the assumption of no intermolecular collisions and specular, elastic reflection at the walls, the motion of a given molecule is as shown in two-dimensional projection in Fig. 2. The molecule in this purely fictitious situation moves at a constant speed C irrespective of its direction. When it collides with the wall, its velocity component perpendicular to the wall is reversed whereas that parallel to the wall is unaltered. If coordinate axes x_1, x_2, x_3 are taken along three edges of the box, the magnitudes of the corresponding velocity components thus have fixed values $|C_1|$, $|C_2|$, $|C_3|$, where $C_1^2 + C_2^2 + C_3^2 = C^2$.

The force exerted on a given wall of the box by the molecule depends on the number of collisions with the wall per unit time and on the change in momentum per collision. Let us consider the force on the wall perpendicular to Ox_1 at the right-hand side of the box. If l is the length of an edge of the box, the time required to traverse the distance between the two walls perpendicular to Ox_1, irrespective of collisions with other walls, is $l/|C_1|$. Since there are two such traverses between each collision with the right-hand wall, the number of collisions per unit time with this wall is therefore $|C_1|/2l$. At each such collision, the molecule, having mass m, suffers a change in x_1-momentum of amount $2m\,|C_1|$, that is, from mC_1 to $-mC_1$. The magnitude of the total change in x_1-momentum per unit time, which is equal to the total force exerted on the wall by the molecule, is thus

$$2m\,|C_1| \times |C_1|/2l = \frac{mC_1^2}{l}.$$

Since the area of the wall is l^2, the corresponding pressure on the wall is

$$\frac{mC_1^2}{l^3} = \frac{mC_1^2}{V},$$

where $V = l^3$ is the volume of the box.

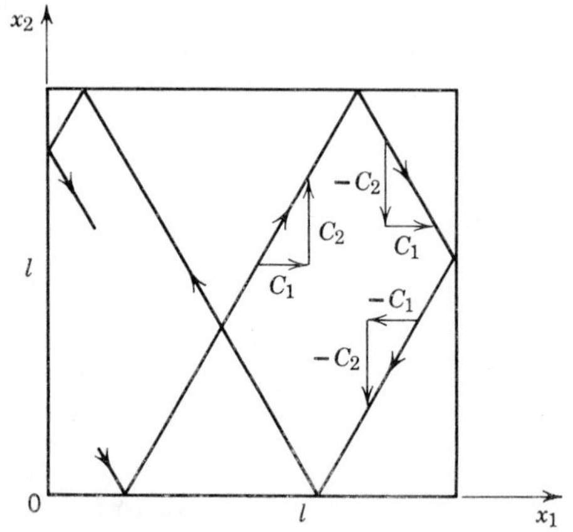

Fig. 2. Projection on the x_1, x_2-plane of the path of a molecule in a cubical box in the ficticious situation of no intermolecular collisions and specular, elastic reflection at the walls.

Now suppose that the gas mixture consists of a large number of molecules of different mass $m_a, m_b, m_c, \ldots$ moving with speeds $C_a, C_b, C_c, \ldots$. The total pressure exerted on the wall perpendicular to Ox_1 by all the molecules is then

$$\frac{1}{V}(m_a C_{a_1}^2 + m_b C_{b_1}^2 + m_c C_{c_1}^2 + \cdots) = \frac{1}{V}\sum_z m_z C_{z_1}^2.$$

But under the assumed conditions of equilibrium the pressure on this wall is the same as on the other walls of the box. Denoting this common pressure by p, we can therefore equate the foregoing pressure to p and write

$$p = \frac{1}{V}\sum_z m_z C_{z_1}^2.$$

Corresponding equations hold for the pressure calculated on the faces perpendicular to Ox_2 and Ox_3. Taking the sum of these equations for the three coordinate directions and dividing by three, we also obtain

$$p = \frac{1}{3V}\sum_z m_z(C_{z_1}^2 + C_{z_2}^2 + C_{z_3}^2) = \frac{1}{3V}\sum_z m_z C_z^2. \tag{3.1}$$

Since the total energy of translation E_{tr} of the molecules is $\frac{1}{2}\sum_z m_z C_z^2$, this equation can be written

$$\boxed{pV = \tfrac{2}{3}E_{tr}}\ . \tag{3.2}$$

The same final results will be rederived in Chapter II on the basis of a more realistic analysis.

The foregoing kinetic-theory equation of state can be compared with the empirically derived equation of state for a thermally perfect gas from classical thermodynamics. This is, in one of its forms,

$$pV = \mathcal{N}\hat{R}T,$$

where $\hat{R}$ is the universal gas constant (i.e., gas constant per mole),[1] $\mathcal{N}$ is the number of moles of gas in the volume V, and T is the gas temperature. We see that the theoretical formula and the empirical equation are identical if we take

$$\boxed{E_{tr} = \tfrac{3}{2}\mathcal{N}\hat{R}T}\ . \tag{3.3a}$$

This equation relates the energy of translation of the molecules in kinetic theory to the absolute temperature as defined in classical thermodynamics. The temperature may thus be interpreted as a measure of the molecular energy. Equation (3.3a) is sometimes taken, from a somewhat different point of view (cf. Chapter IX, Sec. 2), as giving a kinetic *definition* of temperature.

The result (3.3a) can be reexpressed in terms of the average kinetic energy per molecule by dividing both sides by the total number N of molecules. This gives

$$\tilde{e}_{tr} \equiv \frac{E_{tr}}{N} = \frac{3}{2} \times \frac{\mathcal{N}}{N}\hat{R}T = \frac{3}{2} \times \frac{\hat{R}}{\hat{N}}T,$$

where $\hat{N} = N/\mathcal{N}$ is the number of molecules per mole, which is *Avogadro's number*, another universal constant. The quotient $\hat{R}/\hat{N}$, which is in the nature of a "gas constant per molecule," is thus also a universal constant, usually denoted by k and called the *Boltzmann constant*. We thus write

[1] The value of this and other fundamental physical constants is given in the appendix at the end of the book.

for the average kinetic energy per molecule

$$\tilde{e}_{\mathrm{tr}} = \tfrac{3}{2}kT. \tag{3.3b}$$

We can also obtain the specific kinetic energy e_{tr} per unit mass by dividing both sides of equation (3.3a) by the total mass $M = \sum_z m_z$. Noting that the combination $\mathcal{N}\hat{R}/M$ is the ordinary gas constant R (i.e., gas constant per unit mass), we thus obtain

$$e_{\mathrm{tr}} = \tfrac{3}{2}RT. \tag{3.3c}$$

Under our assumption that the molecules have no internal structure, the energy of translation constitutes the entire molecular energy of the gas. It can therefore be taken as equal to the internal energy e of thermodynamics. We thus find for the specific heat at constant volume in the present case

$$c_v = \left(\frac{\partial e}{\partial T}\right)_v = \frac{de_{\mathrm{tr}}}{dT} = \frac{3}{2}R.$$

The specific heat at constant pressure, as given by the thermodynamic relation $c_p - c_v = R$, valid for a thermally perfect gas, is

$$c_p = \tfrac{5}{2}R,$$

and the ratio of specific heats $\gamma \equiv c_p/c_v$ is therefore

$$\gamma = \tfrac{5}{3}.$$

Since our assumed gas model has constant specific heats, it is calorically as well as thermally perfect.[2] The calculated value of c_v is in good agreement with experimental data for monatomic gases at ordinary temperatures. This indicates that our model is a reasonable one for such gases.

The foregoing theory can also be used to obtain an estimate of molecular speeds. For this purpose we return to equation (3.1) and divide both sides by $M = \sum_z m_z$. Introducing the mass density $\rho = M/V$ and a mean-square molecular speed defined by

$$\overline{C^2} \equiv \frac{\sum_z m_z C_z^2}{\sum_z m_z}, \tag{3.4}$$

[2] When a gas is both thermally and calorically perfect, we shall ordinarily describe it by the unmodified term "perfect gas." When we wish to distinguish between thermal perfection (a gas obeying the equation $pV = \mathcal{N}\hat{R}T$) and caloric perfection (a gas with constant specific heats), we shall try to remember to add the adjectives. The reader will recall that a gas can be thermally perfect and calorically imperfect, but not vice versa.

we thus obtain

$$\boxed{\frac{p}{\rho} = \frac{1}{3}\overline{C^2}}. \tag{3.5}$$

This relation allows an estimate of molecular speeds directly from macroscopically measurable quantities. For example, for air[3] at standard conditions (1 atm pressure and 0°C) we have $p = 1.013 \times 10^6$ dyne/cm² and $\rho = 1.288 \times 10^{-3}$ gm/cm³, so that the root-mean-square molecular speed is

$$\sqrt{\overline{C^2}} = \left(3\frac{p}{\rho}\right)^{1/2} = 4.86 \times 10^4 \text{ cm/sec} = 486 \text{ m/sec}.$$

This is of the same order as the corresponding speed of sound (332 m/sec), which is as would be expected since the molecular motion is the real mechanism underlying the propagation of sound. From equation (3.5) and the equation of state in the form $p/\rho = RT$, it follows that the root-mean-square molecular speed, like the speed of sound, is proportional to $T^{1/2}$.

Turning our attention to another matter, let us suppose that the gas is composed of numbers of molecules of distinct species A, B, C, etc. Returning again to equation (3.1) we can then write

$$p = \frac{1}{3V}\sum_{\substack{\text{species}\\ \text{A}}} m_A C_A^2 + \frac{1}{3V}\sum_{\substack{\text{species}\\ \text{B}}} m_B C_B^2 + \cdots.$$

Here the summation for a given species Y is understood to be taken over all molecules of that species; the corresponding mass m_Y is thus a constant but C_Y varies from molecule to molecule. Now, each term of the foregoing equation has itself the same form as equation (3.1), that is, it can be thought of as constituting a partial pressure p_Y calculated as if the molecules of that species filled the container by themselves. We can thus write for a mixture of gases

$$\boxed{p = p_A + p_B + \cdots = \sum_Y p_Y}, \tag{3.6}$$

which is recognized as *Dalton's law of partial pressures* for a mixture of

[3] This use of the theory might be questioned on the grounds that air is composed of diatomic molecules that have an effective internal structure. As will be shown in Chapter IV, however, the internal structure does not alter the relationship between pressure and *translational* energy, so that this application of (3.5) is permissible.

perfect gases. The partial pressure is given analogously to equation (3.2) by

$$p_{\mathrm{Y}}V = \tfrac{2}{3}E_{\mathrm{Y}_{\mathrm{tr}}}, \tag{3.7}$$

where $E_{\mathrm{Y}_{\mathrm{tr}}} = \frac{1}{2}\sum_{\mathrm{Y}} m_{\mathrm{Y}}C_{\mathrm{Y}}^2$ is the total molecular energy of translation of species Y. The corresponding thermodynamic equation here is

$$p_{\mathrm{Y}}V = \mathscr{N}_{\mathrm{Y}}\hat{R}T,$$

where $\mathscr{N}_{\mathrm{Y}}$ is the number of moles of species Y in the mixture, and the common temperature T is used since at equilibrium all gases in the mixture must have the same temperature. Comparing the two equations we obtain, as in equation (3.3a),

$$E_{\mathrm{Y}_{\mathrm{tr}}} = \tfrac{3}{2}\mathscr{N}_{\mathrm{Y}}\hat{R}T. \tag{3.8a}$$

If we now divide by the total number N_{Y} of molecules of species Y and note that $\hat{N} = N_{\mathrm{Y}}/\mathscr{N}_{\mathrm{Y}}$ is again Avogadro's number, we find for the average translational energy per molecule of species Y

$$\tilde{e}_{\mathrm{Y}_{\mathrm{tr}}} = \tfrac{3}{2}kT. \tag{3.8b}$$

The right-hand side of this equation is identical to that of equation (3.3b) for the entire mixture. Equations (3.8b) and (3.3b) thus show that

$$\tilde{e}_{\mathrm{A}_{\mathrm{tr}}} = \tilde{e}_{\mathrm{B}_{\mathrm{tr}}} = \cdots = \tilde{e}_{\mathrm{tr}}. \tag{3.9}$$

That is to say, *when a number of gases are mixed at the same temperature, the average kinetic energy of their molecules is the same.* It follows that the heavy molecules must move, on the average, slower than the light molecules.

One remark needs to be made. The foregoing development has used our prior thermodynamic knowledge of the equation of state of a thermally perfect gas and of the existence of the universal gas constant and Avogadro's number. The appeal to thermodynamics in this restricted way, however, is not essential and has been used here only for the sake of brevity. An alternative—and perhaps logically preferable—approach is given in Jeans' book (pp. 21–30). Proceeding from considerations of the mechanics of molecular collisions for elastic spheres, one can, with a little work, establish the result of equation (3.9) from mechanical considerations alone and with no reference to thermodynamics. The identification of the average kinetic energy with the thermodynamic temperature [equation (3.8b)] is then made from general thermodynamic considerations of internal energy and entropy without recourse to the equation of state. The equation of state for a thermally perfect gas then follows directly from the basic kinetic equation (3.2). From this point of view, the thermodynamic form of the equation of state together with Avogadro's law (which is equivalent to the existence of the universal gas constant) emerge as logical deductions from the kinetic approach rather than as independently introduced empirical findings.

Equations (3.3b) and (3.8b) are a special case of the general *principle of equipartition of energy*, which we shall encounter again later. This principle states that *for any part of the molecular energy that can be expressed as the sum of square terms, each such term contributes an average energy of* $\frac{1}{2}kT$ *per molecule*. By a "square term" we mean a term that is quadratic in some appropriate variable used to describe the energy. In the kinetic energy of translation, for example, the energy of any molecule can be written in the form $\frac{1}{2}m(C_1^2 + C_2^2 + C_3^3)$, which is the sum of three square terms. The average translational energy per molecule is therefore $\frac{3}{2}kT$ in agreement with (3.3b) and (3.8b). For molecules with internal structure, the total number of square terms may be greater. If the total number of such terms is ξ, the average molecular energy, now expressed per unit mass rather than per molecule, is given by

$$e = \xi(\tfrac{1}{2}RT) = \frac{\xi}{2} RT. \tag{3.10}$$

The corresponding specific heats and their ratio γ are then

$$c_v = \frac{\xi}{2} R, \qquad c_p = \frac{\xi + 2}{2} R, \qquad \gamma = \frac{\xi + 2}{\xi}. \tag{3.11}$$

A given type of molecular energy (rotational, vibrational, etc.) cannot always validly be expressed as a sum of square terms. For this to be possible, the type of energy in question must be "fully excited." The meaning of this term will become apparent when we study the effects of internal structure from the quantum-mechanical point of view (see Chapter IV, Sec. 12). For a diatomic gas such as air there are at ordinary temperatures five square terms, three for translation and two for rotation about the two major axes of the molecule. We thus have in this case $\xi = 5$ and hence $\gamma = \frac{7}{5}$, which is a well-known approximate value for diatomic gases.

The treatment leading up to the basic equation (3.1) is unrealistic in that it assumes specular reflection and a special shape of container and ignores the intermolecular collisions that are a characteristic feature of kinetic theory. It also does not demonstrate that the pressure is the same for all regions on the surface of the container. These objections will be overcome in Chapter II. The results, however, will not be altered.

4 MEAN FREE PATH

We now wish to say something about the collisions between molecules, which were ignored in the previous discussion. A concept of fundamental importance here is that of the *mean free path*, which can be defined as the average distance that a molecule travels between successive collisions. We wish in particular to obtain an expression for the mean free path in terms of the quantities that define the gas model.

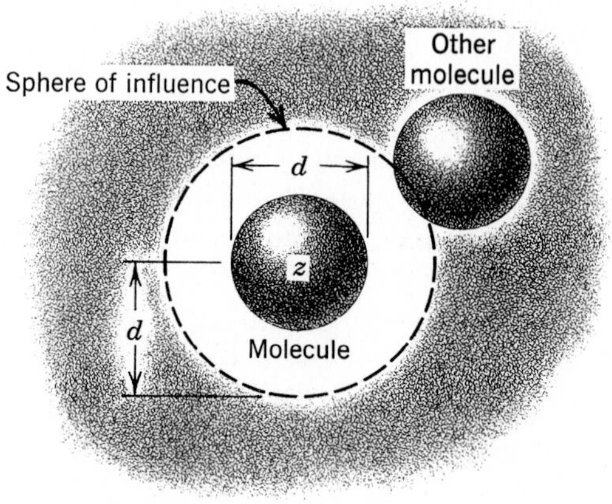

Fig. 3. Sphere of influence of molecule among like molecules.

Let us consider the situation when all of the molecules in an assumed billiard-ball model are of one species and thus have the same diameter d. A given molecule z will then suffer a collision whenever the center of z approaches the distance d from the center of any other molecule. We may thus imagine z as carrying with it a *sphere of influence* of radius d (Fig. 3); a collision will occur whenever the center of another molecule lies on the surface of this sphere of influence.

To make things easy for ourselves, we consider here an oversimplified situation in which z is moving at a uniform speed equal to the mean molecular speed $\bar{C}$ [not to be confused with the slightly different root-mean-square speed $(\overline{C^2})^{1/2}$] and all the other molecules are standing still. The zig-zag path of the molecule and its sphere of influence is then as illustrated in Fig. 4(a), where the molecules with which it collides are indicated by their centers. Ignoring the violation of mechanical principles, imagine that this path is straightened out as shown in Fig. 4(b), with the centers of the target molecules retaining their position relative to the path of the moving molecule prior to the collision. These centers will then all lie within the straight cylindrical volume swept out by the sphere of influence. Since the molecule is traveling at the uniform speed $\bar{C}$, the volume swept out per unit time is obviously $\pi d^2\bar{C}$. If n denotes the number of molecules per unit volume of gas, the number of centers lying within

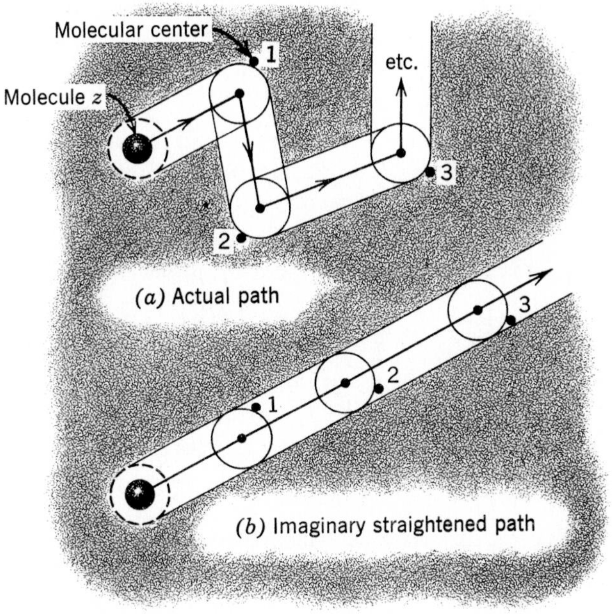

Fig. 4. Path of molecule z among stationary molecules.

the cylindrical volume is $\pi d^2\bar{C}n$. Since each of these centers corresponds to a collision, this product must also represent the number Θ of collisions per unit time for molecule z, that is,

$$\Theta = \pi d^2\bar{C}n. \tag{4.1}$$

But the distance traveled by molecule z per unit time is $\bar{C}$. It follows that the average distance traveled per collision, which is the mean free path λ, is

$$\lambda = \frac{\bar{C}}{\Theta} = \frac{1}{\pi d^2 n}. \tag{4.2}$$

In view of the approximations involved, this gives accurately the mean free path for a particle moving with a speed infinitely greater than that of the other particles. It follows from equation (3.9) that this is approximately the case for a free electron (very small mass) moving in a gas of normal molecular mass.[4] For improved accuracy in purely molecular collisions, the derivation should be modified to account for the motion of the target molecules, and this will be done in Chapter II, Sec. 6. The modified expression for λ turns out to be

$$\boxed{\lambda = \frac{1}{\sqrt{2}\,\pi d^2 n}}\,. \tag{4.3a}$$

The factor $\sqrt{2}$ in the denominator arises essentially from the fact that the correct speed to use for the evaluation of Θ in equation (4.1) is really the mean *relative* speed of the molecules, and this can be shown to be $\sqrt{2}\bar{C}$. Equation (4.3a) can also be written in terms of the mass density $\rho = mn$, where m is the mass of the assumedly identical molecules. This gives

$$\lambda = \frac{m}{\sqrt{2}\,\pi d^2 \rho}\,. \tag{4.3b}$$

Thus for a given value of ρ, the mean free path for the rigid-sphere model is independent of T and depends only on the mass and diameter of the molecules.

Exercise 4.1. The *Knudsen number*, which plays an important role in low-density flow problems, is defined as the dimensionless ratio λ/L, where λ is the mean free path and L is some characteristic length of the boundaries. Flow for which $\lambda/L \geqq 1$ is sometimes called free-molecule flow. Consider a sphere 1 foot in diameter traveling through the atmosphere, and take the diameter of the sphere as the characteristic length. Using the results of this section, find the altitude above which free-molecule flow prevails, assuming that the density of the atmosphere is given to a sufficient approximation by

$$\frac{\rho}{\rho_0} = e^{-\alpha H},$$

where H is the altitude, $\alpha = 4.25 \times 10^{-5}$/ft, and ρ_0 is the sea-level density 1.23×10^{-3} gm/cm^3. The molecular quantities required in the calculation can be taken from Sec. 6.

[This problem is intended only as an illustration, and the results should not be taken literally. To obtain correct results it would be necessary to take account of the variation of the effective molecular diameter with temperature (see final paragraphs of Sec. 5) as well as to use a more accurate density-altitude relation.]

[4] In this case, however, d must be interpreted as the arithmetic mean of the diameters of the electron and molecule, which is very nearly equal to the radius of the molecule.

Exercise 4.2. Consider an equilibrium mixture of two species of molecule A and B with different diameters d_A and d_B. Using the simplest possible methods, find an expression for the collision frequency Θ_A of one A-molecule with all other molecules and for the mean free path λ_A of A-molecules.

5 TRANSPORT PHENOMENA

The phenomena that we have treated so far have had to do with a gas in thermodynamic equilibrium, that is, in which all macroscopic properties are uniform in space and time. When the gas is out of equilibrium by virtue of a nonuniform spatial distribution of some macroscopic quantity (flow velocity, temperature, composition, etc.), additional phenomena arise as a result of the microscopic molecular motion. Very briefly, the molecules in their random thermal movement from one region of the gas to another tend to transport with them the macroscopic properties of the region from which they come. If these macroscopic properties are nonuniform, the molecules thus find themselves out of equilibrium with the properties of the region in which they arrive. The result of these molecular transport processes is the appearance at the macroscopic level of the well-known nonequilibrium phenomena of viscosity, heat conduction, and diffusion. (For a more detailed description of the mechanism of transport processes, see Present, 1958, pp. 38–39.) The situation can be summarized as follows:

Macroscopic Cause	Molecular Transport	Macroscopic Result
Nonuniform flow velocity	Momentum	Viscosity
Nonuniform temperature	Energy	Heat conduction
Nonuniform composition	Mass	Diffusion

For the simplest possible case of nonuniformity in one direction only, macroscopic (i.e., continuum) theory assumes the following relations for the foregoing phenomena, where the symbols are as defined below:

$$\tau = \mu \frac{du_1}{dx_2}, \tag{5.1}$$

$$q = -K \frac{dT}{dx_2}, \tag{5.2}$$

$$\Gamma_A = -D_{AB} \frac{dn_A}{dx_2}. \tag{5.3}$$

Here the nonuniformities are taken to be in the x_2-direction. In equation (5.1), u_1 is the component of flow velocity in the x_1-direction (the other components being zero), μ is the coefficient of viscosity, and τ is the shearing stress. In equation (5.2), T is the temperature, K is the coefficient of thermal conductivity, and q is the heat flow per unit area per unit time, where it is understood that the area is taken normal to the x_2-axis. In equation (5.3), n_A is the number density of molecules of species A in an assumed mixture of species A and B, D_{AB} is the coefficient of diffusion in such a mixture, and Γ_A is the flow of A-molecules per unit area per unit time. We shall now use the methods of kinetic theory in a crude way to

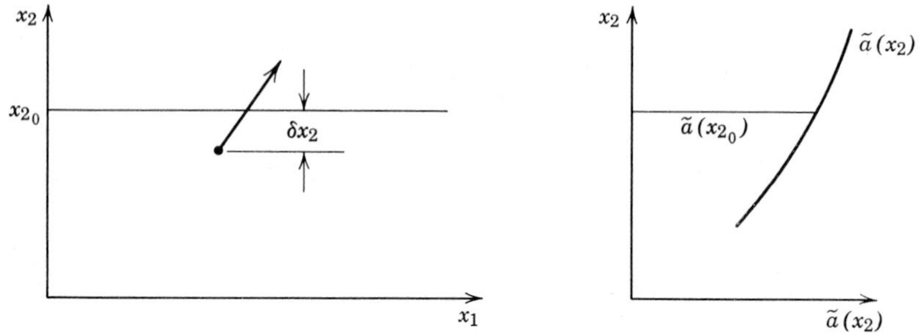

Fig. 5. Assumed situation for calculation of transport properties.

deduce the same equations and to obtain expressions for the transport coefficients. In this we follow essentially the unified treatment given by Present (1958, Chapter 3).

Let $\tilde{a}(x_2)$ denote some mean molecular quantity (measured per molecule), which varies in the x_2-direction only (Fig. 5). When a molecule crosses a plane $x_2 = x_{2_0}$, it transports a value of $\tilde{a}$ characteristic mainly of the place at which it made its last collision but dependent to some extent on the location of previous collisions. Let $\tilde{a}(x_{2_0} - \delta x_2)$ represent the mean value of $\tilde{a}$ transported in the positive x_2-direction by a molecule from below x_{2_0} and $\tilde{a}(x_{2_0} + \delta x_2)$ the value transported in the negative x_2-direction by a molecule from above. In a first approximation, δx_2 would represent the average distance from the plane to the point at which the molecule made its last collision before crossing; in a higher approximation it will depend on the location of previous collisions and on the nature of the quantity being transported. Whatever the approximation, we may expect that its value is roughly the same as that of the local mean free path. We therefore set $\delta x_2 = \alpha_{\tilde{a}}\lambda$, where $\alpha_{\tilde{a}}$ is a number that is only slightly different from unity and varies somewhat depending on the quantity $\tilde{a}$ in question. Furthermore, to a first approximation the average number of molecules crossing the plane per unit area per unit time from either above or below is proportional to $n\bar{C}$, where n is the number density of molecules and $\bar{C}$ is the average speed of the random molecular motion, both evaluated at x_{2_0}. This result can be arrived at by dimensional reasoning or by proportion—doubling either n or $\bar{C}$ would certainly double the flux of molecules. The net amount of $\tilde{a}$ transported in the positive x_2-direction per unit area per unit time, denoted by $\Lambda_{\tilde{a}}$, is therefore

$$\Lambda_{\tilde{a}} = \eta n\bar{C}[\tilde{a}(x_{2_0} - \alpha_{\tilde{a}}\lambda) - \tilde{a}(x_{2_0} + \alpha_{\tilde{a}}\lambda)],$$

where η is a constant of proportionality. Expanding $\tilde{a}$ in a Taylor's series and retaining only first-order terms in λ, we obtain

$$\boxed{\Lambda_{\tilde{a}} = -\beta_{\tilde{a}} n\bar{C}\lambda \frac{d\tilde{a}}{dx_2}}, \tag{5.4}$$

where $\beta_{\tilde{a}} = 2\eta\alpha_{\tilde{a}}$ is a new constant of proportionality and $d\tilde{a}/dx_2$ is the gradient of $\tilde{a}$ evaluated at the plane under consideration. Equation (5.4)

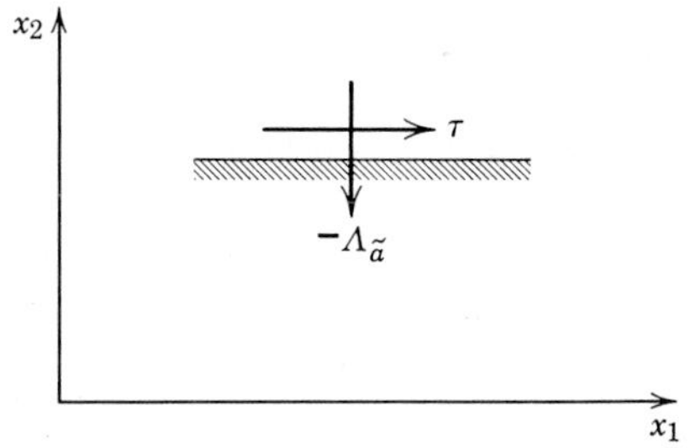

Fig. 6. Correspondence between momentum flux and positive shearing stress.

is the general transport equation. From it three equations equivalent to (5.1) to (5.3) can be obtained as described in the following paragraphs.

(a) Momentum Transport. In this case the quantity being transported is the mean x_1-momentum of the molecules. Since the mean molecular velocity in a moving fluid is equal to the flow velocity, we have specifically $\tilde{a} = mu_1(x_2)$, where m is the mass of one molecule, all of which are here taken to be alike. The resulting flux of $\tilde{a}$ (momentum per unit area per unit time) is then equivalent by Newton's law to a shearing stress (force per unit area). In Fig. 6, for example, a net flux of momentum into the material below the plane in question is equivalent to a shearing stress τ on that material in the positive x_1-direction. If this stress is counted as positive in accord with the usual sign convention for shearing stress, a negative flux then corresponds to a positive stress, and we have specifically

$-\Lambda_{\tilde{a}} = \tau$. Equation (5.4) thus becomes

$$\tau = \beta_\mu nm\bar{C}\lambda \frac{du_1}{dx_2} = \beta_\mu \rho \bar{C}\lambda \frac{du_1}{dx_2},$$

where we have written β_μ for simplicity. Comparison with the phenomenological equation (5.1) shows that

$$\boxed{\mu = \beta_\mu \rho \bar{C}\lambda}. \tag{5.5}$$

We have already seen in equation (4.3b) that λ is inversely proportional to ρ. It follows that for a given value of $\bar{C}$, which like $(\overline{C^2})^{1/2}$ is proportional to $T^{1/2}$, the value of μ is independent of density. This result was first obtained in 1860 by Maxwell, who found it so surprising that he would not rest until he had confirmed it by experiment. Of the dependence on T, we shall have more to say later.

(b) Energy Transport. Here the quantity being transported is the mean energy of the random and internal motions of the molecules. (We take into account here the internal structure.) We thus have, following equations (3.10) and (3.11), $\tilde{a} = \tilde{e} = (\xi/2)kT = (\xi/2)mRT = mc_vT$. The resulting flux $\Lambda_{\tilde{a}}$ is the heat flow q. Equation (5.4) thus becomes

$$q = -\beta_K nm\bar{C}\lambda c_v \frac{dT}{dx_2} = -\beta_K \rho \bar{C}\lambda c_v \frac{dT}{dx_2},$$

and comparison with (5.2) shows that

$$\boxed{K = \beta_K \rho \bar{C}\lambda c_v}. \tag{5.6}$$

(c) Mass Transport. Properly speaking, we are concerned here with a mixture of gases, and the quantity being transported is the mass—and with it the identity—of the different species. It will be sufficient for our purposes, however, to continue to deal with a single species. To have something to be transported, we must then imagine that certain of the molecules are "tagged" in some way that distinguishes them from the others without significantly altering their molecular properties. This would be the case, for example, with radioactive tracer molecules. The transport of such tagged molecules through otherwise identical molecules as the result of a concentration gradient is called *self diffusion*. The property being transported is then the identity of the molecule, that is, the probability that it is a tagged molecule. This probability is expressed quantitatively by the fraction n_A/n, where n_A is the number density of tagged molecules, called A-molecules, and n is the number density of all molecules. We thus have $\tilde{a} = n_A/n$. The corresponding flux is taken as $\Lambda_{\tilde{a}} = \Gamma_A$, where Γ_A is the net transport of A-molecules per unit area per unit time. Putting these expressions into (5.4) we obtain, on the assumption that the total number of molecules n is uniform,[5]

$$\Gamma_A = -\beta_D n \bar{C} \lambda \frac{d(n_A/n)}{dx_2} = -\beta_D \bar{C} \lambda \frac{dn_A}{dx_2}.$$

Comparison with equation (5.3) gives

$$\boxed{D_{AA} = \beta_D \bar{C} \lambda}, \tag{5.7}$$

where we have written D_{AA} for the coefficient of self-diffusion. Equation (5.7) is also a reasonable approximation for the diffusion of one gas through another that is *nearly* identical. This is the situation, for example, with nitrogen and carbon monoxide, which have very similar molecules with nearly the same molecular weight. For the so-called *mutual diffusion* of truly dissimilar molecules, the analysis leading up to equation (5.4) does not apply. In this case we must take account of the fact that the mean free path and mean velocity of the different species are different (see, e.g., Jeans, 1940, Chapter VIII, or Present, 1958, Chapter 4). We shall not attempt to treat this more difficult problem. Further discussion of the motion of mixtures of gases will be given in Chapter IX, Sec. 10.

The precise determination of the constants β_μ, β_K, and β_D is a matter of considerable difficulty. To illustrate the results, we shall mention here only the first two of these, and then only very briefly. (For a more complete discussion see Jeans, 1940, Chapters VI and VII. For a treatment of β_D see the same reference, Chapter VIII.) The simplest possible approximation assumes that: (a) the effective slant path of each molecule as regards the value of $\tilde{a}$ that it transports across x_{2_0} is exactly λ (i.e., the effect of collisions prior to the one before crossing is negligible), and (b) all molecules have precisely the speed $\bar{C}$. This leads to the result that

$$\beta_\mu = \beta_K = \tfrac{1}{2}. \tag{5.8a}$$

Taking account of the actual distribution of molecular speeds about the mean (cf. Chapter II) reduces the value to

$$\beta_\mu = \beta_K = \tfrac{1}{3}. \tag{5.8b}$$

Including the effect of previous collisions ("persistence of velocity") increases the effective distance δx_2 from which the molecules come and

[5] This corresponds to the assumption that the temperature and pressure are uniform.

thus tends to increase β_μ and β_K again. In the transport of energy there is also a correlation between the velocity of the molecule and the amount of energy transported (i.e., molecules with high translational energy tend to come from greater distances away from x_{2_0}) and this increases β_K still further. The final result of including these and other secondary effects for monatomic gases is that

$$\beta_\mu = 0.499 \cong \tfrac{1}{2} \qquad \text{and} \qquad \beta_K \cong \tfrac{5}{4}. \tag{5.8c}$$

For more complex gases no such simple results are available.

The relation between μ and K implied by equations (5.5) and (5.6) is of interest. Combining these equations we obtain

$$K = \frac{\beta_K}{\beta_\mu} \mu c_v.$$

If the transport of momentum and energy followed the same mechanism [equations (5.8a) or (5.8b)], we would have

$$K = \mu c_v. \tag{5.9a}$$

Owing primarily to the velocity correlation in the transport of energy, we actually obtain [equations (5.8c)]

$$K = \frac{\frac{5}{4}}{\frac{1}{2}} \mu c_v = \frac{5}{2} \mu c_v. \tag{5.9b}$$

From this we find for the *Prandtl number* $\mathrm{Pr} \equiv c_p\mu/K$, which is an important dimensionless parameter in viscous heat-transfer problems,

$$\mathrm{Pr} = \frac{c_p}{c_v} \times \frac{c_v\mu}{K} = \frac{2}{5}\gamma. \tag{5.10}$$

This checks well with experiment for a monatomic gas ($\gamma = \frac{5}{3}$ and hence $\mathrm{Pr} = \frac{2}{3}$). For diatomic gases such as air, however, it is known that $\mathrm{Pr} \cong \frac{3}{4}$ and $\gamma = \frac{7}{5}$, and these do not check well with equation (5.10). This is not surprising, since the theoretical values of β_K and β_μ used in equation (5.9b) were those for a monatomic gas.

A crude method for taking account of the effects of internal structure for polyatomic molecules was suggested by Eucken. His method divides the molecular energy into two parts

$$e = e_{\mathrm{tr}} + e_{\mathrm{int}},$$

where e_{tr} is the energy of translational motion and e_{int} is the energy associated with the internal structure. We have correspondingly

$$c_v = c_{v_{\mathrm{tr}}} + c_{v_{\mathrm{int}}}. \tag{5.11}$$

The assumption is now made that transport of translational energy involves correlation with the velocity as for a monatomic gas [equation (5.9b)], but that transport of internal energy involves no correlation and is therefore similar to that of momentum [equation (5.9a)]. We write accordingly

$$K = K_{\text{tr}} + K_{\text{int}} = \tfrac{5}{2}\mu c_{v_{\text{tr}}} + \mu c_{v_{\text{int}}},$$

or, in view of equation (5.11),

$$K = \mu(\tfrac{3}{2}c_{v_{\text{tr}}} + c_v).$$

Using $c_{v_{\text{tr}}} = \tfrac{3}{2}R$, which follows from equation (3.3c), together with the perfect-gas relation $c_v = R/(\gamma - 1)$, we obtain finally for the Prandtl number

$$\text{Pr} = \frac{4\gamma}{9\gamma - 5}. \tag{5.12}$$

This is known as *Eucken's relation*. It gives the same result as the previous relation for a monatomic gas ($\gamma = \tfrac{5}{3}$ and $\text{Pr} = \tfrac{2}{3}$) and is a reasonably good approximation for other gases at ordinary temperatures (e.g., for air $\gamma = \tfrac{7}{5}$ and $\text{Pr} = \tfrac{28}{38} = 0.737 \cong \tfrac{3}{4}$).

We close this section with a brief discussion of the effect of temperature on the coefficient of viscosity μ. This will serve also to illustrate the effect of the intermolecular forces, which we have up till now ignored. The discussion proceeds from equations (5.5) and (4.3b), which show that for a fixed value of m

$$\mu \sim \bar{C}/d^2. \tag{5.13}$$

Since $\bar{C}$ is proportional to $T^{1/2}$ (this will be shown in Chapter II), it follows that for the billiard-ball model we have $\mu \sim T^{1/2}$. Actually, μ is found to vary more rapidly than this, the difference being due to the variation of the intermolecular force with distance.

Consider, for example, the Sutherland model (Fig. 1), which has a rigid-sphere center surrounded by a weak field of attraction. As two spheres of this kind approach each other their paths are deflected even before the spheres actually strike. From the point of view of an observer fixed relative to a given molecule z, the path of a colliding molecule is as shown in Fig. 7. As a result of the attractive force, the center of the colliding molecule follows the curved path PQR, with a collision occurring when the center is at the point Q on the sphere of influence of z. The overall change in direction of the molecule as the result of the encounter is the same as if it had followed the straight-line path $PQ'R$, with the collision occurring at the point Q' on some larger, effective sphere of influence. The gross behavior of the spheres with a weak attractive field is thus equivalent to that of nonattractive spheres of a somewhat larger diameter. The effect of the attractive force can thus be accounted for to a first approximation by replacing d in relation (5.13) by an effective diameter d_{eff}. The larger the relative speed of the molecules, the less time there is for the attractive forces to deflect them and the smaller is d_{eff} relative to d. Detailed analysis of the dynamics of the situation (Loeb,

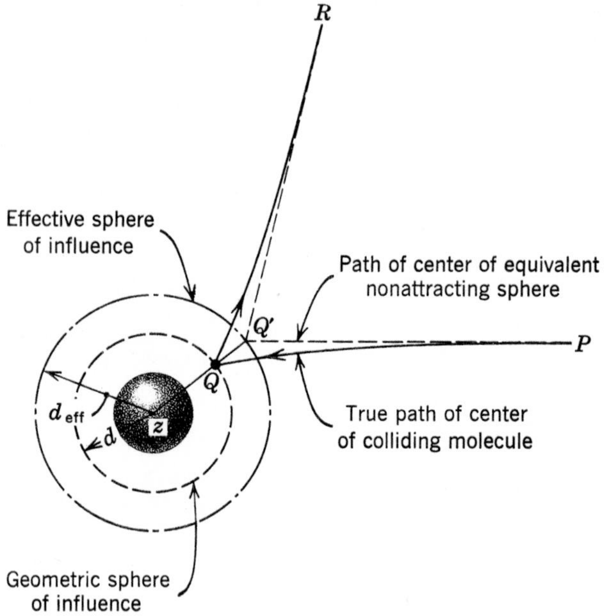

Fig. 7. Collision of two Sutherland molecules as seen by observer fixed relative to molecule z.

1961, p. 221; Kennard, 1938, p. 154) shows that

$$d_{\text{eff}}^2 = d^2\left[1 + \frac{4}{m}\,\overline{1/g^2}\int_d^\infty - F(r)\,dr\right],$$

where $F(r)$ is the intermolecular force as in Fig. 1 and $\overline{1/g^2}$ is the mean of $1/g^2$, where g is the relative speed of the molecules. Since it can be shown that $\overline{1/g^2} \sim 1/T$, this equation has for a given force field the form

$$d_{\text{eff}}^2 = d^2\left(1 + \frac{\chi}{T}\right),$$

where χ is a constant that is positive for attractive forces ($F < 0$). Using

d_{eff}^2 in (5.13) in place of d^2 and noting as before that $\bar{C} \sim T^{1/2}$, we obtain finally

$$\frac{\mu}{\mu_{\text{ref}}} = \left(\frac{T}{T_{\text{ref}}}\right)^{1/2} \frac{1 + \chi/T_{\text{ref}}}{1 + \chi/T}, \tag{5.14}$$

where μ_{ref} is the coefficient of viscosity at some reference temperature T_{ref}. This is *Sutherland's formula.* Although it is still rather crude from a theoretical point of view, it does provide a more rapid variation of μ than the simple $T^{1/2}$. With the constants determined empirically, it gives accurate results for certain gases such as oxygen and nitrogen and is therefore of considerable practical value.

Exercise 5.1. The thickness δ of a laminar boundary layer at a given angular position on a circular cylinder of diameter D varies as $\delta/D \sim 1/\sqrt{\text{Re}}$. Here the Reynolds number Re is defined by $\text{Re} \equiv \rho_\infty U_\infty D/\mu_\infty$, where U_∞ is the flow velocity of the undisturbed stream relative to the cylinder and ρ_∞ and μ_∞ are the corresponding density and coefficient of viscosity. If λ_∞ is the mean free path in the undisturbed stream, find how the ratio λ_∞/δ varies as a function of Re and M, where M is the Mach number $\text{M} \equiv U_\infty/\sqrt{\gamma R T_\infty}$.

Exercise 5.2. In the treatment of energy transport in this section we in effect assumed (by considering only the energy of the random and internal motions) that there was no macroscopic motion of the gas. Consider now a gas with not only a temperature gradient dT/dx_2 but also a gradient of flow velocity du_1/dx_2 as in the treatment of momentum transport. By the mean-free-path methods of this section, show that the flux of molecular energy in the x_2-direction is now given by

$$\Lambda = -K\frac{dT}{dx_2} - u_1\mu\frac{du_1}{dx_2} = q - u_1\tau.$$

The second term on the right is interpreted at the macroscopic level as the work done by the shear stress.

6 MOLECULAR MAGNITUDES

We are now in a position to use macroscopic measurements to make an estimate of the magnitudes involved in molecular phenomena. Among the quantities that can be measured macroscopically are p, ρ, μ, and $\hat{M}$, where $\hat{M}$ is the molecular weight (mass per mole). We should like to find the following four quantities that characterize the molecular model: m, d, n, and $\sqrt{\overline{C^2}}$. Since $m = \hat{M}/\hat{N}$, the determination of m also entails the determination of Avogadro's number $\hat{N}$.

We shall consider air, which has the following macroscopic properties at standard conditions of 1 atm and 0°C:

$$p = 1.013 \times 10^6 \text{ dyne/cm}^2,$$
$$\rho = 1.288 \times 10^{-3} \text{ gm/cm}^3,$$
$$\mu = 171 \times 10^{-6} \text{ gm/cm sec.}$$

We shall ignore the fact that air is a mixture of gases, and consider it to be made up of fictitious identical molecules. Since the molecular weights of N_2 and O_2 are not very different, this assumption is not far from the truth. On this basis the molecular weight is $\hat{M} \cong 28.9$ gm/mole.

The root-mean-square molecular speed at the above conditions has already been found following equation (3.5) and is

$$\sqrt{\overline{C^2}} = \left(3\frac{p}{\rho}\right)^{1/2} \cong \underline{5 \times 10^4 \text{ cm/sec.}} \tag{6.1}$$

We can estimate d and $\hat{N}$ simultaneously by the following procedure. From equations (5.5) and (5.8c) plus equation (4.3b) we can write

$$\mu = \tfrac{1}{2}\rho\bar{C}\lambda = \tfrac{1}{2}\rho\bar{C}\frac{m}{\sqrt{2}\,\pi\, d^2\rho} = \frac{1}{2\sqrt{2}\,\pi} \times \frac{\hat{M}}{\hat{N}} \times \frac{\bar{C}}{d^2},$$

or, if we take $\bar{C} \cong \sqrt{\overline{C^2}}$ and $2\sqrt{2}\,\pi \cong 9$,

$$\hat{N}d^2 \cong \frac{1}{9}\frac{\hat{M}\sqrt{\overline{C^2}}}{\mu}. \tag{6.2}$$

Since the quantities on the right are known, this gives a relation between $\hat{N}$ and d. A second relation can be found by going outside kinetic theory and considering the situation in *liquid* air. We assume as a rough estimate that each molecule in the liquid state occupies a cube of side d. If ρ_L denotes the density of the liquid state, we then have $m = \rho_L d^3$ or, since $m = \hat{M}/\hat{N}$,

$$\hat{N}d^3 \cong \frac{\hat{M}}{\rho_L}. \tag{6.3}$$

Dividing (6.3) by (6.2) and substituting the known values of μ, $\sqrt{\overline{C^2}}$, and $\rho_L = 0.86$ gm/cm³, we find

$$d \cong \frac{9\mu}{\rho_L\sqrt{\overline{C^2}}} \cong \underline{3.7 \times 10^{-8} \text{ cm.}} \tag{6.4}$$

Returning to equation (6.3), we then obtain for Avogadro's number

$$\hat{N} \cong \frac{\hat{M}}{\rho_L d^3} \cong 6.7 \times 10^{23}/\text{mole}$$

The accepted value of this universal constant, from accurate considerations based on measurements of electronic charge, is

$$\hat{N} = \underline{6.02252 \times 10^{23}/\text{mole.}} \qquad (6.5)$$

With the foregoing results, we now find for the mass of one molecule

$$m = \hat{M}/\hat{N} \cong \underline{4.80 \times 10^{-23} \text{ gm}},$$

and for the number density

$$n_0 = \frac{\rho}{m} \cong \underline{2.69 \times 10^{19}/\text{cm}^3}. \qquad (6.6)$$

This last is the number of molecules per cubic centimeter of air at standard conditions of $p = 1$ atm and $T = 0°\text{C}$. Since the perfect-gas equation can be written $p = nkT$, the same number must hold, to the accuracy of perfect-gas theory, for *any* gas at that pressure and temperature. This *standard number density*, which we denote by the addition of the subscript $(\ )_0$, is known also as *Loschmidt's number.*

We have thus found the four quantities that we set out to evaluate. It is also of interest to calculate the average spacing δ between molecules. Since the average volume available per molecule is $1/n_0$, this is given by

$$\delta = \frac{1}{n_0^{1/3}} \cong 3.3 \times 10^{-7} \text{ cm}.$$

We can also find the mean free path, which is

$$\lambda = \frac{1}{\sqrt{2}\pi n_0 d^2} \cong 6 \times 10^{-6} \text{ cm},$$

and the frequency of collisions per molecule, which is

$$\Theta = \frac{\bar{C}}{\lambda} \cong \frac{\sqrt{\overline{C^2}}}{\lambda} \cong 10^{10}/\text{sec}.$$

It is of interest to compare the various microscopic distances that we have obtained. On the basis of the foregoing results we can write

$$\lambda : \delta : d = 6 \times 10^{-6} : 3.3 \times 10^{-7} : 3.7 \times 10^{-8} \cong 170 : 10 : 1.$$

Thus the mean free path is much greater than the average spacing, which in turn is much greater than the molecular diameter. Since the effective range of intermolecular forces is of the order of the diameter, this justifies the assumption that the gas molecules interact only during some kind of collision process of relatively short duration. This idea is implicit in our entire development of kinetic theory. It remains valid so long as the density is not too high.

The foregoing picture, with its minute distances and astronomical number density, would be fantastic if we were not already accustomed to molecular ideas. Some feeling for the wonder of it may be regained with the following quotation from Jeans (1940, p. 32):

> . . . , a man is known to breathe out about 400 c.c. of air at each breath, so that a single breath of air must contain about 10^{22} molecules. The whole atmosphere of the earth consists of about 10^{44} molecules. Thus one molecule bears the same relation to a breath of air as the latter does to the whole atmosphere of the earth. If we assume that the last breath of, say, Julius Caesar has by now become thoroughly scattered through the atmosphere, then the chances are that each of us inhales one molecule of it with every breath we take. A man's lungs hold about 2000 c.c. of air, so that the chances are that in the lungs of each of us there are about five molecules from the last breath of Julius Caesar.

References

Jeans, J., 1940, *An Introduction to the Kinetic Theory of Gases*, Cambridge University Press.

Kennard, E. H., 1938, *Kinetic Theory of Gases*, McGraw-Hill.

Loeb, L. B., 1961, *The Kinetic Theory of Gases*, 3rd ed., Dover.

Present, R. D., 1958, *Kinetic Theory of Gases*, McGraw-Hill.

Slater, J. C., 1939, *Introduction to Chemical Physics*, McGraw-Hill.

Chapter II

Equilibrium Kinetic Theory

1 INTRODUCTION

The mean-free-path methods of Chapter I are as far as we shall go for the present in our discussion of nonequilibrium kinetic theory. We shall return to nonequilibrium theory and transport processes in a more rigorous way in Chapters IX and X. For the time being we restrict our attention to a more detailed and rigorous consideration of the equilibrium state. In this chapter in particular, we shall (1) give a more rigorous discussion of the equation of state of a perfect gas, (2) look into the distribution of molecular velocities and the resulting implications with regard to molecular collisions and the mean free path, and (3) examine the conditions of equilibrium of a reacting gas mixture from the kinetic point of view. Some of the results, although obtained from equilibrium ideas, will also be useful later in certain nonequilibrium situations.

2 VELOCITY DISTRIBUTION FUNCTION

All molecules of a gas do not move with the same velocity, nor does the velocity of a given molecule remain constant with time. For a detailed and rigorous discussion of kinetic theory, we must have some statistical way of specifying this fact. This is provided by the *velocity distribution function.*

The distribution function is a general concept in statistics, and there can be distribution functions for all sorts of quantities. We shall introduce the velocity distribution function by analogy to a kind of distribution function with which the student is presumably familiar but which he may not have recognized as such. This is the mass density ρ in a nonuniform density field.

To begin, let us consider a gas uniformly distributed throughout a container of volume V. We consider all the molecules to be alike of mass m and suppose that there are N of these molecules. The mass density of the gas is thus

$$\rho = m\frac{N}{V} = mn, \qquad (2.1)$$

where $n = N/V$ is the corresponding number density. Since the gas is uniformly distributed, ρ and n each have the same value for all subvolumes of V, provided these volumes contain a sufficient number of molecules.

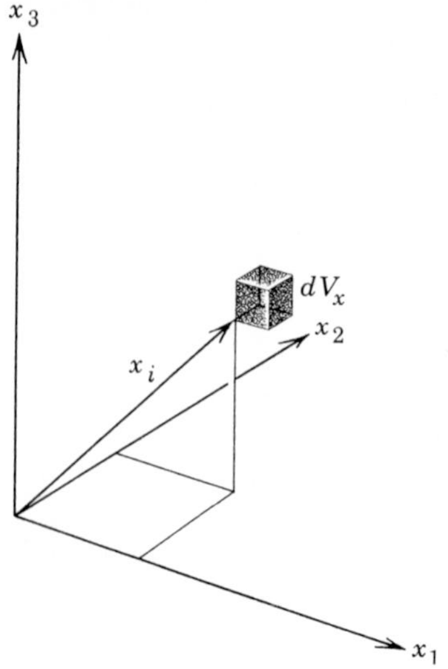

Fig. 1. Volume element in physical space.

If the gas is *non*uniformly distributed in the container, we introduce the idea of a *local* mass density $\rho(x_1, x_2, x_3) = \rho(x_i)$. Here x_i is shorthand notation for the position *vector* with Cartesian components x_1, x_2, x_3 (Fig. 1); this type of notation will be used for vectors throughout the book. To relate the familiar macroscopic idea of a local density to microscopic molecular ideas, we proceed as follows. Let ΔN be the number of molecules contained in the volume ΔV_x located between x_1 and $x_1 + \Delta x_1$, x_2 and $x_2 + \Delta x_2$, x_3 and $x_3 + \Delta x_3$. We then have, analogous to (2.1),

$$\rho(x_i) = \lim_{\Delta V_x \to 0} m\frac{\Delta N}{\Delta V_x} = m \lim_{\Delta V_x \to 0} \frac{\Delta N}{\Delta V_x} = mn(x_i). \qquad (2.2)$$

Here the limit is taken to mean that ΔV_x approaches zero on the scale of the container but remains large on the scale of the molecular spacing. Thus ΔV_x is given a macroscopic interpretation even in the limit. Since the molecular spacing is normally very tiny (see Chapter I, Sec. 6), this is a valid procedure except for gases so rarefied as to be of little practical interest. The local number density $n(x_i) = \lim(\Delta N/\Delta V_x)$ in equation (2.2) gives the number of molecules per unit volume as a function of position. It is thus a measure of the distribution of molecules with regard to position in space—in short, it is a *position distribution function.* If we know this function, the number of molecules dN in the macroscopically infinitesimal element $dV_x = dx_1\,dx_2\,dx_3$ located at the position x_i is then

$$dN = n(x_i)\,dV_x. \qquad (2.3a)$$

This will be the case so long as the dimensions of the subvolumes are large compared with the average spacing between the molecules. Since $\rho(x_i)$ is proportional to $n(x_i)$ by (2.2), the local mass density may also be looked on as a position distribution function, the only difference being that it applies to mass rather than to molecular number. Knowing this distribution function, we can find the mass dM in an infinitesimal volume located at x_i from the relation

$$dM = \rho(x_i)\, dV_x, \tag{2.3b}$$

a procedure that is presumably familiar. Generally—and loosely—speaking, a distribution function gives the concentration of some quantity per unit "volume" as a function of position in some sort of "space." To find the amount of that quantity in an infinitesimal volume located at a given position in that space, we merely proceed as in equations (2.3), that is, we multiply the distribution function by the infinitesimal element of volume.

Although $n(x_i)$ can be taken as the position distribution function for molecules, it is sometimes convenient to define a normalized quantity $\omega(x_i) = n(x_i)/N$, and speak of this function $\omega(x_i)$ as the distribution function. The number of molecules located in the volume dV_x is then

$$dN = N\omega(x_i)\, dV_x. \tag{2.4}$$

The fractional number of molecules within the volume is thus $\omega(x_i)\, dV_x$. This can also be interpreted as the probability that a molecule chosen at random will lie within the given volume. Since the total number of molecules in V is N, $\omega(x_i)$ is subject to the condition that

$$\int_V N\omega(x_i)\, dV_x = N,$$

or

$$\int_V \omega(x_i)\, dV_x = 1, \tag{2.5}$$

where the integration is taken over the volume V.

The foregoing illustrates what we mean by a distribution function. We are interested here, however, not in the spatial distribution of molecules in a nonuniform gas but in the distribution of molecular velocities in a spatially uniform gas.

We return therefore to a consideration of the uniform gas of N molecules in the volume V. At every instant, each of the N molecules will have a velocity that can be completely specified by its three Cartesian components C_1, C_2, C_3. The molecule can thus be represented uniquely by a point in a Cartesian space having the velocity components as coordinates

(Fig. 2). In this so-called *velocity space* the molecule is thus represented by a point located at the end of the velocity vector C_i laid off from the origin. The entire gas in the volume V is represented correspondingly by a cloud of N points in this space. The situation is thus analogous to (but definitely different from) the distribution of the molecules themselves in physical space. As before, a local point density $F(C_i)$—points per unit "volume," analogous to $n(x_i)$—can be introduced. It is defined such that the number of molecular points whose coordinates in velocity space lie in the element of volume $dV_C = dC_1\, dC_2\, dC_3$ located between C_1 and $C_1 + dC_1$, C_2 and $C_2 + dC_2$, and C_3 and $C_3 + dC_3$ is equal to $F(C_i)\, dV_C$. Since each point represents a molecule, this also gives the number of molecules for which the velocity components lie between the same limits, or equivalently, for which the velocity vectors end inside the volume dV_C. The function $F(C_i)$ is referred to as a *velocity distribution function*. As before, it is sometimes convenient to replace $F(C_i)$ by a normalized distribution function, defined by $f(C_i) \equiv F(C_i)/N$. The number of molecules whose components of velocity lie in the range C_1 to $C_1 + dC_1$, C_2 to $C_2 + dC_2$, and C_3 to $C_3 + dC_3$ is then

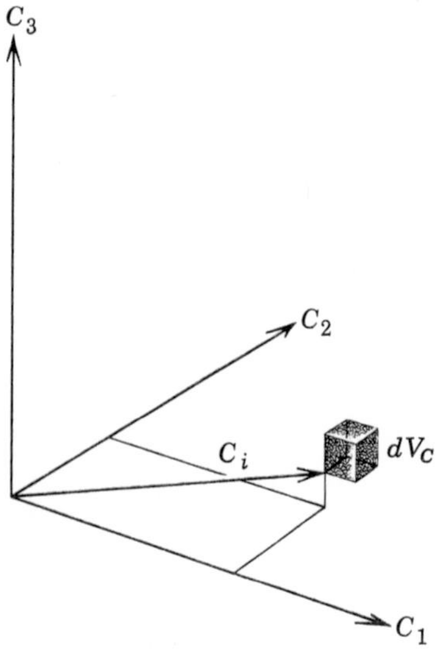

Fig. 2. Volume element in velocity space.

$$\boxed{dN = Nf(C_i)\, dV_C}, \tag{2.6a}$$

or, in terms of the number density $n = N/V$,

$$dn = nf(C_i)\, dV_C. \tag{2.6b}$$

The probability of any given molecule having velocity components within the specified range is $f(C_i)\, dV_C$. Since all the N molecules must lie somewhere in the velocity space, $f(C_i)$ is subject to the condition

$$\int_{-\infty}^{\infty} Nf(C_i)\, dV_C = N,$$

or

$$\boxed{\int_{-\infty}^{\infty} f(C_i)\, dV_C = 1}, \tag{2.7}$$

where the integration extends over all of velocity space, that is, from $-\infty$ to ∞ for each velocity component.

In the work that follows we shall often want to take averages of quantities that depend on the velocity of the molecule. Let $Q = Q(C_i)$ be any quantity (measured per molecule) that is a function of the velocity. The average value of Q for the whole assembly of N molecules is then

$$\bar{Q} = \frac{\int_N Q\,dN}{N} = \frac{\int_{-\infty}^{\infty} Q(C_i)Nf(C_i)\,dV_C}{N},$$

or

$$\boxed{\bar{Q} = \int_{-\infty}^{\infty} Qf\,dV_C}, \tag{2.8}$$

where the functional notation is omitted for brevity.

Throughout this book we shall work with the normalized distribution function, which we have denoted by $f(C_i)$. Some authors prefer to write their equations in terms of the non-normalized function, which we have called $F(C_i)$, but use the symbol $f(C_i)$ to denote this latter function. Their equations are readily obtained from ours by formally replacing $f(C_i)$ in our equations by $f(C_i)/N$ or $f(C_i)/n$.

3 EQUATION OF STATE FOR A PERFECT GAS

With the aid of the distribution function we can now give a derivation of the perfect-gas law that does not depend on the special assumptions used in the earlier analysis. For simplicity, we assume here that all molecules are alike and thus have the same mass m.

We consider a locally planar element dS of a fixed, solid wall and choose an x_1, x_2, x_3 coordinate system with the x_3-axis normal to the element as shown in Fig. 3. Molecules will strike and leave dS in various directions, depending on the direction of the velocity vector of the particular molecule. We assume that each molecule that strikes the element is brought to rest in the positive x_3-direction and then emitted with a component of motion in the negative direction. The molecule loses momentum normal to the wall in being brought to rest and gains negative momentum in being emitted. Each such change in momentum is accompanied by a force on the wall, and this force will fluctuate with time. In a gas at equilibrium the time average of the force, measured per unit area, is the pressure of the gas.

In calculating the force in this instance we make no supposition about the nature of the reflection for a given molecule. In particular, we dispense with the unrealistic assumption, made in Sec. 3 of the preceding chapter, that the reflection is specular and without change in speed. This can be done by considering the incoming and outgoing molecules in two separate groups.

We look first at the incoming molecules, that is, at molecules with positive values of C_3. Of these, let us consider to begin with only those with

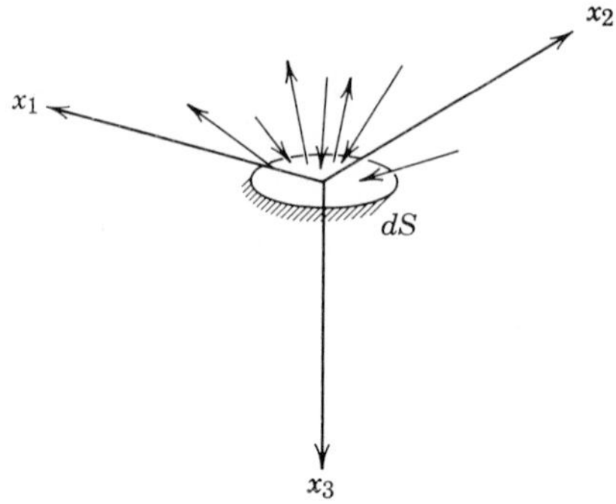

Fig. 3. Element of wall and coordinate system.

velocities infinitesimally close to a specific velocity vector C_i. We consider, in other words, a special class of incoming molecules with velocity components in the range C_1, C_2, C_3 to $C_1 + dC_1$, $C_2 + dC_2$, $C_3 + dC_3$, where C_1, C_2, C_3 have given numerical values. We speak of these as "molecules of class C_i." All classes will be accounted for later by integration. The molecules of class C_i that strike dS in an interval of time dt will be those that, at the beginning of the interval, are contained in the slant cylinder with base dS and surface elements parallel to C_i and of length $C\,dt$ (Fig. 4). Molecules belonging to the class C_i but lying outside the cylinder obviously cannot reach dS in the interval dt; by the same token, all molecules of class C_i that lie inside the cylinder must reach dS in the stated interval.[1]

[1] In making these statements we have ignored the effect of molecular collisions inside the slant cylinder. This is justified if we take dt small enough that the distance $C\,dt$ is small compared with the mean free path. In a sufficiently dilute gas the molecular spacing is sufficiently small compared with the mean free path that this can be done and a large number of molecules still retained in the cylinder. The neglect of collisions can also be argued on equilibrium grounds—that is, at equilibrium as many molecules are knocked into the given class in time dt as are knocked out of it. We prefer the first argument, however, since we shall want to apply certain of the resulting expressions to nonequilibrium situations in Chapter IX.

Molecules of other classes, of course, will also reach dS in time dt, but we are for the time being not interested in these. Now, the altitude of the cylinder in question is $C_3\,dt$, and its volume is therefore $C_3\,dt\,dS$. Furthermore, by expression (2.6b), the number of molecules of class C_i per unit volume in the physical space is $nf(C_i)\,dC_1\,dC_2\,dC_3$. The number of molecules of class C_i striking dS in the interval dt is therefore $nC_3f(C_i)\,dC_1\,dC_2\,dC_3\,dS\,dt$, and the flux of such molecules (number per

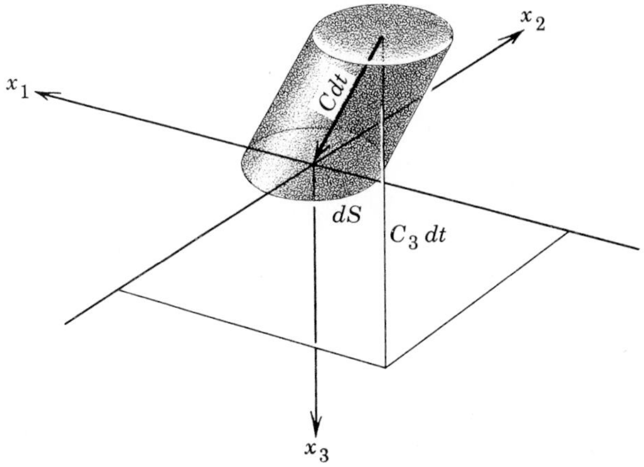

Fig. 4. Slant cylinder containing molecules of class C_i that strike dS in time dt.

unit area per unit time) arriving at dS is correspondingly

$$nC_3f(C_i)\,dC_1\,dC_2\,dC_3. \tag{3.1}$$

To obtain the flux of molecules arriving at an element of area perpendicular to the x_1- or x_2-axes we need merely replace C_3 ahead of f by C_1 or C_2. Expressions of this kind will have considerable use later. If dS is regarded as an open area rather than a wall, expression (3.1) also gives the flux of molecules of class C_i through that area.

The calculation of the normal stress (normal force per unit area) on dS due to the incoming molecules is now simple. Since each such molecule loses normal momentum in the amount mC_3 when it strikes the wall, the total normal momentum lost by incoming molecules of class C_i per unit area per unit time is

$$nmC_3^2f(C_i)\,dC_1\,dC_2\,dC_3.$$

The normal stress σ_I exerted on dS by all incoming molecules is equal to the total loss of normal momentum for all such molecules. This is found by integrating over all classes of incoming molecules as follows:

$$\sigma_I = nm\int_0^\infty\int_{-\infty}^\infty\int_{-\infty}^\infty C_3^2f(C_i)\,dC_1\,dC_2\,dC_3.$$

The integration with respect to C_1 and C_2 extends from $-\infty$ to ∞. That with respect to C_3 extends only from 0 to ∞, as is appropriate for incoming molecules.

The foregoing arguments can be repeated with regard to outgoing molecules. We thus find for the normal stress σ_O exerted on dS by all such molecules

$$\sigma_O = nm\int_{-\infty}^{0}\int_{-\infty}^{\infty}\int_{-\infty}^{\infty} C_3^2 f(C_i)\, dC_1\, dC_2\, dC_3.$$

Here the integration with respect to C_3 extends from $-\infty$ to 0 since outgoing molecules have negative values of C_3.

The complete normal stress σ on the wall is the sum of σ_I and σ_O. We thus have

$$\begin{aligned}\sigma &= nm\int_{0}^{\infty}\int_{-\infty}^{\infty}\int_{-\infty}^{\infty} C_3^2 f(C_i)\, dC_1\, dC_2\, dC_3 \\ &\quad + nm\int_{-\infty}^{0}\int_{-\infty}^{\infty}\int_{-\infty}^{\infty} C_3^2 f(C_i)\, dC_1\, dC_2\, dC_3 \\ &= nm\int_{-\infty}^{\infty} C_3^2 f\, dV_C\end{aligned}$$

or, in view of equation (2.8),

$$\sigma = nm\overline{C_3^2}.$$

No assumption of equilibrium has been made in the development to this point. The foregoing expressions thus hold whether the gas is in equilibrium or not.

We now assume the gas to be in equilibrium, that is, there are no gradients of temperature or mean velocity. It is implicit, of course, that the wall is also in equilibrium with the gas, that is, that it has the same temperature. Since the container is at rest, the molecular velocity C_i must now be entirely random. Under these conditions every direction is statistically like every other direction; or, as stated in Chapter I, Sec. 3, no direction is preferred. We must thus have, in particular, $\overline{C_1^2} = \overline{C_2^2} = \overline{C_3^2}$. But since $C^2 = C_1^2 + C_2^2 + C_3^2$, it follows that

$$\overline{C^2} = \overline{C_1^2} + \overline{C_2^2} + \overline{C_3^2} = 3\overline{C_3^2}. \tag{3.2}$$

Since at equilibrium the normal stress on the wall is equal to the pressure p of the gas, we can thus write finally, with $\rho = nm$,

$$\boxed{\frac{p}{\rho} = \frac{1}{3}\overline{C^2}}\,. \tag{3.3}$$

This is identical to equation (I 3.5), which was obtained from cruder arguments. The results based on the earlier equation, or on the equivalent equation (I 3.2), thus follow as before.

The foregoing result can be interpreted directly in terms of the pressure in the interior of the gas if we define the normal stress as the momentum transport per unit area per unit time across an imaginary element dS within the gas. The stress σ_I is now interpreted as arising from the flux of momentum through dS in one direction and σ_O from the flux in the opposite direction. The pressure p at equilibrium then follows as before.

Throughout this section we have assumed, as justified by the calculations at the end of Chapter I, that the molecules experience a force only during an actual collision. In other words, we assume that the average spacing of the molecules is much greater than the range of the intermolecular forces. As a result the molecules cannot, so to speak, "reach out" across the element dS in the interior of the gas and exert a force on the molecules on the other side. The only way they can exert such a force is to actually pass through dS and undergo a collision. It is thus reasonable to define the pressure at equilibrium solely in terms of a momentum flux. This is crucial to the kinetic-theory derivation of the perfect-gas equation of state. If the density is high enough that the molecular spacing is at all comparable to the range of the intermolecular forces, then the so-called "van der Waals forces" become important and the definition of pressure and the resulting equation of state must be modified accordingly (see Jeans, 1940, Chapter III).

Note that we have nowhere in this section made the explicit assumption that the molecules collide according to the billiard-ball model. All that we have really assumed is that there exists some sort of loosely defined collision process of relatively short duration (cf. Chapter IX, Sec. 3). The perfect-gas result thus corresponds, not to the billiard-ball model alone, but to any gas provided it is sufficiently dilute.

4 MAXWELLIAN DISTRIBUTION—CONDITION FOR EQUILIBRIUM

The result of equation (3.3) was obtained without finding the form of $f(C_i)$. To proceed further we can no longer ignore this question.

As first published by Maxwell in 1860, the velocity distribution function for a gas at equilibrium is

$$f(C_i) = f(C_1, C_2, C_3) = \left(\frac{m}{2\pi kT}\right)^{3/2} \exp\left[-\frac{m}{2kT}(C_1^2 + C_2^2 + C_3^2)\right].$$

Unfortunately, the establishment of this formula by kinetic theory represents a pedagogical dilemma in an elementary discussion. The simple treatments, such as that used by Maxwell himself, are lacking in rigor and physical content; the rigorous treatment devised by Boltzmann, although instructive physically, is long and complicated. We shall give here an abbreviated discussion that follows the general approach of the rigorous treatment but cites some of the details without proof. The proof of these details is provided later in Chapter IX. An alternative derivation by the entirely different methods of statistical mechanics is also given in Chapter IV. With molecular-beam techniques, the distribution of molecular velocities can be measured directly and is found to be in excellent agreement with the above formula. The Maxwellian distribution can thus, if we wish, be taken as a fundamental physical law that requires no theoretical proof. From this point of view, the present treatment could be regarded as a kind of plausibility argument.

Our argument proceeds from a consideration of the rate of change of the number of molecules in a given class C_i as the result of collisions. Each time a molecule of this class collides with another molecule, its velocity is changed and it is removed from the class. Such collisions thus tend to deplete the molecules in the class in question. Conversely, when a collision between two molecules occurs such that one of them emerges with the velocity C_i, a molecule is added to the class. Such collisions act to replenish the number of molecules of the class C_i. At equilibrium the net change from the two types of collisions must be zero, and this fact can be used to find the distribution function.

To make these ideas explicit, we consider first a general collision between two molecules of arbitrary classes U_i and Y_i, that is, with velocity components in the respective ranges U_1, U_2, U_3 to $U_1 + dU_1$, $U_2 + dU_2$, $U_3 + dU_3$ and Y_1, Y_2, Y_3 to $Y_1 + dY_1$, $Y_2 + dY_2$, $Y_3 + dY_3$. The relative-velocity vector of the molecule of class Y_i relative to the molecule of class U_i will be denoted by g_i. It has the magnitude

$$g = [(Y_1 - U_1)^2 + (Y_2 - U_2)^2 + (Y_3 - U_3)^2]^{1/2}. \tag{4.1}$$

Let us temporarily take a molecule of class U_i as the focus of attention. If we assume a billiard-ball model, we can then say, as in Chapter I, Sec. 4, that at the instant of impact the center of the molecule of class Y_i must lie on the sphere of influence of radius d about the center of the molecule of class U_i (Fig. 5). To completely specify the collision, however, it is also necessary to give the direction of the line connecting the centers of the two molecules at the instant of impact, the so-called *line of centers*. Two angles are required for this purpose. A convenient choice for one of these is the acute angle ψ between the line of centers and the line of the vector

g_i. These two lines define a plane, which we call the *plane of the collision.* The direction of the line of centers is then completely specified if we specify the orientation of this plane. This we do by taking as our second angle the angle ε that the plane makes with some arbitrary reference plane parallel to g_i.

We now ask: What is the frequency of collisions occurring between molecules of classes U_i and Y_i and such that the direction of the line of

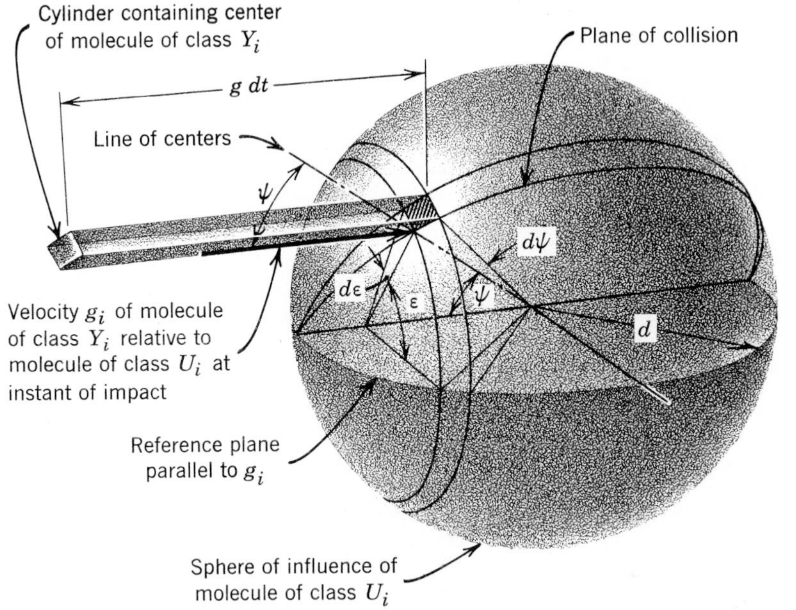

Fig. 5. Construction for calculation of collision rate between molecules of classes U_i and Y_i.

centers lies in the range ψ, ε to $\psi + d\psi$, $\varepsilon + d\varepsilon$? At the instant of this particular type of collision the center of the molecule of class Y_i must lie on the small rectangle cut out on the sphere of influence of the molecule of class U_i by the angles $d\psi$ and $d\varepsilon$ (Fig. 5). The area of this rectangle is $d^2 \sin \psi \, d\psi \, d\varepsilon$. If such a collision is to occur within a small interval of time dt, then at the beginning of that interval the molecule of class Y_i must lie within a slant cylinder based on the small rectangle and having surface elements parallel to g_i and of length $g\,dt$. Since the altitude of this cylinder is $g\,dt \cos \psi$, the volume of the cylinder is $d^2 \sin \psi \cos \psi \, d\psi \, d\varepsilon \, g\,dt$. Now, a cylinder of this kind can be associated with each of the molecules of class U_i. These are present in a unit volume of gas in the number $nf(U_i)\,dV_U$, where n is the number density of molecules in physical space and dV_U is, as before, the volume element $dU_1\,dU_2\,dU_3$ in velocity space. It follows that the total volume of the slant cylinders associated with all the molecules of class U_i taken together is

$$n\,d^2 f(U_i) g \sin \psi \cos \psi \, d\psi \, d\varepsilon \, dV_U \, dt.$$

The number of molecules of class Y_i having centers within this collection of cylinders can be found by multiplying the foregoing expression by the number of molecules of class Y_i per unit volume, which is $nf(Y_i)\,dV_Y$. Since each of these molecules satisfies the conditions for a collision, we thus find for the number of collisions of the specified type per unit volume of gas per unit time

$$n^2 d^2 f(U_i) f(Y_i) g \sin \psi \cos \psi \, d\psi \, d\varepsilon \, dV_U \, dV_Y. \tag{4.2}$$

This is the answer to the question posed at the beginning of the paragraph.

We now wish to apply this formula to the collisions that deplete and replenish the particular class C_i that was originally under discussion. In doing this we shall pair off a specific type of depleting collision with a corresponding type of replenishing collision in a special way. As the typical depleting collision we take those collisions that occur between molecules of the given class C_i and some other specific class Z_i and that have their line of centers in a specified range ψ, ε to $\psi + d\psi$, $\varepsilon + d\varepsilon$. The two molecules entering such a collision do so with initial velocities C_i and Z_i and leave with final velocities denoted respectively by C_i' and Z_i'. We shall see from detailed consideration of collision dynamics in Chapter IX, Sec. 7 that if C_i, Z_i, ψ, and ε are given, then C_i' and Z_i' are completely determined. This is because the four conservation equations (one for energy and three for momentum) plus the two quantities ψ and ε give, in effect, six relations for the determination of the six components of C_i' and Z_i'. We can write accordingly that

$$C_i' = C_i'(C_i, Z_i, \psi, \varepsilon) \quad \text{and} \quad Z_i' = Z_i'(C_i, Z_i, \psi, \varepsilon). \tag{4.3}$$

We now consider the *inverse collision* to the foregoing. The inverse to a given collision is by definition a collision having (a) initial velocities equal to the final velocities of the given collision and (b) the same direction of the line of centers. The inverse to the foregoing collision thus has initial velocities C_i' and Z_i' and the same direction of the line of centers as before. As will be seen from the detailed considerations of collision dynamics in Chapter IX, the final velocities of the molecules leaving this inverse collision are then precisely C_i and Z_i. That is to say, an inverse collision reproduces the velocities at which the molecules entered the corresponding original collision. An example of a collision and its inverse is shown in Fig. 6.[2] Since the inverse collision here produces a molecule with velocity C_i, it constitutes a replenishing collision in the sense originally discussed. The reason for pairing a given type of depleting collision with the inverse replenishing collision is that the initial velocities of the latter can then be expressed in terms of the initial velocities of the former [cf. equations (4.3)]. As will be seen, this facilitates the integration necessary to determine the net change in the number of molecules in the class C_i as the result of all possible depleting and replenishing collisions.

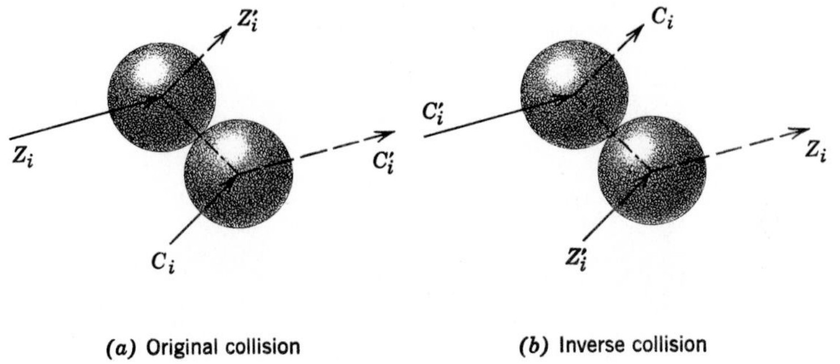

Fig. 6. Example of a collision and its inverse.

We now apply expression (4.2) to the types of collisions just discussed, which are illustrated in terms of relative velocity in Fig. 7. The diagram is drawn here from the point of view of an observer fixed relative to the molecule of class C_i or C_i'. By reference to Fig. 5 and expression (4.2), we write at once for the number of depleting collisions per unit volume per unit time

$$n^2 d^2 f(C_i) f(Z_i) g \sin\psi \cos\psi \, d\psi \, d\varepsilon \, dV_C \, dV_Z.$$

Since each depleting collision removes one molecule from the class C_i, this also gives the rate of removal of molecules by collisions of the given type. To treat the inverse replenishing collisions we note that the relative velocity g_i' leaving the depleting collision lies in the plane of the collision. Furthermore it has a magnitude g' equal to the magnitude g of the incoming velocity and makes an angle ψ' with the line of centers equal to the incoming angle ψ. For the present billiard-ball model, we shall take the truth of the foregoing statements as reasonably obvious; they can be

[2] Note that an inverse collision is not the same as a *reverse* collision, in which the final velocities of the original collision are reversed and the molecules retrace their original paths. In a reverse collision here the initial velocities would be $-C_i'$ and $-Z_i'$ and the final velocities $-C_i$ and $-Z_i$.

demonstrated precisely from the later considerations of Chapter IX, Sec. 7. Now, for the inverse replenishing collision the velocity entering the collision must also be g_i'. Since the line of centers is the same for the two collisions, it follows that this velocity must again lie in the plane of the original collision, that is, the plane of collision for the original and inverse collisions is identical in the present frame of reference. The orientation of this plane is still specified by the same angle ε as originally defined. It

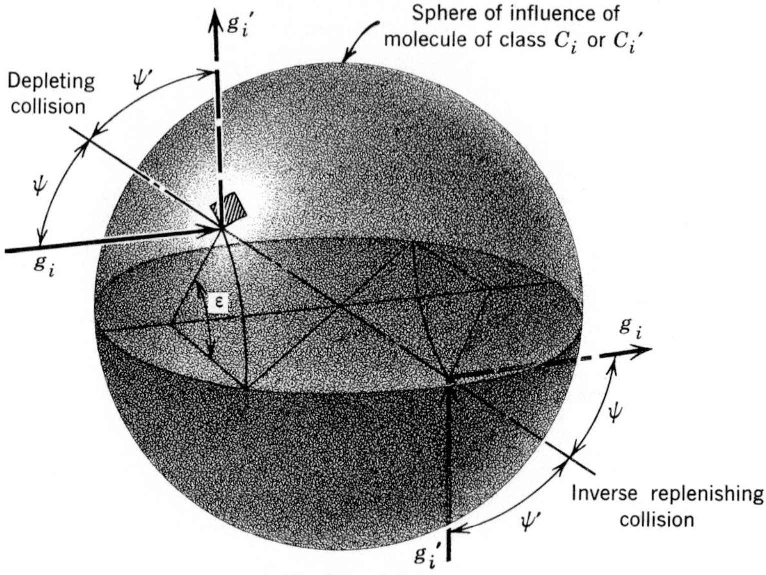

Fig. 7. Details of depleting collision and inverse replenishing collision in terms of relative velocities.

is thus easy to see that (4.2) can be applied to the inverse replenishing collisions with the same $d\varepsilon$ as before. Making this application, we thus write for the number of such collisions per unit volume per unit time

$$n^2d^2f(C_i')f(Z_i')g' \sin \psi' \cos \psi' \, d\psi' \, d\varepsilon \, dV_{C'} \, dV_{Z'}.$$

Each such collision restores one molecule to the class C_i. Subtracting the first of the foregoing expressions from the second and making use of the aforementioned fact that $g' = g$ and $\psi' = \psi$, we then obtain the following expression for the net rate of change of the number of molecules of class C_i per unit volume of gas as a result of a given type of depleting collision and its inverse:

$$n^2d^2[f(C_i')f(Z_i') \, dV_{C'} \, dV_{Z'} - f(C_i)f(Z_i) \, dV_C \, dV_Z]g \sin \psi \cos \psi \, d\psi \, d\varepsilon. \tag{4.4}$$

To find the total effect of all collisions on the number of molecules in class C_i, expression (4.4) must be integrated over all types of depleting collisions, that is, over all values of Z_i for a fixed C_i and over all values of ψ and ε. The replenishing collisions are automatically taken care of in this process, since they have been paired off with the depleting collisions.[3] To actually carry out the integration of the first term in the square brackets it would be necessary to transform the variables of integration from C_i' and Z_i' to Z_i, ψ, and ε. This can be done for the arguments of the functions f by direct substitution from the relations of the form (4.3). The transformation of the differential product $dV_{C'}\, dV_{Z'} = dC_1'\, dC_2'\, dC_3'\, dZ_1'\, dZ_2'\, dZ_3'$ can also be carried out on the basis of these relations; again the details will be left to Chapter IX, Sec. 7. As it turns out, we obtain simply $dV_{C'}\, dV_{Z'} = dV_C\, dV_Z$. (This is a special case of an important general dynamic result known as *Liouville's theorem.*) With these substitutions the limits of integration are then from 0 to $\pi/2$ for ψ and from 0 to 2π for ε (see Fig. 5) and over all of velocity space for Z_i. The result gives the total rate of change of the number of molecules in the given class C_i per unit volume of gas as the result of collisions. Since the number of such molecules is $nf(C_i)\, dV_C$, this quantity can also be written

$$\left\{\frac{\partial[nf(C_i)\, dV_C]}{\partial t}\right\}_{\text{coll}} = \left\{\frac{\partial[nf(C_i)]}{\partial t}\right\}_{\text{coll}} dV_C.$$

Equating the two expressions and cancelling the common differential dV_C, we obtain finally

$$\boxed{\left\{\frac{\partial}{\partial t}[nf(C_i)]\right\}_{\text{coll}} = \int_{-\infty}^{\infty}\int_0^{2\pi}\int_0^{\pi/2} n^2 d^2[f(C_i')f(Z_i') - f(C_i)f(Z_i)]g \sin\psi \cos\psi\, d\psi\, d\varepsilon\, dV_Z}, \tag{4.5}$$

where it is understood that C_i' and Z_i' are given by equations of the form (4.3). This expression will play an important role in the development of nonequilibrium kinetic theory in Chapter IX.

For the time being we are interested only in a gas in equilibrium. In this situation the number of molecules in every velocity class must be constant, that is, $\{\partial[nf(C_i)]/\partial t\}_{\text{coll}} = 0$ for all C_i. Equation (4.5) shows that this will be the case if f is such that

$$\boxed{f(C_i')f(Z_i') - f(C_i)f(Z_i) = 0}\,. \tag{4.6}$$

[3] It is apparent that this accounts for *all* replenishing collisions. If there were others, their inverse would be a depleting collision and these are in fact all taken care of in the integration.

As far as we can tell at the moment, this is only a *sufficient* condition for equilibrium. The vanishing of the integral in (4.5) does not demand the vanishing of the integrand but only the cancellation of positive and negative contributions to the integral from different parts of the region of integration. It will be shown in Chapter IX, Sec. 4, however, that (4.6) is also a *necessary* condition for equilibrium, that is, that it is the *only* condition under which the integral will vanish. This shows that in equilibrium it is not enough for the overall effect of all collision processes to be zero; it is necessary that each detailed process and its inverse be individually in balance.

The foregoing is a special instance of the *principle of detailed balancing*, which we shall have occasion to use in several places later. The principle in its general form is a result of quantum theory, and it can be formulated precisely only within the framework of that theory (see Heitler, 1954, pp. 412–414 or Wu and Ohmura, 1962, pp. 423–424). For purposes of this book it may be taken as requiring that in a system at equilibrium each individual molecular process and its inverse proceed, on the average, at the same rate. By the inverse of a given process we mean in each case the process that differs from the given one by interchange of the initial and final conditions. The principle does not, in fact, apply with absolute generality (see preceding references). The exceptions, however, are of no concern in the processes with which we deal.

To arrive at equation (4.5) and thence the condition (4.6), we have for simplicity assumed a billiard-ball model. As will be explained in Chapter IX, Sec. 3, however, a collision integral with form similar to (4.5) and containing the same bracketed terms can be obtained for any type of molecule with a spherically symmetrical force field. The condition (4.6) thus applies to any dilute gas whose molecules satisfy this requirement.

5 MAXWELLIAN DISTRIBUTION—FINAL RESULTS

The problem of finding the velocity distribution function for a gas in equilibrium is now reduced to that of finding a function f that satisfies equation (4.6). This could be done with formal mathematical methods, but it is easier to proceed from our knowledge of mechanics. To do this we take the logarithm of (4.6) and write

$$\ln f(C_i') + \ln f(Z_i') = \ln f(C_i) + \ln f(Z_i). \tag{5.1}$$

Since a collision changes C_i, Z_i to C_i', Z_i', this equation states that there is a certain function $\ln f$ of the molecular velocity such that the sum of that function for the two molecules in a collision is the same both before and after the collision. From mechanics we know four functions of velocity that have this property: the energy $\frac{1}{2}m(C_1^2 + C_2^2 + C_3^2)$ and the three components of momentum mC_1, mC_2, mC_3. Any linear combination of these four quantities will also have the required property, and it can be shown that this is the most general function of which this is so (see, e.g., Kennard, 1938, pp. 42–45). The general solution of equation (5.1) and hence of equation (4.6) is therefore

$$\ln f(C_i) = b\frac{m}{2}(C_1^2 + C_2^2 + C_3^2) + a_1 mC_1 + a_2 mC_2 + a_3 mC_3 + a_4,$$

where b and the a's are constants. That this is in fact a solution can be verified by substitution into (5.1) and use of the conservation equations of mechanics. The foregoing solution can also be written

$$\ln f(C_i) = b\frac{m}{2}[(C_1 - \alpha_1)^2 + (C_2 - \alpha_2)^2 + (C_3 - \alpha_3)^2] + \alpha_4,$$

or finally

$$f(C_i) = A \exp\left\{-\beta\frac{m}{2}[(C_1 - \alpha_1)^2 + (C_2 - \alpha_2)^2 + (C_3 - \alpha_3)^2]\right\}, \quad (5.2)$$

where A, β, α_1, α_2, and α_3 are new constants. The minus sign is used with β in anticipation of the conditions on A and β that will be satisfied later.

The constants in (5.2) can be evaluated with the aid of known conditions on C_i and $f(C_i)$. We note first that $(C_1 - \alpha_1)$ occurs in (5.2) only through its square. Thus corresponding positive and negative values of this quantity occur with equal probability and its average value must obviously be zero, that is, $\overline{C_1 - \alpha_1} = \overline{C_1} - \overline{\alpha_1} = 0$. It follows that $\alpha_1 = \overline{\alpha_1} = \overline{C_1}$, that is, α_1 is equal to the average value of C_1. But for our equilibrium system at rest, the velocity C_i is random and we have $\overline{C_1} = 0$. We therefore obtain $\alpha_1 = 0$, with a similar result for α_2 and α_3. Equation (5.2) thus reduces to

$$f(C_i) = A \exp\left[-\beta\frac{m}{2}(C_1^2 + C_2^2 + C_3^2)\right]. \quad (5.3)$$

To find the remaining constants A and β, we first obtain a relationship between the two by putting (5.3) into equation (2.7), which requires that $\int_{-\infty}^{\infty} f(C_i)\, dV_C = 1$. This gives

$$A\int_{-\infty}^{\infty} e^{-\beta(m/2)C_1^2}\, dC_1 \int_{-\infty}^{\infty} e^{-\beta(m/2)C_2^2}\, dC_2 \int_{-\infty}^{\infty} e^{-\beta(m/2)C_3^2}\, dC_3 = 1.$$

The integral here is one of a general form that will appear often; it is treated in the appendix at the rear of the book. Using the results given there, we find for the typical integral

$$\int_{-\infty}^{\infty} e^{-\beta(m/2)z^2}\,dz = \left(\frac{2\pi}{\beta m}\right)^{1/2},$$

and hence from the foregoing equation we have

$$A = \left(\frac{\beta m}{2\pi}\right)^{3/2}. \tag{5.4}$$

With this relationship, we can now find β by calculating $\overline{C_3^2}$, which is already known from equations (3.2) and (3.3) to be equal to $\overline{C^2}/3 = p/\rho = RT = (k/m)T$. Using equation (2.8), substituting from (5.3) and (5.4), and evaluating the integrals from the results in the appendix, we find

$$\begin{aligned}\overline{C_3^2} &= \int_{-\infty}^{\infty} C_3^2 f(C_i)\,dV_C \\ &= \left(\frac{\beta m}{2\pi}\right)^{3/2} \int_{-\infty}^{\infty}\int_{-\infty}^{\infty}\int_{-\infty}^{\infty} C_3^2 e^{-\beta(m/2)(C_1^2+C_2^2+C_3^2)}\,dC_1\,dC_2\,dC_3 = \frac{1}{\beta m}.\end{aligned} \tag{5.5}$$

Equating this result to $(k/m)T$ gives β finally as

$$\beta = \frac{1}{kT}. \tag{5.6}$$

The value of A follows from (5.4) and is

$$A = \left(\frac{m}{2\pi kT}\right)^{3/2}. \tag{5.7}$$

With these values of A and β, equation (5.3) becomes finally

$$\boxed{f(C_i) = \left(\frac{m}{2\pi kT}\right)^{3/2} \exp\left[-\frac{m}{2kT}(C_1^2 + C_2^2 + C_3^2)\right]}\,. \tag{5.8}$$

This is the famous *Maxwellian distribution* cited at the beginning of the preceding section. If we define

$$\Phi(C_1) = \left(\frac{m}{2\pi kT}\right)^{1/2} \exp\left[-\frac{m}{2kT}C_1^2\right] \tag{5.9}$$

and similarly for the other directions, the probability that a given molecule chosen at random will have velocity in the range C_1, C_2, C_3 to $C_1 + dC_1$, $C_2 + dC_2$, $C_3 + dC_3$ is

$$f(C_i)\,dC_1\,dC_2\,dC_3 = \Phi(C_1)\,dC_1 \times \Phi(C_2)\,dC_2 \times \Phi(C_3)\,dC_3.$$

The right-hand side of this expression is the product of three independent probabilities, each of which gives the probability that the molecule will have a component of velocity in a specified range in *one* of the coordinate directions. Each of the probabilities is independent of the components in the other directions. (This had been assumed by Maxwell.) The distribution function for a single component as given by (5.9) has the form of Gauss's error-distribution curve. This is illustrated in Fig. 8, which

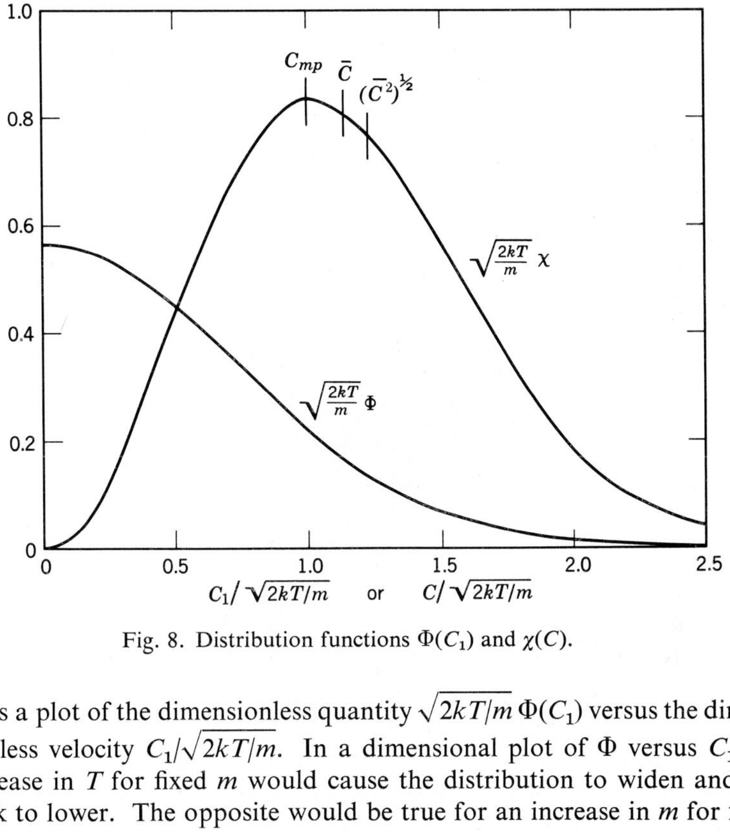

Fig. 8. Distribution functions $\Phi(C_1)$ and $\chi(C)$.

gives a plot of the dimensionless quantity $\sqrt{2kT/m}\,\Phi(C_1)$ versus the dimensionless velocity $C_1/\sqrt{2kT/m}$. In a dimensional plot of Φ versus C_1, an increase in T for fixed m would cause the distribution to widen and the peak to lower. The opposite would be true for an increase in m for fixed T.

It is also useful to know the distribution of the magnitude C of the velocity vector (i.e., the speed) without regard to direction. To obtain this we first transform the expression

$$f(C_i)\,dC_1\,dC_2\,dC_3 = \left(\frac{m}{2\pi kT}\right)^{3/2} e^{-(m/2kT)(C_1^2+C_2^2+C_3^2)}\,dC_1\,dC_2\,dC_3$$

into spherical polar coordinates in velocity space (Fig. 9). The "volume" element on the right then becomes $dC_1\,dC_2\,dC_3 = (C\,d\phi)\;(C\sin\phi\,d\theta)\;dC = C^2\sin\phi\,d\phi\,d\theta\,dC$, and $C_1^2 + C_2^2 + C_3^2$ is replaced by C^2. Redefining the left-hand side, we thus obtain the following expression for the probability that a given molecule has velocity in the range C, ϕ, θ to $C + dC$, $\phi + d\phi$, $\theta + d\theta$:

$$f(C, \phi, \theta)\,dC\,d\phi\,d\theta = \left(\frac{m}{2\pi kT}\right)^{3/2} C^2 e^{-(m/2kT)C^2} \sin\phi\,d\phi\,d\theta\,dC.$$

By integrating over all values of ϕ and θ we find for the probability that a

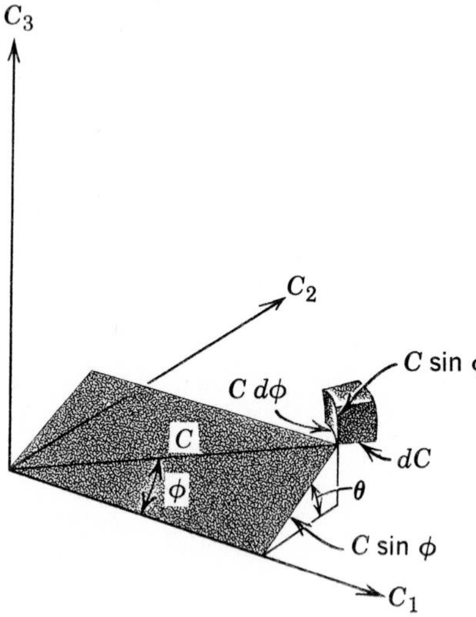

Fig. 9. Polar coordinates and volume element in velocity space.

given molecule has speed between C and $C + dC$ regardless of direction

$$\chi(C)\,dC = \int_0^{\pi} \sin\phi\,d\phi \int_0^{2\pi} d\theta \left(\frac{m}{2\pi kT}\right)^{3/2} C^2 e^{-(m/2kT)C^2}\,dC,$$

whence we obtain

$$\boxed{\chi(C) = 4\pi\left(\frac{m}{2\pi kT}\right)^{3/2} C^2 e^{-(m/2kT)C^2}}. \qquad (5.10)$$

The speed distribution function $\chi(C)$ has the form shown in Fig. 8.

Certain special speeds that can be evaluated with the aid of $\chi(C)$ are of interest:

1. The most probable speed, as determined by the position of the maximum of the distribution curve, is easily found by differentiation and

is

$$C_{mp} = \left(\frac{2kT}{m}\right)^{1/2}. \tag{5.11}$$

2. The average speed $\bar{C}$ is obtained from

$$\bar{C} = \int_0^\infty C\chi(C)\,dC.$$

Using the integrals from the appendix we find

$$\bar{C} = \frac{2}{\pi^{1/2}}\left(\frac{2kT}{m}\right)^{1/2} \cong 1.13C_{mp}. \tag{5.12}$$

3. The root-mean-square speed is obtained from

$$\overline{C^2} = \int_0^\infty C^2\chi(C)\,dC,$$

and turns out to be

$$(\overline{C^2})^{1/2} = \left(\frac{3kT}{m}\right)^{1/2} \cong 1.22C_{mp}. \tag{5.13}$$

This is equivalent to the earlier result of equation (3.3) as it should be.

The position of the foregoing speeds in relation to the curve for $\chi(C)$ is indicated on Fig. 8. They can also be compared with the speed of sound given by $a = (\gamma kT/m)^{1/2}$. For a monatomic gas ($\gamma = \frac{5}{3}$), this gives $a \cong 0.91\, C_{mp}$. The speed of sound is thus of the same order of magnitude as the previous molecular speeds. The value of all these speeds increases as T increases and decreases as m increases.

Exercise 5.1. A perfect gas containing a single species of molecular weight $\hat{M}$ is enclosed in a container at equilibrium at pressure p and temperature T. The gas escapes into vacuum through a small circular hole of area A in the wall of the container. Assume the following ideal conditions: (1) The wall of the container is negligibly thick and is planar in the vicinity of the hole. (2) The diameter of the hole is appreciably smaller than the mean free path, but considerably larger than the molecular diameter.

(a) Show that the number of molecules escaping from the hole per unit area per unit time is given by $n\bar{C}/4$.

(b) Obtain an expression for the rate of mass outflow in terms of (and only of) p, T, $\hat{M}$, A, and $\hat{R}$.

(c) Show that the mean kinetic energy of the escaping molecules is greater than that of the molecules inside the container in the ratio 4/3. Can you give a physical explanation of this result?

Exercise 5.2. Consider a gas of molecules of mass m at equilibrium at temperature T. Let C_t be the magnitude of the component of molecular velocity parallel to one of the coordinate planes in physical space.

(a) Obtain an expression for the fraction of molecules with C_t in the range C_t to $C_t + dC_t$. Check that the integration of this expression over all molecules gives unity.

(b) Show that

$$\frac{(C_t)_{mp}}{C_{mp}} = \left(\frac{1}{2}\right)^{1/2}, \qquad \frac{\bar{C}_t}{\bar{C}} = \frac{\pi}{4}, \qquad \left(\frac{\overline{C_t^2}}{\overline{C^2}}\right)^{1/2} = \left(\frac{2}{3}\right)^{1/2}.$$

Exercise 5.3. Consider again a gas of molecules of mass m at equilibrium at temperature T.

(a) Obtain an expression for the fraction of molecules having kinetic energy $\epsilon \equiv \frac{1}{2}mC^2$ in the range ϵ to $\epsilon + d\epsilon$.

(b) Show that $\epsilon_{mp} = \frac{1}{2}kT$ and $\bar{\epsilon} = \frac{3}{2}kT$.

(c) If the error function (a tabulated function) is defined by

$$\operatorname{erf} x = \frac{2}{\sqrt{\pi}} \int_0^x e^{-z^2}\, dz,$$

show that the fraction of molecules with kinetic energy equal to or greater than a specified value ϵ is given by

$$1 + \frac{2}{\sqrt{\pi}} \sqrt{\frac{\epsilon}{kT}}\, e^{-\epsilon/kT} - \operatorname{erf}\left(\sqrt{\frac{\epsilon}{kT}}\right).$$

(d) What percentage of the total kinetic energy of molecular motion is possessed by the molecules that have an individual kinetic energy less than one-tenth of the average kinetic energy? (Do not attempt to evaluate any integrals in closed form. Use a suitable infinite series instead and retain only the first two terms.) *Ans.:* 0.24%.

6 COLLISION RATE AND MEAN FREE PATH

We are now in a position to make an improved calculation of the frequency of molecular collisions and of the length of the mean free path. The results will be of particular use when we take up the collision theory of chemical reaction rates in Chapter VII. With this in mind, we shall generalize the work somewhat here by considering collisions between two different molecular species A and B of a gas mixture. (The mixture may also contain other species, but that will not concern us for the moment.) We assume once again the billiard-ball model, to which the formulas of this section are as they stand specifically restricted.

Expression (4.2) for the frequency of collisions between billiard-ball molecules of classes U_i and Y_i is easily generalized to collisions between different species. The development is fairly obvious, and we leave the details to the student. In the generalized expression, $f(U_i)$ is replaced by

$f_A(C_i)$, where f_A is the distribution function for molecules of species A and C_i now denotes any velocity for molecules of this particular species. The function $f(Y_i)$ is similarly replaced by $f_B(Z_i)$, where Z_i is now the velocity for molecules of species B. In place of the common diameter d, we obtain the average diameter $d_{AB} \equiv (d_A + d_B)/2$, where d_A and d_B are the diameters of the molecules of the two species. The result is the following expression for the frequency of collisions per unit volume of gas between molecules of species A of class C_i and molecules of species B of class Z_i:

$$n_A n_B \, d_{AB}^2 f_A(C_i) f_B(Z_i) g \sin \psi \cos \psi \, d\psi \, d\varepsilon \, dV_C \, dV_Z. \tag{6.1}$$

Here n_A and n_B are the number density of molecules of the two species, and g is the relative speed

$$g = [(Z_1 - C_1)^2 + (Z_2 - C_2)^2 + (Z_3 - C_3)^2]^{1/2}. \tag{6.2}$$

Expression (6.1) becomes identical to expression (4.2) when only a single species is present. Although we shall not go into the details here, it can be shown (see Jeans, 1940, pp. 114–115) that in a mixture of gases each species individually follows a Maxwellian distribution at the common equilibrium temperature of the mixture. Replacing the f's by means of equation (5.8) we thus obtain in place of (6.1)

$$n_A n_B \frac{(m_A m_B)^{3/2}}{(2\pi kT)^3} d_{AB}^2 g e^{-(1/2kT)(m_A C^2 + m_B Z^2)} \times \sin \psi \cos \psi \, d\psi \, d\varepsilon \, dC_1 \, dC_2 \, dC_3 \, dZ_1 \, dZ_2 \, dZ_3, \tag{6.3}$$

where $C^2 = C_1^2 + C_2^2 + C_3^2$ and $Z^2 = Z_1^2 + Z_2^2 + Z_3^2$ and the volume elements dV_C and dV_Z in velocity space have been written out explicitly.

We now want to find the total frequency of collisions between molecules of species A and B. For this it is necessary to integrate expression (6.3) over all directions of the line of centers and over all velocity classes for both species. The integration with respect to ψ and ε, which defines the direction of the line of centers, is trivial and could be carried out immediately; we choose to defer it, however, since some of the intermediate expressions will be of use in later work. The integration with respect to the velocity classes, on the other hand, is difficult because of the appearance of the components of C_i and Z_i in a complicated way in g.

To circumvent this difficulty, we transform the variables of integration from the components of velocity of the molecules themselves to the components of the relative velocity and of the velocity of the center of mass. The x_1-component of the velocity g_i of molecule B relative to molecule A is

$$g_1 = Z_1 - C_1, \tag{6.4}$$

and similarly for the other components. The x_1-coordinate of the center of mass of the molecule pair is

$$x_{c_1} = \frac{m_A x_{A_1} + m_B x_{B_1}}{m_A + m_B},$$

from which it follows by differentiation that the x_1-component of the velocity W_i of the center of mass is

$$W_1 = \frac{m_A C_1 + m_B Z_1}{m_A + m_B}. \tag{6.5}$$

From equations (6.4) and (6.5) we easily obtain

$$C_1 = W_1 - \frac{m_B}{m_A + m_B} g_1, \qquad Z_1 = W_1 + \frac{m_A}{m_A + m_B} g_1. \tag{6.6}$$

These and the corresponding equations in the other coordinate directions give the desired transformation. By means of these equations we can write

$$m_A C_1^2 + m_B Z_1^2 = (m_A + m_B) W_1^2 + \frac{m_A m_B}{m_A + m_B} g_1^2.$$

Combining equations of this type for each component and defining the *reduced mass*

$$m_{AB}^* \equiv \frac{m_A m_B}{m_A + m_B}, \tag{6.7}$$

we obtain for the sum appearing in the exponential factor of (6.3)

$$\tfrac{1}{2} m_A C^2 + \tfrac{1}{2} m_B Z^2 = \tfrac{1}{2}(m_A + m_B) W^2 + \tfrac{1}{2} m_{AB}^* g^2. \tag{6.8}$$

To transform the differentials we have, for the x_1-direction,

$$dC_1\, dZ_1 = \frac{\partial(C_1, Z_1)}{\partial(W_1, g_1)}\, dW_1\, dg_1, \tag{6.9}$$

where the Jacobian of the transformation is found from (6.6) as

$$\frac{\partial(C_1, Z_1)}{\partial(W_1, g_1)} = \begin{vmatrix} \dfrac{\partial C_1}{\partial W_1} & \dfrac{\partial C_1}{\partial g_1} \\[2ex] \dfrac{\partial Z_1}{\partial W_1} & \dfrac{\partial Z_1}{\partial g_1} \end{vmatrix} = \begin{vmatrix} 1 & -\dfrac{m_B}{m_A + m_B} \\[2ex] 1 & \dfrac{m_A}{m_A + m_B} \end{vmatrix} = 1. \tag{6.10}$$

Corresponding results hold for the other directions. With these relations, the expression (6.3) can be rewritten

$$n_A n_B \frac{(m_A m_B)^{3/2}}{(2\pi k T)^3} d_{AB}^2 g \exp\left\{-\frac{1}{2kT}\left[(m_A + m_B) W^2 + m_{AB}^* g^2\right]\right\}$$
$$\times \sin\psi \cos\psi\, d\psi\, d\varepsilon\, dW_1\, dW_2\, dW_3\, dg_1\, dg_2\, dg_3. \tag{6.11}$$

The next step is to transform the "volume" elements in each of the new velocity spaces W_i and g_i from Cartesian to spherical coordinates by the relations (cf. Sec. 5)

$$\begin{aligned} dW_1\,dW_2\,dW_3 &= W^2 \sin\phi_W\,d\phi_W\,d\theta_W\,dW, \\ dg_1\,dg_2\,dg_3 &= g^2 \sin\phi_g\,d\phi_g\,d\theta_g\,dg. \end{aligned} \tag{6.12}$$

With this transformation, the required integrations can be readily carried out, since the expression to be integrated now has the form of a product of functions of the individual variables of integration.

The integration will first be made over all possible velocities of the center of mass. To do this we integrate with respect to ϕ_W, θ_W, and W from, respectively, 0 to π, 0 to 2π, and 0 to ∞. (The integration with respect to W requires the use of one of the integrals in the appendix.) This leads finally to the expression

$$n_A n_B \left(\frac{\overset{*}{m}_{AB}}{2\pi kT}\right)^{3/2} d_{AB}^2 g^3 e^{-(\overset{*}{m}_{AB}/2kT)g^2} \sin\phi_g \sin\psi\cos\psi\,d\psi\,d\varepsilon\,d\phi_g\,d\theta_g\,dg. \tag{6.13}$$

To obtain all possible velocities of the center of mass we have obviously had to take account of all possible absolute velocities of both species of molecule. The foregoing integration thus, in effect, encompasses all possible velocities of both species. The expression (6.13) is therefore to be interpreted as giving the frequency of collisions per unit volume of gas between *all* molecules of species A and *all* molecules of species B and such that the relative velocity is in the range g, ϕ_g, θ_g to $g + dg$, $\phi_g + d\phi_g$, $\theta_g + d\theta_g$, and the direction of the line of centers in the range ψ, ε to $\psi + d\psi$, $\varepsilon + d\varepsilon$. By "all" molecules of a given species we mean all classes of absolute velocity of that species. This is in contrast to the limited classes of velocity accounted for in the original expression (6.1).

We now take account of all possible directions (but not magnitudes) of the relative velocity by integrating with respect to ϕ_g from 0 to π and with respect to θ_g from 0 to 2π. To obtain an expression that we shall need later, we also take partial account of the possible directions of the line of centers by integrating with respect to ε over its range from 0 to 2π. We thus obtain the following expression for the frequency of collisions per unit volume of gas between all molecules of species A and all molecules of species B and such that the relative speed is in the range g to $g + dg$ and the line of centers lies in the angular range ψ to $\psi + d\psi$ measured from the direction of the relative motion:

$$8\pi^2 n_A n_B \left(\frac{\overset{*}{m}_{AB}}{2\pi kT}\right)^{3/2} d_{AB}^2 g^3 e^{-(\overset{*}{m}_{AB}/2kT)g^2} \sin\psi\cos\psi\,d\psi\,dg. \tag{6.14}$$

This expression will be used in the development of the collision theory of chemical reaction rates in Chapter VII.

Finally, we can find the total frequency of collisions per unit volume of gas between molecules of species A and B by integrating with respect to ψ from 0 to $\pi/2$ and with respect to g from 0 to ∞. This so-called *bimolecular collision rate*, denoted by Z_{AB}, comes out to be

$$Z_{AB} = n_A n_B d_{AB}^2 \left(\frac{8\pi kT}{m_{AB}^*}\right)^{1/2}. \tag{6.15a}$$

Equation (6.15a) as it stands cannot be used to obtain the mutual collision rate within a given species A by formally replacing B with A. In performing the integrations necessary to obtain the foregoing results from the original expression (6.1) we have, as already pointed out, summed over all velocity classes of both collision partners. If we consider two specific velocity classes U_i and Y_i with no regard as to which species is associated with which velocity, a collision between molecules with these velocities will appear in the summation twice, once with $C_i = U_i$, $Z_i = Y_i$ and once with $C_i = Y_i$, $Z_i = U_i$. In collisions of unlike molecules these are two distinct cases. In collisions of like molecules, however, they are identical, that is, interchanging the molecules between the velocity classes does not provide a different collision. This is illustrated in Fig. 10, where in (*b*) only the artifice of different intensities of shading for the two collision partners allows us to recognize that an interchange has been made; without this artifice the interchange is unrecognizable and therefore meaningless. If we had carried through the foregoing analysis for like molecules we would thus have counted each collision twice. To obtain the collision rate within a given species we must therefore divide the integrated results by 2. The same formulas can be made to serve for both cases if we write the bimolecular collision rate (6.15a), for example, as

$$\boxed{Z_{AB} = \frac{n_A n_B}{\sigma} d_{AB}^2 \left(\frac{8\pi kT}{m_{AB}^*}\right)^{1/2}}, \tag{6.15b}$$

where σ is a symmetry factor that is 1 for unlike molecules ($B \not\equiv A$) and 2 for like molecules ($B \equiv A$). It should be noted that the reduced mass m_{AB}^* does not go over into the molecular mass when the results are applied to like molecules. Instead we have $m_{AA}^* = m_A m_A/(m_A + m_A) = m_A/2$. Exercise 6.2 gives an application of the ideas of this paragraph and may be helpful in understanding them.

The foregoing results can be used to find the mean free path as follows. Each collision included in Z_{AB} in equation (6.15a) terminates one free path for one species of molecule—say A—by collision with another species—say B. The formula (6.15b) with $\sigma = 1$ can thus be interpreted as also giving the number of free paths of the n_A molecules of species A per unit volume of gas that are terminated per unit time through collisions with species B. In the case of like molecules, however, each collision terminates *two* free paths for molecules of the given species. To find the number of free paths terminated per unit time by the n_A molecules of

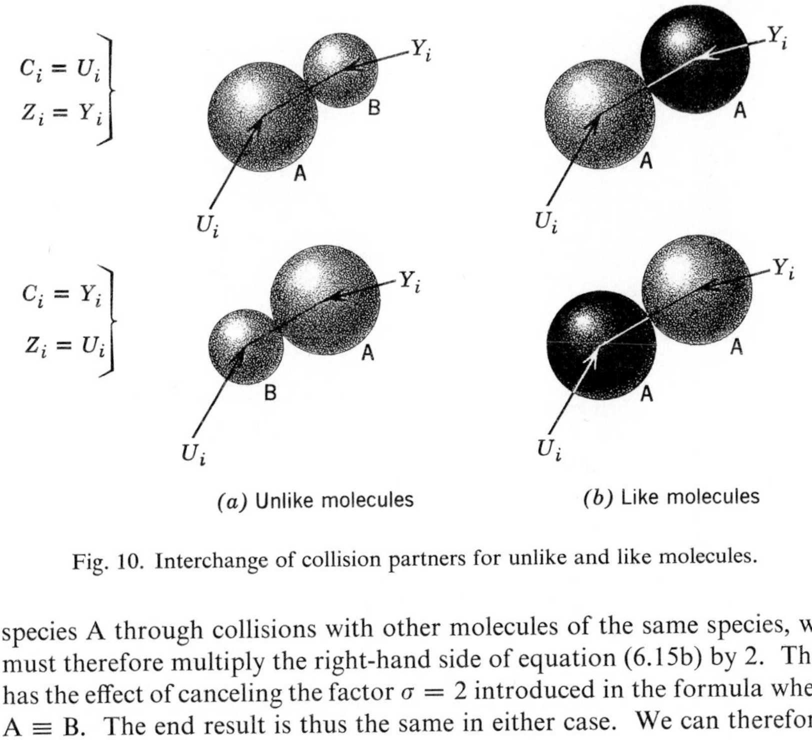

Fig. 10. Interchange of collision partners for unlike and like molecules.

species A through collisions with other molecules of the same species, we must therefore multiply the right-hand side of equation (6.15b) by 2. This has the effect of canceling the factor $\sigma = 2$ introduced in the formula when $A \equiv B$. The end result is thus the same in either case. We can therefore write the common expression

$$n_A n_Y \, d_{AY}^2 \left(\frac{8\pi kT}{m_{AY}^*}\right)^{1/2}$$

for the number of free paths terminated by n_A molecules of species A per unit volume of gas through collisions with some species Y, and this is correct whether Y is species A itself or some other species. The number of free paths terminated per unit time by one of the molecules out of the n_A

can be found by dividing the foregoing result by n_A and is

$$\Theta_{AY} = n_Y\, d_{AY}^2 \left(\frac{8\pi kT}{m_{AY}^*}\right)^{1/2}.$$

Since this equation holds for any target species Y, the number of free paths terminated per unit time by one molecule of species A through collisions with all species in a complex mixture can be found by summation and is

$$\Theta_A = \sum_Y \Theta_{AY} = \sqrt{8\pi kT} \sum_Y \frac{n_Y\, d_{AY}^2}{\sqrt{m_{AY}^*}} = \sqrt{8\pi kT} \sum_Y n_Y\, d_{AY}^2 \sqrt{\frac{1}{m_A} + \frac{1}{m_Y}},$$

where the summation extends over all species in the mixture. Now, the average distance traveled by a molecule of species A per unit time is [see equation (5.12)]

$$\bar{C}_A = \sqrt{8kT/\pi m_A}.$$

It follows that the average distance traveled between each termination of a free path, which is the mean free path for molecules of species A, is

$$\lambda_A = \frac{\bar{C}_A}{\Theta_A} = \frac{1}{\pi \sum_Y n_Y\, d_{AY}^2 \sqrt{1 + (m_A/m_Y)}}. \tag{6.16}$$

When there is only a single species present this reduces to

$$\lambda = \frac{1}{\sqrt{2}\,\pi\, d^2 n},$$

which was cited without proof as equation (I 4.3a).

Exercise 6.1. Measurements made in argon (molecular weight $\simeq$ 40) at a temperature of 300°K and a mass density of 2×10^{-3} gm/cm³ give a coefficient of viscosity of 2100×10^{-7} gm/cm sec. (These values are not strictly correct, having been altered slightly to facilitate the numerical work.) Calculate the rate of collision between pairs of argon molecules. *Ans.:* 1.14×10^{29}/cm³sec.

Exercise 6.2. (a) Consider a gas consisting of a single molecular species with mass m, diameter d, number density n, and temperature T. Write the expression for the rate Z of bimolecular collisions per unit volume.

(b) Consider a gas made up of two molecular species A and B with mass m_A and m_B, diameter d_A and d_B, and number density n_A and n_B, also at the temperature T. Obtain an expression for the *total* rate Z of bimolecular collisions *of all kinds* per unit volume.

(c) Show that the result of (b) reduces to that of (a) if the two species are made identical and the total number density is unaltered.

Exercise 6.3. This exercise will serve to review many of the ideas and procedures of Chapters I and II. Consider a gas whose molecules are flat discs sliding on a frictionless plane (e.g., poker chips sliding on a frictionless billiard table). Assume that the molecules exhibit those properties of the familiar billiard-ball model that are consistent with their two-dimensional motion. Supposing that the gas obeys the equation of state $p/\rho = RT$, where now p is the force per unit *length* and ρ is the density per unit *area*, find the following:

(a) The relation connecting p, ρ, and $\overline{C^2}$ using the methods of Chapter I, Sec. 3. Check whether the principle of equipartition of energy holds.

(b) An approximate expression for the mean free path, using the methods of Chapter I, Sec. 4.

(c) The velocity distribution function at equilibrium. (Use either the arguments of Chapter II, Secs. 4 and 5, or the *functional form* of the three-dimensional distribution function plus some suitable reasoning.) Find $\bar{C}$ and $\overline{C^2}$ and compare with the results of (a).

(d) The bimolecular collision rate for unlike molecules, using the method of this section. Compute the mean free path λ when there is only one species present, and compare with the result of (b).

7 CHEMICAL EQUILIBRIUM AND THE LAW OF MASS ACTION

In the foregoing sections we have supposed that any gas mixtures with which we were concerned were chemically inert, so that no changes in molecular structure took place upon collision. In this section we consider briefly the conditions of equilibrium in a chemically reacting mixture of gases from the kinetic point of view. We shall later treat the same problem more extensively with entirely different methods.

To fix our ideas we take up the problem here in terms of a specific reaction, namely, that for the decomposition of hydrogen iodide. The chemical formula, or *chemical equation*, that describes this reaction is

$$2HI \rightleftharpoons H_2 + I_2. \tag{7.1}$$

The numbers that precede the chemical symbols and that "balance" the equation are called the *stoichiometric coefficients*. The equation expresses the fact that 2 molecules (or moles) of HI must disappear for every molecule (or mole) each of H_2 and I_2 that appears. In any system in equilibrium both the forward ($\rightharpoonup$) and the backward ($\leftharpoondown$) reactions go on simultaneously and continuously. The seemingly constant state of equilibrium represents a dynamic balance between the forward and backward directions. On a time-averaged basis, there are definite proportions of the various species in the equilibrium mixture, and we wish to know what these proportions are.

For a gas reaction to take place, a minimum requirement is that the reacting molecules must be in collision with each other. It is often necessary further that the molecules have considerably more than the average amount of energy, but this latter requirement need not concern us for the time being. Let us consider first the forward reaction. For decomposition of the HI to occur, there must be a bimolecular collision between two HI molecules. We assume that in a certain fraction of these collisions, which fraction may depend on the temperature, a reaction will occur. The number of reacting collisions per unit volume per unit time is therefore proportional to the corresponding rate of bimolecular collisions. Using equation (6.15b) for this collision rate, we can thus write

$$\left.\begin{array}{l}\text{number of decomposing collisions}\\ \text{per unit volume per unit time}\end{array}\right\} = k_f(T)n_{\mathrm{HI}}^2, \tag{7.2}$$

where $k_f(T)$ is a coefficient depending on the temperature and the physical properties of the molecules. That this expression must depend on the square of the number density of HI molecules can also be argued directly on elementary grounds: If we double the number of HI molecules we plainly double the chance that one will be found at the point where the collision will take place, and furthermore we double the chance that, if one is there, another will be there also.

For the formation of HI (i.e., the backward reaction) to take place, there must be a bimolecular collision between a hydrogen and an iodine molecule. On the same assumptions as before regarding reacting collisions, we can write, again with the aid of equation (6.15b),

$$\left.\begin{array}{l}\text{number of forming collisions}\\ \text{per unit volume per unit time}\end{array}\right\} = k_b(T)n_{\mathrm{H_2}}n_{\mathrm{I_2}}, \tag{7.3}$$

where $k_b(T)$ is the coefficient appropriate to the backward reaction.

For equilibrium to be maintained, the number of decomposing collisions (7.2) must just equal the number of forming collisions (7.3). Equating the two expressions, we find the following condition for the particular values of the number densities at equilibrium, denoted by asterisks:

$$\frac{\overset{*}{n}_{\mathrm{H_2}}\overset{*}{n}_{\mathrm{I_2}}}{\overset{*}{n}{}^2_{\mathrm{HI}}} = \frac{k_f(T)}{k_b(T)} \equiv K(T). \tag{7.4}$$

This is a special case of the famous *law of mass action* first formulated in 1867 by the Norwegian scientists Guldberg and Waage. It is more often written in terms of the *concentration* in moles per unit volume, denoted for any arbitrary species Y by [Y]. Since $n_{\mathrm{Y}} = \hat{N}[\mathrm{Y}]$, where $\hat{N}$ is Avogadro's number, we have in these terms

$$\frac{[\mathrm{H_2}]^*[\mathrm{I_2}]^*}{[\mathrm{HI}]^{*2}} = K_c(T). \tag{7.5}$$

Here $K_c(T)$ is the *equilibrium constant* for concentrations. It is called a "constant" because reaction equilibrium is often studied experimentally at a fixed temperature; actually it is a function of T.

Equation (7.5) is not sufficient in itself to provide a solution for the three quantities $[H_2]^*$, $[I_2]^*$, and $[HI]^*$. We require two additional equations, which are given in this case by the fact that the reaction cannot change the total number of hydrogen and iodine atoms in the system. The equations expressing this fact are

$$\begin{aligned}[HI]^* + 2[H_2]^* &= \frac{n_H}{\hat{N}},\\ [HI]^* + 2[I_2]^* &= \frac{n_I}{\hat{N}},\end{aligned} \tag{7.6}$$

where n_H and n_I are the number of hydrogen and iodine atoms per unit volume irrespective of whether they are present in the pure species or in the hydrogen iodide. Their value presumably would be known or calculable from the given conditions in any particular problem.

From the foregoing equations we can find, for example, the effect on the equilibrium of adding—say—H_2 at constant temperature and volume. If the H_2 did not react, then the left-hand side of (7.5) would increase, which would violate the equation at constant T. The only way that the equality can be maintained is for some of the added H_2 to react with the I_2 already present to form additional HI. This will reduce $[H_2]$ and $[I_2]$ and increase [HI] until the equality (7.5) is satisfied. Such changes are seen to be compatible (as indeed they must) with equations (7.6), where n_H would have been increased by the addition of H_2 while n_I remained unchanged.

Equation (7.5) can be rewritten in terms of the equilibrium partial pressures p_Y^* instead of the concentrations by means of the relation

$$p_Y^* = n_Y^* kT = [Y]^* \hat{R} T,$$

which assumes that each gas in the mixture behaves as a perfect gas. The result, which is typical of a form often encountered, is

$$\frac{p_{H_2}^* p_{I_2}^*}{p_{HI}^{*2}} = K_c(T) = K_p(T). \tag{7.7}$$

Other equivalent forms can be obtained, but we shall not go into them here. Equation (7.5) can also be extended without difficulty to more complex chemical reactions involving a greater number of chemical species. We shall defer this, however, until the next chapter, where we take up an alternative derivation by the methods of chemical thermodynamics.

The foregoing derivation does not give us an expression for the function $K_c(T)$. Indeed the derivation of such an expression by purely kinetic methods is impossible, since such a derivation would require knowledge of the probability that a given collision will result in a reaction, and such knowledge is not available on purely theoretical grounds. Fortunately, conditions of equilibrium can be treated by other, more powerful methods that are capable of giving more complete results within their own restricted sphere. We address ourselves to these methods in the next two chapters.

Exercise 7.1. Consider a unit volume containing a mixture of gaseous hydrogen and iodine reacting according to the chemical equation (7.1). The equilibrium composition at $T = 675°K$ was measured by chemical methods and found to consist of 9.664 moles of HI, 0.168 moles of H_2, and 0.168 moles of I_2. If we add one mole of I_2 to the mixture keeping the volume and temperature constant, what will be the composition after equilibrium is reestablished?

References

Heitler, W., 1954, *The Quantum Theory of Radiation*, Oxford University Press.
Jeans, J., 1940, *An Introduction to the Kinetic Theory of Gases*, Cambridge University Press.
Kennard, E. H., 1938, *Kinetic Theory of Gases*, McGraw-Hill.
Wu, T.-Y., and T. Ohmura, 1962, *Quantum Theory of Scattering*, Prentice-Hall.

Chapter III

Chemical Thermodynamics

1 INTRODUCTION

In this chapter we temporarily forsake the molecular approach to take up certain ideas of chemical thermodynamics. Chemical thermodynamics, like all of thermodynamics, treats matter in bulk, that is, with no regard for its detailed molecular structure. Unlike kinetic theory, which can provide a certain amount of information about the rates of chemical processes, chemical thermodynamics can give only the final, equilibrium conditions. Within this restricted range, however, it provides more complete and satisfactory information—for example, it tells us something about the dependence of the equilibrium constant on the temperature. As always with thermodynamics, the results contain certain quantities, such as specific heats, that cannot be obtained from thermodynamics itself and for which we often resort to measurement. Only when we return to the molecular considerations of statistical mechanics in Chapter IV will we be able to obtain all of the bulk properties in terms of a basic molecular model.

There are two somewhat different approaches to chemical thermodynamics: (1) the classical approach developed originally by J. Willard Gibbs in this country in the latter half of the nineteenth century, and (2) the more recent approach of irreversible, or nonequilibrium, thermodynamics as pioneered by De Donder, Prigogine, and their co-workers in Belgium, starting about 1920. We here follow the second approach, more or less as exemplified in the excellent book by Prigogine (1961). (See also Prigogine and Defay, 1954, and Van Rysselberghe, 1963.) This approach gives a straightforward development of chemical thermodynamics and at the same time provides certain ideas from irreversible thermodynamics that we shall need in our later work. Most texts follow the classical approach, and the student will probably have to familiarize himself with this sooner or later. Useful books for this purpose are those of Zemansky (1957), Rossini (1950), and Denbigh (1963). We assume that the reader is acquainted with the basic ideas and results of classical thermodynamics but has no familiarity with chemical thermodynamics as such or with the ideas of nonequilibrium thermodynamics.

2 THERMODYNAMIC SYSTEMS AND KINDS OF EQUILIBRIUM

Application of the methods of thermodynamics requires the choice of a *thermodynamic system*. A system is any quantity of matter separated from the surroundings by an enclosure of macroscopic dimensions. The enclosure can be any closed mathematical surface; it need not be a solid wall. Systems can be usefully classified according to the exchanges of energy (heat and work) and matter that can take place across the enclosure.

1. *Isolated systems* exchange neither energy nor matter with their surroundings.
2. *Closed systems* exchange energy but not matter.
3. *Open systems* exchange both energy and matter.

We shall concern ourselves in this chapter mainly with isolated or closed systems. The extension to open systems will be made only when required.

Classical thermodynamics concerns itself with systems in equilibrium. Here it is useful to think about three distinct kinds of equilibrium.

1. *Mechanical equilibrium* exists when there is no unbalanced force in the interior of a system or between a system and its surroundings. In the absence of body forces this implies a uniform pressure in the system.
2. *Thermal equilibrium* exists when all parts of a system are at the same temperature and this temperature is the same as that of the surroundings.
3. *Chemical equilibrium* exists when a system has no tendency to undergo a spontaneous change in chemical composition, no matter how slow.

When the conditions for all three kinds of equilibrium are simultaneously satisfied, we say that the system is in a state of *thermodynamic equilibrium*. Such states of complete equilibrium can be described in terms of macroscopic coordinates (i.e., state variables) that do not change with time. Classical thermodynamics can make predictions only about states of thermodynamic equilibrium; it can tell us nothing about the rates at which processes take place. When a process is followed by means of classical thermodynamics it must be regarded as consisting of a succession of states of thermodynamic equilibrium, that is, as taking place infinitely slowly. Such processes are necessarily reversible.

In the physical world we must deal with irreversible processes that are a succession of nonequilibrium states. To do this in any detail the notions of classical thermodynamics must be supplemented by other ideas. For systems in mechanical or thermal nonequilibrium we usually proceed by dividing the system into a large number of subsystems that are of infinitesimal size relative to the original system but still of macroscopic size relative to the molecular structure of the medium. By assuming that each subsystem is in local equilibrium internally (but not necessarily in equilibrium with its surrounding subsystems), we can apply equilibrium thermodynamics and the concept of state variables to the subsystems. We can thus build up, by integration, a picture of the behavior of the entire nonequilibrium system. This is what we normally do in gas dynamics and heat transfer—so successfully, in fact, that we rarely give it a second thought. (The limits of applicability of this approach are, as a matter of fact, a subtle matter that has yet to be clarified in any general sense. The microscopic basis for the procedure will be examined for nonuniform gases in Chapters IX and X.)

When we come to the description of a system in chemical nonequilibrium, the procedure must be somewhat different. We shall here assume that the system is in mechanical and thermal equilibrium and that it is homogeneous in space. We assume that the system has a definite volume V and, since it is in mechanical and thermal equilibrium, we can assign to it a definite pressure p and temperature T. Furthermore, since the system is homogeneous, its composition can be specified by giving the number of moles $\mathcal{N}_s$ of each of the constituent chemical species X_s. (We assume only a single phase, i.e., the gaseous one, throughout.) Thus the thermodynamic state of a system of l chemical species can be completely specified by specifying the values of p, V, T, $\mathcal{N}_1, \mathcal{N}_2, \ldots, \mathcal{N}_l$. Similar to the situation in equilibrium thermodynamics, we assume that the foregoing variables are not independent but are connected by an equation of state of the form

$$p = p(V, T, \mathcal{N}_1, \mathcal{N}_2, \ldots, \mathcal{N}_l). \tag{2.1}$$

Consistent with this we also assume that we can assign to the system values of internal energy E and entropy S given by relations of the same form.

The foregoing description holds for a system in chemical nonequilibrium; it assumes only mechanical and thermal equilibrium. During a change of chemical state, the mole numbers $\mathcal{N}_s$, which determine the chemical composition, will change. In general, for a given set of conditions p, V, T of a given system there will be one set of values of the $\mathcal{N}_s$'s for which the system is in chemical and therefore thermodynamic equilibrium (cf.

Chapter II, Sec. 7). The equations (2.1) that relate the state properties of the system when it is out of equilibrium must obviously reduce to those for thermodynamic equilibrium when the equilibrium values of the $\mathcal{N}_s$'s are substituted. To assure that this requirement is satisfied we make the following assumption, which we shall illustrate below:

> *The state equation for any property of a system in chemical nonequilibrium as a function of any other two properties and all of the $\mathcal{N}_s$'s is identical in form to the corresponding equation for the system in thermodynamic equilibrium.*

The foregoing assumption tells us at once how to obtain a state relation for a system in chemical nonequilibrium from the already known relation for the system in complete equilibrium. For example, for a mixture of thermally perfect gases in thermodynamic equilibrium we know that we can write, following Dalton's law of partial pressures,

$$p = \frac{1}{V}\sum_s \mathcal{N}_s^* \hat{R}T. \tag{2.2}$$

Here the mole numbers $\mathcal{N}_s^*$ are the equilibrium values obtained from equations such as (II 7.5) and (II 7.6) with $[X_s]^*$ replaced by $\mathcal{N}_s^*/V$. To represent the pressure of the system in chemical nonequilibrium, the foregoing assumption requires merely that we remove the asterisk. We thus obtain for the nonequilibrium case

$$p = \frac{1}{V}\sum_s \mathcal{N}_s \hat{R}T. \tag{2.3}$$

This equation obviously reduces to the equilibrium relation (2.2) when the $\mathcal{N}_s$'s have their special equilibrium values, that is, equation (2.2) can now be regarded as a special case of the more general equation (2.3).

The foregoing assumption obviously does not exhaust the mathematical possibilities for assuring the required behavior of the relations in the equilibrium situation. There could, for example, be a multiplying factor in the right-hand side of (2.1) that would reduce to unity when $\mathcal{N}_s = \mathcal{N}_s^*$. Experience indicates that this is not the case and that the assumption is a good one so long as only chemical nonequilibrium is present. When other types of nonequilibrium occur simultaneously, the assumption is not always valid.

In the present chapter we shall use the foregoing ideas primarily for the determination of the equilibrium values of the mole numbers, that is, for a thermodynamic derivation of the law of mass action. Later, however, we shall use them for the treatment of nonequilibrium systems in their own right.

3 CONSERVATION OF MASS

In a closed system, which is our main concern in this chapter, the total mass of the contents cannot change. If chemical nonequilibrium exists, however, the amounts of the individual species will vary. These variations are governed by chemical equations such as that given in Chapter II, Sec. 7 for the decomposition of hydrogen iodide:

$$2\mathrm{HI} \rightleftharpoons \mathrm{H}_2 + \mathrm{I}_2. \tag{3.1}$$

The meaning of this equation, as explained previously, is that the changes in the mole numbers of HI, H_2, and I_2 are in the ratio $-2:+1:+1$. If this is the only reaction occurring, we can thus write for infinitesimal changes $d\mathscr{N}_s$

$$d\mathscr{N}_{\mathrm{HI}}:d\mathscr{N}_{\mathrm{H}_2}:d\mathscr{N}_{\mathrm{I}_2} = -2:+1:+1.$$

We can therefore introduce a common, infinitesimal factor of proportionality $d\xi$ such that

$$\frac{d\mathscr{N}_{\mathrm{HI}}}{-2} = \frac{d\mathscr{N}_{\mathrm{H}_2}}{+1} = \frac{d\mathscr{N}_{\mathrm{I}_2}}{+1} = d\xi,$$

or

$$d\mathscr{N}_{\mathrm{HI}} = -2\,d\xi, \qquad d\mathscr{N}_{\mathrm{H}_2} = +1\,d\xi, \qquad d\mathscr{N}_{\mathrm{I}_2} = +1\,d\xi.$$

The quantity ξ so introduced is called the *degree of advancement* of the reaction. Its advantage is that it allows us to keep track of the changes in the mole numbers of the various species in terms of only a single variable.

The chemical equation for an arbitrary reaction involving l species X_s can be written

$$\alpha_1 X_1 + \alpha_2 X_2 + \cdots + \alpha_j X_j \rightleftharpoons \beta_{j+1} X_{j+1} + \cdots + \beta_l X_l, \tag{3.2}$$

where the species on the left are the *reactants* with stoichiometric coefficients α_s and those on the right are the *products* with stoichiometric coefficients β_s. Which are taken as reactants and which as products is purely a matter of choice; the reaction may proceed in either direction and we may interchange the left- and right-hand sides as we wish. Once our choice has been made, however, we must stay with it consistently. To write mole-number relations for equation (3.2), it is convenient to introduce a new stoichiometric coefficient ν_s such that $\nu_s = -\alpha_s$ for reactants and $\nu_s = \beta_s$ for products. The stoichiometric coefficient is thus counted as negative for reactants and positive for products. With this convention the proportionality relation for the infinitesimal changes $d\mathcal{N}_s$ can be generalized to

$$d\mathcal{N}_1 : d\mathcal{N}_2 : \cdots : d\mathcal{N}_l = \nu_1 : \nu_2 : \cdots : \nu_l.$$

From this we obtain, in terms of the common factor of proportionality $d\xi$,

$$\boxed{\frac{d\mathcal{N}_1}{\nu_1} = \frac{d\mathcal{N}_2}{\nu_2} = \cdots = \frac{d\mathcal{N}_l}{\nu_l} = d\xi}, \tag{3.3a}$$

or

$$\begin{aligned} d\mathcal{N}_1 &= \nu_1\, d\xi, \\ &\cdots \\ &\cdots \\ d\mathcal{N}_l &= \nu_l\, d\xi. \end{aligned} \tag{3.3b}$$

If $\mathcal{N}_{s_0}$ denotes the number of moles of the various species at some initial or reference condition at which ξ is counted as zero, equations (3.3b) can be integrated to obtain

$$\begin{aligned} \mathcal{N}_1 - \mathcal{N}_{1_0} &= \nu_1 \xi, \\ &\cdots \\ &\cdots \\ \mathcal{N}_l - \mathcal{N}_{l_0} &= \nu_l \xi. \end{aligned} \tag{3.4}$$

For example, if the initial condition for the reaction (3.1) is one containing only HI, we have

$$\begin{aligned} \mathcal{N}_{HI} &= (\mathcal{N}_{HI})_0 - 2\xi, \\ \mathcal{N}_{H_2} &= \xi, \\ \mathcal{N}_{I_2} &= \xi. \end{aligned} \tag{3.5}$$

It follows from (3.4) that for a closed system in which a single reaction occurs we may replace the $\mathcal{N}_s$'s in the thermodynamic state relations by the quantities $\mathcal{N}_{s_0}$ and the degree of advancement ξ. For a system in which the composition at some reference condition is known, the thermochemical state of the system can thus be specified by giving two of the physical coordinates, say V and T, and the single chemical variable ξ. For example, the state relation (2.1) can be written

$$p = p(V, T, \xi).$$

The quantity ξ can thus itself be regarded as a state variable. Chemical equilibrium for a given V and T will correspond to certain values $\mathcal{N}_s^*$ of $\mathcal{N}_s$ and hence to a specific value ξ^* of ξ.

The foregoing considerations have been for a single reaction. They can easily be extended to multiple, simultaneous reactions by introducing a degree of advancement for each reaction, writing down equations similar to (3.3b) for the contribution of each reaction, and summing over all reactions.

Exercise 3.1. What is the relation of equations (3.5) to the atom-conservation equations (II 7.6)? What is the equation for the equilibrium value $\xi^* = \xi^*(V, T)$ for reaction (3.1)?

4 CONSERVATION OF ENERGY; FIRST LAW

The law of conservation of energy (or first law of thermodynamics) postulates the existence of a function of state called the internal energy E of the system and relates the change of this function to the flow of energy from the surroundings. For a closed system undergoing an infinitesimal reversible process, the law is usually expressed in classical (i.e., equilibrium) thermodynamics by the equation

$$\boxed{dE = dQ - p\,dV}, \tag{4.1}$$

where dQ is the amount of heat received by the system from the surroundings and p and V are the pressure and volume of the system. Here E and p are state functions in the usual sense of equilibrium thermodynamics.[1] For simple systems they are related to the other state variables V and T by state equations of the form

$$p = p(V, T), \qquad E = E(V, T). \tag{4.2}$$

The justification of equation (4.1) lies, of course, in experience.

For the study of chemical nonequilibrium, the law of conservation of energy can still be applied in the form (4.1). It is only necessary that we redefine p and E in the way indicated in Sec. 2. The state equations (4.2) are thus replaced by

$$p = p(V, T, \mathcal{N}_1, \ldots, \mathcal{N}_l), \qquad E = E(V, T, \mathcal{N}_1, \ldots, \mathcal{N}_l), \tag{4.3a}$$

or, for a system with a single reaction, by

$$p = p(V, T, \xi), \qquad E = E(V, T, \xi). \tag{4.3b}$$

[1] The reader is reminded that the change dE between two given adjacent states, being the differential of a state variable, does not depend on the nature of the process; dQ is not the differential of a state variable and does depend on the process.

When the system is in equilibrium, the mole numbers or the degree of advancement are given in turn by equations of the form $\mathcal{N}_s^* = \mathcal{N}_s^*(V, T)$ or $\xi^* = \xi^*(V, T)$. The state equations (4.3) then reduce to the equilibrium relations (4.2). The situation for complete thermodynamic equilibrium thus appears again as a special, limiting case of chemical nonequilibrium. The justification for the extension of equation (4.1) to nonequilibrium situations lies again in experience. (For an interesting discussion of the ideas underlying the first law, see the book by Bridgman, 1952.)

In nonequilibrium thermodynamics, as in equilibrium, it is useful to define the enthalpy given by the equation

$$H \equiv E + pV. \tag{4.4}$$

For chemical nonequilibrium, H is thus also given by a state relation of the form

$$H = H(V, T, \mathcal{N}_1, \ldots, \mathcal{N}_l) \qquad \text{or} \qquad H = H(V, T, \xi). \tag{4.5}$$

With the definition (4.4), equation (4.1) for the first law can be rewritten

$$dH = dQ + V\,dp. \tag{4.6}$$

When applied to a unit mass, as is often necessary in gas dynamics, equations (4.1) and (4.6) can be written

$$de = dq - p\,d\left(\frac{1}{\rho}\right), \tag{4.7}$$

$$dh = dq + \frac{1}{\rho}\,dp, \tag{4.8}$$

where e and h are quantities per unit mass, dq is the heat received per unit mass, and ρ is the mass density.

The actual evaluation of E and H for application of the first law to chemically reacting systems requires some care. This will be taken up in principle in Sec. 8 of this chapter and in detail in Chapter V, Sec. 5.

5 THE SECOND LAW

The second law of thermodynamics postulates the existence of a state function called the entropy and defines the basic properties of this function. One of the basic differences between equilibrium and nonequilibrium thermodynamics is in the way in which the second law is stated.

In classical—or equilibrium—thermodynamics, with which the reader is presumed familiar, the law is often stated in two parts as follows:

There exists a variable of state called the entropy S, such that:

1. For a closed system that undergoes a change from one state of thermodynamic equilibrium A to another such state B, the change in entropy is given by

$$S_B - S_A = \int_A^B \left(\frac{dQ}{T}\right)_{\text{rev}}, \tag{5.1a}$$

where dQ is the heat received and T is the corresponding absolute temperature (see below) for any reversible path leading from A to B.

2. For the same system undergoing an irreversible (i.e., real) process between the same two equilibrium end states A and B we must have

$$S_B - S_A > \int_A^B \left(\frac{dQ}{T}\right)_{\text{irrev}}, \tag{5.1b}$$

where dQ is the heat received in the particular process in question.

The temperature T is in both cases the absolute temperature of the reservoir that supplies the (positive or negative) elements of heat dQ. Since in a reversible process the temperature of the reservoir and the temperature of the system are equal, T in equation (5.1a) is also the temperature of the system to which the heat is supplied. For the irreversible process of (5.1b), the temperature of the system is undefined within the context of classical thermodynamics.

Part 1 of the second law tells us how to evaluate the change in entropy between two equilibrium states; we merely devise a *hypothetical reversible* path between the two states (such paths can always be found and any one will do) and evaluate the integral $\int dQ/T$ over this path. Part 2 puts a restriction on how any irreversible process between two equilibrium end states must proceed; it must proceed such that the integral $\int dQ/T$ evaluated for the actual irreversible process is less than the change in entropy *as calculated by the prescription of part* 1. The justification for this statement of the second law resides in the fact that it can, by a purely logical process, be transformed into equivalent statements about heat flow, perpetual-motion machines, etc. that are in accord with experience.

In irreversible—or nonequilibrium—thermodynamics the second law can be stated as follows:

There exists a variable of state called the entropy S, such that:

1. For any system whatever undergoing any process whatever, the

change in entropy dS can be split into two parts

$$\boxed{dS = d_eS + d_iS}\,, \tag{5.2a}$$

where d_eS is the flow of entropy into the system from the surroundings and d_iS is the production of entropy by irreversible processes within the system.

2. The entropy production d_iS is never negative. It is zero if the system undergoes only reversible processes and positive if it undergoes irreversible processes, that is,

$$\boxed{\begin{array}{ll} d_iS = 0 & \text{(reversible processes)} \\ d_iS > 0 & \text{(irreversible processes)} \end{array}}\,. \tag{5.2b}$$

3. For a *closed* system undergoing any process whatever (reversible or irreversible), d_eS is given by

$$\boxed{d_eS = \frac{dQ}{T} \qquad \text{(closed system, any process)}}\,, \tag{5.2c}$$

where dQ is the heat received by the system at absolute temperature T.

The temperature T is here the absolute temperature of the system itself, which is assumed to be definable in the nonequilibrium situation.

The irreversible statement of the second law is more powerful than the classical statement in that it applies to systems while they are actually in a nonequilibrium state and not merely to equilibrium end states. [This is implied by the differential form of the irreversible equations; strictly speaking, equation (5.1b) of the classical statement cannot logically be written in general in differential form.] That the classical statement is included within the irreversible statement can be seen as follows. For a *closed* system ($d_eS = dQ/T$) undergoing a *reversible* process ($d_iS = 0$), equation (5.2a) reduces to $dS = (dQ/T)_{\text{rev}}$. This can be written after integration as

$$S_B - S_A = \int_A^B \left(\frac{dQ}{T}\right)_{\text{rev}}, \tag{5.3a}$$

which is identical to equation (5.1a) of the classical statement. Part 1 of the classical statement is thus included within the irreversible statement. For a *closed* system undergoing an *irreversible* process, equation (5.2a) becomes, after integration,

$$S_B - S_A = \int_A^B \left(\frac{dQ}{T}\right)_{\text{irrev}} + \int_A^B d_iS. \tag{5.3b}$$

Since for an irreversible process $d_iS > 0$, this equation includes the inequality (5.1b) of the classical statement.[2] Part 2 of the classical statement is thus also included within the irreversible statement. The latter statement, however, is much more powerful than the former in that it replaces an inequality by an equality and thus implies an ability to actually calculate d_iS. A central problem of irreversible thermodynamics, in fact, is the development of formulas for the entropy production d_iS in specific cases. Without them equation (5.3b) would be really no more than part 2 of the classical statement.

The irreversible statement has not been established on as firm a foundation as is possible with the classical statement. It assumes that a meaningful temperature and entropy can be assigned to systems that are out of equilibrium, and that explicit evaluation of the production of entropy can be made under these conditions. For a few cases this procedure can be shown to be completely in accord with the classical ideas; for others (such as heat conduction and viscous flow) it is in accord with well-established practice or with kinetic theory; for still others (explosions, etc.) it is probably not valid. For a critical discussion of the irreversible law and of the formulas for d_iS, the student can refer to the book by Bridgman (1952) and the article by Tolman and Fine (1948). According to Prigogine and Defay (1954, p. xviii), in the present case of partial (i.e., chemical) nonequilibrium the treatment is valid *provided* that the reaction is sufficiently slow that the equilibrium distribution of energy among the component molecules is not disturbed appreciably. The applicability of the irreversible approach is thus *conditional*, in contrast to the classical approach, which is always valid. Thus the irreversible statement, although more powerful, is limited to a more restricted class of problems.

We note that for an isolated system ($dQ = 0$ and hence $d_eS = 0$), equation (5.2a) reduces, with the aid of (5.2b), to

$$dS = d_iS \geqq 0 \quad \text{(isolated system)}. \tag{5.4}$$

This is equivalent to the well-known classical statement that the entropy of an isolated system can never decrease.

[2] A minor difficulty arises here from the fact that T in equation (5.3b) is the temperature T_s of the system whose entropy is being studied whereas T in the inequality (5.1b) is the temperature T_r of the reservoir supplying the heat. This difficulty can be resolved by noting that positive values of dQ must be accompanied by $T_r \geqq T_s$, whereas negative values of dQ must be accompanied by $T_r \leqq T_s$. In either case, it follows that, algebraically speaking, $dQ/T_s \geqq dQ/T_r$, that is, the integrand of the first integral in (5.3b) cannot be smaller than the integrand in (5.1b). The difference in T thus serves to reinforce the statement that (5.1b) is included within (5.3b).

We shall not have space here to go into the various interesting applications of the irreversible form of the second law or to discuss the methods for calculating d_iS for various nonequilibrium phenomena. For this the student should see the article by Tolman and Fine referred to earlier or the book by Prigogine (1961). Here we must restrict ourselves to a discussion of chemical nonequilibrium.

6 THE GIBBS EQUATION FOR A CHEMICALLY REACTING SYSTEM

The calculation of the entropy production in cases of chemical nonequilibrium rests on a formula introduced by Gibbs in his original development of chemical thermodynamics. It can be obtained as a formal, mathematical consequence of the assumption introduced in Sec. 2 with regard to the specification of the state of a system in chemical nonequilibrium. It may be helpful, however, to introduce it on a somewhat more physical basis.

For this purpose it will be convenient to work with a new thermodynamic state variable G, called the *Gibbs free energy*, defined by

$$G \equiv E + pV - TS. \tag{6.1}$$

The Gibbs free energy, like the enthalpy defined in equation (4.4), is a secondary state variable. The entire physical content of thermodynamics has already been embodied in the properties of p, V, T, E, and S, and the development of chemical thermodynamics could be carried out with these primary variables alone. It is purely a matter of convenience that we introduce additional functions. In terms of the enthalpy, G can also be written

$$G = H - TS, \tag{6.2a}$$

or in differential form

$$dG = dH - T\,dS - S\,dT. \tag{6.2b}$$

It follows from these relations that G, like E, V, S, and H, is an extensive property; that is, the total value of the property for the system is equal to the sum of the values for the constituent parts. This is in contrast to the intensive properties such as p and T, which are the same for all parts of a system in mechanical and thermal equilibrium.

On the basis of the first and second laws we can write at once an expression for dG for a closed system undergoing a reversible process.

From the first law for a closed system, equation (4.6) gives $dH = dQ + V\,dp$. With the additional requirement of a reversible process, the second law as stated by equations (5.2) gives $dQ = T\,dS$. Putting these expressions into (6.2b), we obtain

$$dG = V\,dp - S\,dT \quad \text{(closed system, reversible process).} \tag{6.3}$$

We now wish to extend this formula to irreversible chemical processes. To do this let us give up temporarily our restriction to closed systems. We also assume to begin with that no chemical reactions take place, that is, that the various species in the system are inert. Imagine now that infinitesimal amounts $d_e\mathcal{N}_1, d_e\mathcal{N}_2, \ldots, d_e\mathcal{N}_l$ of the various species are

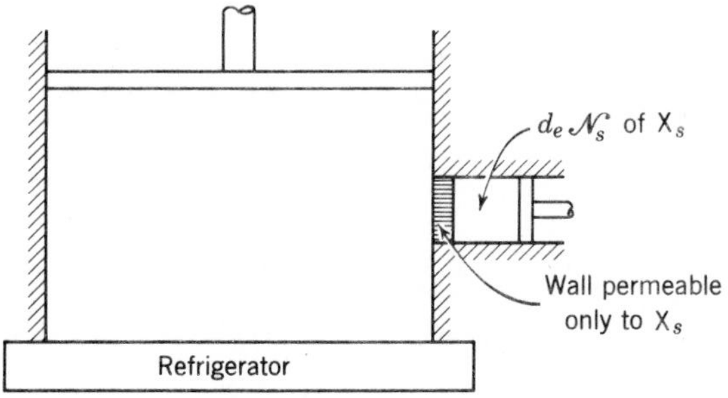

Fig. 1. Device for reversible addition of species at constant p and T.

added *reversibly* from outside the system, which is taken to be the interior of the large cylinder in Fig. 1. This can be accomplished with the aid of a small piston and cylinder, one for each species X_s, separated from the system by a wall permeable only to the given species. (Such *semipermeable membranes* actually exist for some gases and may be imagined generally.) If the infinitesimal amount of gas $d_e\mathcal{N}_s$ in the small cylinder, assumed to be at the same temperature T as the system, is pushed into the system infinitely slowly, the pressure and concentration gradients can be kept as small as we like; and, since the species are inert, the entire addition can be done reversibly.

Now suppose that at the same time a piston is withdrawn slowly in the large cylinder and, if necessary, heat is removed by a refrigerator, both in such a way that the pressure and temperature of the system are held constant. According to equation (6.3) there can then be no changes in G from changes in p and T. Since the system is larger after the mass addition than it was before, however, the value of the extensive property G must obviously have been changed. A term must therefore be added to (6.3) to account for the change. It seems reasonable to assume in doing this that the contribution to dG from the addition of a given species X_s will be proportional to the amount of that species added. If the pertinent factor of proportionality is denoted by $\hat{\mu}_s$, we thus write in place of (6.3)

$$dG = V\,dp - S\,dT + \sum_s \hat{\mu}_s\, d_e\mathscr{N}_s \text{ (open system, reversible process).} \quad (6.4)$$

Equation (6.4) was set up on the assumption that all the species involved were chemically inert and hence that there could be no irreversible processes as the result of chemical reactions within the system. Now let us suppose that chemical reactions occur. In this case the mole numbers may change both by mass addition from outside the system and by

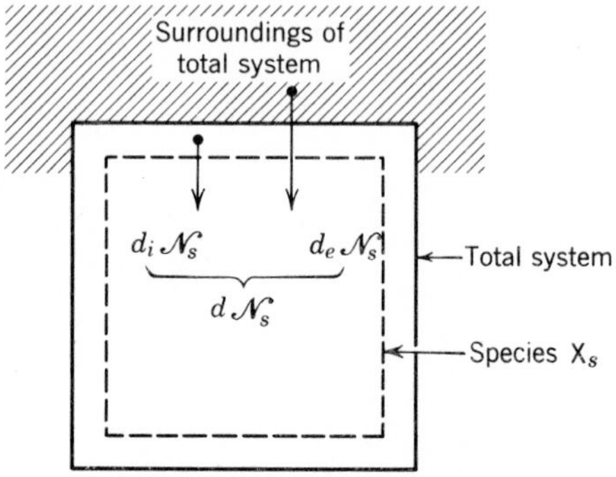

Fig. 2. Total system and coexisting subsystems.

chemical reactions within the system. Under these conditions, an infinitesimal process involving a change in p, T, and the $\mathscr{N}_s$'s will, in general, be irreversible, since chemical equilibrium may not exist. To calculate the corresponding change in G, we assume that equation (6.4) can be used directly. It is only necessary that we drop the subscript e from $d_e\mathscr{N}_s$ in recognition of the fact that the changes in $\mathscr{N}_s$ may come from mass addition to the system or from chemical reactions, or both.

The justification for this extension of (6.4) can be argued on the following grounds. Let us consider that each species X_s of our system constitutes a separate open subsystem contained within, but coextensive with the original total system as shown in Fig. 2. This idea of "separate but coexisting" subsystems is, in a rough sense, similar to our concept of the total pressure of a mixture of perfect gases as being the sum of the partial pressures of the individual gases. According to this concept, the "surroundings" of each subsystem are regarded as consisting of all of the other subsystems plus the surroundings of the original total system. Since each subsystem contains only one species, no chemical reactions can take place *within* it, that is, each such subsystem is internally inert. The chemical reactions of the total system and hence the irreversible processes that accompany them are to be regarded as taking place, so to speak, entirely *between* the subsystems. Assuming that mechanical and thermal equilibrium are maintained at all times, we can thus regard each subsystem as an inert open system with no internal irreversible processes, which means that we can calculate its change in G from equation (6.4) as it stands. We thus write for each species X_s considered as a separate subsystem

$$dG_s = V\,dp_s - S_s\,dT + \hat{\mu}_s\,d\mathcal{N}_s,$$

where now the addition $d\mathcal{N}_s$ from the surroundings of the subsystem will in general include a part $d_e\mathcal{N}_s$ coming from the surroundings of the total system and a part $d_i\mathcal{N}_s$ coming from the other species as a result of the chemical reactions within the total system. We have here in effect assumed a mixture of perfect gases by assuming well-defined partial quantities p_s, etc. for the subsystem independent of the other subsystems. Summing the foregoing equation over all species, that is, over all subsystems, and noting that $G = \sum_s G_s$, $S = \sum_s S_s$, and $p = \sum_s p_s$, we obtain for the total system

$$dG = V\,dp - S\,dT + \sum_s \hat{\mu}_s\,d\mathcal{N}_s \quad \begin{pmatrix}\text{open system, irreversible}\\ \text{chemical processes}\end{pmatrix}. \tag{6.5}$$

This is precisely what is obtained by dropping the subscript e from equation (6.4) in accord with the assumption of the preceding paragraph. Although this plausibility argument for (6.5) has been in terms of a mixture of perfect gases, the equation itself is valid for a homogeneous mixture of any kind.

Equation (6.5) is consistent with our ideas regarding the specification of the state of a system in chemical nonequilibrium. Following the notions in Sec. 2 we can write

$$G = G(p, T, \mathcal{N}_1, \ldots, \mathcal{N}_l), \tag{6.6a}$$

or in differential form

$$dG = \left(\frac{\partial G}{\partial p}\right)_{T,\mathcal{N}_s} dp + \left(\frac{\partial G}{\partial T}\right)_{p,\mathcal{N}_s} dT + \sum_s \left(\frac{\partial G}{\partial \mathcal{N}_s}\right)_{p,T,\mathcal{N}_s'} d\mathcal{N}_s, \tag{6.6b}$$

where the subscript $\mathcal{N}_s'$ in the last term indicates that all mole numbers except $\mathcal{N}_s$ itself are held constant when the derivative is taken. Equation (6.6b) is seen to be of the same form as (6.5). We can therefore write

$$V = \left(\frac{\partial G}{\partial p}\right)_{T,\mathcal{N}_s}, \qquad S = -\left(\frac{\partial G}{\partial T}\right)_{p,\mathcal{N}_s},$$

and most important

$$\boxed{\hat{\mu}_s = \left(\frac{\partial G}{\partial \mathcal{N}_s}\right)_{p,T,\mathcal{N}_s'}} \quad . \tag{6.7}$$

This may, in fact, be taken as the definition of $\hat{\mu}_s$, which was originally introduced as a somewhat indefinitely defined proportionality factor. It is known as the *chemical potential* and plays an important role in chemical thermodynamics. As we shall see, it is an intensive variable and is in general a function of the state of the system as given by—say—p, T, and *all* of the $\mathcal{N}_s$'s. Even though a species may not be present in a system, its chemical potential is nevertheless not zero. There is always the possibility of introducing it into the system, in which case the value of G must be altered and the value of the corresponding $\hat{\mu}_s$ hence must be different from zero.

By taking the differential of the defining equation (6.1) and introducing dG from equation (6.5), we can write

$$\boxed{dS = \frac{1}{T}\,dE + \frac{p}{T}\,dV - \frac{1}{T}\sum_s \hat{\mu}_s\,d\mathcal{N}_s} \quad \begin{pmatrix}\text{open system, irreversible}\\ \text{chemical processes}\end{pmatrix}. \tag{6.8}$$

This may be regarded as the fundamental equation of chemical thermodynamics. We shall refer to it as the *Gibbs equation*, although Gibbs himself introduced it as an expression for dE and had only reversible changes in mind. As written here, it provides the differential form of the state relation

$$S = S(E, V, \mathcal{N}_1, \ldots, \mathcal{N}_l). \tag{6.9}$$

By the procedures of the preceding paragraph, it leads to

$$\hat{\mu}_s = -T\left(\frac{\partial S}{\partial \mathcal{N}_s}\right)_{E,V,\mathcal{N}_s'}, \tag{6.10}$$

and other alternative expressions for the chemical potential can also be obtained (see Exercise 6.1).

One more equation, which follows from (6.6a) and (6.7), will be useful later. Imagine a system at given p and T in which the number of moles of each species, originally of amounts $\mathcal{N}_1, \ldots, \mathcal{N}_l$, is increased by a common factor λ. Since the Gibbs function G is an extensive property, it must also be increased by the same factor, that is,

$$G(p, T, \lambda\mathcal{N}_1, \ldots, \lambda\mathcal{N}_l) = \lambda G(p, T, \mathcal{N}_1, \ldots, \mathcal{N}_l).$$

Differentiating with respect to λ we obtain

$$\frac{\partial G(\ldots, \lambda\mathcal{N}_s, \ldots)}{\partial(\lambda\mathcal{N}_1)} \times \frac{\partial(\lambda\mathcal{N}_1)}{\partial\lambda} + \cdots + \frac{\partial G(\ldots, \lambda\mathcal{N}_s, \ldots)}{\partial(\lambda\mathcal{N}_l)} \times \frac{\partial(\lambda\mathcal{N}_l)}{\partial\lambda} = G,$$

or

$$\frac{\partial G(\ldots, \lambda\mathcal{N}_s, \ldots)}{\partial(\lambda\mathcal{N}_1)}\mathcal{N}_1 + \cdots + \frac{\partial G(\ldots, \lambda\mathcal{N}_s, \ldots)}{\partial(\lambda\mathcal{N}_l)}\mathcal{N}_l = G,$$

where it is understood that p and T are constant. But this equation must be true for all values of λ, in particular for $\lambda = 1$. In this case we obtain

$$\frac{\partial G}{\partial\mathcal{N}_1}\mathcal{N}_1 + \cdots + \frac{\partial G}{\partial\mathcal{N}_l}\mathcal{N}_l = G,$$

or in view of equation (6.7)

$$\boxed{G = \sum_s \hat{\mu}_s \mathcal{N}_s} \quad . \tag{6.11}$$

This result shows clearly that the chemical potentials must be intensive properties. They are in fact the specific Gibbs free energy measured per mole of species and evaluated at the conditions of the mixture. They might therefore be denoted more logically by $\hat{g}_s$, although the notation $\hat{\mu}_s$ is customary. Since $\hat{\mu}_s$ is an intensive property, it can be a function of the mole numbers of the system only through some intensive property such as the corresponding mole fraction $\mathcal{N}_s/\sum_s \mathcal{N}_s$.

Exercise 6.1. Show that the chemical potential is given alternatively by

$$\hat{\mu}_s = \left(\frac{\partial G}{\partial\mathcal{N}_s}\right)_{p,T,\mathcal{N}'_s} = \left(\frac{\partial E}{\partial\mathcal{N}_s}\right)_{S,V,\mathcal{N}'_s} = \left(\frac{\partial H}{\partial\mathcal{N}_s}\right)_{S,p,\mathcal{N}'_s} = \left(\frac{\partial F}{\partial\mathcal{N}_s}\right)_{T,V,\mathcal{N}'_s}.$$

In the last of these, F is the Helmholtz free energy, which is defined by $F \equiv E - TS$ and which will be used in Chapter IV.

7 ENTROPY PRODUCTION IN CHEMICAL NONEQUILIBRIUM; CONDITION FOR REACTION EQUILIBRIUM

With the Gibbs equation (6.8) it is a simple matter to obtain the expression for entropy production in chemical nonequilibrium. Since irreversible chemical processes are included in (6.8), dS can be replaced according to equation (5.2a) of the second law by

$$dS = d_eS + d_iS.$$

If we restrict ourselves once more to closed systems, d_eS is given by equation (5.2c) of the second law as $d_eS = dQ/T$ and dQ is given in turn by the first law (4.1) as $dQ = dE + p\,dV$. We thus have for the flow of entropy into the nonequilibrium system from its surroundings

$$d_eS = \frac{1}{T}\,dE + \frac{p}{T}\,dV.$$

For the closed system the changes in the mole numbers are given entirely by the chemical reactions between the species, so that

$$d\mathcal{N}_s = d_i\mathcal{N}_s.$$

Using these expressions in equation (6.8) we find for the production of entropy by chemical reactions

$$\boxed{d_iS = -\frac{1}{T}\sum_s \hat{\mu}_s\, d_i\mathcal{N}_s}\,. \tag{7.1}$$

For simplicity equation (7.1) has been derived here for a closed system. The identical result can be obtained with slightly more difficulty for chemical reactions in an open system (see Prigogine, 1961, p. 28).

If we restrict ourselves to a single reaction, the changes $d\mathcal{N}_s = d_i\mathcal{N}_s$ for the closed system are given in terms of the change $d\xi$ in the degree of advancement by equation (3.3a), which we now write as

$$d_i\mathcal{N}_s = \nu_s\, d\xi.$$

Again, the same equation can be obtained for a chemical reaction in an open system. Putting this into (7.1) we have in general for the entropy production

$$d_iS = -\frac{1}{T}\left(\sum_s \nu_s\hat{\mu}_s\right) d\xi.$$

Now, the second law requires that d_iS given by this equation be either positive, corresponding to an irreversible process, or zero, corresponding to a reversible process. But saying that a process is reversible in the present context is equivalent to saying that the system is in chemical equilibrium at all times, or, in other words, that the process is one of an infinitely slow succession of states of chemical equilibrium. The condition for chemical equilibrium is therefore that d_iS be zero at all times, which requires that

$$\boxed{\sum_s \nu_s\hat{\mu}_s^* = 0}\,, \tag{7.2}$$

where $\hat{\mu}_s^*$ is the value of the chemical potential at the equilibrium state.

Equation (7.2) is the *equation of reaction equilibrium.* It is not restricted to perfect gases—or even to gases, for that matter—and may be looked on as a universally valid formulation of the law of mass action. The quantity $-\Sigma \nu_s \hat{\mu}_s$ is known in chemical literature as the *affinity* of the chemical reaction and is commonly denoted by the symbol A. The affinity plays a central role in chemical thermodynamics as elaborated by De Donder and his school (Prigogine and Defay, 1954).

8 MIXTURES OF PERFECT GASES

To apply the equations of the preceding section to nonequilibrium mixtures of thermally perfect gases, which are our primary concern in gas dynamics, we must find expressions for $\hat{\mu}_s$ for such mixtures. In line with the basic assumption of Sec. 2, it will suffice for this purpose to study mixtures in thermodynamic equilibrium. But a mixture in a state of thermodynamic equilibrium has a well-defined and fixed composition. For purposes of calculating the state variables (but not necessarily their derivatives) we may therefore ignore the chemical reactions and treat the mixture *as if* it were inert.

The partial pressure of one species in an inert mixture of thermally perfect gases, and hence by the foregoing reasoning in a nonequilibrium reacting mixture of such gases as well, is given by

$$p_s = \frac{1}{V} \mathscr{N}_s \hat{R} T. \tag{8.1}$$

The combined pressure $p = \sum_s p_s$ is then

$$p = \frac{1}{V} \sum_s \mathscr{N}_s \hat{R} T = \frac{1}{V} \mathscr{N} \hat{R} T, \tag{8.2}$$

where we have introduced the notation $\mathscr{N} = \sum_s \mathscr{N}_s$ for the total number of moles. With this notation p_s can also be written

$$p_s = \frac{1}{V} \frac{\mathscr{N}_s}{\mathscr{N}} \mathscr{N} \hat{R} T = \frac{\mathscr{N}_s}{\mathscr{N}} p. \tag{8.3}$$

The *mole fraction* $\mathscr{N}_s/\mathscr{N}$ is sometimes a convenient means of specifying the composition in place of the mole numbers themselves. It is often denoted by the single symbol $x_s \equiv \mathscr{N}_s/\mathscr{N}$, although we shall not do so here. We note that

$$\sum_s \frac{\mathscr{N}_s}{\mathscr{N}} = \frac{1}{\mathscr{N}} \sum_s \mathscr{N}_s = 1,$$

so that if all but one mole fraction is known, the remaining one is determined.

For a perfect gas, the specific energy and enthalpy per mole are functions only of the temperature, which is the same for all gases in the mixture. The combined energy and enthalpy of the mixture are thus easily obtained by summing over the component gases and are

$$E = \sum_s \mathcal{N}_s \hat{e}_s(T) = \sum_s \mathcal{N}_s \left(\int_{T_0}^{T} \hat{c}_{v_s}\, dT + \hat{e}_{s0} \right), \tag{8.4a}$$

$$H = \sum_s \mathcal{N}_s \hat{h}_s(T) = \sum_s \mathcal{N}_s \left(\int_{T_0}^{T} \hat{c}_{p_s}\, dT + \hat{h}_{s0} \right), \tag{8.4b}$$

where $(\hat{\ })_s$ denotes specific quantities measured per mole of species X_s. They are related to the specific quantities measured per unit mass by $(\hat{\ })_s = \hat{M}_s(\ )_s$, where $\hat{M}_s$ is the molecular weight of the species. The quantities $\hat{e}_{s_0}$ and $\hat{h}_{s_0}$ are the values of $\hat{e}_s$ and $\hat{h}_s$ at some reference temperature T_0.

The evaluation of the entropy of the mixture requires more care, since the specific entropy of a perfect gas depends on pressure as well as temperature and the partial pressure is different for the different gases. We might expect that the combined entropy would be the sum of the partial entropies, each calculated in terms of the appropriate partial pressure, that is,

$$S = \sum_s \mathcal{N}_s \hat{s}_s(T, p_s). \tag{8.5}$$

That the foregoing is correct can be seen as follows. Consider a mixture containing only two species X_1 and X_2 and imagine a device capable of separating the mixture *reversibly* into its component species. The device (Fig. 3) consists of two equal compartments each of volume V separated by a fixed wall permeable only to X_1. Two pistons are incorporated, coupled to move together, and made of materials such that one is impermeable and one is permeable only to X_2. The initial state with the mixture in the left-hand compartment is as shown in (*a*). Now suppose that the pistons (which are frictionless) are moved to the right infinitely slowly and that the whole system is kept at a constant temperature. This provides a reversible isothermal process. When the pistons have been moved as far to the right as they will go [Fig. 3(*c*)], the two gases will be completely separated into two compartments, each of which has the same volume as that occupied by the original mixture. Now let us consider an infinitesimal intermediate step in the process [Fig. 3(*b*)]. Since the process is reversible and the gases are inert, the change in entropy is given by the first and second laws as

$$T\, dS = dQ = dE - dW,$$

where dW is the work done on the gas. Since the volume occupied by each gas is constant and the temperature is constant, the pressure of each gas is also constant. Thus at any intermediate position the force on the moving piston, which is permeable to X_2 and hence does not feel the pressure p_2, is p_1 times the piston area. This force is directed toward the left. The force on the solid piston is p_1 times the piston area and is

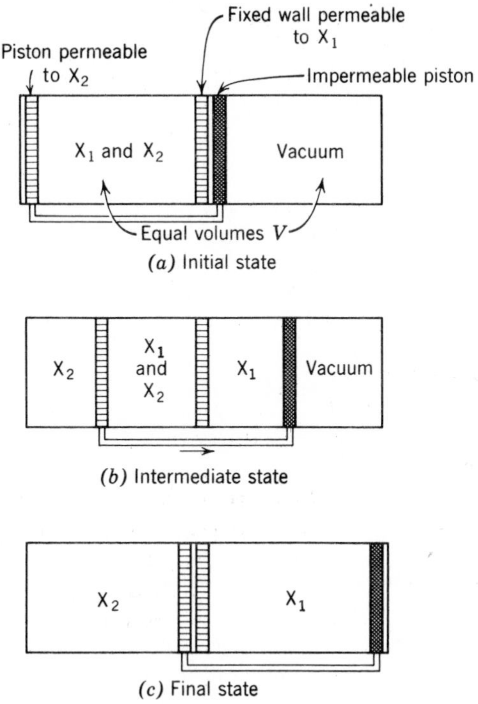

Fig. 3. Reversible separation of two gases.

directed toward the right. It follows that the net force required to move the coupled piston system is zero and hence that $dW = 0$. Also, since the process is isothermal and the internal energy of a perfect gas depends only on T, we have $dE = 0$. It follows from the equation that $dS = 0$ for each infinitesimal step and hence that

$$S_{\text{initial}} = S_{\text{final}}.$$

But S_{initial} is the desired value for the entropy S of the mixture of gases, whereas S_{final} is the sum of the entropies of the component gases each at the common temperature T and at the partial pressure that that gas possesses in the mixture. We therefore have

$$S = S_1(T, p_1) + S_2(T, p_2). \tag{8.6}$$

This is an example of *Gibbs's theorem.* The generalization to any number of component gases is simple. The result, when put in terms of the molar entropies, gives equation (8.5).

By substitution of the well-known expression for the entropy of a thermally perfect gas, equation (8.5) can be written

$$S = \sum_s \mathcal{N}_s\left(\int_{T_0}^{T} \hat{c}_{p_s}\frac{dT}{T} - \hat{R}\ln\frac{p_s}{p_0} + \hat{s}_{s_0}\right), \tag{8.7a}$$

where $\hat{s}_{s_0}$ is the value of $\hat{s}_s$ at the reference temperature T_0 and reference pressure p_0. This result can be put in terms of the complete pressure p by substitution for p_s from equation (8.3). This gives

$$S = \sum_s \mathcal{N}_s\left(\int_{T_0}^{T} \hat{c}_{p_s}\frac{dT}{T} - \hat{R}\ln\frac{p}{p_0} + \hat{s}_{s_0}\right) - \hat{R}\sum_s \mathcal{N}_s \ln\frac{\mathcal{N}_s}{\mathcal{N}}. \tag{8.7b}$$

These are completely equivalent expressions for the entropy of the gas mixture.

The two summations in the expression (8.7b) can be given a physical interpretation. The first summation represents the total entropy that the gases that make up the mixture would have if they were in some way partitioned off in separate compartments and each had the same temperature T and pressure p as was assigned to the mixture. For this to be the case the equation of state requires that each gas would have to occupy a volume $V_s = (\mathcal{N}_s/\mathcal{N})V$, and the sum of the individual volumes would thus equal the volume V occupied by the mixture. The situation is as shown in Fig. 4. The gas mixture, with entropy S as given by (8.7b), can be obtained from this system by removing the partitions and allowing the gases, which are here inert, to diffuse into each other. The second summation on the right must therefore represent the entropy generated by the diffusion process, which is thus seen to be irreversible. Since the mole fractions $\mathcal{N}_s/\mathcal{N}$ are less than one, this term will always be positive, as it must by the second law applied to the present adiabatic process. The magnitude of the term depends only on the mole numbers and is independent of the nature of the gases.

Note that the diffusive mixing described here is quite a different thing from the reversible mixing that would be accomplished by running the device of Fig. 3 from state (*c*) back to state (*a*). The increase in entropy occurs in the diffusive case because the individual gases each occupy a larger volume after mixing than before. When the entropy of a mixture is calculated from (8.7b), the second term is sometimes called the *entropy of mixing*. This does not imply, however, that a diffusive process such as we have described actually took place.

The Gibbs free energy for a mixture of thermally perfect gases follows directly from the relation $G = H - TS$ and the expressions (8.4b) and (8.7b). It can be written

$$G = \sum_s \mathcal{N}_s \left[\hat{\mu}_s^0(T) + \hat{R}T\left(\ln p + \ln \frac{\mathcal{N}_s}{\mathcal{N}} \right) \right], \tag{8.8a}$$

where $\hat{\mu}_s^0(T)$ is a function purely of the temperature and is given by

$$\hat{\mu}_s^0(T) \equiv \int_{T_0}^{T} \hat{c}_{p_s}\, dT - T\int_{T_0}^{T} \hat{c}_{p_s} \frac{dT}{T} + \hat{h}_{s_0} - T\hat{s}_{s_0} - \hat{R}T \ln p_0. \tag{8.8b}$$

Comparison of (8.8a) with equation (6.11) for $G = \sum_s \hat{\mu}_s \mathcal{N}_s$ shows that

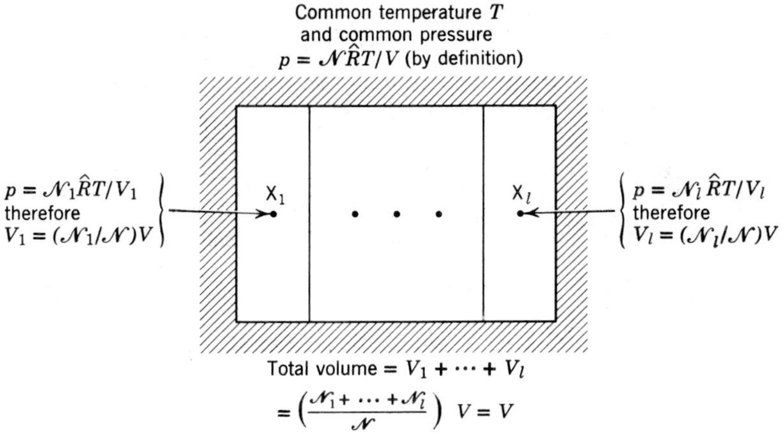

Fig. 4. System of separate gases having combined entropy given by first summation in equation (8.7b).

the chemical potential of one gas in a mixture of thermally perfect gases is

$$\hat{\mu}_s = \hat{\mu}_s^0(T) + \hat{R}T\left(\ln p + \ln \frac{\mathcal{N}_s}{\mathcal{N}} \right). \tag{8.9a}$$

The chemical potential of one species thus depends on p and T of the mixture and, through $\mathcal{N}$, on the mole numbers of *all* of the species. As anticipated following equation (6.11), the mole numbers appear combined into the mole fraction $\mathcal{N}_s/\mathcal{N}$. The meaning of the quantity $\hat{\mu}_s^0(T)$ is also apparent from (8.9a). For a given temperature, it is the chemical potential (or Gibbs free energy per mole) of the species in the pure state ($\mathcal{N}_s = \mathcal{N}$) at unit pressure. Equation (8.9a) can be simplified by rewriting it in terms of the partial pressure $p_s = (\mathcal{N}_s/\mathcal{N})p$, which gives

$$\hat{\mu}_s = \hat{\mu}_s^0(T) + \hat{R}T \ln p_s. \tag{8.9b}$$

9 LAW OF MASS ACTION

To find the equilibrium condition for a mixture of thermally perfect gases it is now only necessary to substitute expression (8.9b) for $\hat{\mu}_s$ into the equation (7.2) for reaction equilibrium: $\sum_s \nu_s \hat{\mu}_s^* = 0$. This gives for the equilibrium partial pressure p_s^*

$$\sum_s \nu_s \hat{\mu}_s^0 + \hat{R}T \sum_s \nu_s \ln p_s^* = 0,$$

or equivalently

$$\boxed{\prod_s (p_s^*)^{\nu_s} = K_p(T)}\,, \tag{9.1a}$$

where

$$K_p(T) \equiv \exp\left(-\sum_s \nu_s \hat{\mu}_s^0 / \hat{R}T\right). \tag{9.1b}$$

This result can also be written in terms of the molar concentrations $[X_s]^*$ by means of the relation $p_s^* = [X_s]^* \hat{R}T$. We thus obtain

$$\boxed{\prod_s [X_s]^{*\nu_s} = K_c(T)}\,, \tag{9.2a}$$

where

$$K_c(T) \equiv K_p(T)/(\hat{R}T)^{\sum_s \nu_s}. \tag{9.2b}$$

Equations (9.1a) and (9.2a) are general forms of the *law of mass action* for mixtures of thermally perfect gases. A specific case of these equations for the reaction $2HI \rightleftharpoons H_2 + I_2$ was obtained from kinetic considerations in Chapter II, Sec. 7. The present thermodynamic derivation has the advantage of telling us something about the form of the equilibrium constant through equations (9.1b) and (9.2b). The function on the right in (9.1b) is still not completely known, of course, since it depends on $\hat{c}_{p_s}$, $\hat{h}_{s_0}$, and $\hat{s}_{s_0}$, and these cannot be found by thermodynamic considerations alone. Within the context of chemical thermodynamics, therefore, the determination of the equilibrium constant must come from experiment. This task can be facilitated, however, by the knowledge embodied in equation (9.1b). A brief indication of this is given in the next section.

Exercise 9.1. Consider an equilibrium mixture of monatomic and diatomic oxygen reacting according to the dissociation-recombination reaction $O_2 \rightleftharpoons 2O$. If the mass density of the mixture is 1×10^{-3} gm/cm³ and the temperature is 3000°K, what is the partial pressure of O, assuming that K_p for the reaction is 15,000 dyne/cm² at this temperature? *Ans.:* 339,000 dyne/cm².

Exercise 9.2. Suppose that four thermally perfect gases react according to the equation

$$\alpha_1 X_1 + \alpha_2 X_2 \rightleftharpoons \beta_3 X_3 + \beta_4 X_4.$$

Consider a stoichiometric mixture consisting initially of only X_1 and X_2. (A mixture is said to be stoichiometric when the mole numbers of the constituents are in exact proportion to their stoichiometric coefficients.)

(a) Show that for such a mixture the progress of the reaction can be specified in terms of the degree of reaction δ defined by

$$\delta \equiv \frac{\mathcal{N}_{1_0} - \mathcal{N}_1}{\mathcal{N}_{1_0}} = \frac{\mathcal{N}_{2_0} - \mathcal{N}_2}{\mathcal{N}_{2_0}}.$$

The quantity δ thus specifies the fraction of the initial constituents that have become products. Its value can vary from 0 to 1.

(b) Show that the equilibrium value of δ is given by

$$\left[\frac{(\beta_3)^{\beta_3}(\beta_4)^{\beta_4}}{(\alpha_1)^{\alpha_1}(\alpha_2)^{\alpha_2}}\right]\frac{(\delta^*)^{\beta_3+\beta_4}}{(1-\delta^*)^{\alpha_1+\alpha_2}}\left[\frac{\alpha_1+\alpha_2-(\alpha_1+\alpha_2-\beta_3-\beta_4)\delta^*}{p}\right]^{\alpha_1+\alpha_2-\beta_3-\beta_4} = K_p(T),$$

where p is the pressure of the mixture.

(c) Obtain an equation equivalent to that of (b) for the general dissociation-recombination reaction $A_2 \rightleftharpoons 2A$. (In this case δ becomes identical to the degree of dissociation α, which we shall encounter again in Chapter V.)

10 HEAT OF REACTION; VAN'T HOFF'S EQUATION

In the experimental study of chemical equilibrium, an important role is played by the *heat of reaction* $\Delta\hat{H}$. This is by definition the heat absorbed by a closed system at constant pressure and temperature per unit change in the degree of advancement ξ, that is,

$$\Delta\hat{H} \equiv \left(\frac{dQ}{d\xi}\right)_{p,T}. \tag{10.1}$$

An expression for $\Delta\hat{H}$ can be found from the first law for a closed system in the form (4.6):

$$dQ = dH - V\,dp.$$

Writing the state relation $H = H(p, T, \mathcal{N}_1, \ldots, \mathcal{N}_l)$ in differential form similar to (6.6b) and substituting into this equation, we obtain, since $d\mathcal{N}_s = \nu_s\,d\xi$,

$$dQ = \left[\left(\frac{\partial H}{\partial p}\right)_{T,\mathcal{N}_s} - V\right]dp + \left(\frac{\partial H}{\partial T}\right)_{p,\mathcal{N}_s} dT + \sum_s \left(\frac{\partial H}{\partial \mathcal{N}_s}\right)_{p,T,\mathcal{N}_s'} \nu_s\,d\xi.$$

From this and (10.1) it follows that

$$\Delta\hat{H} = \sum_s \nu_s \left(\frac{\partial H}{\partial \mathcal{N}_s}\right)_{p,T,\mathcal{N}'_s}.$$

For a mixture of thermally perfect gases H is given by equation (8.4b) as $H = \sum_s \mathcal{N}_s \hat{h}_s(T)$, and hence by differentiation we have

$$\left(\frac{\partial H}{\partial \mathcal{N}_s}\right)_{p,T,\mathcal{N}'_s} = \hat{h}_s(T).$$

We thus find finally for the heat of reaction of a mixture of thermally perfect gases

$$\Delta\hat{H} = \sum_s \nu_s \hat{h}_s(T). \tag{10.2}$$

In this case $\Delta\hat{H}$ is a function only of the temperature.

The heat of reaction can be related to the equilibrium constant by writing equation (9.1b) as

$$\ln K_p(T) = -\sum_s \nu_s \left(\frac{\hat{\mu}_s^0}{\hat{R}T}\right)$$

and differentiating with respect to T to obtain

$$\frac{d \ln K_p}{dT} = -\sum_s \nu_s \frac{d}{dT}\left(\frac{\hat{\mu}_s^0}{\hat{R}T}\right).$$

Here the ordinary derivative is used since the quantities being differentiated are functions only of T. Substituting for $\hat{\mu}_s^0$ from (8.8b) and carrying out the differentiation, we find

$$\frac{d \ln K_p}{dT} = -\sum_s \nu_s \frac{1}{\hat{R}T^2}\left(-\int_{T_0}^{T} \hat{c}_{p_s}\, dT - \hat{h}_{s_0}\right) = \sum_s \nu_s \frac{\hat{h}_s}{\hat{R}T^2},$$

or in view of equation (10.2)

$$\boxed{\frac{d \ln K_p}{dT} = \frac{\Delta\hat{H}}{\hat{R}T^2}}. \tag{10.3a}$$

This is *van't Hoff's equation*, which together with the law of mass action itself, is among the most important equations of chemical thermodynamics. It can also be written

$$\frac{d \ln K_p}{d(1/T)} = -\frac{\Delta\hat{H}}{\hat{R}}. \tag{10.3b}$$

van't Hoff's equation can be used either to calculate $\Delta\hat{H}$ from the measured temperature variation of K_p or alternatively, if $\Delta\hat{H}$ is known from calorimetric measurements, to study the variation of K_p with T. For details of these matters see, for example, Zemansky (1957, pp. 435–440). In either case some measurements of a macroscopic nature are necessary. For a complete determination of K_p in terms of a microscopic molecular model, which means in effect determining $\hat{c}_{p_s}$, $\hat{h}_{s_0}$, and $\hat{s}_{s_0}$ in equation (8.8b), we must go beyond thermodynamics to statistical mechanics. We proceed to this subject in the next chapter.

Exercise 10.1. The forward direction in the gaseous reaction $2HI \rightleftharpoons H_2 + I_2$ is endothermic. That is, heat must be added when the reaction proceeds from left to right at constant pressure and temperature. In which direction will the equilibrium of the reaction be shifted by an increase in temperature at constant pressure?

Exercise 10.2. Consider the decomposition of water vapor according to the equation $2H_2O \rightleftharpoons 2H_2 + O_2$. The heat of reaction $\Delta\hat{H}$ was found experimentally to be practically constant at the value 119,600 cal/mol in the temperature range from 1500 to 2200°K at a pressure of 1 atm. The equilibrium constant K_p as measured at the same pressure and at a temperature of 1500°K was found to be 3.80×10^{-12} atm. If we start with 10 moles of pure H_2O vapor in a container at $T = 1900$°K and $p = 1$ atm, what will be the value of the degree of advancement ξ at equilibrium at the same temperature and pressure?

Ans.: 1.65×10^{-2} moles.

References

Bridgman, P. W., 1952, *The Nature of Some of Our Physical Concepts*, Philosophical Library.

Denbigh, K., 1963, *The Principles of Chemical Equilibrium*, Cambridge University Press.

Prigogine, I., 1961, *Thermodynamics of Irreversible Processes*, 2nd ed., Interscience.

Prigogine, I., and R. Defay, 1954, *Chemical Thermodynamics*, Longmans, Green.

Rossini, F. D., 1950, *Chemical Thermodynamics*, Wiley.

Tolman, R. C., and P. C. Fine, 1948, On the Irreversible Production of Entropy. *Revs. Mod. Phys.*, vol. 20, no. 1, p. 51.

Van Rysselberghe, P., 1963, *Thermodynamics of Irreversible Processes*, Blaisdell.

Zemansky, M. W., 1957, *Heat and Thermodynamics*, 4th ed., McGraw-Hill.

Chapter IV

Statistical Mechanics

1 INTRODUCTION

We now return to the microscopic—or molecular—approach and take up the study of statistical mechanics. Statistical mechanics, like kinetic theory, goes beyond thermodynamics in that its aim is to interpret and, insofar as possible, predict the bulk properties of matter in terms of the microscopic particles of which it is composed. The most obvious difference between the two disciplines lies in the manner in which the interaction of the particles is treated. In kinetic theory we concern ourselves, sometimes at considerable length, with the detailed mechanics of the collision process; in statistical mechanics we are able to avoid these details by means of certain powerful and general statistical ideas. By virtue of this simplification, statistical mechanics can penetrate more deeply than does kinetic theory. Historically, however, this greater depth has been paid for by the restriction of statistical mechanics primarily to equilibrium situations. This restriction, as we shall see in Chapters VII and IX, does not apply to kinetic ideas. Only in recent years has the statistical mechanics of nonequilibrium phenomena been developed in any systematic way (see, e.g., Prigogine, 1962).

The primary aim of this chapter will be the very practical one of using the methods of statistical mechanics to calculate the equilibrium properties of matter that are needed in the application of macroscopic thermodynamics. The required connection that this implies between statistical mechanics and thermodynamics can be approached from two points of view. On the one hand, we could develop statistical mechanics initially with no reference to thermodynamic ideas. With this done, we could then show that certain mathematical expressions that appear in the statistical development have the same properties as the macroscopic quantities of thermodynamics and can therefore be taken as statistical definitions of these quantities, for example, statistical temperature, and statistical entropy. From this point of view, thermodynamics is merely statistical mechanics at the macroscopic level. As an alternative to this approach, we can consider the definitions and relationships of equilibrium thermodynamics as given and use this knowledge wherever appropriate in the development of the statistical ideas. This point of view, which is similar to that followed in Chapters I and II, will be adopted here as probably the simpler for a beginning treatment. It is also perhaps the more natural for an engineer, who will often deal interchangeably with statistical and thermodynamic ideas in practical applications.

The formal connection between statistical mechanics and thermodynamics will be provided by the following important relation, which is postulated for reasons to be explained later:

$$S = k \ln \Omega. \tag{1.1}$$

The quantity S on the left is the entropy as defined in macroscopic thermodynamics. It will be noted that this quantity played no part in the equilibrium kinetic theory of Chapter II; it occupies, however, a central role here. The factor k on the right is a constant that need not concern us for the present. Statistical considerations at the microscopic level enter through the variable Ω, which is a measure of the "randomness" or "disorder" of the collection of particles that constitute the system under consideration. More exactly, it is the number of ways in which the microscopic particles of the system can be arranged consistent with the nature of the particles and with the given macroscopic circumstances. The precise definition and determination of Ω, and the calculation thence of S, is the central problem of this chapter.

Conceptually, statistical mechanics, lacking as it does the concrete collisional models of kinetic theory, is an abstract and subtle science. Furthermore, there exist alternative approaches to various aspects of the subject. These, when combined in various possible ways, lead to different presentations that may have little superficial resemblance, especially at the outset. This is sometimes a source of difficulty for the beginning student. We shall attempt to mention certain of these alternatives as we go along. In the interest of getting on to the applications, however, we shall have to ignore a number of fundamental difficulties. We shall also restrict ourselves purely to gaseous systems, in fact, to perfect gases or mixtures thereof. Readers interested in pursuing the subject more deeply may consult the texts by Rushbrooke (1949) and Huang (1963) or the more advanced treatises by Mayer and Mayer (1940), Fowler and Guggenheim (1946), or Tolman (1938).

2 MACROSCOPIC AND MICROSCOPIC DESCRIPTIONS

We consider then an assembly of a very large number of microscopic particles, which taken together constitute a macroscopic thermodynamic system. The particles with which we deal will be mainly atoms and molecules, but they could equally well be ions, electrons, etc. We assume for the time being that all particles of the system are identical, although this assumption will be relaxed later. Under this assumption, the macroscopic, that is, thermodynamic, state of the system is specified if we know the total number N of the given particles, their total energy E, and the volume V of the system. (The total energy of the particles is, of course, identical to the thermodynamic internal energy E.)

As in kinetic theory, we assume that the microscopic particles of the system interact with each other, either through direct collisions or some other process. These interactions give rise to an exchange of energy between the particles. As a result, the particles will at different times have different states and arrangements on the microscopic scale, all corresponding to the same macroscopic state of the overall system. To study the situation on the microscopic level by equilibrium statistical mechanics, we ignore the details of the interaction process and proceed directly from considerations of the energy of the particles.

In specifying the energy of the particles we shall here take the point of view of quantum mechanics. The distinctive feature of quantum mechanics, which we shall examine more closely in the next section, is that for any one microscopic particle there exists a unique set of permissible energy states. These may be identified conveniently in sequence by the corresponding value of the energy—say—$\epsilon_1, \epsilon_2, \ldots, \epsilon_i, \ldots$, where $\epsilon_{i+1} \geqq \epsilon_i$. These energy values are discrete, although they may be very close together, and no particle may exist in more than one energy state at any one time. (The equal sign in the foregoing relation has been included to allow for the possibility that more than one state may have the same value of the energy. We shall refer to this so-called "degenerate" situation in more detail later.) In ascribing energy states to the individual particles rather than to the overall system, we have tacitly assumed that the particles are only weakly interacting, that is, that at any given instant the state of any one particle is unaffected by the state of any of the other particles. This implies that the only mechanism for energy interchange is by direct collision and that the time occupied by such collisions is negligible compared with the time between collisions. The interdependence of the particles is then reflected only in the requirement that the collisions shall not affect the total energy E of the system.

For many purposes of this chapter we could equally well describe the microscopic state of the particles in terms of classical mechanics. This was, in fact, the procedure that had to be adopted by both Boltzmann and Gibbs, who founded statistical mechanics in the late nineteenth century before quantum theory was known. In the classical description, the microscopic state of a particle is described in terms not of discrete energy states but of continuous position coordinates and the corresponding momenta (or velocities) of the particle. The energy then appears as a continuous function of these variables. This classical point of view was a natural one in our discussion of kinetic theory, and we shall continue to find it useful on occasion for descriptive purposes. For actual analysis, however, we shall from the outset adopt the quantum formulation as being both simpler and more general—and, for certain quantities such as the vibrational specific heat of a diatomic gas, essential for correct and useful results. Where the classical formulation is valid, a rigorous relationship can be established between the two descriptions. We shall not have space, however, to go into this in any general way. For a careful and detailed discussion, the reader can consult Chapter IV of Rushbrooke (1949).

3 QUANTUM ENERGY STATES

Before proceeding with the statistical mechanics itself, it will be helpful to consider briefly some of the ideas underlying the quantum-state description.

The existence of quantum energy states is bound up intimately with *Heisenberg's uncertainty relation*, which can be written

$$|\Delta q|\,|\Delta p| \cong h. \tag{3.1}$$

Here $|\Delta q|$ and $|\Delta p|$ are the uncertainties in any positional coordinate and the corresponding momentum, both in a generalized sense. The quantity h is *Planck's constant*, which has the very small value of 6.6256×10^{-27} ergsec.

From the experimental point of view, relation (3.1) shows that there are limits to the accuracy with which the position and momentum of a particle can be simultaneously measured. For example, if we wish to measure a position coordinate within a specified accuracy $|\Delta q|$, we must tolerate an inaccuracy of $|\Delta p| \cong h/|\Delta q|$ in a simultaneous measurement of the corresponding momentum. This is so because the very act of measuring the position will alter the momentum from what it would otherwise have been. More fundamentally, however, the uncertainty relation shows that, even from the theoretical point of view, it makes no sense to speak of particles as having absolutely defined values of position and momentum. If we wish to regard a particle, even theoretically, as having an absolutely defined position (i.e., $|\Delta q| = 0$), then according to relation (3.1) we must regard the momentum as indeterminate (i.e., $|\Delta p| = \infty$). In other words, it would make no sense to even think of the particle as having a definable momentum. This comes about because the particle description of matter is not absolutely valid at the microscopic level. It must be regarded rather as a kind of analogy, not to be taken too literally, but useful for visualization and calculation. If we wish to use it, we must then allow a certain tolerance or "fuzziness" in our specification of the state of the particles. In particular, we cannot from the quantum-mechanical point of view speak of a given particle as having an absolute position but only of having a certain probability of lying in the vicinity of that position.

The specification of the foregoing probability is made in quantum mechanics in terms of a *wave function* ψ. This function has no direct physical significance itself, but is defined such that the product $|\psi|^2 \, dV_x$ gives the probability that a given particle lies in a given differential element of volume dV_x. (We speak henceforth only of a single particle in true physical space, which will be sufficient for our purposes.) The wave function thus plays a role here roughly similar to that occupied in fluid mechanics by the potential function φ, which has no physical significance itself but which serves when operated on suitably (by differentiation in this case) to give information about the fluid velocity. The governing equation for ψ is provided by the *Schrödinger wave equation*, which replaces Newton's equation at the microscopic level. For a single, structureless particle of mass m moving in a potential field, this equation, when specialized to a so-called "stationary state" (or "standing wave"), can be written in Cartesian space coordinates as

$$\boxed{\frac{h^2}{8\pi^2 m}\left(\frac{\partial^2\psi}{\partial x_1^2} + \frac{\partial^2\psi}{\partial x_2^2} + \frac{\partial^2\psi}{\partial x_3^2}\right) + (\epsilon - \epsilon_p)\psi = 0} \, . \qquad (3.2)$$

Here ϵ is the total energy of the particle and ϵ_p its potential energy. The function ψ is furthermore subject to the conditions that it shall be everywhere finite and continuous. This is necessary so that the probability $|\psi|^2 \, dV_x$ of finding the particle in dV_x will also be finite and will not undergo a discontinuity within dV_x. The value of ψ must also be everywhere unique, since the probability of finding a particle in dV_x cannot be multivalued.

There are also conditions on the derivatives of ψ, but they need not concern us here. Since there is a certainty, that is, a probability of 1, that the particle will be found *somewhere*, ψ is also subject to the "normalization" condition

$$\int |\psi|^2 \, dV_x = 1, \tag{3.3}$$

where the integration is carried out through all space. Like Newton's equation and other basic equations of physics, Schrödinger's equation cannot be derived but must be postulated. Its validity rests on the success of its results.

We now apply equation (3.2) to a particle moving in field-free space inside a rectangular box with impenetrable walls. This will illustrate the occurrence of discrete energy states as well as provide results needed in our later analysis of the gas properties associated with molecular translation. Since the space is field-free, the potential energy ϵ_p interior to the box is zero and the total energy of the particle is equal to the kinetic energy, that is,

$$\epsilon = \tfrac{1}{2}m(C_1^2 + C_2^2 + C_3^2) = \frac{1}{2m}(p_1^2 + p_2^2 + p_3^2), \tag{3.4}$$

where C_1, C_2, C_3 are the velocity components and $p_1 = mC_1$, etc. are the corresponding momenta. Equation (3.2) thus becomes

$$\frac{h^2}{8\pi^2 m}\left(\frac{\partial^2 \psi}{\partial x_1^2} + \frac{\partial^2 \psi}{\partial x_2^2} + \frac{\partial^2 \psi}{\partial x_3^2}\right) + \frac{1}{2m}(p_1^2 + p_2^2 + p_3^2)\psi = 0. \tag{3.5}$$

Since the probability of finding the particle outside the box is nil, ψ must be zero everywhere external to the box; and, since ψ must be continuous, it must therefore be zero at the walls of the box as well. This supplies the required boundary condition on ψ.

To solve equation (3.5) we separate variables by assuming ψ in the form

$$\psi(x_1, x_2, x_3) = \psi_1(x_1)\psi_2(x_2)\psi_3(x_3). \tag{3.6}$$

Substituting this into equation (3.5) we see that the equation will be satisfied if we have

$$\frac{h^2}{4\pi^2}\frac{\partial^2 \psi_1}{\partial x_1^2} + p_1^2 \psi_1 = 0, \tag{3.7}$$

with similar equations in the x_2- and x_3-directions. The general solution of this equation can be written at once as

$$\psi_1 = A_1 \sin 2\pi \frac{|p_1|}{h} x_1 + B_1 \cos 2\pi \frac{|p_1|}{h} x_1, \tag{3.8}$$

where A_1 and B_1 are constants to be determined.[1] We now suppose that the sides of the box are of length a_1, a_2, a_3 in the coordinate directions. If the box is then placed entirely in the positive octant of the coordinate system, with one corner at the origin, the boundary conditions on ψ_1 become

$$\psi_1(0) = 0 \qquad \text{and} \qquad \psi_1(a_1) = 0. \tag{3.9}$$

The first of these conditions requires that $B_1 = 0$. The second cannot then be satisfied by adjusting A_1, but only by setting $2\pi(|p_1|/h)a_1 = n_1\pi$, where n_1 is a positive integer.[2] We thus find that a solution of (3.7) subject to the boundary conditions (3.9) is not possible for arbitrary values of the momentum p_1, but only for discrete values such that

$$|p_1| = \frac{h}{2} \times \frac{n_1}{a_1} \quad \text{where } n_1 = 1, 2, 3, \ldots. \tag{3.10}$$

These are the so-called *eigenvalues* of the differential equation (3.7), and corresponding eigenvalues exist for the other two coordinate directions. From this and the relation (3.4), it follows that solutions of the original differential equation (3.5) subject to the given boundary conditions will exist only for the discrete values of the energy ϵ given by the equation

$$\epsilon_{n_1,n_2,n_3} = \frac{h^2}{8m}\left(\frac{n_1^2}{a_1^2} + \frac{n_2^2}{a_2^2} + \frac{n_3^2}{a_3^2}\right) \quad \text{where } \left.\begin{matrix} n_1 \\ n_2 \\ n_3 \end{matrix}\right\} = 1, 2, 3, \ldots. \tag{3.11}$$

These then are the permissible *quantum energy states* for a particle in a rectangular box. The integer numbers n_1, n_2, n_3 that define the states are called *quantum numbers*. With the foregoing results we could also write the complete wave function from equation (3.6) and determine the remaining constant $A = A_1A_2A_3$ from the normalization condition (3.3). The integration would now be limited to the interior of the box since only there is ψ different from zero. We will not concern ourselves with this, however, since only the energy states are of interest here.

Before returning to statistical mechanics, one more point is worth noting. It is well known that electron and molecular beams exhibit the same interference and diffraction phenomena as shown by electromagnetic

[1] We include only the absolute value of p, since it is only $|\psi|^2$ that is really of importance. In other words, the difference between solutions with positive and negative values of p is of no consequence, since they provide identical information about the position of the particle.

[2] We rule out the trivial solution given by $n_1 = 0$, since this would correspond to $\psi_1 = 0$ and hence $|\psi|^2\, dV_x = 0$, that is, the probability of finding the particle anywhere is zero, hence no particle.

radiation. These phenomena can be accounted for if matter, like light, is imagined as having a wave as well as a particle nature. Specifically, matter behaves as if it has, associated with each component of momentum p, a wavelength Λ given by

$$\boxed{\Lambda = \frac{h}{|p|}}\ . \tag{3.12}$$

This is the *de Broglie relation*, and Λ is the *de Broglie wavelength*. Combining this equation with the result (3.10), we obtain the following relationship between the dimension of the box and the corresponding de Broglie wavelength:

$$a_1 = n_1 \frac{\Lambda_1}{2}\ . \tag{3.13}$$

Thus, for a permissible energy state, the state of motion must be such that one-half of the de Broglie wavelength in a given coordinate direction fits into the box in that direction an integral number of times. This is a reflection, of course, of the fact that the de Broglie wavelength (3.12) is here identical to the wavelength of the wave function ψ_1 given by equation (3.8).

The foregoing results were derived for the translational motion of a single particle. For a weakly interacting system as assumed in Sec. 2, however, equation (3.11) will also give the translational energy states for each individual particle in the complete system. Results for the permissible energy states associated with other types of motion, as for example, the rotation and vibration of diatomic molecules, can be obtained by solving the appropriate Schrödinger equation. For want of space, we shall merely state these results as we need them. For a complete treatment, the reader should consult any standard text on quantum mechanics (e.g., Pauling and Wilson, 1935, or Leighton, 1959).

4 ENUMERATION OF MICROSTATES

We now return to the mainstream of statistical mechanics. The situation that we have postulated can be summarized as follows: We have a macroscopic system of N identical microscopic particles, each of which occupies at any instant one of its permissible energy states $\epsilon_1, \epsilon_2, \ldots, \epsilon_i, \ldots$ (considered now in a general sense). The total energy of the system is E. We now ask the following question that will be fundamental to our considerations: In how many different ways can the foregoing situation occur? Or, putting it differently, in how many different ways can the N particles be distributed over the given permissible energy states, consistent with a specified value of the total energy E? Following the usual terminology, we shall refer to each of these possible distributions as a *microstate* of the system. In other words, a microstate is any distribution of the microscopic particles over the available quantum energy states in such a way as to be consistent with the macroscopic state of the system. Our question then amounts to asking briefly, how many microstates are there corresponding to the given macroscopic situation?

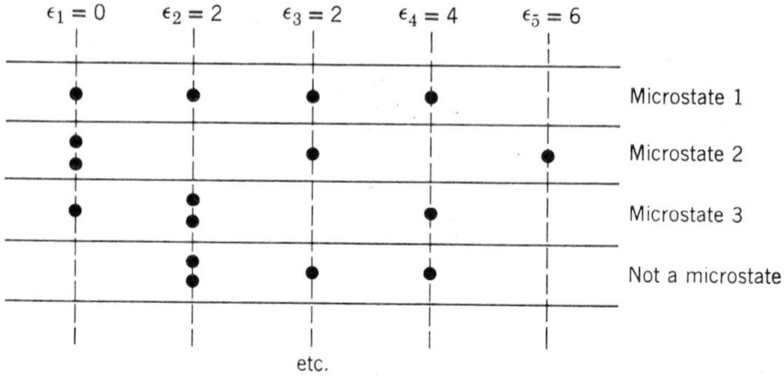

Fig. 1. Partial enumeration of microstates for $N = 4$, $E = 8$.

The reason for asking the foregoing question is that the answer, for reasons that will be explained later, will be identified with the quantity Ω in our postulated equation (1.1). This answer will come out, of course, as a function of the given quantities N and E, as well as of the volume V of the system. This last enters, as we shall see, through the energy states ϵ_i. We thus arrive at a *statistical* relationship of the form $\Omega = \Omega(E, V, N)$, and the postulated equation (1.1) then leads in turn to the result that $S = S(E, V, N)$. This is consistent with the fact that S, introduced originally as a *thermodynamic* state variable, must depend on the macroscopic state of the system, which was taken as specified (see opening paragraph of Sec. 2) by the variables E, V, N. The entire formulation is thus self-consistent, as indeed it must be.

The answer to the question of our first paragraph could be obtained, in principle, by simple trial and error. We could imagine ourselves, for example, as apportioning the N particles out to the given energy states in successively different ways, checking each time to see if the total energy has the required value E. This is illustrated in Fig. 1 for the grossly oversimplified case of $N = 4$ and $E = 8$, with five energy states given by $\epsilon_1 = 0$, $\epsilon_2 = 2$, $\epsilon_3 = 2$, $\epsilon_4 = 4$, and $\epsilon_5 = 6$. The first three trials are seen to fulfill the conditions for a microstate of the system; the fourth trial does not. The reader can satisfy himself that there are a total of eleven microstates for the given system. If N_i denotes the number of particles in the arbitrary energy state ϵ_i, the foregoing process can be indicated formally by the summation

$$\Omega = \sum_{\substack{\text{over all sets of} \\ N_i \text{ such that} \\ \left(\sum_i N_i = N \atop \sum_i N_i \epsilon_i = E\right)}} 1. \tag{4.1}$$

The meaning of this perhaps curious looking equation is that each set of the N_i's that is found to satisfy the restrictive conditions in the parentheses constitutes a microstate and is therefore tallied as contributing 1 toward the total value of Ω. In addition to the restrictive condition on the total energy, represented by the equation $\sum_i N_i \epsilon_i = E$, the specification of the number of particles N must also be regarded as supplying the condition $\sum_i N_i = N$.

In carrying out the foregoing enumeration we have tacitly imposed still another condition. This is that the particles with which we deal are *completely indistinguishable*. This means that it is only the number of particles in a given energy state that counts for our enumeration, not the identity of the particles, which is in fact nonexistent. Thus, for example, in Fig. 1, the interchange of the particles between states ϵ_3 and ϵ_4 in microstate 1 does not create a new microstate. Since the particles are indistinguishable, the resulting situation is in no way different from the original one. The reasons for this assumption of indistinguishability lie basically in quantum-mechanical considerations of the overall wave function for collections of particles. Such matters are beyond the scope of our treatment. (For a brief discussion, see Fast, 1962, pp. 185–192.) We can only observe that microscopic particles of a given kind possess of themselves no "labels" (i.e., color, number, etc.) that can be used to tell them apart. Thus, aside from their energy state, which is not an intrinsic characteristic of the particle itself, an observer will find it impossible to distinguish between two particles of the same kind. The assumption of indistinguishability is therefore essential.

We have also assumed in the foregoing that there is no restriction on the number N_i of particles that can be in any one energy state at a given instant. Quantum mechanics indicates that this is in fact the case for atoms and molecules made up of an even number of elementary material particles (e.g., electrons, protons, neutrons). Systems composed of such atoms or molecules are said to be governed by *Bose-Einstein statistics*. For systems composed either of the elementary material particles themselves (e.g., an electron gas) or of atoms and molecules made up of an odd number of such particles, a further condition must be imposed. This is given by the *Pauli exclusion principle*, which states that no quantum energy state may contain more than one particle, that is, that all the N_i's must be either 0 or 1. In counting up the microstates of such systems, this must also be added as a restrictive condition to the summation (4.1). With this further restriction, the example of Fig. 1 would have only one possible microstate, namely, "microstate 1." The statistics of systems subject to the exclusion principle are referred to as *Fermi-Dirac statistics*. The difference between the two types of systems lies basically in the symmetry properties of the associated quantum-mechanical wave functions. These details are again beyond the present treatment.

Let us now return to the problem of carrying out the enumeration indicated by equation (4.1). The trial-and-error method of counting up the microstates, although possible in principle, becomes absurdly impractical when the number of particles and energy states is large, as is the case for systems of practical interest. Fortunately, however, this difficulty can be overcome. To do this we make use of the usual fact that the quantum energy states, in terms of their energy values, lie very close together. That this is so can be shown by considering the energy states for weakly interacting particles in a box as given by equation (3.11). For simplicity, we assume that the box is a cube of volume V, so that we have $a_1^2 = a_2^2 = a_3^2 = V^{2/3}$, and equation (3.11) becomes

$$\epsilon_{n_1,n_2,n_3} = \frac{h^2}{8mV^{2/3}}(n_1^2 + n_2^2 + n_3^2) = \frac{h^2}{8mV^{2/3}}\, n^2, \qquad (4.2)$$

where we have introduced the notation $n^2 = n_1^2 + n_2^2 + n_3^2$. To get a rough idea of the spacing of the energy states, we inquire how many of the states have an energy less than a certain value ϵ, to be chosen later. These states must satisfy the condition $\epsilon_{n_1,n_2,n_3} < \epsilon$, which reduces with the aid of the foregoing equation to

$$n < 2\,\frac{V^{1/3}}{h}\,(2m\epsilon)^{1/2}.$$

Let us consider the octant of a Cartesian space defined by the necessarily positive quantum numbers n_1, n_2, n_3 (see Fig. 2). The foregoing inequality shows that in this "quantum-number space" the energy states in question must lie within the spherical surface having a radius n_s given by the right-hand side of the inequality. But, since the quantum numbers must be integers, each cube of unit volume in the Cartesian space will contain, in effect, just one energy state. If the radius of the sphere is sufficiently large, the number of energy states that we seek will therefore be very nearly equal to the volume of the one-eighth sphere of radius shown. We thus find for the number Γ of states with energy less than ϵ

$$\Gamma = \tfrac{1}{8} \times \tfrac{4}{3}\pi\left[2\,\frac{V^{1/3}}{h}(2m\epsilon)^{1/2}\right]^3 = \tfrac{4}{3}\pi\,\frac{V}{h^3}(2m\epsilon)^{3/2}. \tag{4.3}$$

For the sake of argument, let us take ϵ equal to the average kinetic energy of the particles, which we know from kinetic theory to be $\tfrac{3}{2}kT$ [cf. equation

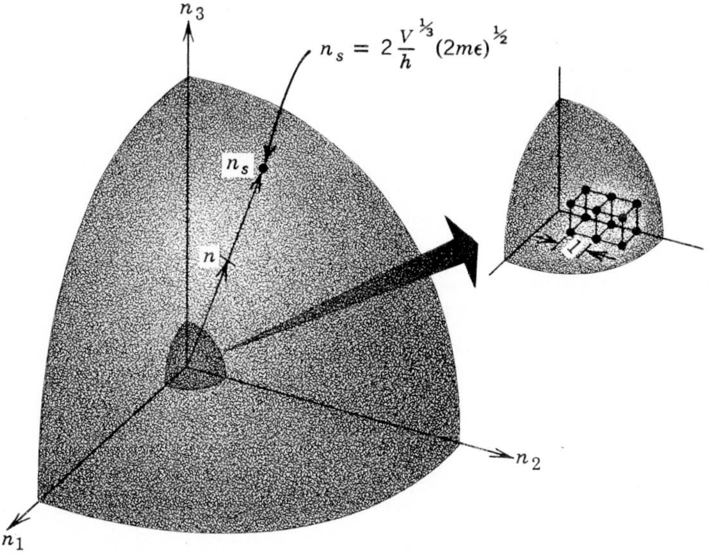

Fig. 2. Quantum-number space. Each dot represents a particular combination of integer values of n_1, n_2, n_3 (i.e., a particular quantum state) and thus occupies the center of a cube of unit volume.

(I 3.3b)]. Then, for nitrogen molecules ($m = 4.65 \times 10^{-23}$ gm) in a volume of 1 cm^3 at a temperature of 0°C, equation (4.3) gives a value of approximately 2×10^{26}. Although the quantum states will not be equally spaced in terms of their energy, it is apparent that their average spacing must be very small indeed. It is also pertinent to note that the number of molecules in the foregoing volume at 1 atm pressure, as given by Loschmidt's number (see Chapter I, Sec. 6), is about 2.7×10^{19}. Even if *all* these molecules are in the 2×10^{26} states that have less than the average energy (which is, of course, impossible), the distribution over the available states will still be exceedingly sparse—so sparse, in fact, that the great majority of the states are empty and only about one in every 10^7 contains a particle. This fact will be of importance later.

In view of the very close spacing of the quantum energy states it is possible and, as we shall see, useful to divide the states into groups, each of which is regarded as characterized by a unique value of the energy. To be specific, we conceive of the scale of energies ϵ_i as divided into successive regions, denoted by subscript j, and each containing a range of energy values $(\delta\epsilon_i)_j$. The values of the $(\delta\epsilon_i)_j$ are taken large enough to encompass many energy states but small compared with the total energy E. This is possible because of the close spacing of the states. Since the $(\delta\epsilon_i)_j$ are small in comparison with E, all the states in a given group, although still individually identifiable, can be treated to a sufficient accuracy as having the same energy ϵ_j. This could be, for example, the mean energy of the group, although precise definition is not important. We denote the number of energy states in a group by C_j. The number of particles in the group is then denoted by N_j. Once the manner of subdivision is chosen, the ϵ_j and C_j are fixed quantities and only the N_j are variable. In the usual terminology each set of the distribution numbers N_j is said to define a *macrostate* of the system. A change of a particle from one energy state to another within a specified group, although it will alter the microstate of the system, will not affect the value of N_j for that group. Each macrostate may thus comprise many microstates. In explicit terms, a macrostate may therefore be defined as a set of all those microstates that have the same specific values for the numbers N_j of particles in the various groups into which the range of quantum energy states has been divided.

Our task of enumerating the total number of microstates Ω can now be carried out in two steps as follows: (1) Find the number of microstates for a given fixed macrostate, that is, for given values of the N_j's. (2) Sum these numbers over all possible macrostates, that is, over all values of the N_j's consistent with the thermodynamic state of the system.

To carry out step (1), we concentrate our attention first on a *single* group of C_j specified states containing a given number N_j of indistinguishable particles. We then ask in how many ways the N_j particles can be distributed over the C_j states. This is equivalent to asking how many ways there are to place N_j indistinguishable particles into C_j identifiable containers. The answer to this question will, of course, be different depending on whether we employ Bose-Einstein or Fermi-Dirac statistics.

Bose-Einstein Statistics

Here there is no limit to the number of particles in any one container. Let us imagine a straight-line array of N_j particles (represented by x's) and $C_j - 1$ vertical partitions as follows:

$$\begin{array}{cccccccccccc} \text{xx} & | & \text{xxx} & | & \text{x} & | & & | & \text{xx} & | \ldots | & \text{xx.} \\ 1 & & 2 & & 3 & & 4 & & 5 & & C_j \end{array}$$

The spaces between the partitions and to the left and right of the final partitions at each end are taken to represent the C_j containers. The identity of the containers is specified by numbering their position from left to right in the array, that is, 1, 2, 3, . . . , C_j. Let us begin by thinking of the N_j particles and the $C_j - 1$ partitions as in some way labeled and distinguishable. The number of distinct ways in which the sequence of particles and partitions can be arranged, proceeding again from left to right, is then $(N_j + C_j - 1)!$. This is so because the first in the sequence can be chosen from a total of $(N_j + C_j - 1)$ objects, the second from the remaining $(N_j + C_j - 2)$ objects, the third from the $(N_j + C_j - 3)$ objects that then remain, and so on, giving altogether $(N_j + C_j - 1)(N_j + C_j - 2)(N_j + C_j - 3) \cdots 2 \times 1 = (N_j + C_j - 1)!$ ways of choosing the sequence. We observe, however, that considering the partitions as distinguishable is actually improper for our purposes. That is, the identity of the containers, which is the only thing of consequence, is fixed purely by their position from left to right in the array and is in no way dependent on the assumed identity of the partitions. Thus, the mere interchange of partitions does not in fact lead to different situations. Since the distinguishable partitions, considered alone, can be arranged in $(C_j - 1)!$ different ways, this means that for a given distribution of the particles to the containers we have counted $(C_j - 1)!$ times too many situations by the foregoing scheme. We must therefore divide our result by $(C_j - 1)!$ to correct for the assumed distinguishability of the partitions. We observe, furthermore, that contrary to our original assumption the particles are also indistinguishable. Since N_j distinguishable particles can by themselves be arranged in $N_j!$ different ways, we must therefore, by reasoning similar to the foregoing, divide by $N_j!$ to correct for this assumed distinguishability. Thus we arrive at the expression

$$\frac{(N_j + C_j - 1)!}{(C_j - 1)!\, N_j!} \tag{4.4a}$$

for the number of ways in which N_j indistinguishable particles can be placed in C_j identifiable containers with no restriction on the number of particles in each container.

Fermi-Dirac Statistics

Here there can be no more than one particle in each container, which means that $N_j \leqq C_j$. The reasoning is then simpler than that given previously. We think of the N_j particles at first as being distinguishable—say by numbering I, II, III, . . . , N_j—and imagine them laid out in numerical order from left to right as follows:

$$\begin{array}{ccccc} \text{x} & \text{x} & \text{x} & \cdots & \text{x.} \\ \text{I} & \text{II} & \text{III} & & N_j \end{array}$$

We then imagine that N_j of the C_j identifiable containers are filled by placing them one by one over the individual particles, proceeding from left to right. There are $C_j!/(C_j - N_j)!$ distinct ways of doing this, since the first can be chosen from C_j containers, the second from $(C_j - 1)$, and so on, with the last being chosen from $[C_j - (N_j - 1)]$. The total number of ways of choosing the sequence of containers is thus $C_j(C_j - 1) \cdots (C_j - N_j + 1)$, which is equal to $C_j!/(C_j - N_j)!$. The particles, however, are not in fact distinguishable, that is, do not have identity I, II, ... A mere rearrangement on the particles of one of the foregoing choices of N_j containers thus does not really constitute a different situation. Since there are $N_j!$ ways in which this rearrangement can be performed, we have therefore counted $N_j!$ times too many situations by the above scheme. We must therefore divide by $N_j!$, which gives

$$\frac{C_j!}{(C_j - N_j)!\, N_j!} \tag{4.4b}$$

for the number of ways in which N_j indistinguishable particles can be placed in C_j identifiable containers with not more than one particle to a container.

The foregoing expressions give the number of arrangements possible for the N_j particles in a single group of C_j states. This number for any one group is independent of the arrangement in the other groups. To find the total number of combined arrangements for the complete system for a given set of the N_j's, we therefore take the product of these expressions over all of the groups. We thus obtain the following expressions for the number W of microstates in a given macrostate:

For Bose-Einstein statistics:

$$\boxed{W(N_j) = \prod_j \frac{(N_j + C_j - 1)!}{(C_j - 1)!\, N_j!}}\;; \tag{4.5a}$$

For Fermi-Dirac statistics:

$$\boxed{W(N_j) = \prod_j \frac{C_j!}{(C_j - N_j)!\, N_j!}}\;. \tag{4.5b}$$

The notation $W(N_j)$ indicates that the value of W will differ depending on the value of the N_j's for the macrostate in question. This completes step (1).

We now proceed to step (2) and sum the foregoing quantities over all possible macrostates, that is, over all possible sets of the N_j's. We thus obtain finally for the total number of microstates of the system

$$\boxed{\Omega = \sum_{\substack{\text{over all sets of} \\ N_j \text{ such that} \\ \left(\substack{\sum_j N_j = N \\ \sum_j N_j \epsilon_j = E}\right)}} W(N_j)} \, . \tag{4.6}$$

The restrictive conditions that determine the possible macrostates appear here, of course, in terms of the N_j's instead of the N_i's.

Exercise 4.1. Show that equations (4.5) and (4.6) reduce formally to equation (4.1) if the groups of energy states are chosen so that the microscopic representation of the system reduces to that employed before the groups were introduced.

5 DISTRIBUTION OVER ENERGY STATES—GENERAL CASE

We should now like to evaluate the sum indicated in equation (4.6). The distribution numbers N_j, of course, will not appear in the answer, since the sum is taken over all possible values of these numbers. Actually, the sum (4.6) cannot be evaluated in closed algebraic form unless N is very large. In this circumstance, however—and this will be the case in the systems with which we deal—we can take advantage of certain mathematical approximations that lead fairly simply to the desired result. These approximations appear to be crude and open to objection. A rigorous but more difficult method due to Fowler and Darwin, however, leads to precisely the same results when N is very large (see Fowler and Guggenheim, 1956). As in most elementary discussions, therefore, we shall feel justified in using the simpler treatment, which is perhaps also the more illuminating for the newcomer to the subject.

We begin by making the seemingly drastic assumption (which can, however, be justified for very large N) that *only the largest term in the sum* (4.6) *makes any effective contribution to* Ω. We shall return to an examination of this assumption at the end of the next section, when the necessary formulas are available. For the time being let us accept it and replace the sum (4.6) accordingly by

$$\Omega = W_{\text{max}}, \tag{5.1}$$

where W_{max} is the largest of the individual terms defined by the expressions (4.5), subject to the restrictive conditions appended to equation (4.6).

The problem is now one of picking out $W_{\max}$. For this purpose we find it convenient to deal, not with W itself, but with $\ln W$ and to look for a maximum of this latter function. Accordingly, we write, from equations (4.5), for Bose-Einstein statistics,

$$\ln W = \sum_j [\ln (N_j + C_j - 1)! - \ln (C_j - 1)! - \ln N_j!], \quad (5.2a)$$

and for Fermi-Dirac statistics,

$$\ln W = \sum_j [\ln C_j! - \ln (C_j - N_j)! - \ln N_j!]. \quad (5.2b)$$

To find the maximum of $\ln W$, we make use of two additional assumptions, namely, that *the factorial functions in* (5.2) *may be approximated by means of Stirling's formula* (see below) and that *ln W may then be treated as a continuous function of the N_j's.* Stirling's formula states that, for sufficiently large values of the argument z, the value of $\ln z!$ is given approximately by

$$\ln z! \cong z \ln z - z + \tfrac{1}{2} \ln 2\pi z. \quad (5.3a)$$

For the very large values of z encountered in statistical mechanics this may be further approximated to the following form, which we shall use throughout:

$$\boxed{\ln z! \cong z \ln z - z}\,. \quad (5.3b)$$

(For a derivation of Stirling's formula, see, e.g., Wilks, 1962, pp. 175–177. The reader may find it instructive to test the accuracy of equations (5.3) by substitution of increasing numerical values of z.) The application of Stirling's formula to equations (5.2) implies that, not only are the C_j's large as assumed, but all of the N_j's as well. Strictly speaking, this is impossible, since the finiteness of the total energy must obviously require that $N_j \to 0$ as $\epsilon_j \to \infty$. The groups for which N_j is small, however, contribute little to the value of $\ln W$, so that the ensuing error is negligible. Applying Stirling's formula (5.3b) to equations (5.2), we then obtain, after some regrouping of terms,

$$\ln W = \sum_j \left[\pm C_j \ln \left(1 \pm \frac{N_j}{C_j}\right) + N_j \ln \left(\frac{C_j}{N_j} \pm 1\right)\right], \quad (5.4)$$

where it is understood that the upper sign in the $\pm$ refers to Bose-Einstein statistics and the lower to Fermi-Dirac statistics. To obtain the Bose-Einstein result, unity in equation (5.2a) has been justifiably neglected in comparison with C_j.

The condition for $\ln W$ to be a maximum is that the small change $\delta(\ln W)$ corresponding to any small changes δN_j must be equal to zero.

Since we assume that ln W may now be treated as a continuous function, this condition can be written

$$\delta(\ln W) = \sum_j \frac{\partial(\ln W)}{\partial N_j} \delta N_j = 0.$$

Substitution for ln W from equation (5.4) and simplification of the resulting equation then leads to the following condition on the N_j's:

$$\sum_j \left[\ln\left(\frac{C_j}{N_j} \pm 1\right)\right] \delta N_j = 0. \tag{5.5}$$

The small changes δN_j, of course, are not independent, but are related by the following equations obtained from the restrictive conditions on equation (4.6):

$$\sum_j \delta N_j = 0, \qquad \sum_j \epsilon_j \, \delta N_j = 0. \tag{5.6}$$

A convenient method for solving an equation such as (5.5) subject to restrictive conditions such as (5.6) is given by *Lagrange's method of undetermined multipliers.* (For a development of this method see, e.g., Rushbrooke, 1949, pp. 18–19, or Mayer and Mayer, 1940, Appendix A VI.) According to Lagrange's method, the most general value that ln $(C_j/N_j \pm 1)$ can have so as to satisfy equation (5.5) subject to the conditions (5.6) is the following linear combination of the coefficients unity and ϵ_j of the δN_j in conditions (5.6):

$$\ln\left(\frac{C_j}{N_j} \pm 1\right) = \alpha \times 1 + \beta \times \epsilon_j = \alpha + \beta\epsilon_j. \tag{5.7}$$

Here α and β are undetermined multipliers whose value is to be found later. That this is indeed a solution can be verified by substitution into the left-hand side of (5.5) to obtain

$$\sum_j \left[\ln\left(\frac{C_j}{N_j} \pm 1\right)\right] \delta N_j = \sum_j (\alpha + \beta\epsilon_j)\, \delta N_j = \alpha \sum_j \delta N_j + \beta \sum_j \epsilon_j \, \delta N_j,$$

which is zero for all values of δN_j for which the conditions (5.6) are satisfied. The particular values of N_j that make ln W and hence W a maximum (it can be shown that there is no real minimum) are thus, from equation (5.7),

$$\boxed{\frac{N_j^*}{C_j} = \frac{1}{e^{\alpha+\beta\epsilon_j} \mp 1}}, \tag{5.8}$$

where the minus sign is for the Bose-Einstein statistics and the plus sign for the Fermi-Dirac statistics. The asterisk has been added to these particular values of the N_j's for clarity. We shall sometimes drop it later when there is no possibility of confusion.

The values of α and β could, in principle, be found from the following equations obtained by substituting (5.8) into the restrictive conditions $\sum_j N_j = N$ and $\sum_j N_j\epsilon_j = E$:

$$\sum_j \frac{C_j}{e^{\alpha+\beta\epsilon_j} \mp 1} = N, \qquad \sum_j \frac{\epsilon_j C_j}{e^{\alpha+\beta\epsilon_j} \mp 1} = E. \tag{5.9}$$

These equations supply, in principle, two simultaneous equations for the determination of α and β as functions of N and E and the given C_j and ϵ_j. With α and β known, the expressions (5.8) could then be substituted into equation (5.4) to find $\ln W_{max}$, and our task would be completed. Unfortunately, the determination of α and β directly from equations (5.9) is not possible in general.

6 DISTRIBUTION OVER ENERGY STATES—LIMITING CASE

The determination of α and β and the completion of the analysis is, in fact, simple only in a limiting situation, which fortunately covers the cases of practical interest for our purposes. This situation is obtained when the temperature is sufficiently high so that the particles are distributed widely over the permissible energy states. As was exemplified in Sec. 4, the energy levels will then be sparsely populated, and we can write that $C_j \gg N_j$. The only way that equation (5.8) can be consistent with this result is for the exponential term in the denominator to be very large. In this situation, unity can be neglected in comparison with the exponential term, and we obtain from both types of statistics the common limiting result

$$\frac{N_j^*}{C_j} = e^{-\alpha-\beta\epsilon_j}. \tag{6.1}$$

Similarly, by applying the condition $C_j \gg N_j$ in equation (5.4) and using the approximation $\ln(1 \pm x) \cong \pm x$ for $x \ll 1$, we obtain the following common limiting equation for $\ln W$:

$$\ln W = \sum_j \left[N_j + N_j \ln \frac{C_j}{N_j}\right] = \sum_j N_j\left(\ln \frac{C_j}{N_j} + 1\right). \tag{6.2}$$

A common result is obtained because the energy states under normal conditions are so sparsely populated that most of them are empty. In the typical case considered in Sec. 4, for example, we saw that only about one in every 10^7 states contains a particle. The likelihood that a state will contain more than one particle, even if permitted, is thus practically nil. It is to be expected therefore that the Bose-Einstein and Fermi-Dirac statistics will lead to identical results in the limit.

The foregoing limiting results can also be obtained directly on the basis of the so-called *Boltzmann statistics*. This approach, which was appropriate to the classical methods used before quantum statistical mechanics was known, assumes in counting up the microstates that the particles are truly distinguishable. This leads without difficulty to equation (6.1), but an approximate and much argued over correction to account for the actual indistinguishability is needed to arrive at equation (6.2). Because of this difficulty we have chosen to avoid the Boltzmann statistics here and obtain our results solely as the limit of the correct quantum statistics. For convenience we shall refer to these results as corresponding to the *Boltzmann limit*.

Since ϵ_j has a wide range of values, including possibly zero, the only way that the exponential term in equation (5.8) can be large for all groups of states is to have $\alpha \gg 1$. It can be shown (see Exercise 10.1) that this condition for the validity of the Boltzmann limit is exactly equivalent to the condition[3]

$$\frac{V}{N} \times \frac{(2\pi mkT)^{3/2}}{h^3} \gg 1. \tag{6.3}$$

This requirement would be violated, for example, when the particle mass m is very small and the number density N/V is sufficiently large, as is the case for an electron gas in metals. It is also violated for normal molecular values of m and N/V when T is very low. In this situation the particles tend to crowd together in the lower energy states, and it obviously makes considerable difference whether more than one particle is allowed in a given state. There is then a significant difference between the Bose-Einstein and Fermi-Dirac statistics.[4] Whenever condition (6.3) is violated, the gas is said to be *degenerate* and care must be taken to use the appropriate statistics. For most systems with which we deal in gas dynamics, the condition is satisfied, and the Boltzmann limit is sufficient. We shall assume that this is the case throughout the subsequent work unless stated otherwise.

[3] Note that this condition is, except for a constant factor, the same as the approximate condition that would be obtained by requiring that the number of quantum states with energy less than $\frac{3}{2}kT$, as given by equation (4.3), be much greater than the number of particles N.

[4] Even before this happened, however, the gas would in most cases cease to be a system of weakly interacting particles and this would also have to be taken into account.

After these digressions, we can return to equations (6.1) and (6.2). Elimination of α is now a simple matter. Neglecting unity in comparison with the exponential term in the first of equations (5.9) we obtain

$$e^{-\alpha} = \frac{N}{\sum_j C_j e^{-\beta\epsilon_j}}. \tag{6.4}$$

Putting this into (6.1) we then find

$$N_j^* = N \frac{C_j e^{-\beta\epsilon_j}}{\sum_j C_j e^{-\beta\epsilon_j}}. \tag{6.5}$$

The elimination of β is not so simple, since the second of equations (5.9) reduces in the limit, and with the aid of equation (6.4), to

$$N \frac{\sum_j \epsilon_j C_j e^{-\beta\epsilon_j}}{\sum_j C_j e^{-\beta\epsilon_j}} = E. \tag{6.6}$$

The solution of this equation for β is still not possible in practice. Fortunately, there are easier ways to find β, as we shall see later.

Equation (6.5) gives the values N_j^* that maximize ln W in the Boltzmann limit. Substitution into equation (6.2) therefore gives the value of ln W_{max}, and the assumption (5.1) that $\Omega = W_{\text{max}}$ then leads to the value of ln Ω. We thus obtain finally, with the aid of the equations $\sum_j N_j = N$ and $\sum_j N_j\epsilon_j = E$,

$$\boxed{\ln \Omega = N\left(\ln \frac{\sum_j C_j e^{-\beta\epsilon_j}}{N} + 1\right) + \beta E}. \tag{6.7}$$

This result provides the answer, in the Boltzmann limit, to the question posed at the beginning of Sec. 4, namely, what is the total number Ω of microstates consistent with the overall specification of the system?

Up to this point we have regarded the distribution numbers N_j^* given by equation (6.5) as merely intermediate values to be used in the determination of W_{max} and thence of Ω. They can, however, be given a direct physical meaning. This arises out of the fact that all the terms $W(N_j)$ that *actually* contribute significantly to the sum (4.6)—we here revert to the true situation that $\Omega \neq W_{\text{max}}$—correspond to values of N_j that lie very close to N_j^*. That is to say, the maximum of W as a function of the N_j's is a sharp one. To see this let us calculate, for the Boltzmann limit, the number of ways W in which a distribution can occur for values of N_j close to N_j^*. To do this we set $N_j = N_j^* + \Delta N_j$ and substitute into equation

(6.2) to obtain

$$\ln W = \sum_j (N_j^* + \Delta N_j)[-\ln (N_j^* + \Delta N_j) + \ln C_j] + N.$$

If we assume $|\Delta N_j|/N_j^* \ll 1$, we can expand $\ln (N_j^* + \Delta N_j)$ as

$$\ln (N_j^* + \Delta N_j) = \ln N_j^* + \ln \left(1 + \frac{\Delta N_j}{N_j^*}\right)$$

$$= \ln N_j^* + \frac{\Delta N_j}{N_j^*} - \frac{1}{2}\left(\frac{\Delta N_j}{N_j^*}\right)^2 + \cdots.$$

Putting this into the previous equation, we obtain, to the second order in the ΔN_j's,

$$\ln W = \sum_j N_j^* \ln \frac{C_j}{N_j^*} + N - \sum_j \left[\Delta N_j \left(1 - \ln \frac{C_j}{N_j^*}\right) + \frac{1}{2}\frac{(\Delta N_j)^2}{N_j^*} + \cdots\right].$$

But the condition on the total number of particles gives $\sum_j \Delta N_j = 0$ and equation (5.5) provides the result, for $C_j \gg N_j$, that $\sum_j [\ln (C_j/N_j^*)]\Delta N_j = 0$. We thus find, with the aid of equation (6.2) written for N_j^*,

$$\ln W = \ln W_{\max} - \sum_j \left[\frac{1}{2}\frac{(\Delta N_j)^2}{N_j^*} + \cdots\right],$$

or finally, to the second order,

$$\ln \frac{W}{W_{\max}} = -\frac{1}{2}\sum_j \left(\frac{\Delta N_j}{N_j^*}\right)^2 N_j^*.$$

(We note in passing that the right-hand side of this equation is always negative, showing that the extremum previously calculated is indeed a maximum.) Consider 1 cm³ of gas at 1 atm pressure and 0°C, for which (see Chapter I, Sec. 6) $N = \Sigma N_j^* \cong 2 \times 10^{19}$. Suppose that we have an average deviation for all values of j of $(|\Delta N_j|/N_j^*)_{\text{av}} = 10^{-3}$. The foregoing equation then gives

$$\ln \frac{W}{W_{\max}} = -\frac{1}{2}\left(\frac{|\Delta N_j|}{N_j^*}\right)^2_{\text{av}} \sum_j N_j^* = -\tfrac{1}{2} \times 10^{-6} \times 2 \times 10^{19} = -10^{13},$$

or

$$\frac{W}{W_{\max}} = e^{-10^{13}},$$

which is an extremely small number indeed. This shows that in a system with a large number of particles, even a small (e.g., 0.1%) deviation in the distribution numbers from the special values N_j^* leads to an enormous reduction in the number of microstates corresponding to a given macrostate. Conversely, the macrostates for which the number of microstates is at all comparable to $W_{\max}$ must have values of N_j extremely close to N_j^*.

We now introduce the fundamental physical assumption of statistical mechanics, namely, that

all possible microstates of the system corresponding to given values of N, E, and the ϵ_j's are a priori equally probable.

That is to say, *every one* of the possible microstates that goes to make up the number Ω is equally likely. Or, in more physical terms, the system, if left alone, will in time explore without favoritism all possible microstates consistent with its prescribed values of N, E, and ϵ_j. We make this assumption since there is no reason to do otherwise, that is, there is no apparent reason why nature should prefer one microstate to another. It is, of course, a physical assumption and its final justification must rest on the success of the theory built on it.

The situation is thus as follows:

1. According to our assumption, the system will, as the result of collisions between the particles, pass through all possible microstates with equal frequency.
2. By the preceding analysis, the overwhelming majority of the possible microstates corresponds to distribution numbers N_j in the immediate vicinity of N_j^*.

It follows from these two statements that the system must spend most of the time in the macrostate corresponding to N_j^*, or some macrostate only negligibly different from it. In other words, if we could in some way measure the number of particles with energy ϵ_j, we would expect to find essentially the value N_j^*. The fact that any measurement must be an average over some period of time only serves to strengthen this expectation. Although the discussion here has been in terms of the Boltzmann limit, the conclusions are equally true in the Bose-Einstein and Fermi-Dirac statistics.

The foregoing discussion gives a physical as well as mathematical meaning to the macrostate described by the numbers [cf. equation (6.5)]

$$\boxed{N_j^* = N\frac{C_j e^{-\epsilon_j/kT}}{\sum\limits_j C_j e^{-\epsilon_j/kT}}}, \tag{6.8}$$

where we have, for convenience of reference, anticipated the later result that $\beta = 1/kT$. It is, for all practical purposes, the actual distribution over energy groups of the particles in a system in equilibrium. This important distribution is known as the *Boltzmann distribution*. The sum in the denominator of (6.8) is called the *molecular partition function* and will be denoted here by

$$\boxed{Q \equiv \sum_j C_j e^{-\epsilon_j/kT}}, \tag{6.9}$$

or, written out,

$$Q \equiv C_1 e^{-\epsilon_1/kT} + C_2 e^{-\epsilon_2/kT} + \cdots.$$

We see that the successive terms in the partition function are proportional to the number of particles in the corresponding group of energy states. That is, from equation (6.8), for two groups j and k we have

$$\frac{N_j^*}{N_k^*} = \frac{C_j e^{-\epsilon_j/kT}}{C_k e^{-\epsilon_k/kT}}.$$

The reason for the name of the partition function is thus apparent: it tells us how the particles are partitioned, or divided up, among the energy groups.

The foregoing results are in terms of groups of energy states, all states in a given group being treated as having the same energy ϵ_j. It is sometimes desirable to revert back to our original description in terms of individual states, all with perhaps different energies ϵ_i. Since the energies ϵ_i of the states in a given group are only very slightly different from the value of ϵ_j for that group, and since there are C_j such states in the group, we can write the partition function (6.9) within the accuracy of our formulation as

$$Q \equiv \sum_j C_j e^{-\epsilon_j/kT} = \sum_i e^{-\epsilon_i/kT}, \tag{6.10}$$

where the second summation is now taken over all energy *states*. Turning to equation (6.8), let us write this equation in the form

$$\frac{N_j^*}{C_j} = N \frac{e^{-\epsilon_j/kT}}{\sum_j C_j e^{-\epsilon_j/kT}}, \tag{6.11}$$

where N_j^*/C_j now gives the average number of particles in each energy state within the given group j. This, together with equation (6.10), suggests writing for the population of the individual states the equation

$$N_i^{*\prime} = N \frac{e^{-\epsilon_i/kT}}{\sum_i e^{-\epsilon_i/kT}}. \tag{6.12}$$

This is a reasonable extension since summing this equation over all states i within a given group j leads, to the accuracy of our formulation, directly back to equation (6.11). Equation (6.12), which will be useful later, in effect defines the population numbers $N_i^{*\prime}$. Since $e^{-\epsilon_i/kT}$ is approximately equal to $e^{-\epsilon_j/kT}$ within a given group, it follows from equations (6.11) and

(6.12), together with equation (6.10), that $N_i^{*\prime} \cong N_j^*/C_j$. Furthermore, since $C_j \gg N_j^*$, the numbers $N_i^{*\prime}$ are smaller than unity. They cannot therefore represent the number of particles in a given state at a particular instant and are not to be confused with the numbers N_i of equation (4.1), which are necessarily integers. They are nevertheless useful as giving a kind of measure of the average population (over a time interval) of a given state.

It is also convenient at times to have these results in terms of energy *levels* instead of energy states. An energy level is defined as consisting of all energy states having *identical* values of the energy ϵ_i. A level containing more than one state is then said to be *degenerate*. We denote the number of states contained in a given level of common energy ϵ_l by the symbol g_l, commonly called the *degeneracy* of that level. To write equations (6.10) and (6.12) in terms of levels, we need only collect the energy states together into groups for which $\epsilon_i = \epsilon_l$. We thus obtain

$$Q = \sum_l g_l e^{-\epsilon_l/kT}, \tag{6.13}$$

and

$$N_l^{*\prime} = N \frac{g_l e^{-\epsilon_l/kT}}{\sum_l g_l e^{-\epsilon_l/kT}}, \tag{6.14}$$

where the summations are now taken over all energy levels. These equations have a formal resemblance to equations (6.9) and (6.8). They are, however, essentially different. The values of the energy for the states in a given *level* in (6.13) and (6.14) are identically equal, whereas those for the states in a given *group* in (6.9) and (6.8) are only approximately so. Each group of states, in fact, will often contain a considerable number of levels.

We note that in the forms (6.12) and (6.14) the numbers C_j, which represent the number of states in our earlier groups, have been eliminated from the final results. This is reasonable since the original division of the energy states into groups was an artificial process introduced for analytical purposes. The numbers C_j thus have no inherent physical significance and should be expected to disappear eventually. In some instances, however, it will still be convenient to employ the results in the form (6.11).

Before moving on to other matters, we can now re-examine our seemingly drastic assumption that $\Omega = W_{\max}$. We may justifiably ask whether there are not other terms $W(N_j)$ in the vicinity of $W_{\max}$ that will contribute to the sum (4.6) for the exact value of Ω. The answer is that there are, but that it does not matter. This is because it is not Ω that we are really interested in, but $\ln \Omega$. For example, suppose for the sake of discussion that there are as many as N terms in the sum all equal to $W_{\max}$ and that all other terms are negligible. We then have $\Omega = NW_{max}$, from which we can obtain

$$\frac{\ln \Omega}{\ln W_{max}} = 1 + \frac{\ln N}{\ln W_{max}}.$$

The value of $\ln W_{max}$ is known from equation (6.7), which was derived on the assumption that $\Omega = W_{max}$. Putting that result into the foregoing relation we can write

$$\frac{\ln \Omega}{\ln W_{max}} = 1 + \frac{\ln N}{N(\ln Q - \ln N + 1) + \beta E}.$$

If we assume for simplicity that the particles of our system have only translational energy, we can write from our earlier results of kinetic theory that $E = N(\frac{3}{2}kT)$. We thus finally obtain

$$\frac{\ln \Omega}{\ln W_{max}} = 1 + \frac{\ln N}{N(\ln Q - \ln N + 1 + \frac{3}{2}\beta kT)}.$$

Since Q and β are independent of N, the second term on the right will approach zero as $N \to \infty$, and we find accordingly that $\ln \Omega \to \ln W_{max}$. This rough example suggests that our assumption that $\Omega = W_{max}$ is reasonable for sufficiently large values of N. A further circumstance favoring the present method is the fact that the use of Stirling's formula (5.3b) in the evaluation of W_{max} leads to a value that is larger than that obtained when more accurate methods are used. This error tends to cancel the previous error made by omitting terms other than W_{max} from the correct sum (4.6) for Ω.[5] These remarks should be regarded, however, only as plausible explanations for the success of our seemingly approximate methods. A firmer justification lies, as mentioned at the beginning of Sec. 5, in the fact that our special assumptions, which were purely mathematical, are not essential but can be removed by more rigorous analysis. There is no change in the final results.

Exercise 6.1. Consider helium at a pressure of 1 atm in a volume of 1 cm^3. Assuming that the helium may be regarded as a thermally perfect gas, check the degree to which the inequality (6.3) for the validity of the Boltzmann limit is satisfied (a) at 273°K and (b) at 4.2°K. (The latter is the boiling point of liquid helium, at which the Boltzmann limit would not be expected to apply.)

Exercise 6.2. It is known that the sum

$$\sum_{n=0}^{N} \frac{N!}{n!\,(N-n)!}$$

[5] For a detailed discussion of these questions see Davidson (1962), pp. 302–308, or Wilson (1957), pp. 90–94.

has the exact value 2^N. Evaluate the sum approximately by (a) assuming that the entire sum is equal to the largest term and (b) picking out this largest term with the aid of Stirling's formula. What does your result show regarding the effect of approximations (a) and (b) in this instance?

7 RELATION TO THERMODYNAMICS; BOLTZMANN'S RELATION

Our main interest in statistical mechanics for the present book lies in its ability to throw light on and in some instances predict the macroscopic properties of matter with which we deal in thermodynamics. For these purposes we must find some link between the variables of thermodynamics and the statistical quantities that we have discussed. This link can be made in various ways. The one that we shall use has already been set down as equation (1.1), and we shall now attempt to make this important equation plausible.

The basis for equation (1.1) lies in a recognition of the analogy that exists between the entropy of a system and the degree of molecular disorder or randomness of the system. This can best be illustrated by several examples. For convenience we revert to the classical description in terms of the position and velocity of the particles, although this is not essential. We deal in each case with a closed system, that is, a system in which the number N of particles is fixed.

Expansion into a Vacuum. Let us consider the expansion of a gas from a high-pressure chamber, through a stopcock, and into another chamber originally at a vacuum, the entire system being thermally insulated. We know from thermodynamics that the equalization of pressure in the two chambers after the stopcock is opened is accompanied by an increase in entropy. It is also obviously accompanied by an increase in our degree of uncertainty about the collection of molecules that make up the gas. Specifically, we know less than we did about the possible location of any given molecule, since each molecule now has a greater volume available to it than it did originally. Since disorder and uncertainty go hand in hand, the increase in entropy of the assembly is thus accompanied by an increase in disorder. (The molecular velocity, i.e., temperature, is the same before and after the expansion, so there is no change in our uncertainty in this regard.)

Mixing of Two Gases by Diffusion. We consider two different gases originally at the same temperature and pressure and separated from each other by a removable partition. As we have seen in Chapter III, Sec. 8, the diffusive mixing that follows removal of the partition is accompanied by an increase in entropy. It is also accompanied by an increase in uncertainty or disorder, since the mixing reduces our knowledge regarding the possible location of a given molecule of either kind. (The molecular velocities are unchanged by the mixing, so again this does not enter the picture.)

Reversible Heating of a Gas. We consider the reversible heating of a gas in a fixed volume. This is known from thermodynamics to involve an increase in entropy. Since the volume is fixed there is no increase in uncertainty regarding the location of the gas molecules. There is, however, an increase in temperature, which is accompanied by a widening of the Maxwellian velocity distribution [cf. eq. (II 5.10)] and a resulting increase in uncertainty about the velocity of any given molecule.

We thus see that in all three processes that we have examined for a closed system, two irreversible and one reversible, an increase in entropy corresponds to an increase in molecular disorder. The same can be found in other examples. We may therefore surmise that the entropy of a system is in some way a measure of the degree of disorder of the system.

To make the foregoing idea useful, we must have some quantitative measure of the disorder. In the discussion of the three examples that we have considered, we implicitly regarded the microscopic state of the system at any instant as specified in terms of the position and velocity of every molecule at that instant. We then assessed the degree of disorder of the system by, in effect, assessing the number of different microscopic states that the assembly could attain consistent with its macroscopic situation, that is, the more possible positions or velocities in which the molecules could find themselves, other things being equal, the greater the degree of disorder. This leads us to define the disorder of an assembly as measured by the number of different microscopic states in which the assembly can be arranged consistent with its overall situation. If this number of microscopic states is denoted as before by Ω, the conclusion of the preceding paragraph can then be expressed by a relation of the form

$$S = \phi(\Omega), \tag{7.1}$$

where ϕ is an unknown, universal function of the single argument Ω for a system of fixed N.

It should be noted that the quantity Ω is to be regarded here in a very general sense. It will take different forms for different types of systems, and it may be evaluated in terms of positions and velocities from the point of view of classical mechanics (as in this section) or in terms of energy states from the point of view of quantum mechanics (as in our preceding work). This will depend on what is appropriate to the problem at hand.

The expressions for Ω given in the preceding sections apply primarily to gaseous systems treated from the quantum-mechanical point of view.

The properties of S and Ω allow an immediate solution for the form of the function ϕ in the relation (7.1). Let us consider two systems with numbers of particles N_1 and N_2 and having entropies S_1 and S_2 and numbers of microstates Ω_1 and Ω_2. We then consider the combined system formed by taking the two systems together (but not intermixed). Since entropy is additive, the entropy S_{12} of the combined system with number of particles $N_{12} = N_1 + N_2$ must be

$$S_{12} = S_1 + S_2. \tag{7.2}$$

Furthermore, since every microstate of the first system can exist in conjunction with Ω_2 microstates of the second system, the number of microstates Ω_{12} of the combined system is

$$\Omega_{12} = \Omega_1\Omega_2. \tag{7.3}$$

Putting the relation (7.1) into (7.2) and substituting from equation (7.3), we obtain the following relation governing the function ϕ:

$$\phi(\Omega_1\Omega_2) = \phi(\Omega_1) + \phi(\Omega_2). \tag{7.4}$$

A differential equation for ϕ can now be obtained by differentiating this relation successively with respect to Ω_1 and Ω_2. The first differentiation gives

$$\Omega_2\phi'(\Omega_1\Omega_2) = \phi'(\Omega_1),$$

and from the second we have

$$\Omega_1\Omega_2\phi''(\Omega_1\Omega_2) + \phi'(\Omega_1\Omega_2) = 0,$$

or, in terms of the general argument Ω,

$$\Omega\phi''(\Omega) + \phi'(\Omega) = 0.$$

The general solution of this second-order differential equation is

$$\phi(\Omega) = A(N)\ln\Omega + B(N), \tag{7.5}$$

where $A(N)$ and $B(N)$ are constants for a given system of fixed N but may vary from one system to another. The forms of A and B can be found by rewriting the original governing relation (7.4) with the aid of this result. This gives

$$A(N_1 + N_2)\ln\Omega_1 + A(N_1 + N_2)\ln\Omega_2 + B(N_1 + N_2)$$
$$= A(N_1)\ln\Omega_1 + B(N_1) + A(N_2)\ln\Omega_2 + B(N_2).$$

For this equation to be satisfied, A must be an absolute constant—say

k—and B must have the linear form $B(N) = bN$, where b is a second constant. This completes the solution for ϕ.

With equation (7.5) and the foregoing results for A and B, equation (7.1) becomes

$$S = k \ln \Omega + S_0,$$

where $S_0 \equiv bN$ is a constant for a given system. It is convenient (see below) to take $S = 0$ for every system in a completely ordered state

Fig. 3. Memorial to Ludwig Boltzmann, Central Cemetery, Vienna (photo by Gerard Van Hoven).

(i.e., $\Omega = 1$). We thus set $S_0 = 0$ for *all* systems and write, as anticipated by our earlier equation (1.1),

$$\boxed{S = k \ln \Omega}\ . \tag{7.6}$$

This beautifully simple relation is commonly known as *Boltzmann's relation*, although Boltzmann himself never wrote it in precisely this form. In recognition of its importance, it stands carved on the memorial over Boltzmann's grave in the Central Cemetery in Vienna (Fig. 3).

The theoretical reasons for setting $S_0 = 0$ to obtain equation (7.6) are somewhat subtle (see Wilks, 1961, pp. 88–89, and Schrödinger, 1952, pp. 16–17). They boil down to the fact that the really important thing is that S_0 is a constant for a given system and that there is no need ever to know its value. The sensible and consistent thing therefore is to set it equal to zero for all systems. With this choice, the entropy of all systems will tend to zero as the absolute temperature tends to zero if the number of possible microstates Ω tends to one, that is, if the limit $T \to 0$ actually leads to the completely ordered state. This behavior has been demonstrated on the basis of quantum mechanics for certain particular systems, but no general proof has been derived. Writers on the subject differ as to whether such a proof is theoretically possible.

The choice of $S_0 = 0$ and the conjecture that $\Omega \to 1$ as $T \to 0$ is the statistical-mechanical counterpart of *Nernst's heat theorem* of macroscopic thermodynamics, also known as the *third law of thermodynamics.* Nernst's theorem, which is a generalization from experimental observations, asserts in one of its possible forms (that due to Planck) that the entropy of all substances in complete thermodynamic equilibrium tends to a common value as $T \to 0$; this common value is arbitrarily taken as zero. Various other statements are also given in the literature, particularly to allow for the existence at low temperatures of states that are only partially in equilibrium. Since we shall not have occasion to use Nernst's theorem, we shall not go into it further. For an up-to-date discussion of the subject, which has been notable for the disagreement that it has evoked in the past, the reader can consult Wilks (1961) or Wilson (1957), Chapter 7.

The discussion of this section is not to be regarded as a derivation of Boltzmann's relation, but merely as a plausibility argument for the existence of such a relation. The relation will be taken here as an hypothesis whose validity is to be judged on the basis of the success of the theory erected upon it. The constant k in the equation will be identified later with Boltzmann's constant, but it is to be regarded for the time being merely as an unspecified constant of proportionality.

Exercise 7.1. Give a possible interpretation, in molecular terms similar to those used in the three examples at the beginning of this section, of the fact that when reversible work is done by changing the volume of a gas in a perfectly insulated container, the entropy remains constant.

Exercise 7.2. Consider a gaseous system composed of a fixed number of identical molecules N contained in a variable volume V. For a given, *fixed* temperature (and hence a fixed distribution of the molecules as regards velocity), the number of microstates can be assessed classically in terms of the position of the molecules by dividing the volume up into z small cells of identical size such that $z \gg N$. The size of the cells is fixed and does not change as V changes.

(a) Write an expression for the number of microstates Ω, where Ω now has the special meaning of the number of ways in which the N indistinguishable molecules can be placed in the z cells. Since $z \gg N$, you may ignore the possibility that two molecules are in the same cell.

(b) Using Stirling's formula and the approximation that $\ln(1 - x) \cong -x$ for $x \ll 1$, show that

$$\ln \Omega = N \ln \frac{z}{N} + N.$$

(c) If two volumes are in the ratio $r = V_2/V_1$ so that accordingly $z_2 = rz_1$, find an expression for $\ln (\Omega_2/\Omega_1)$ in terms of V_2/V_1 and N.

(d) Using only your knowledge from macroscopic thermodynamics, write an expression for the entropy difference $S_2 - S_1$. Assume that the gas is perfect and recall that the temperature has been assumed to have a fixed value independent of the volume.

(e) By comparing the results of (c) and (d), show that for this specific example, at least,

$$S_2 - S_1 = k \ln \Omega_2 - k \ln \Omega_1,$$

or

$$S = k \ln \Omega.$$

8 THERMODYNAMIC PROPERTIES

With Boltzmann's relation we can now proceed to the statistical calculation of thermodynamic properties. Substituting the earlier results of equation (6.7) into Boltzmann's relation, we obtain at once for the entropy of a system in the Boltzmann limit

$$S = k\left[N\left(\ln \frac{\sum_j C_j e^{-\beta\epsilon_j}}{N} + 1\right) + \beta E\right]. \tag{8.1}$$

In this result, β is still unknown. We can evaluate it, however, by reference to the general thermodynamic relation (III 6.8). Written for a single-component gas, this relation is

$$dS = \frac{1}{T}(dE + p\,dV - \tilde{\mu}\,dN), \tag{8.2}$$

where we have used $\tilde{\mu}$, the chemical potential per molecule, in place of $\hat{\mu}$, the chemical potential per mole. From this we see that we must have

$$\left(\frac{\partial S}{\partial E}\right)_{V,N} = \frac{1}{T}. \tag{8.3}$$

Differentiating (8.1), and noting that β is a function of E as dictated by equation (6.6), we obtain

$$\left(\frac{\partial S}{\partial E}\right)_{V,N} = k\left\{\beta + \frac{\partial}{\partial \beta}\left[N\left(\ln \frac{\sum_j C_j e^{-\beta\epsilon_j}}{N} + 1\right) + \beta E\right]\frac{\partial \beta}{\partial E}\right\}.$$

In the partial derivative with respect to β (with everything else fixed, of course), we have

$$\frac{\partial}{\partial\beta}\left[N\left(\ln\frac{\sum_j C_j e^{-\beta\epsilon_j}}{N}+1\right)+\beta E\right]=N\frac{-\sum_j \epsilon_j C_j e^{-\beta\epsilon_j}}{\sum_j C_j e^{-\beta\epsilon_j}}+E$$

$$=-N\frac{E}{N}+E=0,$$

where use has been made of equation (6.6). We thus obtain

$$\left(\frac{\partial S}{\partial E}\right)_{V,N}=k\beta.$$

Comparison with equation (8.3) shows that

$$\beta=\frac{1}{kT}.$$

Introducing this and the relation (6.10) for the partition function into equation (8.1), we can thus write for the entropy

$$\boxed{\begin{aligned} S &= Nk\left(\ln\frac{Q}{N}+1\right)+\frac{E}{T} \\ Q &= \sum_j C_j e^{-\epsilon_j/kT}=\sum_i e^{-\epsilon_i/kT} \end{aligned}} \tag{8.4}$$

where

It is convenient at this point to introduce the thermodynamic state variable called the *Helmholtz free energy F*, which is defined by the equation

$$F\equiv E-TS. \tag{8.5}$$

The useful properties of this variable can be found by differentiating this definition, which gives

$$dF=dE-T\,dS-S\,dT,$$

and then eliminating $dE - T\,dS$ between this relation and equation (8.2) to obtain[6]

$$dF=-S\,dT-p\,dV+\tilde{\mu}\,dN.$$

We see from this equation that

$$S=-\left(\frac{\partial F}{\partial T}\right)_{V,N},\qquad p=-\left(\frac{\partial F}{\partial V}\right)_{T,N},\qquad \tilde{\mu}=\left(\frac{\partial F}{\partial N}\right)_{T,V}, \tag{8.6a}$$

[6] Comparison of this equation with equation (III 6.5) shows that F is the natural variable to use when the conditions of the system are given in terms of T and V, whereas G is the natural variable when they are given in terms of T and p. This illustrates the convenience of defining different thermodynamic variables for different purposes.

and hence from the definition (8.5) that

$$E = F - T\left(\frac{\partial F}{\partial T}\right)_{V,N} = -T^2\left(\frac{\partial(F/T)}{\partial T}\right)_{V,N}. \tag{8.6b}$$

We thus see that if F is known as a function of T, V, and N, all other variables of interest can be found from equations (8.6).

The value of F for the present system can be found from equations (8.4) and (8.5). The result can then be substituted into equations (8.6) to find the other thermodynamic variables. We thus obtain finally the following expressions for the thermodynamic properties of our system (where note has been taken that the ϵ_i's and hence Q are independent of N):

$$\boxed{\begin{aligned} F &= -NkT\left(\ln\frac{Q}{N} + 1\right) \\ S &= Nk\left[\ln\frac{Q}{N} + 1 + T\frac{\partial(\ln Q)}{\partial T}\right] \\ E &= NkT^2\frac{\partial(\ln Q)}{\partial T} \\ p &= NkT\frac{\partial(\ln Q)}{\partial V} \\ \tilde{\mu} &= -kT\ln\frac{Q}{N} \end{aligned}}\ . \tag{8.7}$$

We see that the partition function is the key to the problem; if it can be found as a function of T and V, the thermodynamic state variables follow at once from the foregoing equations.

Before we proceed to the problem of finding Q, an alternative approach to statistical mechanics should be mentioned. In the foregoing development we have treated the equilibrium properties of a system by direct consideration of the quantum energy states of the individual particles. This method is essentially a quantum-mechanical extension of the pioneering classical work of Boltzmann. It is perfectly valid for weakly interacting particles, and, as we shall see, leads to correct results for a thermally perfect gas or for mixtures of such gases. The method fails, however, for particles that are more than weakly interacting, and it cannot therefore be applied to imperfect gases. This is due to the fact that the quantum energy states of the system as a whole cannot then be described in terms of the states of the individual particles but must be considered directly. This difficulty can be overcome by adopting an alternative method due originally to Gibbs. In this method one abandons the attempt to study a collection of particles and studies instead a collection or *ensemble* of systems of structure similar to the system of actual interest and distributed over a range of different possible quantum states of the system as a whole. By evaluating the average behavior of the systems in an appropriate ensemble, one can then predict what is to be expected *on the average* for the single system of actual interest. The ensembles may be of several kinds, depending on one's knowledge of the system, and the averaging may be done in various ways—these are matters of detail. When a certain type of ensemble (the so-called *microcanonical ensemble*) is applied to weakly interacting particles, the method becomes essentially equivalent to what we have done and the same results are obtained. Since in gas dynamics we are concerned mostly with cases of weak interaction, we shall not pursue the matter further. For particularly lucid accounts of the ensemble method, the reader may consult Denbigh (1955), Chapters 11 and 12, or Andrews (1963). In the following sections we shall direct ourselves to the determination of the partition function and thence the thermodynamic properties of our system of weakly interacting particles.

Exercise 8.1. Since one part of the energy, the potential energy, may be measured from any zero that we wish, the values ϵ_i of the energy are not in general absolute but may be changed by a constant. Let us arbitrarily decide to measure the particle energies from a new origin such that the new energy ϵ_i' is greater than the old by a fixed amount ϵ^*, that is, $\epsilon_i' = \epsilon_i + \epsilon^*$. If the new partition function is evaluated formally in terms of ϵ_i' rather than ϵ_i, show by means of equations (8.7) that the new thermodynamic properties (indicated by primes) are related to the old by

$$F' = F + N\epsilon^*, \qquad S' = S, \qquad E' = E + N\epsilon^*, \qquad p' = p, \qquad \tilde{\mu}' = \tilde{\mu} + \epsilon^*.$$

Thus the choice of the zero point for the energy affects F, E, and $\tilde{\mu}$ but not S and p.

9 PROPERTIES ASSOCIATED WITH TRANSLATIONAL ENERGY

We shall first apply the foregoing results to calculate the thermodynamic properties of a system of N structureless molecules moving freely in a volume V. By structureless we mean that the molecules have mass but no internal structure. The only energy that they can possess is therefore translational. This, with the assumption of weak interaction (i.e., only upon collision), is the same model that led in our discussions of kinetic theory to the relations for a thermally and calorically perfect monatomic gas. We should expect therefore, provided our entire structure of statistical mechanics is valid, to obtain once again the relations for such a gas.

If the volume V is a rectangular box having sides of length a_1, a_2, and a_3, then according to our earlier result (3.11) the possible energy states of a particle of mass m moving in this box are

$$\epsilon_i = \epsilon_{n_1,n_2,n_3} = \frac{h^2}{8m}\left(\frac{n_1^2}{a_1^2} + \frac{n_2^2}{a_2^2} + \frac{n_3^2}{a_3^2}\right) \qquad \text{where} \qquad \left.\begin{matrix} n_1 \\ n_2 \\ n_3 \end{matrix}\right\} = 1, 2, 3, \ldots .$$

The partition function $Q = \sum_i e^{-\epsilon_i/kT}$ is therefore as follows, where we

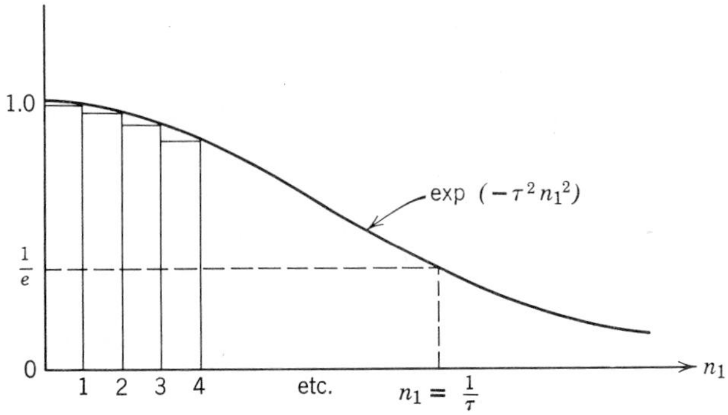

Fig. 4. Determination of the sum $\sum_{n_1=1}^{\infty} e^{-\tau^2 n_1^2}$.

must sum over all values of n_1, n_2, and n_3 to include all the energy states:

$$Q_{\text{tr}} = \sum_{n_1=1}^{\infty} \sum_{n_2=1}^{\infty} \sum_{n_3=1}^{\infty} \exp\left[-\left(\frac{n_1^2}{a_1^2} + \frac{n_2^2}{a_2^2} + \frac{n_3^2}{a_3^2}\right)\frac{h^2}{8mkT}\right]$$

$$= \sum_{n_1=1}^{\infty} \exp\left(-\frac{n_1^2}{a_1^2}\frac{h^2}{8mkT}\right) \sum_{n_2=1}^{\infty} \exp\left(-\frac{n_2^2}{a_2^2}\frac{h^2}{8mkT}\right) \sum_{n_3=1}^{\infty} \exp\left(-\frac{n_3^2}{a_3^2}\frac{h^2}{8mkT}\right).$$

With later use in mind, we append the subscript tr to denote quantities associated with translation. Let us consider one representative sum of this triple product, say

$$\sum_{n_1=1}^{\infty} \exp\left(-\frac{n_1^2}{a_1^2}\frac{h^2}{8mkT}\right) = \sum_{n_1=1}^{\infty} e^{-\tau^2 n_1^2},$$

where $\tau^2 = h^2/8ma_1^2kT$. When τ^2 is very small (e.g., 10^{-16}, as would be the case for a hydrogen atom, $m = 1.67 \times 10^{-24}$ gm, with $a_1 = 1$ cm and $T = 0°\text{C}$), the curve of $e^{-\tau^2 n_1^2}$ will appear as shown in Fig. 4. The value of τ^2 determines the horizontal spread of the curve; when τ^2 is small, the curve falls very gradually. Now, it can be seen that the sum

$$\sum_{n_1=1}^{\infty} e^{-\tau^2 n_1^2} = \sum_{n_1=1}^{\infty} (e^{-\tau^2 n_1^2} \times 1)$$

represents the area under the stepped curve in the figure; and if τ^2 is very small this area will be sensibly equal to the area under the exponential curve. For very small τ^2 we may therefore, with only negligible error, replace the summation by an integral and write

$$\sum_{n_1=1}^{\infty} e^{-\tau^2 n_1^2} = \int_0^{\infty} e^{-\tau^2 n_1^2}\, dn_1 = \frac{\sqrt{\pi}}{2\tau} = a_1 \frac{\sqrt{2\pi mkT}}{h},$$

where the value of the integral has been obtained from the formulas in the appendix. With this result we can then write

$$Q_{tr} = a_1 \frac{\sqrt{2\pi mkT}}{h} a_2 \frac{\sqrt{2\pi mkT}}{h} a_3 \frac{\sqrt{2\pi mkT}}{h},$$

or finally, since $a_1 a_2 a_3 = V$,

$$\boxed{Q_{tr} = V\left(\frac{2\pi mkT}{h^2}\right)^{3/2}}. \tag{9.1}$$

The same result can be obtained on the basis of a more general volume than a rectangular box.

The thermodynamic properties associated with Q_{tr} are now readily found. Substitution into the first of equations (8.7) gives for F_{tr}

$$F_{tr} = -NkT\left[\ln\frac{V}{N} + \tfrac{3}{2}\ln T + \ln\left(\frac{2\pi mk}{h^2}\right)^{3/2} + 1\right]. \tag{9.2}$$

Differentiating according to the second of equations (8.6a), we then find for the pressure

$$p = -\left(\frac{\partial F_{tr}}{\partial V}\right)_{T,N} = NkT \times \frac{1}{V},$$

or

$$\boxed{pV = NkT}. \tag{9.3}$$

We have thus derived the equation of state for a thermally perfect gas by statistical-mechanical methods, provided we identify the constant of proportionality k, which was introduced originally in the equation $S = k \ln \Omega$, as Boltzmann's constant. This identification can be made on more general grounds than this, but we shall not have space to go into the matter here.

The entropy can be found from equation (9.2) and the first of equations (8.6a) and is

$$S_{tr} = -\left(\frac{\partial F_{tr}}{\partial T}\right)_{V,N} = Nk\left[\ln\frac{V}{N} + \tfrac{3}{2}\ln T + \ln\left(\frac{2\pi mk}{h^2}\right)^{3/2} + \tfrac{5}{2}\right], \quad (9.4)$$

or, after substitution from equation (9.3) in the form $\ln (V/N) = \ln T - \ln p + \ln k$,

$$S_{tr} = Nk\left\{\tfrac{5}{2}\ln T - \ln p + \ln\left[\left(\frac{2\pi m}{h^2}\right)^{3/2} k^{5/2}\right] + \tfrac{5}{2}\right\}. \quad (9.5a)$$

This is the famous *Sackur-Tetrode equation* for the absolute translational entropy. It can be written in terms of entropy per unit mass $s_{tr} \equiv S_{tr}/mN$ as

$$\boxed{s_{tr} = \tfrac{5}{2}R\ln T - R\ln p + R\left\{\ln\left[\left(\frac{2\pi m}{h^2}\right)^{3/2} k^{5/2}\right] + \tfrac{5}{2}\right\}}, \quad (9.5b)$$

where $R = k/m$ is the gas constant per unit mass. This is seen to be identical in form to the well-known thermodynamic result

$$s = c_p \ln T - R \ln p + \text{const}$$

for the specific entropy of a thermally and calorically perfect gas. The present results agree with our earlier findings from kinetic theory that for a monatomic gas $c_p = c_{p_{tr}} = \frac{5}{2}R$. They go much farther, however, in that they also provide us with a value for the constant, which both thermodynamics and kinetic theory were powerless to evaluate.

A partial check on the validity of our entire theoretical structure can be obtained from the foregoing results. One first calculates the entropy of a suitable monatomic gas, say gaseous mercury, at a given temperature and pressure by means of equation (9.5a). This equation was derived purely from molecular considerations of the gaseous state and requires no knowledge of the fact that the substance in question is liquid or solid at lower temperatures. One can also calculate the entropy quite independently by integrating the classical thermodynamic formula $dS = (dQ/T)_{rev}$ from absolute zero, making use of experimental values of the specific heat in the solid, liquid, and gaseous states and of the heats of fusion and vaporization. It is also necessary to invoke Nernst's heat theorem that $S = 0$ at $T = 0$ (see Sec. 7) and to extrapolate the experimental data in the vicinity of absolute zero. This calculation thus makes detailed use of the entire "history" of the substance, but requires no knowledge whatsoever of the existence of molecules. When the calculations are made and the two values are compared, the agreement is found to be excellent (see, e.g., Wilson, 1957, p. 197). This is a remarkably convincing result in view of the completely different methods used in obtaining the two values.[7]

The fact that equation (9.5) shows the entropy approaching an infinite negative value as $T \rightarrow 0$, although it appears to conflict with Nernst's theorem, does not actually do so. As we have seen earlier, the assumptions on which the equation is based become invalid at very low temperatures. Under such conditions the Boltzmann limit must be replaced by the correct Bose-Einstein or Fermi-Dirac statistics. This does not affect the ability of equation (9.5) to give reliable values of the absolute entropy at the higher temperatures, however, since the situation at low temperatures is in no way involved in its derivation.

To complete the picture, the remaining thermodynamic properties can also be obtained. The internal energy $E_{tr} = F_{tr} + TS_{tr}$ is found from equations (9.2) and (9.4) to be

$$E_{tr} = \tfrac{3}{2}NkT, \tag{9.6a}$$

or, in terms of specific energy,

$$\boxed{e_{tr} = \tfrac{3}{2}RT}\ . \tag{9.6b}$$

From this we obtain

$$\boxed{c_{v_{tr}} = \left(\frac{\partial e_{tr}}{\partial T}\right)_v = \tfrac{3}{2}R}\ , \tag{9.7}$$

which is also in accord with our earlier results from kinetic theory.

The chemical potential is obtained with the aid of the last of equations (8.6a). We thus find

$$\tilde{\mu}_{tr} = \left(\frac{\partial F_{tr}}{\partial N}\right)_{T,V} = -kT\left[\ln\frac{V}{N} + \tfrac{3}{2}\ln T + \ln\left(\frac{2\pi mk}{h^2}\right)^{3/2}\right].$$

For the chemical potential per unit mass $\mu_{tr} = (R/k)\tilde{\mu}_{tr}$, written in terms of the pressure, we then have

$$\boxed{\mu_{tr} = RT\ln p - RT\left\{\tfrac{5}{2}\ln T + \ln\left[\left(\frac{2\pi m}{h^2}\right)^{3/2} k^{5/2}\right]\right\}}\ . \tag{9.8}$$

[7] When similar methods are applied to polyatomic gases and other substances, it is sometimes found that the "statistical entropy" is significantly higher than the "calorimetric entropy." Plausible reasons can be given for supposing that in these cases the experimental data were not carried to sufficiently low temperatures or that the substance does not stay in thermodynamic equilibrium as the temperature is reduced, with the result that Nernst's theorem does not apply. For a discussion of these matters, see Wilson (1957), pp. 196–205, or Wilks (1961), Chapter V.

This agrees with equation (III 8.9a) for $\hat{\mu} = (\hat{R}/R)\mu$ when that equation is specialized to a single species ($\mathcal{N}_s = \mathcal{N}$). Statistical mechanics, however, provides a complete expression for the function $\hat{\mu}^0(T) = (\hat{R}/R)\mu^0(T)$, which was previously only partially defined.[8]

It remains to examine the distribution of molecules over the energy states. This is given in general by equation (6.8):

$$N_j^* = N\frac{C_j e^{-\epsilon_j/kT}}{Q}. \tag{9.9}$$

The value of Q for the present case of translational energy is known from equation (9.1). We consider the general group of states j to consist of all those quantum states having energy in the small range from ϵ to $\epsilon + d\epsilon$, where ϵ is a continuous variable. If $\xi(\epsilon)$ is the distribution function for molecular energy, we can then write for the number N_j^* of molecules in the group

$$N_j^* = N\xi(\epsilon)\,d\epsilon.$$

The number C_j of energy states in the group can be found with the aid of equation (4.3), which gave the total number Γ of states with energy less than ϵ. By differentiation, we have specifically[9]

$$C_j = \frac{d\Gamma}{d\epsilon}\,d\epsilon = 2\pi\frac{V}{h^3}(2m)^{3/2}\epsilon^{1/2}\,d\epsilon. \tag{9.10}$$

Substituting these two expressions together with Q_{tr} from equation (9.1) into equation (9.9), and setting $\epsilon_j = \epsilon$ in accord with our choice of the group j, we obtain the following expression for the distribution of molecular energy:

$$\xi(\epsilon)\,d\epsilon = 2\pi\frac{\epsilon^{1/2}e^{-\epsilon/kT}}{(\pi kT)^{3/2}}\,d\epsilon.$$

The right-hand side of this equation can be re-expressed in terms of the velocity magnitude C by means of the relations $\epsilon = mC^2/2$ and $d\epsilon = mC\,dC$. Redefining the transformed result as $\chi(C)\,dC$, we thus find for the distribution of molecular speed

$$\chi(C)\,dC = 4\pi\left(\frac{m}{2\pi kT}\right)^{3/2}C^2e^{-mC^2/2kT}\,dC. \tag{9.11}$$

[8] Note that this expression cannot be obtained by carrying out the integration of equation (III 8.8b) with $\hat{c}_p = \frac{5}{2}\hat{R}$, since $\hat{c}_p$ is not in fact a constant over the full range of temperature.

[9] With reference to Fig. 2, C_j can be interpreted as one-eighth of the volume of a spherical shell having radius n_s and thickness dn_s, where n_s is related to ϵ as shown in the figure.

This is identical to the result (II 5.10) previously obtained for the Maxwellian velocity distribution by kinetic means. We thus see that the Maxwellian distribution is the special case for translational energy of the more general Boltzmann distribution over energy states.

Exercise 9.1. Verify equations (9.3), (9.4), and (9.6a) by direct substitution of Q_{tr} into the appropriate equation (8.7) rather than by derivation from F_{tr}.

Exercise 9.2. Consider a perfect monatomic gas whose molecules are constrained to move in the 1-direction only, so that the permissible energy levels are given by

$$\epsilon_{n_1} = \frac{h^2}{8ma_1^2} n_1^2.$$

(a) Obtain expressions for Q_{tr}, p, S_{tr}, and E_{tr} in terms of a_1 and T.
(b) Show that

$$C_j = \frac{a_1}{h} (2m)^{1/2} \epsilon^{-1/2} \, d\epsilon.$$

(c) Obtain an expression for the distribution function for the molecular speed.

10 CONTRIBUTION OF INTERNAL STRUCTURE

In the preceding section we considered a system of molecules that were independent and structureless and therefore could possess only translational energy. This model was seen to correspond to a thermally and calorically perfect monatomic gas. If the molecules are considered to have an internal structure, as in the case, for example, of a diatomic molecule or of a monatomic molecule when the orbiting electrons are considered, then energy other than translational comes into play. We now consider the contribution of this so-called "internal energy." (This molecular usage of the term internal energy should not be confused with the usage from thermodynamics.)

To begin, suppose that each molecule can have several independent types of energy ϵ', ϵ'', ϵ''', . . . , to be specified later. The total energy of the molecule is then

$$\epsilon = \epsilon' + \epsilon'' + \epsilon''' + \cdots.$$

In keeping with quantum theory, we suppose that each of the various types of energy is quantized, that is, ϵ' can have values ϵ_1', ϵ_2', . . . , ϵ_m', . . . ; ϵ'' can have values ϵ_1'', ϵ_2'', . . . , ϵ_n'', . . . ; etc. Since any energy state of one type can be taken in conjunction with any energy state of another type, the permissible values of ϵ are then given by

$$\epsilon_{m,n,p,\ldots} = \epsilon'_m + \epsilon''_n + \epsilon'''_p + \cdots.$$

Substituting this expression into the relation $Q = \sum_i e^{-\epsilon_i/kT}$ for the partition function, we obtain

$$Q = \sum_{m,n,p,\ldots} \exp\left(\frac{-\epsilon_{m,n,p,\ldots}}{kT}\right) = \sum_{m,n,p,\ldots} \exp\left[-\frac{(\epsilon'_m + \epsilon''_n + \epsilon'''_p + \cdots)}{kT}\right]$$

$$= \sum_m e^{-\epsilon'_m/kT} \sum_n e^{-\epsilon''_n/kT} \sum_p e^{-\epsilon'''_p/kT} \cdots$$

$$= Q'Q''Q''' \cdots. \tag{10.1}$$

We thus find the important *factorization property* of the partition function. That is, the complete partition function for a molecule is expressible as the product of individual partition functions for each kind of independent energy.

The extent to which the energy may be broken down into independent kinds depends on the accuracy required. For gases with weakly interacting particles, the separation into a translational energy ϵ_{tr} and an internal energy ϵ_{int} is completely appropriate. Detailed study shows that at all but very low temperatures ϵ_{int} may with excellent accuracy be divided in turn into two terms, one due to electronic excitation and one due to the combined effects of molecular vibration and rotation (present, of course, only in molecules made up of more than one atom). To a somewhat poorer approximation, the latter term may also be separated. The inaccuracy in this step is caused by the fact that the rotation of the molecule gives rise to a centrifugal field that affects the vibration, while the vibration is accompanied by changes in moment of inertia that affect the rotation. For most engineering purposes, however, these influences may be neglected. We therefore write for the total energy of the molecule

$$\epsilon = \epsilon_{tr} + \epsilon_{rot} + \epsilon_{vib} + \epsilon_{el}, \tag{10.2}$$

where ϵ_{rot}, ϵ_{vib}, and ϵ_{el} are the energies associated with rotation, vibration, and electronic excitation respectively. In accordance with equation (10.1), the partition function can then be written as

$$Q = Q_{tr}Q_{rot}Q_{vib}Q_{el} = Q_{tr} \times \prod_{int} Q_{int}, \tag{10.3}$$

where the product is to be taken over the various internal partition functions.

In the foregoing we have ignored the contributions of the nucleus, in particular the nuclear spin. For chemical purposes this is justified, since, except for the special case of hydrogen where the spin is important at low temperatures, the nucleus plays no part in normal chemical processes. The possible effects of isotope mixing are similarly ignored on the grounds that the isotopic composition usually remains almost constant. In both cases, therefore, the corresponding partition functions are effectively set equal to 1. As a result, the entropies so calculated are not in fact absolute values. This is of no practical consequence, however, since the foregoing effects cancel when we consider *changes* in entropy, which are the quantities of practical importance. In the foregoing we have also ignored the effects of ionization and chemical reaction. These require special treatment since they involve mixtures of more than one kind of particle. These matters are discussed later in Secs. 13 and 14.

Expressions for the thermodynamic variables in terms of the individual partition functions can now be obtained by substituting (10.3) into the equations (8.7). For the energy, written now in terms of the specific energy $e = E/mN$, we thus find, since $R = k/m$,

$$\begin{aligned} e &= RT^2 \frac{\partial}{\partial T} \ln Q \\ &= RT^2 \frac{\partial}{\partial T} \ln Q_{\text{tr}} + \sum_{\text{int}} RT^2 \frac{\partial}{\partial T} \ln Q_{\text{int}} \\ &= e_{\text{tr}} + \sum_{\text{int}} e_{\text{int}}. \end{aligned} \tag{10.4}$$

Hence for the specific heat $c_v = (\partial e/\partial T)_v$, we have

$$c_v = c_{v_{\text{tr}}} + \sum_{\text{int}} c_{v_{\text{int}}}. \tag{10.5}$$

The thermodynamic internal energies and the specific heats are thus additive, as would be expected from the original equation (10.2).

For the specific entropy $s = S/mN$ we find from the second of equations (8.7)

$$\begin{aligned} s &= R\left[\ln \frac{Q}{N} + 1 + T\frac{\partial}{\partial T} \ln Q\right] \\ &= R\left[\ln \frac{Q_{\text{tr}}}{N} + 1 + T\frac{\partial}{\partial T} \ln Q_{\text{tr}}\right] + \sum_{\text{int}} R\left[\ln Q_{\text{int}} + T\frac{\partial}{\partial T} \ln Q_{\text{int}}\right] \\ &= s_{\text{tr}} + \sum_{\text{int}} s_{\text{int}}. \end{aligned} \tag{10.6}$$

Note here that the expressions thus defined for s_{tr} and s_{int} have a different form. The factor N and the term 1 must be included with s_{tr} to give agreement with the results for the translational entropy from Sec. 9 [cf. equations (9.1) and (9.4)].

Turning to the pressure p, we obtain from the fourth of equations (8.7)

$$p = MRT\frac{\partial}{\partial V} \ln Q = MRT\frac{\partial}{\partial V} \ln Q_{\text{tr}} + \sum_{\text{int}} MRT\frac{\partial}{\partial V} \ln Q_{\text{int}}.$$

As we shall see, the Q_{int} are all independent of the volume V. This is a result of the fact that for weakly interacting particles the permissible internal quantum states of the particles have no relation to the geometry of the container. As a result, the only contribution to the pressure is from Q_{tr}. As in Sec. 9, therefore, we reproduce the equation of state for a thermally perfect gas. The gas with internal contributions to the molecular energy is thus still thermally perfect. Since the Q_{int} will be functions of T, however, the specific heat as given by equation (10.5) will vary with temperature. The gas is therefore no longer calorically perfect.

When the ideas of this section are applied, the convention is usually adopted of measuring each of the various kinds of internal energy in equation (10.2) from the lowest or so-called *ground state* of that particular kind of energy. (This is permissible, since the energies are independent.) Thus ϵ_{rot} is counted as zero when the molecule is in its rotational ground state, and similarly for ϵ_{vib} and ϵ_{el}. When all three kinds of internal energy are in their ground state simultaneously, the molecule as a whole is in its internal ground state, and the total internal energy ϵ_{int} is counted as zero. The translational energy ϵ_{tr} is taken as zero, not in the lowest translational state [$n_1 = n_2 = n_3 = 1$ in equation (3.11)], but in some hypothetical—and, according to equations (3.10) and (3.11), unattainable —state of zero translational motion.[10] This we refer to as a *state of rest*. On the basis of these conventions, each molecule may be thought of, relative to its own energy scale, as having zero total energy when it is in its internal ground state and at rest. The specific energy e as calculated for a system of identical molecules from equation (10.4) is then reckoned on the same basis.

Exercise 10.1. Using results from this and the preceding section, show that the undetermined multiplier α of Sec. 6 is given in the Boltzmann limit by

$$\alpha = \ln\left[\frac{V}{N}\left(\frac{2\pi mkT}{h^2}\right)^{3/2}\right] + \sum_{\text{int}} \ln Q_{\text{int}}.$$

It follows from the results of subsequent sections that the Q_{int} are not $\gg 1$. This equation thus confirms *a posteriori* that the condition $\alpha \gg 1$ for the validity of the Boltzmann limit is equivalent to the condition (6.3) previously cited.

11 MONATOMIC GASES

In this and the following section, we consider the properties of calorically imperfect monatomic and diatomic gases. For brevity we confine our

[10] The value of ϵ_{tr} for $n_1 = n_2 = n_3 = 1$ is, however, very small.

attention to the internal energy and the specific heat, although the other properties can be found if required.

For a monatomic gas, rotation and vibration are absent (i.e., $\epsilon_{\text{rot}} = \epsilon_{\text{vib}} = 0$). The partition function as given by equation (10.3) then reduces to simply $Q = Q_{\text{tr}}Q_{\text{el}}$, with corresponding simplification in the equations for e and c_v. The contribution of molecular translation has already been evaluated in equations (9.1), (9.6), and (9.7). We need consider here only the effect of the electrons that are present in the atom.

For many gases and many purposes, the quantum energy levels of the electrons can be ignored. In some cases, however, the levels are degenerate so that even an unexcited atom has a choice of electron states. In others the lower excited levels are sufficiently close to the unexcited (or ground) level that they can become occupied even at practical temperatures. The electronic effects may then be important and must at times be taken into account.

Since the electron states group into degenerate levels, it is convenient to write the electronic partition function, following equation (6.13), as

$$Q_{\text{el}} = \sum_l g_l\, e^{-\epsilon_l/kT} = g_0\, e^{-\epsilon_0/kT} + g_1 e^{-\epsilon_1/kT} + g_2 e^{-\epsilon_2/kT} + \cdots,$$

where $g_0, g_1, g_2, \ldots$ are the degeneracy factors of the levels of energy $\epsilon_0, \epsilon_1, \epsilon_2, \ldots$. In accordance with our convention, the energy of the ground level is taken as zero (i.e., $\epsilon_0 = 0$) and $\epsilon_1, \epsilon_2, \ldots$ are measured from that datum. If the quantities $\epsilon_1/k, \epsilon_2/k, \ldots$, which have the dimensions of temperature, are taken to define the *characteristic temperatures for electronic excitation*, denoted by $\Theta_1, \Theta_2, \ldots$, we then have

$$Q_{\text{el}} = g_0 + g_1 e^{-\Theta_1/T} + g_2 e^{-\Theta_2/T} + \cdots. \tag{11.1}$$

In most cases the Θ_l are large enough that only the first one, two, or three terms make any effective contribution for practical values of T.[11] Physically speaking, this means that only the first one, two, or three energy levels are significantly populated.

Let us assume, as is sometimes the case, that all beyond the first two terms are negligible, so that

$$\boxed{Q_{\text{el}} = g_0 + g_1 e^{-\Theta_1/T}}. \tag{11.2}$$

Differentiating this in accordance with equation (10.4), we find for the

[11] For a discussion of atomic energy levels and of the origin of the terms in equation (11.1), see Herzberg (1944).

electronic contribution to the internal energy

$$e_{el} = R\Theta_1 \frac{(g_1/g_0)e^{-\Theta_1/T}}{1 + (g_1/g_0)e^{-\Theta_1/T}}, \tag{11.3}$$

and hence for the corresponding specific heat

$$c_{v_{el}} = R\left(\frac{\Theta_1}{T}\right)^2 \frac{(g_1/g_0)e^{-\Theta_1/T}}{[1 + (g_1/g_0)e^{-\Theta_1/T}]^2}. \tag{11.4}$$

The resulting variation of $c_{v_{el}}/R$ as a function of T/Θ_1 is shown in Fig. 5 for two values of g_1/g_0. The value of $c_{v_{el}}$ vanishes for both large and small

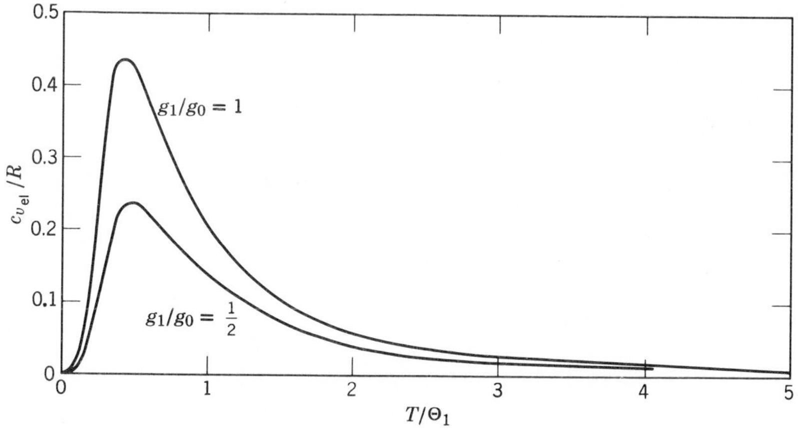

Fig. 5. Variation of electronic specific heat as a function of temperature.

values of T/Θ_1 and has a sharp maximum near $T/\Theta_1 \simeq 0.4$. The height of the maximum is strongly dependent on g_1/g_0. The shape of the curves for large values of T/Θ_1 should not, of course, be taken literally. As T increases for fixed values of Θ_1, Θ_2, etc. in equation (11.1), the higher energy levels eventually become populated, and the terms of Q_{el} beyond the first two become important. The variation of $c_{v_{el}}$ at the larger values of T is then different from that shown in the figure and may possess several maxima.

The values of the constants in equation (11.1) are found by spectroscopic study of the transitions between the various energy levels. (For tabulated data see Moore, 1949, et seq.) From such results we can write, for example, for atomic oxygen, if T is in °K,

$$Q_{el}^{O} = 5 + 3e^{-228/T} + e^{-326/T}, \tag{11.5a}$$

where terms of order $e^{-23,000/T}$ have been neglected, and for atomic nitrogen

$$Q_{\text{el}}^{\text{N}} = 4, \tag{11.5b}$$

where terms of order $e^{-28,000/T}$ have been neglected. Although Q_{el}^{O} consists of three terms in (11.5a), the values of Θ_l for the second and third terms are sufficiently close that $c_{v_{\text{el}}}^{\text{O}}$ will behave more or less as given by equation (11.4) for a value of Θ_1 of about 270 and $g_1 = 3 + 1 = 4$. The maximum will thus occur at about 100°K. At temperatures high enough for atomic oxygen to be formed (i.e., above about 3000°K), the exponential factors in (11.5a) are essentially unity. If the temperature remains low enough that the neglected terms are in fact negligible, it is then sufficiently accurate for practical purposes to set $Q_{\text{el}}^{\text{O}} = 9$. Similar remarks apply to other species with low characteristic temperatures for electronic excitation (e.g., nitric oxide in Sec. 12).

A partition function consisting of only the constant term, as in equation (11.5b) for Q_{el}^{N}, contributes nothing toward e and c_v, since these quantities depend only on the derivatives of Q. It is of importance, however, in the calculation of absolute entropy and in the treatment of chemical equilibrium, where the value of Q itself appears (see Sec. 13).

12 DIATOMIC GASES

For diatomic gases the molecular rotation and vibration enter the picture and, from the practical point of view, overshadow the electronic effects.

We consider first the rotation. Here, in view of our separation of the energy into various kinds, it is sufficient to regard the molecule as a rigid "dumbell," characterized by a constant moment of inertia I. From solution of the appropriate Schrödinger equation it is found that such a molecule can exist in degenerate energy levels, the value of whose energy is given by

$$\epsilon_l = \frac{h^2}{8\pi^2 I} l(l + 1) \qquad \text{where } l = 0, 1, 2, 3, \ldots. \tag{12.1}$$

The existence of the degeneracy, that is, of a number of states with identical energy, arises from the possibility of different directions of the axis of rotation. The degeneracy is given here by $g_l = 2l + 1$. We therefore have for the rotational partition function, from equation (6.13),

$$Q_{\text{rot}} = \sum_{l=0}^{\infty} (2l + 1) \exp\left[\frac{-l(l + 1)h^2}{8\pi^2 IkT}\right], \tag{12.2}$$

or

$$Q_{\text{rot}} = 1 + 3e^{-2\Theta_r/T} + 5e^{-6\Theta_r/T} + \cdots, \tag{12.3}$$

where Θ_r is a *characteristic temperature for rotation* given by

$$\Theta_r \equiv \frac{h^2}{8\pi^2 Ik}. \tag{12.4}$$

Numerical values of Θ_r must be found again by spectroscopic study. (For spectroscopic data on diatomic molecules see, e.g., Herzberg, 1950.) For $T \ll \Theta_r$ we have from equation (12.3), $Q_{\text{rot}} \simeq 1$. We also find correspondingly that $e_{\text{rot}} \simeq 0$ and $c_{v_{\text{rot}}} \simeq 0$, that is, at sufficiently low temperatures the contribution from the rotation vanishes.

Unfortunately, the right-hand side of equation (12.3) cannot be summed explicitly for arbitrary T. In most cases, however, Θ_r is near absolute zero (e.g., 2.1°K for O_2, 2.9°K for N_2, and 2.5°K for NO), so that at ordinary temperatures the ratio Θ_r/T is very small. Under these conditions the sum can be replaced by an integral on the basis of an argument similar to that used in Sec. 9 in connection with Q_{tr}, that is, the sum is essentially equal to the area under a curve of $(2l + 1) \exp[-l(l + 1)\Theta_r/T]$ versus l. We thus find, with the aid of the substitution $z = l(l + 1)$,

$$Q_{\text{rot}} = \int_0^\infty (2l + 1)e^{-l(l+1)\Theta_r/T}\, dl = \int_0^\infty e^{-z\Theta_r/T}\, dz,$$

from which finally

$$Q_{\text{rot}} = \frac{T}{\Theta_r}. \tag{12.5}$$

From this we find, on the basis of equation (10.4),

$$\boxed{e_{\text{rot}} = RT}, \tag{12.6}$$

and hence

$$\boxed{c_{v_{\text{rot}}} = R}. \tag{12.7}$$

Thus at sufficiently high temperatures the rotation contributes RT to the internal energy per unit mass. The rotation is then said to be *fully excited.* This is another example of the equipartition of energy (cf. Chapter I, Sec. 3). That is, when the rotational energy is fully excited it can validly be expressed on the basis of classical mechanics as the sum of two square terms and each square term then contributes $\frac{1}{2}RT$ to the internal energy. The term "fully excited" is perhaps unfortunate in that it suggests that no further energy can then be taken up by the motion under consideration and that the corresponding internal energy is therefore constant. What is really meant is that the energy increases linearly with T; it is c_v that is constant. The increase of $c_{v_{rot}}$ from the zero value applicable at very low temperatures to the fully excited value R takes place at temperatures in the vicinity of Θ_r and is therefore of no consequence in practical gas dynamics.

Equation (12.5) for Q_{rot} is actually correct only for heteronuclear molecules, that is, for diatomic molecules in which the two atoms are dissimilar. For homonuclear molecules, for example, O_2 and N_2, the previous result must be divided by 2. The true reasons for this have to do with symmetry considerations of the quantum-mechanical wave functions and are too lengthy to go into here (see, e.g., Fast, 1962, pp. 275–280). From the classical point of view, it may be thought of as due to the fact that rotation of a homonuclear molecule through 180° results in a spatial situation that is in no way distinguishable from the original one. In any event, we must in general write

$$\boxed{Q_{rot} = \frac{1}{\sigma}\left(\frac{T}{\Theta_r}\right)}, \tag{12.8}$$

where σ is a *symmetry factor* that is 2 for homonuclear and 1 for heteronuclear molecules. This modification does not affect the results for e_{rot} and $c_{v_{rot}}$, which depend only on the derivative of $\ln Q_{rot}$.

We now turn to the effects of molecular vibration. Here the molecule may be regarded to a good approximation as a harmonic oscillator of frequency ν. According to quantum mechanics, the permissible energy states—there is no question here of degenerate levels—are given by

$$\epsilon_i = (i + \tfrac{1}{2})h\nu \qquad \text{where} \quad i = 0, 1, 2, 3, \ldots. \tag{12.9a}$$

Even in the vibrational ground state ($i = 0$) the molecule possesses a non-zero vibrational energy $\epsilon_0 = \frac{1}{2}h\nu$, the so-called *zero-point energy*. Consistent with the convention adopted in Sec. 10, we omit this zero-point energy and measure the vibrational energies from the ground state. This is permissible, since the zero-point energy is a constant and plays no part in the changes of state of the system. We thus write in place of (12.9a)

$$\epsilon_i = ih\nu \qquad \text{where} \quad i = 0, 1, 2, 3, \ldots. \tag{12.9b}$$

Substituting this expression into equation (6.10) we obtain for the vibrational partition function

$$Q_{vib} = \sum_{i=0}^{\infty} e^{-ih\nu/kT}.$$

We recognize this sum as the simple geometrical series

$$\frac{1}{1-x} = 1 + x + x^2 + \cdots = \sum_{i=0}^{\infty} x^i,$$

with $x = e^{-h\nu/kT}$. If we define a *characteristic temperature for vibration* by

$$\Theta_v \equiv \frac{h\nu}{k}, \tag{12.10}$$

we thus obtain finally

$$\boxed{Q_{\text{vib}} = \frac{1}{1 - e^{-\Theta_v/T}}}. \tag{12.11}$$

From this we find, again on the basis of equation (10.4),

$$\boxed{e_{\text{vib}} = \frac{R\Theta_v}{e^{\Theta_v/T} - 1}}, \tag{12.12}$$

and hence[12]

$$\boxed{c_{v_{\text{vib}}} = R\left(\frac{\Theta_v}{T}\right)^2 \frac{e^{\Theta_v/T}}{(e^{\Theta_v/T} - 1)^2} = R\left[\frac{\Theta_v/2T}{\sinh(\Theta_v/2T)}\right]^2}. \tag{12.13}$$

It can be seen from these formulas that as T/Θ_v tends to zero the contribution of vibration to both the internal energy and the specific heat vanishes. When T/Θ_v becomes very large, it is readily shown that e_{vib} tends to RT and $c_{v_{\text{vib}}}$ tends to R. The vibration is then fully excited, and the results are again in accord with the equipartition of energy. (The energy of vibration consists, classically speaking, of two square terms, one from the kinetic energy and one from the potential energy.) Between the two limits the variation of $c_{v_{\text{vib}}}$ is as shown in Fig. 6. In contrast to the situation for rotation, the value of Θ_v, as obtained from spectroscopic data, may be large (e.g., 2270°K for O_2, 3390°K for N_2, and 2740°K for NO). The variation of $c_{v_{\text{vib}}}$ with temperature may thus be significant at temperatures encountered in modern gas dynamics.

The contribution of electronic effects has already been discussed in connection with monatomic gases. The same general considerations apply here. Typical expressions for the electronic partition function for diatomic gases, to be compared with the previous equations (11.5), are as follows for oxygen, nitrogen, and nitric oxide:

$$Q_{\text{el}}^{O_2} = 3 + 2e^{-11,390/T}, \tag{12.14a}$$

$$Q_{\text{el}}^{N_2} = 1, \tag{12.14b}$$

$$Q_{\text{el}}^{NO} = 2 + 2e^{-174/T}, \tag{12.14c}$$

[12] If the zero-point energy had been retained, as is sometimes done, we would have obtained a factor $\exp(-\Theta_v/2T)$ in place of unity in the numerator of equation (12.11), with a subsequent addition of a constant term in e_{vib} but no change in $c_{v_{\text{vib}}}$.

Here terms of order $e^{-19,000/T}$, $e^{-100,000/T}$, and $e^{-65,000/T}$ respectively have been neglected. From these and the results of Fig. 5 we see that the contribution to the specific heat will have a maximum for O_2 at about 4500°K and for NO at about 70°K. For N_2 the contribution is everywhere zero to the present accuracy.

Since the electronic contribution in the range of temperatures of interest is often relatively small (cf. Figs. 5 and 6), it may often be ignored in practical applications. The total specific heat for diatomic gases is then

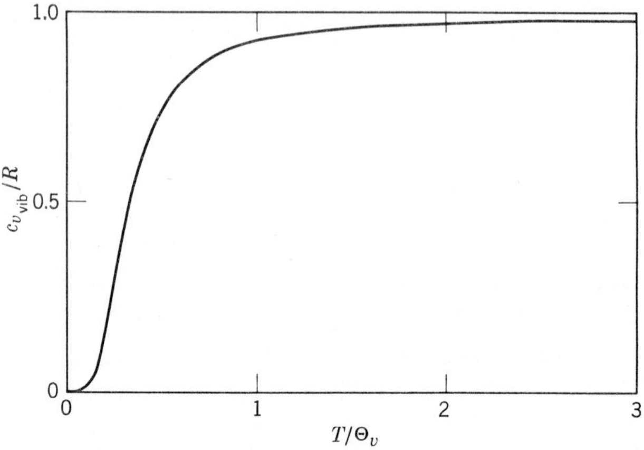

Fig. 6. Variation of vibrational specific heat as a function of temperature.

found by summing the translational, rotational, and vibrational contributions. We thus find, from equations (9.7), (12.7), and (12.13),

$$\boxed{c_v = R\left\{\frac{5}{2} + \left[\frac{\Theta_v/2T}{\sinh(\Theta_v/2T)}\right]^2\right\}} \quad . \tag{12.15}$$

This simple formula is useful for many practical purposes. The contribution of $\frac{5}{2}R$ due to molecular translation and rotation can be—and indeed originally was—obtained on purely classical grounds, that is, with no reference to quantum mechanics. The contribution of vibration, however, depends essentially on quantum-mechanical concepts. Classical statistical mechanics is capable of giving only the high-temperature value of $c_{v_{\text{vib}}} = R$. It thus provides no variation of c_v with T, a result in notorious contradiction to the experimental evidence. Success in accounting for the variation of c_v with T was, in fact, one of the early triumphs of quantum statistical mechanics.

In the foregoing, the vibrating molecule has been regarded as a simple harmonic oscillator. This is, of course, actually not the case, and a brief mention of the so-called *anharmonic* effects is appropriate. The situation can be visualized with the aid of Fig. 7, in which the horizontal scale is the distance r between the centers of the two atoms that make up the diatomic molecule. As this distance changes due to vibration, work must be done against the interatomic forces, attractive at large distances and repulsive at short distances. This results in a change in the internal potential energy

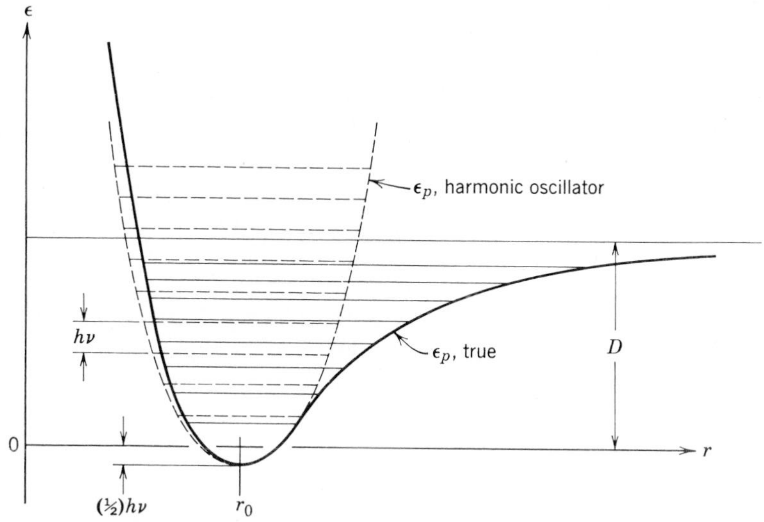

Fig. 7. Potential energy and vibrational levels for a diatomic molecule.

ϵ_p of the molecule. At some distance r_0, where the force $F = -d\epsilon_p/dr$ is zero, the energy ϵ_p is a minimum, and the atoms would be able to remain at rest for an unlimited period. The true value of ϵ_p, shown by the solid curve, increases rapidly as r becomes smaller than r_0, and approaches asymptotically to a certain *dissociation energy* D as r becomes very large, that is, as the atoms of the diatomic molecule become dissociated. Use of the harmonic-oscillator approximation corresponds to replacing this true curve by a parabola (shown dashed) whose curvature matches that of the true curve at r_0. The vibrational energy states given by equation (12.9), and shown by horizontal lines in the figure, are then symmetrical horizontally about r_0. They are infinite in number and are spaced vertically at uniform increments $h\nu$, with a zero-point energy of $\frac{1}{2}h\nu$.[13] In the true case

[13] Note that the energy zero in the figure is taken at the ground vibrational state in accord with our earlier convention, and D is measured accordingly. The figure is only schematic and the spacings shown are not to scale.

the vibrational states are not in fact symmetrical about r_0 and their spacing decreases as ϵ_p increases. The zero-point energy is also changed slightly, and there is a finite limit to the number of states as dissociation is approached.

Despite the qualitative inaccuracies of the harmonic-oscillator approximation, the results that it gives are sufficiently accurate for most practical purposes. For precise work, however, or for temperatures high enough that the upper vibrational states are appreciably populated, the anharmonic effects must be taken into account. When this is done, the coupling between vibration and rotation must be included, since the two effects are in principle of the same order of magnitude. Landau and Lifshitz (1958, p. 145) give an approximate method for this, based on assumptions of classical statistical mechanics valid at sufficiently high temperatures. They find that the anharmonic and coupling effects result in an additive correction to the specific internal energy of magnitude

$$e_{\text{anh}} = \frac{1}{2}\left(1 + 3\alpha - \frac{3}{2}\beta + \frac{15}{8}\alpha^2\right) R \frac{k}{I\nu^2} T^2,$$

where α and β are constants that can be expressed in terms of the spectroscopic constants of the molecule. With the previous definitions $\Theta_v \equiv h\nu/k$ and $\Theta_r \equiv h^2/8\pi^2 Ik$, this can be written

$$e_{\text{anh}} = CR \frac{\Theta_r}{\Theta_v^2} T^2, \tag{12.16}$$

where C is a constant for a given molecule. The corresponding correction to be added to the specific heat is therefore

$$c_{v_{\text{anh}}} = 2CR \frac{\Theta_r}{\Theta_v^2} T. \tag{12.17}$$

In view of the relative magnitudes of Θ_r and Θ_v as previously mentioned, this contribution will generally be small. For a complete quantum-mechanical treatment of the anharmonic and coupling effects, the reader can consult Mayer and Mayer (1940), pp. 160–166.

Exercise 12.1. Show that the specific rotational entropy when the rotation is fully excited is given by

$$s_{\text{rot}} = R\left[\ln \frac{T}{\sigma\Theta_r} + 1\right].$$

Exercise 12.2. Show that the specific vibrational entropy is given by

$$s_{\text{vib}} = R\left[-\ln\left(1 - e^{-\Theta_v/T}\right) + \frac{\Theta_v/T}{e^{\Theta_v/T} - 1}\right].$$

What is the limiting form of this result for $T \gg \Theta_v$?

Exercise 12.3. Show that the specific electronic entropy corresponding to the two-term approximation (11.2) for Q_{el} is

$$s_{el} = R\left[\ln g_0 + \ln\left(1 + \frac{g_1}{g_0}e^{-\Theta_1/T}\right) + \frac{(g_1/g_0)(\Theta_1/T)e^{-\Theta_1/T}}{1 + (g_1/g_0)e^{-\Theta_1/T}}\right].$$

Exercise 12.4. Using the foregoing results, calculate the percentage contribution of the various types of energy to the total specific entropy s for diatomic oxygen at a temperature of 1000°K and a pressure of 1 at m.

Exercise 12.5. Consider nitric oxide (NO) in equilibrium at 850°K. What numerical percentage of the molecules have *both* (i.e., simultaneously) a rotational energy corresponding to the quantum level $l = 1$ and a vibrational energy corresponding to the quantum state $i = 0$, irrespective of the orientation of the axis of rotation or of the state of translational energy or electronic excitation?

Exercise 12.6. At temperatures sufficiently high that dissociation occurs, it is sometimes desired to take account of the fact that the number of vibrational states is actually finite. A crude way of doing this is to retain the harmonic-oscillator model, with its uniform spacing of vibrational states, but arbitrarily truncate the summation for Q_{vib} at the vibrational state immediately below dissociation. We thus write

$$Q_{vib} = \sum_{i=0}^{n-1} e^{-ih\nu/kT},$$

where n is related to the dissociation energy D by $n = D/h\nu$. Show that this leads to

$$Q_{vib} = \frac{1 - e^{-\Theta_d/T}}{1 - e^{-\Theta_v/T}},$$

where $\Theta_d \equiv D/k$ is a characteristic temperature for dissociation. Find the corresponding expressions for e_{vib} and $c_{v_{vib}}$.

13 CHEMICALLY REACTING SYSTEMS AND LAW OF MASS ACTION

In all the foregoing work we have considered systems of N identical particles. Here we shall extend our considerations to a system made up of more than one type of particle. To be specific, we consider in particular a gaseous system in a specified volume V and containing N_A A-particles and N_B B-particles. These particles could be either atoms or molecules, but we shall speak of them as atoms. We suppose further that these atoms can exist separately as A- and B-atoms or can interact chemically to form

AB-molecules according to the chemical equation

$$A + B \rightleftharpoons AB. \tag{13.1}$$

We wish to inquire into the equilibrium composition and thermodynamic properties of the system. We shall see that this inquiry leads us to a statistical-mechanical derivation of the law of mass action, which we have obtained in Chapter II on kinetic grounds and in Chapter III on thermodynamic grounds.

As before we assume that the particles of the system are weakly interacting, that is, that they exchange energy or react chemically only upon actual collision. We suppose further that each A-atom has possible energy states $\epsilon_1^a, \epsilon_2^a, \epsilon_3^a, \ldots, \epsilon_i^a, \ldots$, each B-atom has possible energy states $\epsilon_1^b, \epsilon_2^b, \epsilon_3^b, \ldots, \epsilon_i^b, \ldots$, and each AB-molecule has possible energy states $\epsilon_1^{ab}, \epsilon_2^{ab}, \epsilon_3^{ab}, \ldots, \epsilon_i^{ab}, \ldots$. As before we then divide the energy states for A-atoms into groups such that each of the C_j^a states in a given group has approximately the same energy ϵ_j^a. The number of atoms in the group is denoted by N_j^a. The states for B-atoms and AB-molecules are divided in similar fashion. Once this division is made, a macrostate of the system is defined by the values of the distribution numbers

$$\left.\begin{array}{l} N_1^a, N_2^a, \ldots, N_j^a, \ldots \\ N_1^b, N_2^b, \ldots, N_j^b, \ldots \\ N_1^{ab}, N_2^{ab}, \ldots, N_j^{ab}, \ldots \end{array}\right\}.$$

To evaluate the total number of microstates Ω of the system, we first suppose that the distribution numbers have certain specific values and ask how many microstates exist for this specific macrostate. Considering the A-atoms *alone*, and referring back to the discussion leading up to equations (4.5), we write for the number of ways W^a that the A-atoms can be arranged over *their* energy states

$$W^a = W^a(N_j^a),$$

where the right-hand side is given in terms of the specified N_j^a and the chosen C_j^a by equations (4.5a) or (4.5b), depending on whether the Bose-Einstein or Fermi-Dirac statistics are appropriate. Similar expressions can be written for correspondingly defined quantities W^b and W^{ab}. But each of the arrangements that go to make up W^a can occur in combination with any one of the arrangements that go to make up W^b, and similarly with regard to W^{ab}. The number of combined arrangements, which gives the number of microstates W corresponding to the specified macrostate, is therefore

$$W(N_j^a, N_j^b, N_j^{ab}) = W^a(N_j^a) \times W^b(N_j^b) \times W^{ab}(N_j^{ab}). \tag{13.2}$$

This product replaces the single expression (4.5) that applied for a system of identical particles. To find the *total* number of microstates we must, as before, take the sum of the values of W for all possible sets of the distribution numbers, that is,

$$\Omega = \sum_{\left(\substack{\text{over all possible} \\ \text{sets of} \\ N_j^a, N_j^b, N_j^{ab}}\right)} W^a(N_j^a) \times W^b(N_j^b) \times W^{ab}(N_j^{ab}). \tag{13.3}$$

The possible sets of the distribution numbers are restricted by the requirements that (a) the total number of A-atoms and B-atoms in the system, combined or uncombined, shall be N_A and N_B respectively, and (b) the total energy of the system shall be E. The first restriction can be expressed at once by the *atom-conservation equations*

$$\begin{aligned} \sum_j N_j^a + \sum_j N_j^{ab} &= N_A, \\ \sum_j N_j^b + \sum_j N_j^{ab} &= N_B. \end{aligned} \tag{13.4}$$

The expression of the second restriction requires some care with regard to the zero from which the energies are measured. Following the convention adopted in Sec. 10, we presume, as is usually (but not always) done, that the energies ϵ_j^a, ϵ_j^b, ϵ_j^{ab} are each measured from the ground state of the corresponding particle at rest. As we have seen in connection with Fig. 7, however, the dissociation of an AB-molecule from its ground state into an A-atom and B-atom, both in their ground state, requires the expenditure of an energy of dissociation D. That is, an AB-molecule in its ground state and at rest has an energy D less than that of an A-atom and a B-atom both in their ground state and at rest. Thus the zero for the measurement of ϵ_j^{ab} according to our convention is D below the zero for ϵ_j^a and ϵ_j^b, as illustrated in Fig. 8.[14] To express the total energy of the system, however, we must choose some *arbitrary common* zero, or reference state of the system, from which all energies will be measured. For reasons that we shall not go into here, it is logical, though not essential, to take this zero as identical with the energy zero for atoms. This corresponds to saying that the energy E of the system is taken arbitrarily as zero when all constituents are in the atomic form and are in their ground state and at rest (i.e., at temperature absolute zero); the energy of other states of the assembly is then measured relative to this reference state. With this choice of a

[14] The student may find it helpful to think of D as the energy required to dissociate an AB-molecule at absolute-zero temperature into an A-atom and a B-atom, also at absolute-zero temperature.

common zero, the energy contributed to the system by the several species is

$$E^a = \sum_j N_j^a \epsilon_j^a, \qquad E^b = \sum_j N_j^b \epsilon_j^b, \qquad E^{ab} = \sum_j N_j^{ab}(\epsilon_j^{ab} - D),$$

and the total energy E is then the sum of these, that is,

$$\sum_j N_j^a \epsilon_j^a + \sum_j N_j^b \epsilon_j^b + \sum_j N_j^{ab}(\epsilon_j^{ab} - D) = E. \tag{13.5}$$

This equation, with E constant, expresses the second restriction on the distribution numbers.

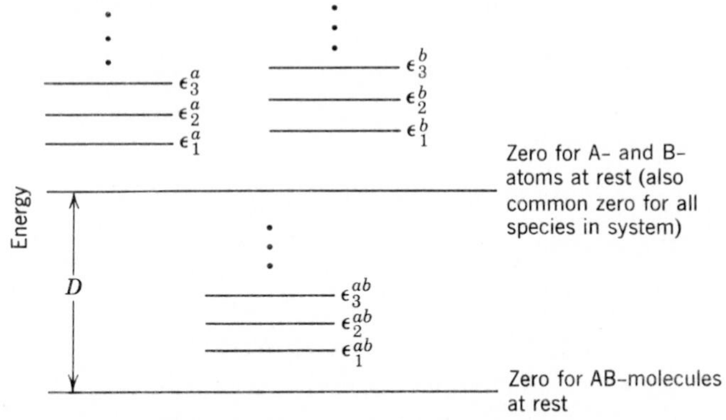

Fig. 8. Energy zeros for species in a reacting system.

To evaluate Ω from equation (13.3) subject to the restrictions (13.4) and (13.5), we proceed as in Sec. 5, that is, we replace (13.3) by the approximation $\Omega = W_{\text{max}}$ and pick out W_{max} with the aid of Stirling's approximate formula for the factorial function. As before, we find it convenient to deal with $\ln W$ rather than W itself. We therefore write from equation (13.2)

$$\ln W = \ln W^a + \ln W^b + \ln W^{ab}. \tag{13.6}$$

By application of Stirling's formula to appropriate expressions of the form (4.5) for W^a, W^b, and W^{ab}, each term on the right could now be replaced by a summation of the form (5.4) for either Bose-Einstein or Fermi-Dirac statistics. For present purposes, however, we may as well proceed directly to the Boltzmann limit at this point rather than wait until the end of the analysis. We therefore use the common limiting form (6.2), which is valid when $C_j \gg N_j$, and write for $\ln W^a$

$$\ln W^a = \sum_j N_j^a \left(\ln \frac{C_j^a}{N_j^a} + 1 \right), \tag{13.7}$$

with similar expressions for $\ln W^b$ and $\ln W^{ab}$. Putting these expressions into (13.6) we then have in the Boltzmann limit

$$\ln W = \sum_j N_j^a\left(\ln \frac{C_j^a}{N_j^a} + 1\right) + \sum_j N_j^b\left(\ln \frac{C_j^b}{N_j^b} + 1\right) + \sum_j N_j^{ab}\left(\ln \frac{C_j^{ab}}{N_j^{ab}} + 1\right). \tag{13.8}$$

To find the maximum value of ln W we proceed as before and set the small change $\delta(\ln W)$ corresponding to any small changes δN_j^a, δN_j^b, δN_j^{ab} equal to zero as follows:

$$\delta(\ln W) = \sum_j \frac{\partial(\ln W)}{\partial N_j^a}\,\delta N_j^a + \sum_j \frac{\partial(\ln W)}{\partial N_j^b}\,\delta N_j^b + \sum_j \frac{\partial(\ln W)}{\partial N_j^{ab}}\,\delta N_j^{ab} = 0.$$

Substitution from (13.8) then leads immediately to the condition

$$\sum_j \ln \frac{C_j^a}{N_j^a}\,\delta N_j^a + \sum_j \ln \frac{C_j^b}{N_j^b}\,\delta N_j^b + \sum_j \ln \frac{C_j^{ab}}{N_j^{ab}}\,\delta N_j^{ab} = 0. \tag{13.9}$$

The changes δN_j are not independent but must be related by the following equations obtained from the restrictive conditions (13.4) and (13.5):

$$\begin{aligned} \sum_j \delta N_j^a + \sum_j \delta N_j^{ab} &= 0, \\ \sum_j \delta N_j^b + \sum_j \delta N_j^{ab} &= 0, \\ \sum_j \epsilon_j^a\,\delta N_j^a + \sum_j \epsilon_j^b\,\delta N_j^b + \sum_j (\epsilon_j^{ab} - D)\,\delta N_j^{ab} &= 0. \end{aligned} \tag{13.10}$$

It can be shown by Lagrange's method of undetermined multipliers that the most general solution of equation (13.9) subject to the conditions (13.10) is

$$\begin{aligned} \ln \frac{C_j^a}{N_j^a} &= \alpha_a + \beta\epsilon_j^a, \\ \ln \frac{C_j^b}{N_j^b} &= \alpha_b + \beta\epsilon_j^b, \\ \ln \frac{C_j^{ab}}{N_j^{ab}} &= \alpha_a + \alpha_b + \beta(\epsilon_j^{ab} - D), \end{aligned} \tag{13.11}$$

where α_a, α_b, and β are three undetermined multipliers. That this is indeed a solution of (13.9) can be verified by substitution as before [cf. material following equation (5.7)]. We thus find for the values of the distribution numbers that make ln W a maximum

$$\begin{aligned} N_j^{a*} &= C_j^a e^{-\alpha_a - \beta\epsilon_j^a}, \\ N_j^{b*} &= C_j^b e^{-\alpha_b - \beta\epsilon_j^b}, \\ N_j^{ab*} &= C_j^{ab} e^{-\alpha_a - \alpha_b + \beta D - \beta\epsilon_j^{ab}}. \end{aligned} \tag{13.12}$$

Summing over the distribution numbers for each species and defining $N^{a^*} \equiv \sum_j N_j^{a^*}$, $N^{b^*} \equiv \sum_j N_j^{b^*}$, $N^{ab^*} \equiv \sum_j N_j^{ab^*}$, we can write

$$
\begin{aligned}
N^{a^*} &= e^{-\alpha_a} \sum_j C_j^a e^{-\beta \epsilon_j^a}, \\
N^{b^*} &= e^{-\alpha_b} \sum_j C_j^b e^{-\beta \epsilon_j^b}, \\
N^{ab^*} &= e^{-\alpha_a - \alpha_b + \beta D} \sum_j C_j^{ab} e^{-\beta \epsilon_j^{ab}}.
\end{aligned} \tag{13.13}
$$

These equations can be used respectively to eliminate $-\alpha_a$, $-\alpha_b$, and $-\alpha_a - \alpha_b + \beta D$ from equations (13.12). Anticipating again the result that $\beta = 1/kT$ and introducing a partition function for each species according to the general definition $Q \equiv \sum_j C_j e^{-\epsilon_j/kT}$ [cf. equation (6.9)], we thus obtain finally

$$
\boxed{
\begin{gathered}
N_j^{a^*} = N^{a^*} \frac{C_j^a e^{-\epsilon_j^a/kT}}{Q^a}, \qquad N_j^{b^*} = N^{b^*} \frac{C_j^b e^{-\epsilon_j^b/kT}}{Q^b}, \\
N_j^{ab^*} = N^{ab^*} \frac{C_j^{ab} e^{-\epsilon_j^{ab}/kT}}{Q^{ab}}
\end{gathered}
} . \tag{13.14}
$$

These results are analogous to equation (6.8) for a simple system and show that each species follows a Boltzmann distribution within itself. The total number of particles of each species, as given by the quantities N^{a^*}, N^{b^*}, and N^{ab^*} that appear on the right-hand side of these equations, are not known *a priori*, however. They are related by the following equation obtained by eliminating α_a and α_b between the three equations (13.13):

$$
\boxed{\frac{N^{ab^*}}{N^{a^*} N^{b^*}} = \frac{Q^{ab}}{Q^a Q^b} e^{D/kT}} . \tag{13.15}
$$

This equation, plus the atom-conservation equations (13.4) rewritten in the form

$$
\boxed{
\begin{aligned}
N^{a^*} + N^{ab^*} &= N_A \\
N^{b^*} + N^{ab^*} &= N_B
\end{aligned}
} \tag{13.16}
$$

provide three equations from which the values of N^{a^*}, N^{b^*}, and N^{ab^*} can be calculated if the partition functions are known. The individual distribution numbers can then be calculated from equations (13.14). The partition function is thus again the key to the problem.

The distribution numbers given by the foregoing equations were derived to make $\ln W$ a maximum. As before, however, they can be given a physical as well as a mathematical meaning. They provide, for all practical purposes, the actual distribution of the species in a system in chemical equilibrium.

To find the thermodynamic properties of the system, we equate the entropy S to $k \ln \Omega$ in accord with Boltzmann's relation and set $\ln \Omega = \ln W_{\max}$ by our assumption. To find $\ln W_{\max}$ we substitute the particular distribution numbers (13.14) into equation (13.8). We thus obtain, since $N^{a*} = \sum_j N_j^{a*}$, etc.,

$$S = k\Big[N^{a^*}\Big(\ln \frac{Q^a}{N^{a^*}} + 1\Big) + \beta \sum_j N_j^{a^*}\epsilon_j^a + N^{b^*}\Big(\ln \frac{Q^b}{N^{b^*}} + 1\Big) + \beta \sum_j N_j^{b^*}\epsilon_j^b + N^{ab^*}\Big(\ln \frac{Q^{ab}}{N^{ab^*}} + 1\Big) + \beta \sum_j N_j^{ab^*}\epsilon_j^{ab}\Big], \tag{13.17a}$$

where we have replaced $1/kT$ by β, which is in fact still unknown. With the aid of equation (13.5), this can also be written

$$S = k\Big[N^{a^*}\Big(\ln \frac{Q^a}{N^{a^*}} + 1\Big) + N^{b^*}\Big(\ln \frac{Q^b}{N^{b^*}} + 1\Big) + N^{ab^*}\Big(\ln \frac{Q^{ab}}{N^{ab^*}} + 1\Big) + \beta(E + N^{ab^*}D)\Big]. \tag{13.17b}$$

These equations can be compared with their counterpart (8.1) for the single-species system. We thus see, as might be expected from our assumption of weak interaction, that the entropy of the reacting system is the sum of the entropies of the various species considered separately.[15] For this to be true, however, it must be understood that the individual entropies are all evaluated at the same V and T, both of which enter into the partition functions. This is in agreement with Gibbs's theorem of Chapter III, Sec. 8.

As before, β can now be found by reference to the Gibbs equation (III 6.8). This equation, when written for the present system, is

$$dS = \frac{1}{T}(dE + p\,dV - \tilde{\mu}^a\,dN^a - \tilde{\mu}^b\,dN^b - \tilde{\mu}^{ab}\,dN^{ab}), \tag{13.18}$$

[15] The presence of the term $N^{ab^*}D$ in (13.17b) arises merely out of the choice of origin for measuring the energy E.

from which it follows that

$$\left(\frac{\partial S}{\partial E}\right)_{V,N^a,N^b,N^{ab}} = \frac{1}{T}.$$

Differentiation of (13.17) and comparison of the final result with this relation shows again that $\beta = 1/kT$.[16]

With the foregoing results, the Helmholtz free energy $F \equiv E - TS$ can be written as

$$\boxed{F = -kT\left[N^{a*}\left(\ln\frac{Q^a}{N^{a*}} + 1\right) + N^{b*}\left(\ln\frac{Q^b}{N^{b*}} + 1\right) + N^{ab*}\left(\ln\frac{Q^{ab}}{N^{ab*}} + 1\right)\right] - N^{ab*}D}. \quad (13.19)$$

As in our previous work, all the thermodynamic properties of the system can be derived from the expression for F. We shall carry out the derivation only for the pressure. This can be shown by means of the Gibbs equation (13.18) and the definition of F, to be given by

$$p = -\left(\frac{\partial F}{\partial V}\right)_{T,N^a,N^b,N^{ab}}. \quad (13.20)$$

To carry out the differentiation we observe that V, which enters the analysis only through the contribution of the translational motion, appears as a multiplying factor in the partition function Q [cf. equations (9.1) and (10.3)]. We therefore have, for each of the species,

$$\left(\frac{\partial \ln Q}{\partial V}\right)_{T,N^a,N^b,N^{ab}} = \frac{d \ln V}{dV} = \frac{1}{V}.$$

Substituting equation (13.19) into (13.20) and using the foregoing result, we readily obtain

$$\boxed{pV = (N^{a*} + N^{b*} + N^{ab*})kT}. \quad (13.21)$$

This result is reminiscent of the thermal equation of state (9.3) for a perfect gas. The sum in parentheses, however, is not a constant since the total number of molecules changes as the chemical balance changes. It must be calculated in any given case from equations (13.15) and (13.16).

[16] Care must again be taken to recognize, as in the development following equation (8.3), that the Q's are functions of β and hence of E.

Equation (13.15), which is the crucial equation here, can be recognized as the *law of mass action* in a somewhat new disguise. For comparison with earlier forms, we introduce the equilibrium partial pressures of the three species according to the equations

$$\overset{*}{p}_a V = N^{a^*} kT, \qquad \overset{*}{p}_b V = N^{b^*} kT, \qquad \overset{*}{p}_{ab} V = N^{ab^*} kT.$$

Equation (13.15) can then be written

$$\frac{\overset{*}{p}_{ab}}{\overset{*}{p}_a \overset{*}{p}_b} = \frac{V}{kT} \times \frac{Q^{ab}}{Q^a Q^b} e^{D/kT}. \tag{13.22a}$$

Since V appears as a multiplying factor in each of the partition functions Q, the right-hand side of this equation is a function of T only, and we can write

$$\frac{\overset{*}{p}_{ab}}{\overset{*}{p}_a \overset{*}{p}_b} = K_p(T). \tag{13.22b}$$

This agrees precisely with equation (III 9.1a)—which was derived from thermodynamic considerations—when that equation is specialized to the reaction $A + B \rightleftharpoons AB$. As usual, however, statistical mechanics goes beyond thermodynamics by giving us a complete recipe for calculating $K_p(T)$.

Exercise 13.1. Carry through an analysis similar to the one given in this section but for a mixture of three *inert* gases, that is, for three gases that do not react chemically. In particular, obtain expressions for S, F, and p and compare these with equations (13.17b), (13.19), and (13.21).

Exercise 13.2. Assume, consistent with our assumption of Chapter III, Sec. 2, that F for a system in mechanical and thermal equilibrium but chemical non-equilibrium is still given by equation (13.19) if the asterisks are removed. This amounts to assuming that each species individually still has a Boltzmann distribution as given by equation (13.14) but that equation (13.15) defining the chemical equilibrium no longer holds. By suitable differentiation as implied by equation (13.18) and the definition of F, obtain expressions for the chemical potentials $\tilde{\mu}^a$, $\tilde{\mu}^b$, and $\tilde{\mu}^{ab}$. Show that *at chemical equilibrium* these chemical potentials satisfy the equation of reaction equilibrium $\sum_s \nu_s \tilde{\mu}_s^* = 0$, which was obtained from thermodynamic considerations in Chapter III, Sec. 7.

Exercise 13.3. We can also define chemical potentials $\tilde{\mu}_A$ and $\tilde{\mu}_B$ by writing the Gibbs equation as

$$dS = \frac{1}{T}(dE + p\,dV - \tilde{\mu}_A\,dN_A - \tilde{\mu}_B\,dN_B).$$

These potentials are thus the chemical potentials for changes in the number of A- or B-atoms in the system with no regard to whether they are combined or uncombined; the potentials $\tilde{\mu}^a$ and $\tilde{\mu}^b$ by contrast are the potentials for changes in the number of atoms present in the uncombined form. The foregoing equation, together with the definition of F, implies that

$$\tilde{\mu}_A = \left(\frac{\partial F}{\partial N_A}\right)_{T,V,N_B}, \qquad \tilde{\mu}_B = \left(\frac{\partial F}{\partial N_B}\right)_{T,V,N_A}.$$

On the basis of equation (13.19), and noting that N^{a*}, N^{b*}, and N^{ab*} are functions of N_A and N_B, obtain the simplest possible expressions for $\tilde{\mu}_A$ and $\tilde{\mu}_B$ for a system in chemical equilibrium. Your results should show, when compared with those of Exercise 13.2, that at chemical equilibrium $\tilde{\mu}^a = \tilde{\mu}_A$ and $\tilde{\mu}^b = \tilde{\mu}_B$, that is, the chemical potential of a given type of atom is the same whether it is uncombined or combined with another type of atom.

14 DISSOCIATION-RECOMBINATION OF SYMMETRICAL DIATOMIC GAS

Before leaving this chapter, it is pertinent to consider the dissociation-recombination reaction for a symmetrical (or homonuclear) diatomic gas A_2:

$$A + A \rightleftharpoons A_2. \tag{14.1}$$

It might seem at first that results for this reaction could be obtained at once from those for the reaction $A + B \rightleftharpoons AB$ by simply identifying B-atoms as A-atoms. The situation, however, is not that simple.

The necessary development is similar to that of the preceding section and will be given here only in broad outline. The student should fill in the steps to make sure that he has mastered the procedures. We consider a system of N_A A-particles, which may exist either as A-atoms or A_2-molecules. We suppose that each A-atom has possible energy states $\epsilon_1^a, \epsilon_2^a, \epsilon_3^a, \ldots, \epsilon_i^a, \ldots$, and each A_2-molecule has possible energy states $\epsilon_1^{aa}, \epsilon_2^{aa}, \epsilon_3^{aa}, \ldots, \epsilon_i^{aa}, \ldots$. As before we divide the energy states into groups j with the number of atoms or molecules in each group denoted by the distribution numbers N_j^a and N_j^{aa}. The total number of microstates of the system is then given by

$$\Omega = \sum_{\left(\substack{\text{over all possible}\\ \text{sets of}\\ N_j^a,\, N_j^{aa}}\right)} W^a(N_j^a) \times W^{aa}(N_j^{aa}), \tag{14.2}$$

where the expressions for $W(N_j)$ are as given by equations (4.5). The possible sets of N_j^a, N_j^{aa} are those compatible with the atom-conservation equation

$$\sum_j N_j^a + 2\sum_j N_j^{aa} = N_A, \tag{14.3}$$

and the energy equation

$$\sum_j N_j^a \epsilon_j^a + \sum_j N_j^{aa}(\epsilon_j^{aa} - D) = E. \tag{14.4}$$

As in all our work, we evaluate Ω by equating it to the largest term in the sum (14.2), after finding this term with the aid of Stirling's formula. The condition for the largest term, in the Boltzmann limit, comes out to be [cf. equation (13.9)]

$$\delta(\ln W) = \sum_j \ln \frac{C_j^a}{N_j^a}\,\delta N_j^a + \sum_j \ln \frac{C_j^{aa}}{N_j^{aa}}\,\delta N_j^{aa} = 0. \tag{14.5}$$

The restrictive conditions as obtained from equation (14.3) and (14.4) are

$$\sum_j \delta N_j^a + 2\sum_j \delta N_j^{aa} = 0, \tag{14.6}$$

$$\sum_j \epsilon_j^a\,\delta N_j^a + \sum_j (\epsilon_j^{aa} - D)\,\delta N_j^{aa} = 0. \tag{14.7}$$

These equations can be solved as before by the method of undetermined multipliers. Omitting the details, we write the final result as

$$\boxed{N_j^{a*} = N^{a*}\frac{C_j^a e^{-\epsilon_j^a/kT}}{Q^a}, \qquad N_j^{aa*} = N^{aa*}\frac{C_j^{aa} e^{-\epsilon_j^{aa}/kT}}{Q^{aa}}}\,, \tag{14.8}$$

where β has been given its usual value $1/kT$. The equilibrium composition turns out to be given by

$$\frac{N^{aa*}}{(N^{a*})^2} = \frac{Q^{aa}}{(Q^a)^2}\,e^{D/kT} \tag{14.9}$$

$$N^{a*} + 2N^{aa*} = N_A\,. \tag{14.10}$$

With the foregoing results, the entropy can be found from $S = k \ln W_{\max}$ and the free energy from $F \equiv E - TS$. We thus obtain finally

$$\boxed{F = -kT\left[N^{a*}\left(\ln\frac{Q^a}{N^{a*}} + 1\right) + N^{aa*}\left(\ln\frac{Q^{aa}}{N^{aa*}} + 1\right)\right] - N^{aa*}D}\,. \tag{14.11}$$

Comparison of these results with equations (13.15), (13.16), and (13.19) for the reaction $A + B \rightleftharpoons AB$ shows that (14.9) could be obtained from the earlier results by formally replacing b by a, but (14.10) and (14.11) could not.

Exercise 14.1. Consider the chemical reaction

$$A + \nu B \rightleftharpoons AB_\nu.$$

For simplicity, we denote the compound AB_ν by C. By following procedures similar to those of this and the preceding section, show that the law of mass action for this reaction is

$$\frac{N^{c*}}{N^{a*}(N^{b*})^\nu} = \frac{Q^c}{Q^a(Q^b)^\nu} e^{D/kT},$$

where D is the energy required to dissociate an AB_ν-molecule into an A-atom and ν B-atoms.

Exercise 14.2. Suppose that N_{AB} AB-molecules and N_{CD} CD-molecules are placed in a container of volume V, where they can react according to the formula

$$AB + CD \rightleftharpoons AC + BD.$$

By the methods of this chapter (which assume, of course, that only the gaseous phase is present) prove that the equilibrium composition at temperature T is given by the system of equations

$$\frac{N^{ac*}N^{bd*}}{N^{ab*}N^{cd*}} = \frac{Q^{ac}Q^{bd}}{Q^{ab}Q^{cd}} \exp\left[(D^{ac} + D^{bd} - D^{ab} - D^{cd})/kT\right],$$

$$N^{ab*} + N^{ac*} = N_{AB},$$

$$N^{cd*} + N^{bd*} = N_{CD},$$

$$N^{ac*} = N^{bd*},$$

where D^{ac}, D^{bd}, etc. are the energies required to dissociate the various molecules into their constituent atoms. Note that in this case the equilibrium composition is independent of the volume V.

References

Andrews, F. C., 1963, *Equilibrium Statistical Mechanics*, Wiley.
Davidson, N., 1962, *Statistical Mechanics*, McGraw-Hill.
Denbigh, K., 1955, *The Principles of Chemical Equilibrium*, Cambridge University Press.
Fast, J. D., 1962, *Entropy*, McGraw-Hill.
Fowler, R., and E. A. Guggenheim, 1956, *Statistical Thermodynamics*, Cambridge University Press.
Herzberg, G., 1944, *Atomic Spectra and Atomic Structure*, 2nd ed., Dover.
Herzberg, G., 1950, *Molecular Spectra and Molecular Structure. I. Spectra of Diatomic Molecules*, 2nd ed., Van Nostrand.
Huang, K., 1963, *Statistical Mechanics*, Wiley.
Landau, L. D., and E. M. Lifshitz, 1958, *Statistical Physics*, Addison-Wesley.
Leighton, R. B., 1959, *Principles of Modern Physics*, McGraw-Hill.
Mayer, J. E., and M. G. Mayer, 1940, *Statistical Mechanics*, Wiley.
Moore, C. E., 1949, 1952, 1958, Atomic Energy Levels. *Natl. Bur. Standards Circ.* 467, vols. I, II, III.

Pauling, L., and E. B. Wilson, 1935, *Introduction to Quantum Mechanics*, McGraw-Hill.
Prigogine, I., 1962, *Non-equilibrium Statistical Mechanics*, Interscience.
Rushbrooke, G. S., 1949, *Introduction to Statistical Mechanics*, Oxford University Press.
Schrödinger, E., 1952, *Statistical Thermodynamics*, 2nd ed., Cambridge University Press.
Tolman, R. C., 1938, *The Principles of Statistical Mechanics*, Oxford University Press.
Wilks, J., 1961, *The Third Law of Thermodynamics*, Oxford University Press.
Wilks, S. S., 1962, *Mathematical Statistics*, Wiley.
Wilson, A. H., 1957, *Thermodynamics and Statistical Mechanics*, Cambridge University Press.

Chapter V

Equilibrium Gas Properties

1 INTRODUCTION

In this chapter we illustrate how the methods of the preceding chapters are used to calculate the equilibrium properties of gases of engineering interest. As a typical example—and because of its obvious importance in aerodynamics—we have chosen to concentrate on air. Since air before dissociation is a mixture primarily of O_2 and N_2, we begin by extending our treatment of the symmetrical diatomic gas. In this connection we consider in particular the useful approximation of the "ideal dissociating gas." Since ionization may play an important role in some problems, a treatment is also given of ionization equilibrium in a simple case. Finally, a discussion is provided of how the properties of air are calculated on the basis of more realistic models involving a considerable number of species. The species that make up the gas mixtures are assumed in all cases to behave individually as thermally perfect gases.

2 SYMMETRICAL DIATOMIC GAS

The results of Chapter IV, Sec. 14 are not in a form convenient for use in gas dynamics. To write them more conveniently we introduce a new variable α called the *degree of dissociation* and defined by

$$\alpha \equiv \frac{N^a}{N_A}. \tag{2.1}$$

The value of α thus gives the number of dissociated A-atoms expressed as a fraction of the total number of A-atoms present in the mixture. If

m is the mass of an A-atom, we can also write

$$\alpha = \frac{mN^a}{mN_A} = \frac{\text{mass of dissociated A-atoms}}{\text{total mass of gas}}, \tag{2.2}$$

so that α may also be thought of as the mass fraction of dissociated gas. The value of α will vary from 0 for the completely combined gas ($N^a = 0$) to 1 for the completely dissociated gas ($N^a = N_A$). On the basis of equation (2.1) and equation (IV 14.10), which apply whether the system is in equilibrium or not, we obtain N^a and N^{aa} in terms of the single variable α as follows:

$$N^a = \alpha N_A, \qquad N^{aa} = \frac{1-\alpha}{2} N_A. \tag{2.3}$$

Substitution of these relations into equation (IV 14.9), then gives for the degree of dissociation α^* at equilibrium

$$\frac{1-\alpha^*}{\alpha^{*2}} \times \frac{1}{2N_A} = \frac{Q^{aa}}{(Q^a)^2} e^{D/kT}.$$

It is convenient at this point to introduce a *characteristic temperature for dissociation* defined by

$$\Theta_d \equiv \frac{D}{k}. \tag{2.4}$$

Taking the reciprocal of the previous equation and setting $N_A = \rho V/m$, where ρ is the mass density of the gas mixture, we thus obtain the following convenient form of the law of mass action for the symmetrical diatomic gas:

$$\boxed{\frac{\alpha^{*2}}{1-\alpha^*} = \frac{m}{2\rho V} \times \frac{(Q^a)^2}{Q^{aa}} e^{-\Theta_d/T}}. \tag{2.5}$$

Since the Q's are functions of V and T, with V contained as a multiplying factor, the right-hand side of this equation is of the form $f(T)/\rho$. Equation (2.5) thus serves to give $\alpha^* = \alpha^*(\rho, T)$ for a symmetrical diatomic gas in chemical equilibrium. The quantity α^*, being a function of the two intensive state variables ρ and T, is itself an intensive state variable. Typical values of Θ_d are 59,500°K for oxygen and 113,000°K for nitrogen.

The thermal equation of state for the gas mixture can be found from equation (IV 14.11), and the relation $p = -(\partial F/\partial V)_{T,N^a,N^{aa}}$. This leads, as we might expect, to the equation

$$pV = (N^{a*} + N^{aa*})kT.$$

Substituting from equations (2.3), we obtain, again with the aid of the relation $N_A = \rho V/m$, the following thermal equation of state for the symmetrical diatomic gas:

$$\boxed{\frac{p}{\rho} = (1 + \alpha^*)R_{A_2}T} . \tag{2.6}$$

Here $R_{A_2} = k/2m$ is the gas constant per unit mass for the diatomic species A_2. The thermal equation of state for an imperfect gas is often written formally as

$$\frac{p}{\rho} = Z(\rho, T)R_0T, \tag{2.7}$$

where Z is known as the *compressibility factor* and R_0 is the ordinary gas constant at standard low temperatures. Since p for a gas in equilibrium must be a function of the two variables ρ and T, this equation in effect defines Z. Comparison of (2.6) and (2.7) shows that for the special case of the symmetrical diatomic gas, Z has the analytical form $Z(\rho, T) = 1 + \alpha^*(\rho, T)$. Its value will vary from 1 (no dissociation) to 2 (complete dissociation).

The caloric equation of state could be found, as in Chapter IV, Sec. 8, from equation (IV 14.11) and the thermodynamic relation

$$E = F - T\left(\frac{\partial F}{\partial T}\right)_{V,N^a,N^{aa}} = -T^2\left(\frac{\partial(F/T)}{\partial T}\right)_{V,N^a,N^{aa}}.$$

To illustrate an alternative method, however, we shall proceed here directly from equation (IV 14.4):

$$E = \sum_j N_j^{a*}\epsilon_j^a + \sum_j N_j^{aa*}(\epsilon_j^{aa} - D). \tag{2.8}$$

With the aid of equations (IV 14.8) for N_j^{a*} and N_j^{aa*}, this relation can be written

$$E = N^{a*}\frac{\sum_j \epsilon_j^a C_j^a e^{-\epsilon_j^a/kT}}{Q^a} + N^{aa*}\frac{\sum_j \epsilon_j^{aa} C_j^{aa} e^{-\epsilon_j^{aa}/kT}}{Q^{aa}} - N^{aa*}D.$$

But we can write in general

$$\frac{\partial}{\partial T}\ln Q = \frac{\partial}{\partial T}\ln \sum_j C_j e^{-\epsilon_j/kT} = \frac{1}{kT^2}\frac{\sum_j \epsilon_j C_j e^{-\epsilon_j/kT}}{Q}.$$

We thus obtain

$$E = kT^2\left[N^{a*}\frac{\partial}{\partial T}\ln Q^a + N^{aa*}\frac{\partial}{\partial T}\ln Q^{aa}\right] - N^{aa*}D.$$

By means of equations (2.3) and the relations $N_A = \rho V/m$, $R_{A_2} = k/2m$, and $\Theta_d = D/k$, this can be written in terms of the specific internal energy $e = E/\rho V$ as

$$\boxed{e = R_{A_2}T^2\left[2\alpha^* \frac{\partial}{\partial T}\ln Q^a + (1-\alpha^*)\frac{\partial}{\partial T}\ln Q^{aa}\right] - (1-\alpha^*)R_{A_2}\Theta_d} \;. \tag{2.9}$$

Since the partition functions are of the form $V \times \text{funct}(T)$, the factors $\frac{\partial}{\partial T}\ln Q$ are functions of T only. Thus, since $\alpha^* = \alpha^*(\rho, T)$, equation (2.9) gives $e = e(\rho, T)$. This is in contrast to the situation for a non-reacting perfect gas, where $e = e(T)$.

Equations (2.5), (2.6), and (2.9), with the partition functions expressed in terms of V and T, give three equations for the five quantities α^*, ρ, T, p, and e. It follows that if any two of these quantities are given, the other three are thereby determined. Thus for the reacting diatomic gas in equilibrium, as for a simple nonreacting gas, the specification of two intensive state variables fixes the state of the gas. This is an example of a general result for reacting gases in equilibrium.

For a given species Y, the energy per unit mass *of that species* can be obtained from [cf. equation (IV 10.4)]

$$e_Y = R_Y T^2 \frac{\partial}{\partial T}\ln Q^y,$$

where R_Y is the gas constant per unit mass of the species and Q^y is the partition function for the species. With this and the relation $R_A = k/m = 2(k/2m) = 2R_{A_2}$, equation (2.9) can also be written

$$\boxed{e = \alpha^* e_A + (1-\alpha^*)e_{A_2} - (1-\alpha^*)R_{A_2}\Theta_d} \;. \tag{2.10}$$

The first two terms on the right are both of the form

$$\frac{\text{mass of species Y}}{\text{unit mass of mixture}} \times \frac{\text{energy of species Y}}{\text{unit mass of species Y}} = \frac{\text{energy of species Y}}{\text{unit mass of mixture}} .$$

Their sum would give the total energy per unit mass of mixture if the energies of the individual species were measured from the same zero. The third term is necessary to take care of the fact that the zeros from which the individual energies are measured are displaced by the dissociation energy D. When the energies of the species are already known, an equation such as (2.10) can be used instead of the more fundamental equation (2.9).

As a simple example of the application of equation (2.10), suppose we are concerned with the temperature range where the rotational and vibrational energy of the A_2 are both fully excited. If the electronic energy is assumed negligible, we can write from the results of Chapter IV

$$e_{A_2} = (e_{A_2})_{tr} + (e_{A_2})_{rot} + (e_{A_2})_{vib} = \tfrac{3}{2}R_{A_2}T + R_{A_2}T + R_{A_2}T = \tfrac{7}{2}R_{A_2}T, \tag{2.11a}$$

and

$$e_A = (e_A)_{tr} = \tfrac{3}{2}R_A T = 3R_{A_2}T. \tag{2.11b}$$

Putting these expressions into equation (2.10), we obtain in this case

$$e = R_{A_2}\left[\frac{7 - \alpha^*}{2}T - (1 - \alpha^*)\Theta_d\right]. \tag{2.12}$$

In all the foregoing, the energy of the mixture has been reckoned from the ground state of A-atoms at rest. Equation (2.10) accordingly gives $e = 0$ for the gas in the completely atomic state ($\alpha^* = 1$) at a temperature of absolute zero ($e_A = 0$). Such a state is, of course, impossible but may be imagined as a convenient reference. An alternative procedure would be to take as the reference state of zero energy the completely diatomic state at absolute zero. Our results can be re-expressed relative to this reference by observing that the transition from the diatomic to the atomic state would require the dissociation of $N_A/2$ molecules, each with dissociation energy D. The atomic reference state thus has energy above that of the diatomic reference state by the amount $\Delta E = (N_A/2)D$, or, in terms of specific energy, $\Delta e = R_{A_2}\Theta_d$. If e' denotes the energy measured from the diatomic reference state, we thus have

$$e' = e + R_{A_2}\Theta_d,$$

or, after substitution from equation (2.10),

$$e' = \alpha^* e_A + (1 - \alpha^*)e_{A_2} + \alpha^* R_{A_2}\Theta_d. \tag{2.13}$$

This equation properly gives $e' = 0$ for the gas in the completely diatomic state ($\alpha^* = 0$) at a temperature of absolute zero ($e_{A_2} = 0$).

The choice of reference is determined by individual preference or particular requirements. The completely monatomic state leads to the simplest "bookkeeping" in calculations for complex gas mixtures. For certain gas-dynamic purposes, however, the completely diatomic state is preferable, since it leads to near-zero values for the energy of the equilibrium mixture at low temperatures and avoids the occurrence of negative values entirely.

3 IDEAL DISSOCIATING GAS

The results for the symmetrical diatomic gas, as given by equations (2.5), (2.6), and (2.9) plus the appropriate expressions for the partition functions, are still somewhat complicated for easy use in gas dynamics. For this reason Lighthill (1957) proposed a simplified model, which he termed the *ideal dissociating gas*. This model has proved useful in the approximate study of dissociating gas flow.

The simplification starts from the law of mass action, equation (2.5). By the factorization property of the partition function (Chapter IV, Sec. 10), we write

$$Q^a = Q^a_{\text{tr}} Q^a_{\text{el}},$$

$$Q^{aa} = Q^{aa}_{\text{tr}} Q^{aa}_{\text{rot}} Q^{aa}_{\text{vib}} Q^{aa}_{\text{el}}.$$

The translational partition function for A-atoms is given by equation (IV 9.1) as

$$Q^a_{\text{tr}} = V\left(\frac{2\pi mkT}{h^2}\right)^{3/2},$$

where m is again the mass of an A-atom. Obviously, Q^{aa}_{tr} is the same except that m is replaced by $2m$. The rotational partition function is given by equation (IV 12.8) and is

$$Q^{aa}_{\text{rot}} = \frac{1}{2}\left(\frac{T}{\Theta_r}\right),$$

where the symmetry factor is taken as 2 for the symmetrical gas. The vibrational partition function reckoned from the ground state of A_2-molecules is provided by equation (IV 12.11) as

$$Q^{aa}_{\text{vib}} = \frac{1}{1 - e^{-\Theta_v/T}}.$$

The electronic partition function has been discussed in general in Chapter IV, Sec. 11; specific expressions are given for O and N in equations (IV 11.5) and for O_2 and N_2 in equations (IV 12.14). Putting the foregoing expressions into equation (2.5), we obtain for the law of mass action of a dissociating gas

$$\frac{\alpha^{*2}}{1 - \alpha^*} = \frac{e^{-\Theta_d/T}}{\rho}\left[m\left(\frac{\pi mk}{h^2}\right)^{3/2}\Theta_r\sqrt{T}(1 - e^{-\Theta_v/T})\frac{(Q^a_{\text{el}})^2}{Q^{aa}_{\text{el}}}\right]. \tag{3.1}$$

The function of T on the right-hand side of this equation is too complicated for easy use. Lighthill noted, however, that in the range of T of frequent interest in gas dynamics (1000 to 7000°K) the variation of the bracketed factor is given primarily by the quantities $\sqrt{T}$ and $(1 - e^{-\Theta_v/T})$, and the effects of these two quantities tend to compensate. Calculated values of the factor in units of gm/cm³ are given for oxygen and nitrogen in the following table taken from Lighthill's work:

T, °K	1000	2000	3000	4000	5000	6000	7000
Oxygen ($\Theta_d = 59{,}500$°K)	145	170	166	156	144	133	123
Nitrogen ($\Theta_d = 113{,}000$°K)	113	135	136	133	128	123	118

The variation is seen to be small, especially when compared with the enormous variation of $e^{-\Theta_d/T}$ over the same range. Lighthill therefore suggested that the factor in brackets be taken as a constant. Since it has the dimensions of a density, he calls it the *characteristic density for dissociation*, denoted by ρ_d. The law of mass action for the resulting ideal dissociating gas is therefore

$$\boxed{\frac{\alpha^{*2}}{1-\alpha^*} = \frac{\rho_d}{\rho} e^{-\Theta_d/T}} \; . \tag{3.2}$$

Suggested values of ρ_d are 150 gm/cm³ for oxygen and 130 gm/cm³ for nitrogen. The fact that these values are so nearly equal is a coincidence and has no significance.

Setting the bracketed expression in (3.1) equal to a constant is equivalent to making an approximation in the temperature dependence of the partition functions. A consistent approximation must be made in the other equations for the gas. In the thermal equation of state this makes no difference, since p depends [see derivation of equation (2.6)] only on the derivatives with respect to V of $\ln Q = \ln [V \times \text{funct}(T)]$. The result is therefore independent of any approximation with regard to the temperature dependence. Equation (2.6) thus gives as it stands the thermal equation of state of the ideal dissociating gas:

$$\boxed{\frac{p}{\rho} = (1 + \alpha^*)R_{A_2}T} \; . \tag{3.3}$$

The caloric equation of state involves the derivatives of $\ln Q$ with respect to T and is therefore affected by the approximation. Comparing equations (2.5) and (3.2), we see that the approximation made in obtaining the latter is equivalent to setting

$$\frac{m(Q^a)^2}{2VQ^{aa}} = \rho_d = \text{const},$$

from which it follows by differentiation that

$$\frac{\partial}{\partial T}\ln Q^{aa} = 2\frac{\partial}{\partial T}\ln Q^a.$$

With this relation equation (2.9) can be written, consistent with our present approximation, as

$$e = 2R_{A_2}T^2\frac{\partial}{\partial T}\ln Q^a - (1-\alpha^*)R_{A_2}\Theta_d.$$

In the range of T of interest the variation of Q^a is given almost solely by Q^a_{tr} and the contribution from Q^a_{el} is negligible. We therefore set $Q^a_{el} =$ constant and write

$$Q^a = \text{const} \times Q^a_{tr} = \text{const} \times VT^{3/2},$$

from which

$$\frac{\partial}{\partial T}\ln Q^a = \frac{3}{2}\times\frac{1}{T}.$$

We thus obtain finally for the caloric equation of state of the ideal dissociating gas

$$\boxed{e = R_{A_2}[3T - (1-\alpha^*)\Theta_d]}. \qquad (3.4)$$

This result can be interpreted physically by returning to equation (2.10) for the internal energy in the general case. Neglecting electronic contributions, we have as in equation (2.11b)

$$e_A = (e_A)_{tr} = \tfrac{3}{2}R_AT = 3R_{A_2}T.$$

If we assume arbitrarily that the energy of molecular vibration of A_2 has one-half its fully excited value, we have

$$e_{A_2} = (e_{A_2})_{tr} + (e_{A_2})_{rot} + (e_{A_2})_{vib} = \tfrac{3}{2}R_{A_2}T + R_{A_2}T + \tfrac{1}{2}R_{A_2}T = 3R_{A_2}T.$$

With these values for e_A and e_{A_2}, equation (2.10) for e goes over precisely into the present equation (3.4). The ideal dissociating gas may thus be thought of as having the vibrational energy of its diatomic component always "half excited." It follows that for the diatomic component, we have $(c_v)_{A_2} = 3R_{A_2}$ and $\gamma_{A_2} = [(c_v)_{A_2} + R_{A_2}]/(c_v)_{A_2} = \frac{4}{3}$.

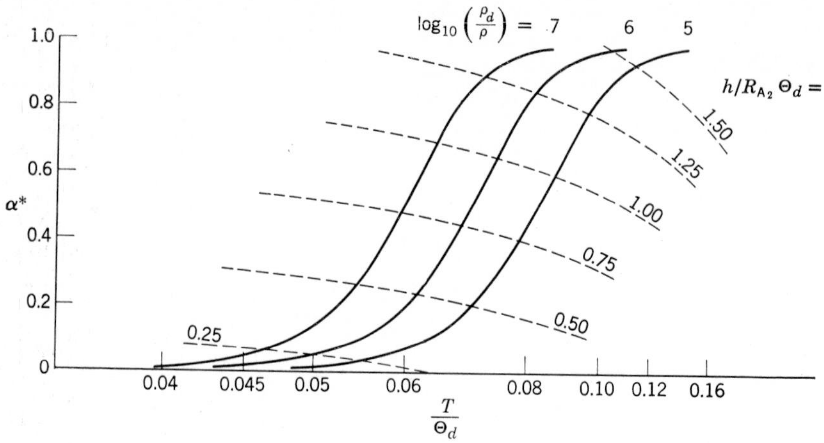

Fig. 1. Degree of dissociation for ideal dissociating gas (after Lighthill, 1957).

Lighthill himself takes as the reference for zero energy the completely diatomic state at absolute zero. He therefore obtains in place of equation (3.4)

$$e = R_{A_2}[3T + \alpha^*\Theta_d]. \tag{3.5}$$

The properties of the ideal dissociating gas are summarized in Figs. 1 and 2, which are taken from Lighthill's paper. Figure 1 shows the degree

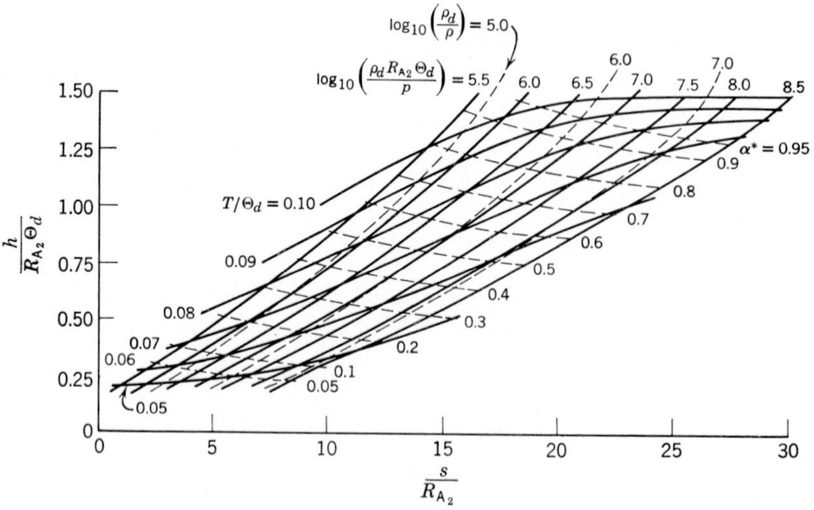

Fig. 2. Mollier diagram for ideal dissociating gas (after Lighthill, 1957).

of dissociation as a function of temperature for three values of the density. We see that for $\rho_d/\rho = 10^5$, which is typical of atmospheric densities, the gas is almost completely dissociated when T is still only about 0.12 of Θ_d. Curves of constant enthalpy are also included in Fig. 1. These are calculated from the following relation obtained from equations (3.3) and (3.5):

$$h \equiv e + \frac{p}{\rho} = R_{A_2}[(4 + \alpha^*)T + \alpha^*\Theta_d]. \tag{3.6}$$

Figure 2 is a classical enthalpy-entropy plot (Mollier diagram) with lines of constant T, p, ρ, and α^*. The entropy here was calculated from the equation of Exercise 3.2 with the constant taken arbitrarily as zero. Note that the enthalpy is not constant for a given temperature as with a perfect gas but rises as the gas dissociates. Figure 2 is qualitatively typical of the Mollier diagram for a mixture of dissociating gases such as air.

Exercise 3.1. Consider a mixture of molecular and atomic oxygen.

(a) Calculate and plot the degree of dissociation as a function of temperature for a pressure of one atmosphere. Do *not* assume an ideal dissociating gas.

(b) What is the density in gm/cm³ of atomic oxygen in the mixture at a temperature of 4500°K and a pressure of one atmosphere?

Exercise 3.2. Obtain the following equation for the specific entropy of the ideal dissociating gas:

$$\frac{s}{R_{A_2}} = 3 \ln \frac{T}{\Theta_d} + \alpha^*(1 - 2 \ln \alpha^*) - (1 - \alpha^*) \ln (1 - \alpha^*) - (1 + \alpha^*) \ln \frac{\rho}{\rho_d} + \text{const.}$$

Exercise 3.3. Derive an expression for the specific heat c_p at constant pressure for an ideal dissociating gas as a function of the degree of dissociation α^* and the temperature T. This expression will contain certain appropriate constants, but should not contain any state variables other than α^* and T.

Exercise 3.4. From thermodynamics it is known that the thermal and caloric equations of state are not independent but must satisfy the so-called "reciprocity relations." One form of these relations (see, e.g., Liepmann and Roshko, 1957, pp. 22–23) is

$$\left(\frac{\partial h}{\partial p}\right)_T = v - T\left(\frac{\partial v}{\partial T}\right)_p,$$

where h is the specific enthalpy and $v \equiv 1/\rho$.

(a) If the approximations for the ideal dissociating gas have been made consistently, the final relations for this gas should satisfy the foregoing equation. Verify that this is indeed the case.

(b) Since the contribution from Q^a_{el} has been arbitrarily neglected in obtaining e for the ideal dissociating gas, a question might arise as to whether in fact the approximations *were* made consistently. Why does this arbitrary approximation have no effect insofar as the satisfaction of the foregoing equation is concerned? (Hint: What would happen if the approximation were not made?)

Exercise 3.5. Consider pure oxygen in equilibrium at a temperature of 5900°K and a pressure of 31.4×10^7 dynes/cm². If the oxygen is regarded as an ideal dissociating gas and the zero for internal energy is taken as corresponding to the completely atomic state at 0°K, what is the specific enthalpy of the gas?

Ans.: -7.66×10^9 dyne cm/gm.

4 IONIZATION EQUILIBRIUM; THE SAHA EQUATION

At sufficiently high temperatures an atom or molecule may lose one or more electrons and become ionized. The equilibrium properties of the resulting mixture of ions, electrons, and neutral particles can be analyzed by means of the law of mass action. Since ionization reactions have certain special features, it will be useful to set the development down in some detail.

We consider the simple single-ionization reaction

$$\mathrm{Z}^+ + \mathrm{e} \rightleftharpoons \mathrm{Z}, \tag{4.1}$$

where Z represents a neutral atom or molecule, Z^+ is the corresponding singly-ionized particle, and e is an electron. This reaction is formally identical to the reaction $\mathrm{A} + \mathrm{B} \rightleftharpoons \mathrm{AB}$ of Chapter IV, Sec. 13, with Z^+ and e playing the roles of the separate particles A and B and Z corresponding to the combination AB. The results of that section can therefore be applied directly. Thus, for the law of mass action we find, from equation (IV 13.15),

$$\frac{N^z}{N^{z^+}N^e} = \frac{Q^z}{Q^{z^+}Q^e}\, e^{I/kT}, \tag{4.2}$$

where I is the ionization energy. For simplicity we have dropped the asterisks denoting the equilibrium composition. Proceeding formally from the conservation equations (IV 13.16), we also have

$$N^{z^+} + N^z = \text{const},$$
$$N^e + N^z = \text{const},$$

where the appropriate values for the constants are perhaps not immediately obvious. They can be evaluated by supposing that our mixture developed from an originally low-temperature state at which there is no ionization.

If this state is denoted by $(\ )_0$, we thus have $N_0^{z^+} = 0$, $N_0^e = 0$, and the foregoing equations become

$$N^{z^+} + N^z = N_0^z, \tag{4.3a}$$

$$N^e + N^z = N_0^z. \tag{4.3b}$$

These equations are formally identical to equations (IV 13.16) except that here $N_A = N_B$. Subtracting the second from the first, we thus find

$$N^{z^+} = N^e. \tag{4.4}$$

This equation, which will be used in place of (4.3b), expresses the fact that the net electrical charge on the originally neutral gas must always remain zero. Equation (4.3a) expresses the associated fact that the total number of atomic nuclei or heavy particles must remain constant. These equations could, of course, have been written from physical principles. We have chosen to proceed formally, however, since the derivation of the law of mass action in Chapter IV depended on two equations of the form (4.3).

Equation (4.2) can now be put in a more convenient form. We define the *degree of ionization* ϕ as the fraction of the originally neutral particles that have become ionized, that is,

$$\phi \equiv \frac{N^{z^+}}{N_0^z}. \tag{4.5}$$

From this and equations (4.4) and (4.3a) we have

$$N^{z^+} = \phi N_0^z, \qquad N^e = \phi N_0^z, \qquad N^z = (1 - \phi)N_0^z. \tag{4.6}$$

Substituting into the reciprocal of equation (4.2) and using the relation $N_0^z = \rho V/m_Z$, where m_Z is the mass of a Z-particle, we obtain

$$\frac{\phi^2}{1 - \phi} = \frac{m_Z}{\rho V} \times \frac{Q^{z^+} Q^e}{Q^z} e^{-\Theta_i/T}, \tag{4.7}$$

where Θ_i is a *characteristic temperature for ionization* defined by

$$\Theta_i \equiv \frac{I}{k}. \tag{4.8}$$

In accord with equation (IV 10.3), the partition functions can be written

$$Q^e = Q_{tr}^e \prod_{int} Q_{int}^e, \qquad Q^{z^+} = Q_{tr}^{z^+} \prod_{int} Q_{int}^{z^+},$$

$$Q^z = Q_{tr}^z \prod_{int} Q_{int}^z.$$

The only internal energy possessed by an electron alone is that associated with its spin. This gives rise to two permissible quantum energy states, both lying in a single ground level of degeneracy 2. We therefore have for the internal contribution of the electron

$$\prod_{\text{int}} Q^e_{\text{int}} = Q^e_{\text{spin}} = 2.$$

The translational partition function for the electron is obtained from equation (IV 9.1) and is

$$Q^e_{\text{tr}} = V\left(\frac{2\pi m_e kT}{h^2}\right)^{3/2},$$

where m_e is the mass of the electron. The masses of the ion and the neutral particle differ only by the relatively very small mass of the electron and may be taken equal. We thus have to a very good approximation

$$Q^{z^+}_{\text{tr}} = Q^z_{\text{tr}}.$$

Putting the foregoing expressions into (4.7), we obtain

$$\boxed{\frac{\phi^2}{1-\phi} = \frac{m_Z}{\rho}\left(\frac{2\pi m_e kT}{h^2}\right)^{3/2} \frac{2\prod_{\text{int}} Q^{z^+}_{\text{int}}}{\prod_{\text{int}} Q^z_{\text{int}}} e^{-\Theta_i/T}} \;. \qquad (4.9)$$

This equation gives $\phi = \phi(\rho, T)$.

Equation (4.9) can be put in terms of the pressure by means of equation (IV 13.21), which becomes in this case

$$pV = (N^{z^+} + N^e + N^z)kT.$$

With equations (4.6) and the relation $N^z_0 = \rho V/m_Z$, this equation can be written

$$\frac{p}{\rho} = (1 + \phi)\frac{k}{m_Z}T. \qquad (4.10)$$

Using this to eliminate ρ from (4.9), we obtain

$$\boxed{\frac{\phi^2}{1-\phi^2} = \frac{1}{p}\left(\frac{2\pi m_e}{h^2}\right)^{3/2}(kT)^{5/2} \frac{2\prod_{\text{int}} Q^{z^+}_{\text{int}}}{\prod_{\text{int}} Q^z_{\text{int}}} e^{-\Theta_i/T}} \;. \qquad (4.11)$$

An equation of the type (4.9) or (4.11) is referred to loosely as the *Saha equation*. When the species Z is a monatomic gas, the internal partition functions are merely those due to electronic excitation (see Chapter IV, Sec. 11). If excitation above the ground level is ignored, they become merely constants, and equation (4.11) can be written

$$\frac{\phi^2}{1-\phi^2} = C\,\frac{T^{5/2}}{p}\,e^{-\Theta_i/T}, \tag{4.12}$$

where C is a constant. It is in essentially this form that Saha (1920) first derived the equation on the basis of classical thermodynamics. Approximations for an ionizing gas similar to those made by Lighthill for the dissociating gas have been proposed by several authors (see, e.g., Laporte, 1958, and Duclos et al., 1962).

Results similar to those of this section can also be derived for reactions involving doubly-ionized particles Z^{++} (see, e.g., Cambel et al., 1963, pp. 48–55).

5 MIXTURE OF GASES

The foregoing work has illustrated the methods necessary to calculate the equilibrium properties of mixtures simple enough to be governed by a single reaction equation. Mixtures with enough species to involve more than a single reaction can be handled by an extension of the same methods. Air at high temperatures forms a typical example of particular interest.

For gas-dynamic purposes, air at temperatures up to about 8000°K may be regarded (see Sec. 6) as composed of essentially five chemical species: O_2, O, N_2, N, and NO. Although ionization will normally have little effect on the energy balance of fluid flow, the presence of free electrons may be important for electromagnetic reasons. Of the five species listed, NO has the lowest ionization energy and is the first to produce electrons as the temperature is increased. If electromagnetic considerations are important, we may therefore add NO^+ and e to the list of species. The equilibrium of the resulting mixture of seven species can be taken to be controlled by the following four reactions, identified by letter r:

$$r = 1: \quad O_2 \rightleftharpoons O + O, \qquad 5.12 \text{ ev} \tag{5.1a}$$

$$2: \quad N_2 \rightleftharpoons N + N, \qquad 9.76 \text{ ev} \tag{5.1b}$$

$$3: \quad NO \rightleftharpoons N + O, \qquad 6.49 \text{ ev} \tag{5.1c}$$

$$4: \quad NO \rightleftharpoons NO^+ + e. \qquad 9.25 \text{ ev} \tag{5.1d}$$

The dissociation or ionization energy in electron-volts is given in each case on the right.[1] The foregoing is far from exhausting the list of reactions than can actually occur between the species. As we shall see later, however, there is no need to introduce all possible reactions when we are interested only in the equilibrium properties of the mixture.

In order for the mixture to be in overall equilibrium, that is, for the concentration of each species to be a constant, each of the four reactions must be in equilibrium individually. This could be shown by setting the problem up in general statistical-mechanical terms, parallel to the treatment in Chapter IV. It also follows at once from the principle of detailed balancing, which states that at equilibrium every molecular process and its inverse must be individually in balance (cf. Chapter II, Sec. 4). Whatever the reasoning, the equilibrium composition is to be found by applying the law of mass action individually to the various reactions.

In applying the law of mass action to problems of this kind, it is often the practice to proceed, not from the law in the statistical form as has been used here, but in the equivalent chemical form derived in Chapter III. The relation between the two forms is easily seen. For reaction (5.1a), for example, the statistical form as given by equation (IV 14.9) can be written

$$\frac{(N^o)^2}{N^{oo}} = \frac{(Q^o)^2}{Q^{oo}} e^{-D_{O_2}/kT}.$$

If [Y] is the concentration of any given species Y in moles per unit volume and $\hat{N}$ is Avogadro's number, we have $N^y = \hat{N}V[\mathrm{Y}]$. The foregoing equation in terms of molar concentrations is therefore

$$\frac{[\mathrm{O}]^2}{[\mathrm{O}_2]} = \frac{(Q^o)^2}{\hat{N}VQ^{oo}} e^{-D_{O_2}/kT},$$

where the right-hand side is now a function only of T. This is a special case of the law of mass action in the general chemical form obtained in equation (III 9.2a):

$$\prod_s [\mathrm{X}_s]^{\nu_s} = K_c(T). \tag{5.2}$$

In this equation X_s denotes the chemical species, differentiated one from another by the subscript $s = 1, 2, 3, \ldots, l$, and ν_s is the corresponding stoichiometric coefficient in the reaction equation (cf. equation (III 3.2), et seq.). As before, $K_c(T)$ is the equilibrium constant for concentrations. The general equation (5.2) was derived in Chapter III by thermodynamic

[1] An electron-volt, which is a convenient unit of energy on the microscopic scale, is the amount of energy acquired by an electron when accelerated through a potential difference of 1 volt. One electron-volt is equivalent to a characteristic temperature Θ of 11,600°K.

methods. With the aid of statistical mechanics the equilibrium constant can always be expressed in terms of the appropriate partition functions.

The molar concentration is an awkward variable in gas dynamics, where specific quantities measured per unit mass of fluid are more natural. A convenient variable here is the number of moles per unit mass of mixture, called the *mole-mass ratio* and denoted by the symbol $\mathfrak{n}$. This is related to the concentration by $[X_s] = \rho \mathfrak{n}_s$. In terms of this new variable the law of mass action has the form

$$\boxed{\prod_s (\mathfrak{n}_s)^{\nu_s} = \frac{1}{\rho^{\sum_s \nu_s}} K_c(T)} \; . \tag{5.3}$$

We now apply equation (5.3) successively to the four reactions (5.1). We thus obtain for the conditions of equilibrium of the mixture

$$\frac{\mathfrak{n}_{\mathrm{O}}^2}{\mathfrak{n}_{\mathrm{O}_2}} = \frac{1}{\rho} K_{c,1}(T), \qquad \frac{\mathfrak{n}_{\mathrm{N}}^2}{\mathfrak{n}_{\mathrm{N}_2}} = \frac{1}{\rho} K_{c,2}(T), \qquad \frac{\mathfrak{n}_{\mathrm{N}}\mathfrak{n}_{\mathrm{O}}}{\mathfrak{n}_{\mathrm{NO}}} = \frac{1}{\rho} K_{c,3}(T),$$
$$\frac{\mathfrak{n}_{\mathrm{e}}\mathfrak{n}_{\mathrm{NO}^+}}{\mathfrak{n}_{\mathrm{NO}}} = \frac{1}{\rho} K_{c,4}(T), \tag{5.4}$$

where for convenience in the specific example we have reverted to our frequent use of the chemical symbol itself as the subscript. In these equations $K_{c,r}(T)$ is the equilibrium constant for reaction r. The foregoing equations, which are four equations in seven unknowns, must be supplemented by three equations expressing the conservation of oxygen and nitrogen nuclei and the fact of zero net charge. These are, in terms of the mole-mass ratios,

$$2\mathfrak{n}_{\mathrm{O}_2} + \mathfrak{n}_{\mathrm{O}} + \mathfrak{n}_{\mathrm{NO}} + \mathfrak{n}_{\mathrm{NO}^+} = 2(\mathfrak{n}_{\mathrm{O}_2})_0, \tag{5.5a}$$
$$2\mathfrak{n}_{\mathrm{N}_2} + \mathfrak{n}_{\mathrm{N}} + \mathfrak{n}_{\mathrm{NO}} + \mathfrak{n}_{\mathrm{NO}^+} = 2(\mathfrak{n}_{\mathrm{N}_2})_0, \tag{5.5b}$$
$$\mathfrak{n}_{\mathrm{NO}^+} - \mathfrak{n}_{\mathrm{e}} = 0, \tag{5.5c}$$

where $(\mathfrak{n}_{\mathrm{Y}})_0$ pertains to the composition at low temperatures where the mixture is completely O_2 and N_2. (These equations are related to earlier equations in terms of numbers of particles by $\mathfrak{n}_{\mathrm{Y}} = N^y/\hat{N}\rho V$.) Equations (5.4) and (5.5) constitute a nonlinear set of seven equations containing seven mole-mass ratios $\mathfrak{n}_{\mathrm{Y}}$. When ρ and T and the low-temperature composition are given, the set can be solved to find $\mathfrak{n}_{\mathrm{Y}} = \mathfrak{n}_{\mathrm{Y}}(\rho, T)$. Because of the nonlinear nature of the law of mass action, the actual solution for the equilibrium composition of a complex gas mixture is a difficult problem. Various special methods have been developed for handling the problem on electronic computing machines (see, e.g., Gordon, Zeleznik, and Huff, 1959, and Hilsenrath, Klein, and Sumida, 1959).

For engineering purposes the equilibrium constants in equations (5.4) can be approximated over a limited range of temperature by

$$K_c = C_c T^{\eta_c} e^{-\Theta/T}, \tag{5.6}$$

where C_c, η_c, and Θ are constants. Values of these constants for the four reactions (5.1) are given in Table A. The units are such that K_c for these reactions is in moles/cm³. The values in the table were obtained by Wray (1962) to give the best fit to the partition-function results over the range from 3000 to 8000°K. They yield the correct equilibrium constants to

Table A Values of Constants in Equation (5.6) *for* K_c

Reaction r	C_c	η_c	Θ, °K
1	1.2×10^3	−0.5	59,500
2	18	0	113,000
3	4.0	0	75,500
4	14.4×10^{-10}	+1.5	107,000

within 10% over this range. Values of $(\mathfrak{n}_Y)_0$ that are a good approximation to real air are $(\mathfrak{n}_{O_2})_0 = 0.00733$ mole/gm, $(\mathfrak{n}_{N_2})_0 = 0.0273$ mole/gm.

It was remarked earlier that the chemical equations (5.1) do not exhaust the possible reactions between the seven species in the mixture. The reader may wonder, therefore, what would happen if we were to add a fifth reaction to the system, say

$$r = 5: \quad O_2 + N_2 \rightleftharpoons 2NO. \quad 1.90 \text{ ev} \tag{5.7}$$

For this reaction to be in equilibrium, equation (5.3) gives the condition

$$\frac{\mathfrak{n}_{NO}^2}{\mathfrak{n}_{O_2}\mathfrak{n}_{N_2}} = K_{c,5}(T), \tag{5.8}$$

and the addition of this equation to the system (5.4) and (5.5) might appear to make the problem overdetermined, that is, we would have eight equations for the seven quantities $\mathfrak{n}_Y$.[2] But we can write alternatively, with the aid of equations (5.4),

$$\frac{\mathfrak{n}_{NO}^2}{\mathfrak{n}_{O_2}\mathfrak{n}_{N_2}} = \left(\frac{\mathfrak{n}_{NO}}{\mathfrak{n}_O\mathfrak{n}_N}\right)^2 \frac{\mathfrak{n}_O^2}{\mathfrak{n}_{O_2}} \frac{\mathfrak{n}_N^2}{\mathfrak{n}_{N_2}} = \frac{K_{c,1}K_{c,2}}{K_{c,3}^2},$$

[2] Note that the dimensions of the equilibrium constant here are different from those of the equilibrium constants in equations (5.4). This is a common situation and calls for care in actual calculations.

and it can further be shown from the general thermodynamic expressions (III 9.2b) and (III 9.1b) for $K_c(T)$ that $K_{c,1}K_{c,2}/K_{c,3}^2 \equiv K_{c,5}$. It follows that equation (5.8) will automatically be satisfied if the first three of equations (5.4) are satisfied. In other words, the first three of equations (5.4) and equation (5.8) are not independent, but any one of them can be derived from the other three. As a result the system is not really over-determined by the addition of the reaction (5.7). This is a consequence of the fact that the reaction equations (5.1) and (5.7) are themselves not independent but can be linearly related, that is, (5.7) can be obtained from the linear combination (5.1a) + (5.1b) − 2(5.1c). To calculate the equilibrium composition in any given case we need only include as many of the possible reactions as are necessary to constitute a complete, linearly independent set. The corresponding laws of mass action plus the equations expressing the conservation of atomic nuclei and net charge will then always be sufficient to determine the composition for given values of ρ and T.

With values of $\mathfrak{n}_s$ known, the thermodynamic properties of the mixture are easily found. Since we assume a mixture of perfect gases, the partial pressure of each species is given by an equation of the form

$$p_s = \rho \mathfrak{n}_s \hat{R} T,$$

where $\hat{R}$ is the universal gas constant (gas constant per mole). Summing over all species then gives for the thermal equation of state of the mixture

$$\boxed{\frac{p}{\rho} = \sum_s \mathfrak{n}_s \hat{R} T} \, . \tag{5.9}$$

This equation, like the equations that follow, is not limited to the seven-species mixture discussed earlier.

If the effect of electronic excitation is ignored, the specific internal energy *per mole* for each species is given by the following equation, whose individual terms we shall take up in order:

$$\hat{e}_s = \hat{e}_{s_{\text{tr+rot}}} + \hat{e}_{s_{\text{vib}}} + \Delta \hat{e}_s. \tag{5.10}$$

On the assumption that the rotational energy is fully excited, the contribution of translation and rotation $\hat{e}_{s_{\text{tr+rot}}}$ is $\frac{3}{2}\hat{R}T$ for the monatomic species and the electrons and $\frac{5}{2}\hat{R}T$ for the diatomic species. The vibrational energy $\hat{e}_{s_{\text{vib}}}$, which is of course zero for the monatomic species and electrons, is given for the diatomic species by equation (IV 12.12):

$$\hat{e}_{s_{\text{vib}}} = \frac{\hat{R}\Theta_{v_s}}{e^{\Theta_{v_s}/T} - 1} \, .$$

The quantity $\Delta\hat{e}_s$ is included to refer the energy of all species to a common zero; its value depends on the reference state chosen. For a complex gas mixture, the simplest choice for the reference state of zero energy—and the choice that we shall make for the common zero for the individual species—is the completely atomic state at 0°K, that is, with all the atoms in their ground state and at rest (cf. Chapter IV, Secs. 13 and 10). The value of $\Delta\hat{e}_s$ for a given species is then the energy required (positive if added, negative if subtracted) to form one mole of that species at 0°K from its constituent atoms, also at 0°K. For neutral monatomic species we therefore have $\Delta\hat{e}_s = 0$, and for neutral diatomic species $\Delta\hat{e}_s = -\hat{D}_s$, where $\hat{D}_s$ is the dissociation energy per mole. For ionized monatomic species we set $\Delta\hat{e}_s = \hat{I}_s$, where $\hat{I}_s$ is the ionization energy per mole, and for ionized diatomic species we set $\Delta\hat{e}_s = -\hat{D}_s + \hat{I}_s$. Since the ionization energy is thus taken to go entirely into forming the ions, the value of $\Delta\hat{e}_s$ for the electrons is taken as zero. This is because in a complex mixture there is no unique way of forming electrons, that is, they may originate from the ionization of various species, all with different ionization energies. The caloric equation of state of the mixture can be obtained by multiplying equation (5.10) by $\mathfrak{n}_s$ and summing over all species. Generalizing the result to allow for the use of *any desired reference state for the mixture*, we thus write for the specific energy per unit mass of gas

$$\boxed{e = \sum_s \mathfrak{n}_s(\hat{e}_{s_{\mathrm{tr+rot}}} + \hat{e}_{s_{\mathrm{vib}}} + \Delta\hat{e}_s) + \Delta e}\,. \tag{5.11}$$

The quantity Δe, which makes possible the change of reference state, is the specific energy required to form the completely atomic reference state at 0°K from the constituents of the new reference state, also at 0°K. A choice often made for a new reference state is the nonionized, completely diatomic state at 0°K. In this case we have

$$\Delta e = \sum_d (\mathfrak{n}_d)_0\,(\hat{D})_d,$$

where the subscript d pertains to the nonionized diatomic species only and the $(\mathfrak{n}_d)_0$ specify the composition of the completely diatomic state. Equations (2.10) and (2.13) for the symmetrical diatomic gas are, in effect, special cases of equation (5.11).

The laws of mass action (5.3), the associated equations for conservation of nuclei and net charge, and the equations of state (5.9) and (5.11) are $l + 2$ equations for the $l + 4$ quantities $\mathfrak{n}_1, \ldots, \mathfrak{n}_l$, ρ, T, p, and e (where l in the example given above was 7). The specification of any two of these quantities is therefore sufficient to determine the state of the system.

The entropy, if required, can be obtained either by integration of the general equilibrium relation $T\,ds = de + p\,d(1/\rho)$ or by means of the thermodynamic equation (III 8.7) for a mixture of thermally perfect gases (which can be shown to be consistent with the statistical-mechanical formulas). All the thermodynamic properties could also be obtained, of course, from the statistical-mechanical formulas involving the partition functions, but this is unnecessarily cumbersome when the specific properties of the individual species are already known, as is often the case.

Although we have used air as an example, the ideas of this section can be applied to any mixture of thermally perfect gases. The thermochemical data necessary for the purpose are found in a wide variety of sources. Latest and most inclusive among the general tabulations are the cumulative, looseleaf JANAF Tables (Stull et al., 1960). Other tabulations are given, for example, by Rossini, Wagman et al. (1952), Rossini, Pitzer et al. (1953), and McBride et al. (1963). For a listing of sources as of 1955, as well as an explanation of the use of the data in its standard chemical form, see also Rossini (1955), pp. 103–108.

Exercise 5.1. On the basis of the general thermodynamic equations (III 9.2b) and (III 9.1b) for $K_c(T)$, show as stated following equation (5.8) that for the chemical reactions (5.1) and (5.7) we have $K_{c,1}K_{c,2}/K_{c,3}^2 \equiv K_{c,5}$.

Exercise 5.2. On the basis of the equations of this section, obtain an expression giving the heat of reaction $\Delta\hat{H}$ (defined in Chapter III, Sec. 10) as a function of T for the reaction $2A \rightleftharpoons A_2$.

6 PROPERTIES OF EQUILIBRIUM AIR

More complicated models for equilibrium high-temperature air have been treated by a number of workers in the United States and abroad. Their results are essentially the same. For a summary of the situation as of this writing, the reader should consult the review article by Hochstim (1963) or the book by Cambel et al. (1963). Here we shall mention only the results of Hilsenrath and his co-workers at the National Bureau of Standards (NBS), which are widely used in the United States.

The NBS work has appeared in a series of three reports. In all of these, tabular values are given for the thermodynamic properties as functions of density and temperature for the temperature range from 1500 to 15000°K. The calculations are based on highly refined evaluations of the partition functions, including anharmonic effects as well as the effects of high-energy states that are only infrequently excited even at high temperatures. In all cases the chemical composition at low temperatures is chosen to approximate that of sea-level air. The three reports differ in the number of species considered and in the omission or inclusion of imperfect-gas effects. Hilsenrath and Beckett (1956), which was in the nature of a preliminary report, considered sixteen species derived from oxygen and nitrogen and treated each species as a thermally perfect gas. Hilsenrath, Klein, and Woolley (1959) improved the gas model by including twelve additional species mainly involving the minor elements carbon, argon, and neon, again on the perfect-gas assumption. The most recent work, by Hilsenrath and Klein (1963), uses the same gas model as the second report but includes certain of the imperfect-gas effects due to interparticle forces. These effects become practically significant at densities greater than atmospheric. Values of specific heat and speed of sound based on the data of Hilsenrath and Klein have been given by Lewis and Neel (1964). The properties of air at temperatures below those considered in the above reports are tabulated by Hilsenrath, Beckett, Benedict et al. (1960) and by Humphrey and Neel (1961).[3] All these reports give results at constant values of temperature. The high-temperature data of Hilsenrath and Klein and the low-temperature data of Humphrey and Neel have been interpolated and retabulated for constant values of (a) entropy and (b) pressure by Neel and Lewis (1964a, 1964b).

Typical results taken from Hilsenrath, Klein, and Woolley (1959) are shown in Figs. 3 and 4. The gas model here consists of the following 28 species:

$$\mathrm{O, N, C, Ar, Ne, e, N_2, O_2, NO, CO, CO_2, NO_2,}$$
$$\mathrm{N_2O, O^-, O^+, O^{++}, N^+, N^{++}, C^+, C^{++}, Ar^+, Ar^{++},}$$
$$\mathrm{Ne^+, O_2^-, O_2^+, N_2^+, NO^+, CO^+.}$$

This list gives some measure of the complexity of the work: 28 partition functions had to be calculated and a nonlinear set of 28 simultaneous equations solved for each set of values of ρ and T.

Figure 3 shows the compressibility factor Z as defined by the equation

$$\frac{p}{\rho} = ZR_0T,$$

where R_0 is the ordinary gas constant for the mixture at standard low-temperature conditions (273°K and atmospheric density). For a mixture

[3] Certain tabulations made prior to about 1955 are in error because of the use of the previously accepted but incorrect value of 7.37 ev for the dissociation energy of nitrogen. The correct value is now accepted to be 9.76 ev as listed with equation (5.1b). Care must be taken to check as to which value was employed before using any of the older tabulations.

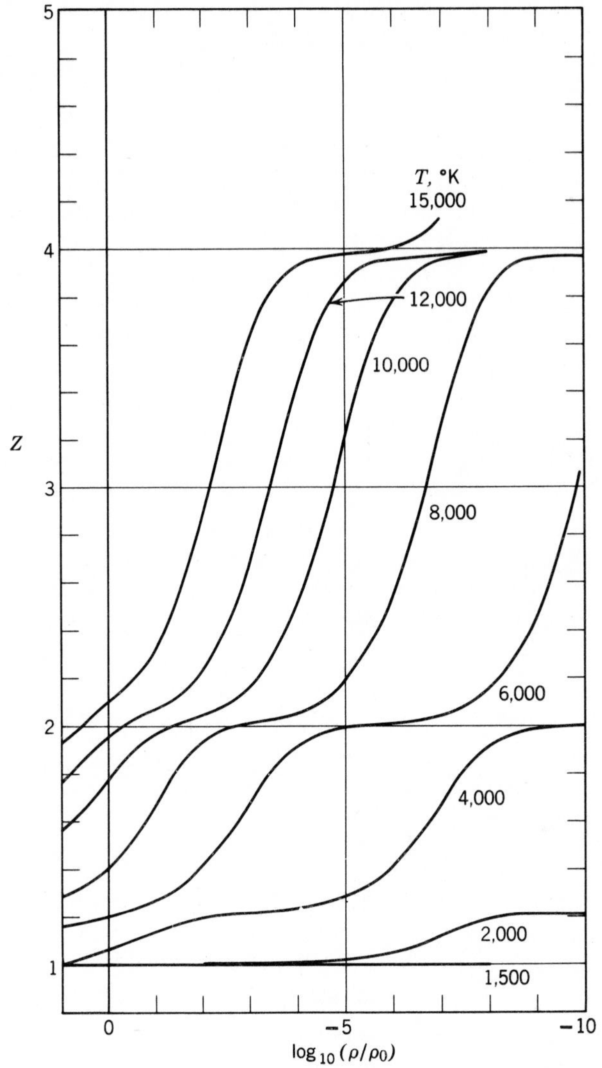

Fig. 3. Compressibility factor for high-temperature air; $\rho_0 = 1.293 \times 10^{-3}$ gm/cm^3 (after Hilsenrath, Klein, and Woolley, 1959).

of perfect gases, as is assumed by Hilsenrath, Klein, and Woolley, Z is equal to the number of moles of mixture at the given conditions for each mole at standard conditions (see Chapter VI, Sec. 5).

Figure 4 shows typical results for the detailed composition as a function of temperature. The dependent variable here is the mole percent, which is the number of moles of the particular species at the given conditions expressed as a percentage of the total number of moles at the same conditions. The results shown are for $\frac{1}{100}$ of standard atmospheric density, but other densities follow more or less the same qualitative trends. We see

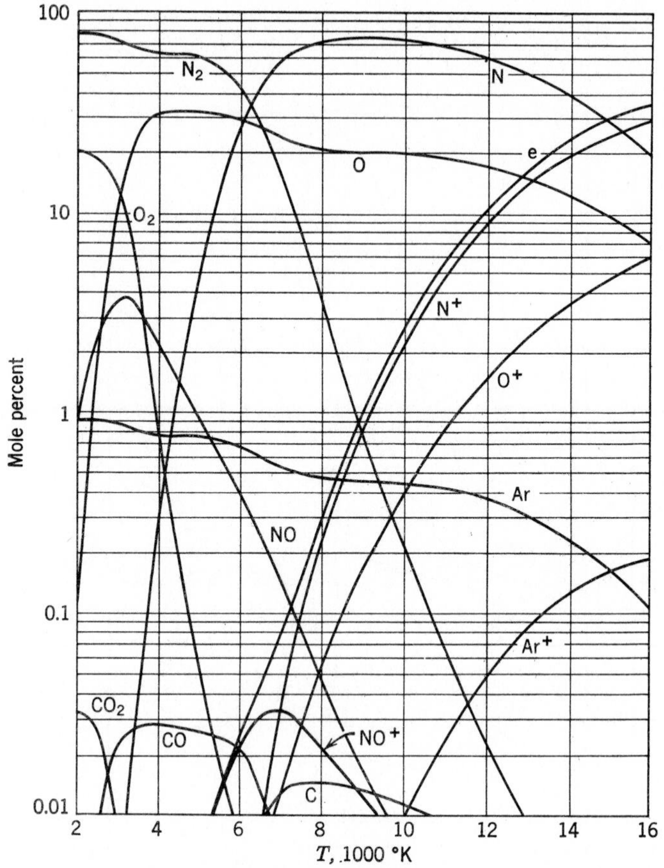

Fig. 4. Equilibrium composition of air at density of 10^{-2} atm (after Hilsenrath, Klein, and Woolley, 1959).

that in the temperature range considered, the only species that appear in amounts of more than one percent are O_2, O, N_2, N, NO, O^+, N^+, and e. At temperatures below about 8000°K, only the first five of these species appear in this amount at the present density. This is the justification for the choice of neutral species used in Sec. 5. The rise of NO to a maximum at relatively low temperatures and its subsequent disappearance is characteristic. The first species to become ionized and thus produce electrons as the temperature increases is seen to be NO, which was also recognized in the gas model of Sec. 5. The NO^+, however, like the neutral NO, soon disappears, and N^+, O^+, and later Ar^+ take over as the predominating ions. The simple model of Sec. 5 would have to be modified accordingly at the higher temperatures. At first glance, Fig. 4 seems to show the amount of inert argon decreasing with temperature even before it begins to ionize significantly. This is only because the results have been presented in mole *percent*; the number of moles of argon, even when constant, becomes continually smaller relative to the total number of moles, which is continually increasing as the result of dissociation.

The references cited earlier give results primarily in tabular form. For gas-dynamic calculations it is often useful to have them in the form of graphs. A large Mollier diagram (enthalpy versus entropy) based on the data of Hilsenrath and Beckett (1956) has been prepared by Feldman (1957). This diagram is particularly convenient and has been widely used. The same data are presented in a plot of enthalpy versus pressure by Moeckel and Weston (1958). A Mollier diagram based primarily on the data of Hilsenrath and Klein (1963) is given in sectional form by Little (1963); this diagram is also available as a large single sheet (see reference).

For engineering calculations and for use in computer subroutines, it is useful to have results in closed analytical form. This has been accomplished by Hansen (1959; see also Hansen and Hodge, 1961) for a gas model consisting of O_2, O, N_2, N, N^+, O^+, and e. The results are derived by using simplified expressions for the partition functions and by dividing the range of temperatures into three regions in which different effects predominate. Hansen's method has been refined and extended by Hochstim (1960, 1961). A second method for obtaining closed-form results is by fitting empirical equations to the tabular data of the earlier references. This has been done by various workers (e.g., Lewis and Burgess, 1963).

Exercise 6.1. Write the equations, similar to equations (5.4) and (5.5), that are necessary to determine the equilibrium composition of a gas mixture composed of the species that appear in Fig. 4.

Exercise 6.2. In a so-called "hotshot" wind tunnel, the air in a closed reservoir is heated to a high temperature and pressure by the discharge of electrical energy (which has been stored in a bank of capacitors) across an arc gap in the reservoir. Suppose that the volume of the reservoir is 400 cm^3, the temperature of the air in the reservoir before discharge is 300°K, and the density in the reservoir before discharge is 50 times standard atmospheric density. If 250,000 joules of energy are stored in the capacitor bank and if the effective efficiency of delivery of this energy to the air is 80%, what is the temperature (in °R), the enthalpy (in Btu/lb), and the pressure (in psia) of the air in the reservoir after the discharge? Use the Mollier diagram of Feldman (1957) or some other suitable chart for the calculations. Consider the air before discharge to be a perfect gas with $\gamma = \frac{7}{5}$.

Ans. (as obtained from Mollier diagram of Feldman): $T = 11{,}250°R$, $h = 4{,}280$ Btu/lb, $p = 18{,}510$ psia.

References

Cambel, A. B., D. P. Duclos, and T. P. Anderson, 1963, *Real Gases*, Academic.

Duclos, D. P., D. P. Aeschliman, and A. B. Cambel, 1962, Approximate Equation for Perfect Gas Plasma. *ARS Jour.*, vol. 32, no. 4, p. 641.

Feldman, S., 1957, Hypersonic Gas Dynamic Charts for Equilibrium Air. *AVCO Res. Lab.*, Everett, Mass.

Gordon, S., F. J. Zeleznik, and V. N. Huff, 1959, A General Method for Automatic Computation of Equilibrium Compositions and Theoretical Performance of Propellants. *NASA*, TN D–132.

Hansen, C. F., 1959, Approximations for the Thermodynamic and Transport Properties of High-Temperature Air. *NASA*, TR R-50.

Hansen, C. F., and M. E. Hodge, 1961, Constant Entropy Properties for an Approximate Model of Equilibrium Air. *NASA*, TN D-352.

Hilsenrath, J., and C. W. Beckett, 1956, Tables of Thermodynamic Properties of Argon-Free Air to 15,000°K. *Arnold Engr. Devel. Center*, AEDC-TN-56-12.

Hilsenrath, J., C. W. Beckett, W. S. Benedict et al., 1960, *Tables of Thermodynamic Properties of Air, Argon, Carbon Dioxide, Carbon Monoxide, Hydrogen, Nitrogen, Oxygen, and Steam*, Pergamon.

Hilsenrath, J., and M. Klein, 1963, Tables of Thermodynamic Properties of Air in Chemical Equilibrium including Second Virial Corrections from 1,500°K to 15,000°K. *Arnold Engr. Devel. Center*, AEDC-TDR-63-161. (Superceded in 1965 by AEDC-TR-65-58 by same authors and under same title.)

Hilsenrath, J., M. Klein, and D. Y. Sumida, 1959, Mechanized Computation of Thermodynamic Tables at the National Bureau of Standards: The Calculation of the Equilibrium Composition and Thermodynamic Properties of Dissociated and Ionized Gaseous Systems; in *Thermodynamic and Transport Properties of Gases, Liquids and Solids*, Am. Soc. Mech. Engrs.

Hilsenrath, J., M. Klein, and H. W. Woolley, 1959, Tables of Thermodynamic Properties of Air Including Dissociation and Ionization from 1,500°K to 15,000°K. *Arnold Engr. Devel. Center*, AEDC-TR-59-20.

Hochstim, A. R., 1960, Approximations to High Temperature Thermodynamics of Air in Closed Form; in *Kinetics, Equilibria and Performance of High Temperature Systems*, Butterworths.

Hochstim, A. R., 1961, Electron Concentration in Closed Form for High Temperature Air and Air with Additives; in *Planetary and Space Science, Vol.* 6, *Electromagnetic Effects of Re-entry*, Pergamon.

Hochstim, A. R., 1963, Theoretical Calculations of Thermodynamic Properties of Air; in *Combustion and Propulsion, Fifth AGARD Colloquium*, ed. by R. R. Hagerty et. al., Macmillan.

Humphrey, R. L., and C. A. Neel, 1961, Tables of the Thermodynamic Properties of Air from 90 to 1500°K. *Arnold Engr. Devel. Center*, AEDC-TN-61-103.

Laporte, O., 1958, High-Temperature Shock Waves; in *Combustion and Propulsion, Third AGARD Colloquium*, Pergamon.

Lewis, C. H., and E. G. Burgess, III, 1963, Empirical Equations for the Thermodynamic Properties of Air and Nitrogen to 15,000°K. *Arnold Engr. Devel. Center*, AEDC-TDR-63-138.

Lewis, C. H., and C. A. Neel, 1964, Specific Heat and Speed of Sound Data for Imperfect Air. *Arnold Engr. Devel. Center*, AEDC-TDR-64-36.

Liepmann, H. W., and A. Roshko, 1957, *Elements of Gasdynamics*, Wiley.

Lighthill, M. J., 1957, Dynamics of a Dissociating Gas. Part I. Equilibrium Flow. *Jour. Fluid Mech.*, vol. 2, pt. 1, p. 1.

Little, W. J., 1963, Mollier Diagram for Air. *Arnold Engr. Devel. Center*, AEDC-TDR-63-190. (Available also in a large single sheet from AEDC, Arnold Air Force Station, Tenn.)

McBride, B. J., S. Heimel, J. G. Ehlers, and S. Gordon, 1963, Thermodynamic Properties to 6,000°K for 210 Substances involving the First 18 Elements. *NASA*, SP-3001.

Moeckel, W. E., and K. C. Weston, 1958, Composition and Thermodynamic Properties of Air in Chemical Equilibrium. *NACA*, TN 4265.

Neel, C. A., and C. H. Lewis, 1964a, Interpolations of Imperfect Air Thermodynamic Data I. At Constant Entropy. *Arnold Engr. Devel. Center*, AEDC-TDR-64-183.

Neel, C. A., and C. H. Lewis, 1964b, Interpolations of Imperfect Air Thermodynamic Data II. At Constant Pressure. *Arnold Engr. Devel. Center*, AEDC-TDR-64-184.

Rossini, F. D., 1955, Fundamentals of Thermodynamics; in *Thermodynamics and Physics of Matter*, ed. by F. D. Rossini, Princeton University Press.

Rossini, F. D., K. S. Pitzer, R. L. Arnett, R. M. Braun, and G. C. Pimentel, 1953, *Selected Values of Physical and Thermodynamic Properties of Hydrocarbons and Related Compounds*, Carnegie Press.

Rossini, F. D., D. D. Wagman, W. H. Evans, S. Levine and I. Jaffe, 1952, Selected Values of Chemical Thermodynamic Properties. *Natl. Bur. Stand. Circ.* 500, U.S. Govt. Print. Off.

Saha, M. N., 1920, Ionization in the Solar Chromosphere. *Phil. Mag.*, vol. 40, no. 238, p. 472.

Stull, D. R. et al., 1960, *JANAF Thermochemical Tables*, Dow Chemical Co., Midland, Mich.

Wray, K. L. 1962, Chemical Kinetics of High Temperature Air; in *Hypersonic Flow Research*, ed. by F. R. Riddell, Academic.

Chapter VI

Equilibrium Flow

1 INTRODUCTION

With the thermodynamic properties of the gas (e.g., air) known, we are in a position to make calculations for various types of flow. To this end we assume that an element of fluid moving through a nonuniform flow field obeys the same thermodynamic relations that we have obtained for a stationary system in complete thermodynamic equilibrium. This allows us to incorporate the equilibrium thermodynamic relations directly into the structure of gas dynamics, as is required to relate the various state variables that appear in the conservation equations. In writing these equations we assume that the effects of molecular transport—that is, viscosity, thermal conductivity, and diffusion—are negligible. Since these are manifestations of nonequilibrium in the translational and rotational motion of the molecules (see Chapters IX and X), this amounts to an assumption of local equilibrium in these motions. The resulting flow, which is thus one of complete local thermodynamic equilibrium, is referred to for brevity as *equilibrium flow*. As we shall see in Chapters VIII and X, the assumption of complete local equilibrium implies that all molecular processes take place within the gas infinitely rapidly, that is, that the gas can adjust instantaneously to changes in its environment.

General relations and conclusions for the equilibrium flow of imperfect as well as perfect gases are given in most textbooks on gas dynamics (e.g., Liepmann and Roshko, 1957, Chapters 2 and 7). For example, for the adiabatic case it is shown that (1) the entropy of a fluid element is constant except during passage through a shock wave and (2) the total enthalpy in steady flow without body forces is constant along a streamline. When specialized to a perfect gas, these conclusions lead to simple analytical results for a number of problems. Because of the more complicated thermodynamic relations that obtain for imperfect gases, however, no simple specific results are possible in the general case. In the range of temperatures below dissociation (less than about 3000°K for air), the analytical results can be modified to account for thermal imperfections due to interparticle forces and for caloric imperfections due to molecular vibration. (For useful results see Eggers, 1950, and Ames Research Staff, 1953.) The resulting equations, however, are far from simple. Once dissociation has set in, even this is no longer possible, and graphical or numerical procedures must be employed.

Since the general conclusions are well known and specific results at high temperatures require specific calculations, we shall not attempt an extended treatment of equilibrium flow. Instead we shall treat only a few specific problems with an eye primarily toward illustrating the complications and differences brought about by the introduction of the imperfect gas. We assume that the reader is acquainted with the qualitative conclusions for imperfect gases as cited earlier and with the specific results for perfect gases as given in any standard textbook on gas dynamics.

2 STEADY SHOCK WAVES

We consider first a standing normal shock wave as shown in Fig. 1. If u denotes the fluid speed, the pertinent conservation equations of fluid flow are then as follows (Liepmann and Roshko, 1957, p. 56):

$$\text{Mass:} \qquad \rho_b u_b = \rho_a u_a, \tag{2.1}$$

$$\text{Momentum:} \quad p_b + \rho_b u_b^2 = p_a + \rho_a u_a^2, \tag{2.2}$$

$$\text{Energy:} \qquad h_b + \tfrac{1}{2}u_b^2 = h_a + \tfrac{1}{2}u_a^2, \tag{2.3}$$

where a and b denote conditions upstream and downstream of the shock. If the conditions a are known, these constitute three equations for the four unknowns u_b, ρ_b, p_b, h_b. The system can be completed by assuming the gas to be in thermodynamic equilibrium on each side of the wave, which supplies the state equation

$$\rho = \rho(p, h). \tag{2.4}$$

To the foregoing we add the requirement, given by the second law of thermodynamics, that $s_b > s_a$. This is necessary to assure the uniqueness of the solution.

In the case of a perfect gas, equation (2.4) has the analytical form $h = [\gamma/(\gamma - 1)]p/\rho$, where γ is the ratio of specific heats. The system can then be solved explicitly, as is done in any standard text on gas dynamics.

In the case of a general gas, where the functional relation (2.4) may be given by a table or chart, we can obtain a numerical solution as follows: We first use equation (2.1) to rewrite equations (2.2) and (2.3) in the form

$$p_b = p_a + \rho_a u_a^2\left(1 - \frac{\rho_a}{\rho_b}\right), \tag{2.5}$$

$$h_b = h_a + \frac{u_a^2}{2}\left[1 - \left(\frac{\rho_a}{\rho_b}\right)^2\right]. \tag{2.6}$$

These supply a convenient system for the iterative calculation of normal shock waves. We first guess a value of ρ_a/ρ_b—say zero—and calculate trial values of p_b and h_b from (2.5) and (2.6). We then find the corresponding value of ρ_b from the state relation (2.4) in tabular or graphical form, as given for example in the references cited in Chapter V, Sec. 6. With this improved approximation to ρ_b known, we can then calculate a revised value of ρ_a/ρ_b. The cycle is then repeated successively until the initial and final values agree. Once the final value of ρ_b is known, u_b can be calculated from (2.1). The convergence of the process is usually fairly rapid (about three cycles for ordinary engineering accuracy), especially at hypersonic speeds where $\rho_a/\rho_b \ll 1$.[1]

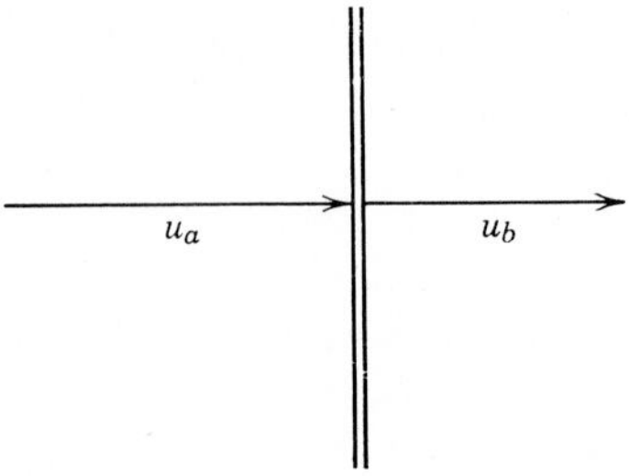

Fig. 1. Standing normal shock wave.

The results of the calculation just described can be represented, for a given gas, by the functional relation

$$\frac{p_b}{p_a} = \text{funct}\,(p_a, \rho_a, u_a). \tag{2.7}$$

For a perfect gas the independent variables can be combined into a single parameter such that

$$\frac{p_b}{p_a} = \text{funct}\left[\frac{u_a}{\sqrt{\gamma(p_a/\rho_a)}}\right] = \text{funct}\,(M_a),$$

where M_a is the Mach number upstream of the shock. In the general case no such simplification is possible, and any collection of numerical results must be given in terms of the three variables p_a, ρ_a, u_a. This leads to extensive sets of tables or charts. To obtain some reduction, results are often given in terms of flight at a given speed u_a at a given altitude H.

[1] For other methods of solution of the normal-shock relations, see Treanor (1960) and Pridmore Brown (1962).

Since $p_a = p_a(H)$ and $\rho_a = \rho_a(H)$, the relation (2.7) can then be written

$$\frac{p_b}{p_a} = \text{funct}\,(H, u_a),$$

and representation on a single graph becomes possible. Typical results of this kind are given by Feldman (1957) and Huber (1958). Tables providing more general results as functions of three variables (ρ_a, T_a, and u_a) for normal undissociated air ahead of the wave are given by Crutchfield (1957).

Because of the complexities in the imperfect-gas relations, it is difficult to make general quantitative statements about shock waves in air at very high speeds. We can say something, however, about how p_b, h_b, ρ_b, etc. vary relative to what they would be if the gas downstream of the wave behaved as a perfect gas (i.e., no dissociation or vibration). This can be examined conveniently in terms of the density ratio ρ_a/ρ_b, which plays a key role in problems in hypersonic aerodynamics. As can be seen from equations (2.5) and (2.6), the thermodynamic characteristics of the gas downstream of the wave influence p_b and h_b only through their effect on this ratio. To study this effect we consider a very strong shock wave (u_a very large) in an oncoming stream at atmospheric conditions. In this case p_a is small compared with the second term in (2.5); and, if the zero reference state for energy and enthalpy is taken as the diatomic state at 0°K, h_a is small compared with the second term in (2.6). We can therefore write to a good approximation in the strong-shock case

$$p_b = \rho_a u_a^2\left(1 - \frac{\rho_a}{\rho_b}\right), \tag{2.8}$$

$$h_b = \frac{u_a^2}{2}\left[1 - \left(\frac{\rho_a}{\rho_b}\right)^2\right]. \tag{2.9}$$

Elimination of u_a^2 between these two equations then leads to the following relation for the density ratio in terms of quantities on the downstream side of the shock:

$$\frac{\rho_a}{\rho_b} = \frac{1}{2h_b\dfrac{\rho_b}{p_b} - 1}. \tag{2.10}$$

For the special case of a perfect gas, where $h = [\gamma/(\gamma - 1)]p/\rho$, this gives

$$\frac{\rho_a}{\rho_b} = \frac{\gamma - 1}{\gamma + 1}, \tag{2.11}$$

which is the well-known result for the limiting density ratio across a normal shock in a perfect gas as $M_a \to \infty$ (cf. Liepmann and Roshko, 1957, p. 59).

For an example of imperfect-gas effects, we can consider the ideal dissociating gas. Here the thermal and caloric equations of state, as obtained from equations (V 3.3) and (V 3.6), are

$$\frac{p}{\rho} = (1 + \alpha^*)R_{A_2}T, \qquad h = R_{A_2}[(4 + \alpha^*)T + \alpha^*\Theta_d]. \tag{2.12}$$

Substitution into (2.10) gives, after some rearrangement,

$$\frac{\rho_a}{\rho_b} = \frac{1}{7} \times \frac{1 + \alpha_b^*}{1 + \frac{1}{7}[1 + 2(\Theta_d/T_b)]\alpha_b^*}. \tag{2.13}$$

If there were no dissociation through the shock wave ($\alpha_b^* = 0$), the result would be independent of T_b, and we would obtain $\rho_a/\rho_b = \frac{1}{7}$. This agrees with what can be found from the perfect-gas equation (2.11) with $\gamma = \frac{4}{3}$, which is the value appropriate to the diatomic component of the ideal dissociating gas (see Chapter V, Sec. 3). Whether dissociation will increase or decrease the density ratio depends on the relative values of the coefficients of α_b^* in the numerator and denominator of (2.13). For strong shock waves the temperature ratio T_b/T_a is typically of the order of 25, so that for atmospheric conditions (T_a approximately 300°K), we have $T_b \cong 7500$°K. We thus find, considering air as pure nitrogen for a rough approximation,

$$\frac{1}{7}\left(1 + 2\frac{\Theta_d}{T_b}\right) = \frac{1}{7}\left(1 + 2\,\frac{113{,}000}{7500}\right) \cong 4.5.$$

It follows from equation (2.13) that the effect of dissociation ($\alpha_b^* > 0$) is to decrease ρ_a/ρ_b relative to what it would be in the absence of dissociation. This is typical of a general result, to wit: at hypersonic flight speeds in atmospheric air the imperfect-gas effects tend to decrease the density ratio relative to what it would be for a perfect gas. As a result ρ_a/ρ_b decreases from the value of $\frac{1}{6}$ for air considered as a perfect gas ($\gamma = \frac{7}{5}$) to values of $\frac{1}{15}$ or less (see, e.g., Feldman, 1957).

In view of the small values of ρ_a/ρ_b, we can go a step further and neglect this ratio in (2.8) and (2.9). This gives

$$p_b = \rho_a u_a^2 \qquad \text{and} \qquad h_b = u_a^2/2.$$

To this rough approximation the values of p_b and h_b are completely independent of the behavior of the gas behind the shock. To see the effect of dissociation on the temperature, we can write from the second of equations (2.12)

$$T_b = \frac{h_b/R_{A_2} - \alpha_b^*\Theta_d}{4 + \alpha_b^*}.$$

Since h_b is essentially the same with or without dissociation, the effect of dissociation must be to decrease the temperature T_b. This effect is as we might expect; the dissociation soaks up energy that would otherwise go into translation, and the temperature behind the shock drops accordingly. The same can also be said with regard to molecular vibration.

The methods and results for normal shock waves can be extended readily to oblique waves by superposing a uniform velocity parallel to the wave in the usual fashion (Liepmann and Roshko, 1957, p. 85). Results for oblique waves are given, for example, by Feldman (1957), Moeckel (1957), and Trimpi and Jones (1960).

Exercise 2.1. Consider the strong normal shock wave directly ahead of a blunt body flying at very high speed and high altitude. Conditions are such that, immediately behind the shock wave, molecular vibration of the molecules of air is in equilibrium in a partially excited state, but no dissociation occurs. Show how the fluid density immediately behind the shock wave will differ from what it would be if no molecular vibration occurred.

3 STEADY NOZZLE FLOW

In this section we consider the steady quasi-one-dimensional adiabatic flow out of a reservoir and through a converging-diverging nozzle of given area distribution $A(x)$ (Fig. 2). We suppose that conditions in the reservoir are such that supercritical (i.e., subsonic-supersonic) flow exists in the nozzle and that the assumption of a perfect gas is not admissible. Assuming thermodynamic equilibrium, we wish to calculate the conditions along the nozzle, as specified by the speed $u(x)$ and any two thermodynamic state variables, say $h(x)$ and $\rho(x)$. We assume that the flow is free of shock waves, as will be the case if the pressure at the nozzle exit is sufficiently low.

The conservation equations for steady quasi-one-dimensional adiabatic flow of a fluid in thermodynamic equilibrium and without shock waves can be taken as follows (Liepmann and Roshko, 1957, pp. 40–49):

$$\text{Mass:} \qquad \rho u A = \text{const}, \tag{3.1}$$

$$\text{Energy:} \quad h + \tfrac{1}{2}u^2 = \text{const}, \tag{3.2}$$

$$\text{Entropy:} \qquad s = \text{const}. \tag{3.3}$$

The third of these equations, which is used here in place of the momentum equation, reflects the fact that the entropy of a fluid element is everywhere the same in the absence of shock waves. It can be obtained as a formal consequence of the momentum and energy equations in differential form; it is also an immediate logical consequence of our basic assumptions that the flow is adiabatic and in equilibrium (i.e., reversible). A determinate set of four equations in four unknowns can be obtained by adding the state relation

$$\rho = \rho(h, s), \tag{3.4}$$

which is most conveniently employed here in the form of a Mollier diagram.

Our first task is to find the constants in equations (3.1), (3.2), and (3.3). These are determined by the conditions of the problem, which are as follows:

1. A specified thermodynamic state exists in the reservoir, that is,

$$h = h_0 \qquad \text{and} \qquad s = s_0 \qquad \text{at} \qquad u = 0. \tag{3.5}$$

These conditions are usually given in terms of other state variables, such

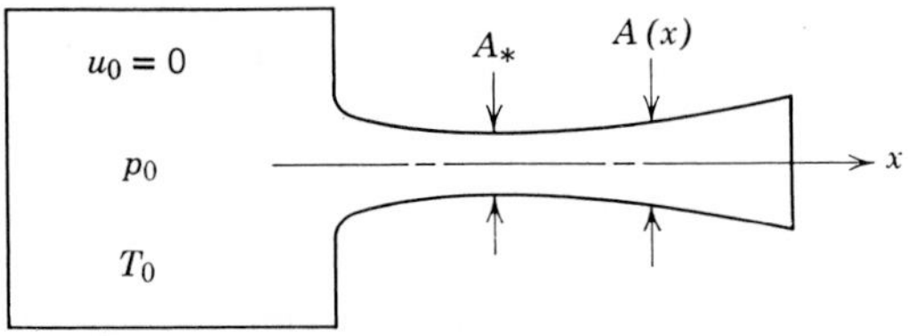

Fig. 2. Reservoir and nozzle.

as p_0 and T_0; we must then find h_0 and s_0 from, say, the Mollier diagram according to the state relations $h_0 = h(p_0, T_0)$, $s_0 = s(p_0, T_0)$.

2. The speed of flow is equal to the speed of sound at the throat, that is, $u_* = a_*$ at $A = A_*$. This is in accord with our basic supposition of supercritical flow and follows from the usual differential area-velocity relation, which is valid for any gas in thermodynamic equilibrium (see Liepmann and Roshko, 1957, p. 52). Since the speed of sound a is itself a thermodynamic state variable, that is, $a \equiv \sqrt{(\partial p/\partial \rho)_s} = a(h, s)$, we can write this condition as

$$u_* = a(h_*, s_*) \qquad \text{at} \qquad A = A_*. \tag{3.6}$$

This specifies a relation at the throat between the kinematic state of the fluid (as given by u) and the thermodynamic state (as given by h and s).

The constants in the energy and entropy equations follow immediately from the conditions (3.5). We thus obtain at once for equations (3.2) and (3.3)

$$h + \tfrac{1}{2}u^2 = h_0 \qquad \text{and} \qquad s = s_0.$$

The determination of the constant in equation (3.1) is a bit more complicated, and proceeds as follows: With the aid of the two foregoing equations, the condition (3.6) can be written

$$\sqrt{2(h_0 - h_*)} = a(h_*, s_0), \tag{3.7}$$

which is a single relation for h_*. The condition at the throat thus resolves itself into a determination of h_* such as to satisfy this relation. If a Mollier diagram is available with lines of constant a (as in the chart of Feldman, 1957), this can be done by trial and error, with a being obtained as a

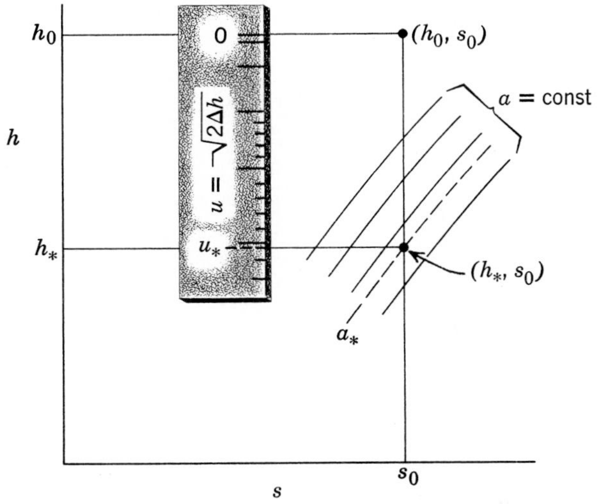

Fig. 3. Determination of condition $u_* = a_*$ from Mollier diagram.

function of h and s from the diagram. It can also be done by first preparing an auxiliary moveable scale graduated in values of u, with the graduations determined from the relation $u = \sqrt{2\,\Delta h}$ (see Fig. 3). The graduations of this scale must, of course, be compatible with the scale to which h is plotted in the Mollier diagram. We then go to the line $s = s_0$ in the Mollier diagram and place the scale vertically with origin at h_0. The desired value of h_* is then found at the point where u on the vertical scale is equal to a as read from the Mollier diagram along the line $s = s_0$.[2] Equation (3.7) shows that h_* is, in effect, fixed once h_0 and s_0 are given, that is, $h_* = h_*(h_0, s_0)$.

[2] If a Mollier diagram is not available with lines of constant a, h_* can be found from the condition that ρu is a maximum at a sonic throat. This is done by calculating $\rho u = \rho\sqrt{2(h_0 - h)}$ from the diagram as a function of h along the line $s = s_0$ and finding the maximum by plotting the results.

From the foregoing determination, we also know $a_* = a(h_*, s_*) = a_*(h_0, s_0)$ and can read correspondingly $\rho_* = \rho(h_*, s_*) = \rho_*(h_0, s_0)$. As with a perfect gas, the throat conditions are thus unique functions of the conditions in the reservoir, although the functional relationships are now more complicated. The constant in the continuity equation is then found by applying that equation at the throat to obtain

$$\rho u A = \rho_*(h_0, s_0) a_*(h_0, s_0) A_* = f(h_0, s_0) A_*.$$

The value of the constant thus depends on h_0, s_0, and A_*.

With the constants determined, the system of equations can now be written

$$\rho u A = f(h_0, s_0) A_*, \tag{3.8}$$

$$h + \tfrac{1}{2}u^2 = h_0, \tag{3.9}$$

$$s = s_0, \tag{3.10}$$

$$\rho = \rho(h, s) = \rho(h, s_0), \tag{3.11}$$

where fA_*, h_0, and s_0 are now known numbers. With these equations we can find the flow quantities along the nozzle. This must be done by trial-and-error or iteration if we want the flow quantities for some given $A(x)$. If we follow the indirect procedure of finding x for a given fluid state, however, the solution can proceed along the line $s = s_0$ in the Mollier diagram as follows:

1. Choose a value of h.
2. Calculate u from equation (3.9): $u = \sqrt{2(h_0 - h)}$.
3. Read ρ from the Mollier diagram according to equation (3.11).
4. Calculate A from equation (3.8): $A = (f/\rho u)A_*$.
5. Find x from the inverse relation $x = x(A)$.

In place of the Mollier diagram, we can also use the constant-entropy properties tabulated for air by Hansen and Hodge (1961) and by Neel and Lewis (1964).

As should be apparent from the foregoing, the results here follow a functional relationship of the form

$$\frac{A}{A_*} = \text{funct}\,(h_0, s_0; u).$$

Since $a_* = a_*(h_0, s_0)$, this can also be written

$$\frac{A}{A_*} = \text{funct}\left(h_0, s_0; \frac{u}{a_*}\right). \tag{3.12}$$

In the case of a perfect gas, this relation reduces to the well-known form $A/A_* = \text{funct}\,(u/a_*)$, and the reservoir conditions appear only implicitly through their influence on a_*. For the imperfect gas, however, no such simplification is possible, and the explicit dependence on the reservoir conditions cannot be eliminated. Useful results for the flow of air in hypersonic nozzles have been given by a number of authors, for example, Erickson and Creekmore (1960), Yoshikawa and Katzen (1961), and Jorgensen and Baum (1962).

Exercise 3.1. Consider the steady flow of chemically reacting air through the converging-diverging nozzle of a hypersonic wind tunnel. Assume that the flow is in complete thermodynamic equilibrium. From measurements of temperature and pressure in the reservoir of the tunnel, the specific enthalpy and entropy at that location are found to be 5.35×10^{10} cm²/sec² and 8.07×10^7 cm²/sec² °K, respectively. The cross-sectional area of the nozzle at the throat is 3.710×10^{-2} cm², and the density and speed of flow at the throat are known to be 5.49×10^{-2} gm/cm³ and 1.13×10^5 cm/sec. The cross-sectional area at the test section is 576 cm². With the aid of the following table, which gives the density of air as a function of enthalpy for the foregoing value of the entropy, calculate the dynamic pressure of the flow in the test section.

Enthalpy, cm²/sec²	Density, gm/cm³
0.25×10^{10}	12.2×10^{-6}
0.20×10^{10}	6.97×10^{-6}
0.15×10^{10}	3.42×10^{-6}
0.10×10^{10}	1.23×10^{-6}
0.05×10^{10}	0.22×10^{-6}

4 PRANDTL-MEYER FLOW

As an example of a typical Prandtl-Meyer flow, we consider the semi-infinite flow field above a convex wall as shown in Fig. 4. Here an originally uniform, parallel supersonic flow is expanded to higher speeds by an increase in the local flow angle θ (measured clockwise from some convenient reference direction). As in supersonic flow generally, a key role is played here by the Mach lines, which are lines inclined relative to the local flow direction at the Mach angle

$$\mu = \tan^{-1} \frac{1}{\sqrt{u^2/a^2 - 1}}. \tag{4.1}$$

Like all Prandtl-Meyer flows, the flow of Fig. 4 has the property that the Mach lines emanating from the wall are straight and the flow variables—u,

θ, h, a, μ, etc.—are constant along these lines (see Liepmann and Roshko, 1957, pp. 93–98). The problem of finding the flow over the wall thus resolves itself into finding the speed u and the thermodynamic state h, a, etc., as functions of the initial conditions $(\ )_a$ and the change in flow direction imposed at the wall. The entire flow field can then be constructed by drawing the Mach lines at the appropriate angle from the wall and thus finding the location of the deduced values of u, h, a, etc. in the field.

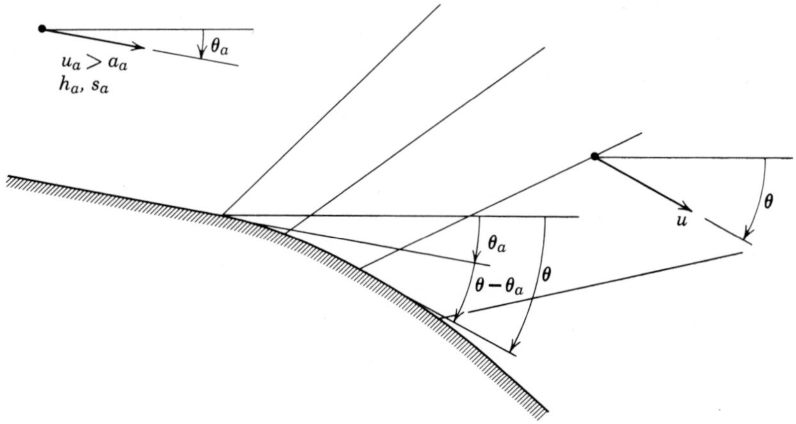

Fig. 4. Prandtl-Meyer flow.

The governing differential equation for Prandtl-Meyer flow can be found by considering the differential velocity change across a Mach line, regarded as a weak, discrete expansion wave. Since the wave can exert no force parallel to itself, the vectorial change δu_i in the velocity vector u_i in passing from one side of the wave to the other must be normal to the wave direction.[3] The situation must therefore be as shown in Fig. 5. The dynamics of the problem have been put in by drawing this figure and writing equation (4.1); from here on the derivation is a matter of geometry. From the angular relationships indicated in the figure, we write, using simple trigonometry,

$$\frac{u + du}{u} = \frac{\sin\left(\frac{\pi}{2} + \mu\right)}{\sin\left(\frac{\pi}{2} - \mu - d\theta\right)}$$

$$= \frac{\cos \mu}{\cos(\mu + d\theta)} = \frac{\cos \mu}{\cos \mu \cos d\theta - \sin \mu \sin d\theta}.$$

[3] Here, as throughout the book, we employ the Cartesian vector notation introduced in Chapter II, Sec. 2.

For vanishingly small $d\theta$ this becomes

$$1 + \frac{du}{u} = \frac{\cos\mu}{\cos\mu - \sin\mu \times d\theta} = \frac{1}{1 - \tan\mu \times d\theta} = 1 + \tan\mu \times d\theta,$$

and we obtain finally with equation (4.1)

$$d\theta = \sqrt{u^2/a^2 - 1}\,\frac{du}{u}. \tag{4.2}$$

This is the well-known differential equation for Prandtl-Meyer flow. In the present derivation no assumption has been made as to a perfect gas.

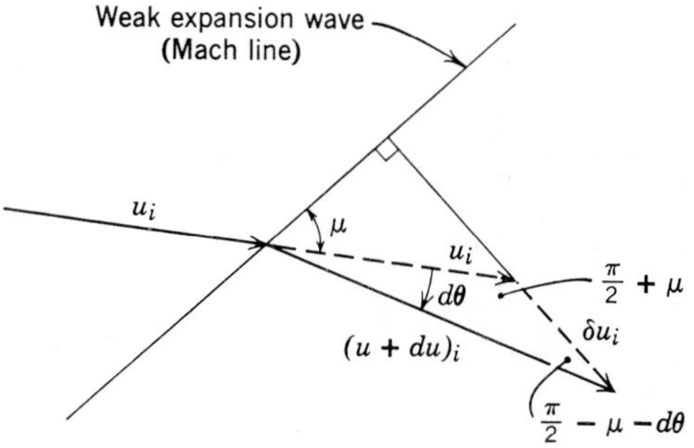

Fig. 5. Velocity change across a weak expansion wave.

To obtain a solution, the foregoing equation must be supplemented by the adiabatic energy equation

$$h + \tfrac{1}{2}u^2 = h_a + \tfrac{1}{2}u_a^2 \equiv h_{t_a}, \tag{4.3}$$

where h_t is the total enthalpy, plus the thermodynamic state relation

$$a = a(h, s) = a(h, s_a). \tag{4.4}$$

In equation (4.4) recognition has again been taken of the fact that the entropy of a fluid element is constant, so that $s = s_a$. With conditions in the initial stream given, equations (4.2), (4.3), and (4.4) constitute three equations for the four quantities θ, u, a, and h. We can thus find any three of these—say, u, a, and h—as a function of the fourth—say θ.

To carry out the solution for a general gas, it is convenient to define a new quantity

$$F \equiv h_{t_a} - h = \frac{u^2}{2}. \tag{4.5}$$

With this, equation (4.2) becomes upon integration

$$\Delta\theta \equiv \theta - \theta_a = \frac{1}{2}\int_{F_a}^{F} \frac{1}{F}\sqrt{\frac{2F}{a^2} - 1}\, dF. \tag{4.6}$$

The integral in this equation can be evaluated numerically or graphically with the aid of a Mollier diagram. We merely proceed downward along the line $s = s_a$ with h as the independent variable, calculating F from (4.5) and reading $a = a(h, s_a)$ from the diagram. The integration must, of course, be performed anew for each new set of initial conditions. It follows from the equations that the results have the functional form

$$u = \sqrt{2F} = u(u_a, h_a, s_a; \Delta\theta),$$
$$h = h(u_a, h_a, s_a; \Delta\theta),$$
$$a = a(u_a, h_a, s_a; \Delta\theta).$$

The corresponding p, T, and ρ can be found from the Mollier diagram according to the state relations $p = p(h, a)$, etc., and the local Mach number is given by

$$M \equiv \frac{u}{a} = M(u_a, h_a, s_a; \Delta\theta). \tag{4.7}$$

The Mach angle can be found from equation (4.1), and the entire flow field can be constructed by laying off the straight Mach lines from the wall at the location of the corresponding value of $\Delta\theta$. In the special case of a perfect gas the initial conditions in equation (4.7) combine, in effect, into a single variable, the initial Mach number M_a. We then have the classical Prandtl-Meyer results according to the relationship $M = M(M_a; \Delta\theta)$, which can be obtained in implicit analytical form (see Liepmann and Roshko, 1957, pp. 98–100). In the general case no such simplification is possible.[4]

Typical results for a Prandtl-Meyer expansion as calculated by Heims (1958) are shown in Fig. 6. The gas here is the argon-free model of air for which thermodynamic calculations were made by Hilsenrath and Beckett (1956). The initial conditions for this particular case are $M_a = 1.00$, $T_a = 6140°\text{K}$, and $p_a = 1.2$ atm. Results are shown by solid lines for equilibrium flow on the basis of the methods just outlined. Results are also shown by dashed lines under the assumption that the chemical composition and vibrational energy are arbitrarily fixed at the initial state and do not participate in the energy exchange during the flow process. This so-called *frozen flow* is thus a perfect-gas flow with an appropriate fixed value of γ (obtained as explained in the next section). The results are calculated accordingly on the basis of the usual perfect-gas formulas.

As can be seen from Fig. 6, the pressure is here more sensitive to the

[4] For alternative formulations of the Prandtl-Meyer flow, which are instructive in showing the relationship to the solution for a perfect gas, see Hayes and Probstein (1959), pp. 259–262.

thermodynamic behavior of the gas than is the density. This is in contrast to the situation in the compression through a shock wave, where, as we saw, the pressure behind the wave is insensitive to the assumption regarding the gas model whereas the density is not. As we would expect, the temperature in the frozen expansion is much lower than in the equilibrium expansion. This is because in the frozen-flow case the energy originally present in dissociation and vibration is locked in and is thus unavailable

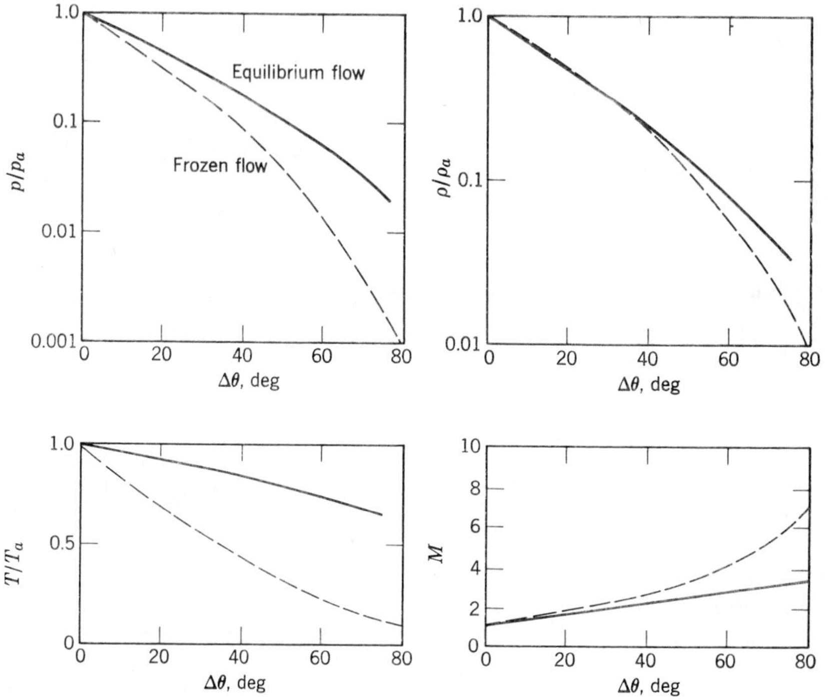

Fig. 6. Results for Prandtl-Meyer expansion; $M_a = 1.00$, $T_a = 6140°K$, $p_a = 1.2$ atm (after Heims, 1958).

for transformation to random translational energy as the gas expands. The generally lower temperatures in the frozen-flow case are reflected also in generally higher Mach numbers, although the speed of flow has a secondary influence here.

5 FROZEN FLOW

It is sometimes useful, as in the previous section, to make flow calculations for a partially dissociated gas under the assumption that the gas composition is arbitrarily fixed at some initial value. This would be the situation if the chemical reactions could in some way suddenly be inhibited, so that, although molecular collisions still take place, there is no accompanying change in molecular structure. We assume that molecular translation of all species and molecular rotation of polyatomic species are at all times in equilibrium. The vibrational energy of the polyatomic species can be taken to be either (a) fixed at some initial value or (b) in equilibrium at its fully excited value. With these assumptions our mixture of perfect gases of now fixed composition will behave as a thermally and calorically perfect gas. Flow calculations are then readily carried out on the basis of the well-known equations for a perfect gas provided R, c_v, c_p, and γ are given the constant values appropriate to the assumed situation. As we shall see in the discussion of nonequilibrium nozzle flow in Chapter VIII, Sec. 13, circumstances do occur in which such frozen flow does effectively exist. Frozen flow is also useful as giving one bound (equilibrium flow being the other) on the flow variables in a general nonequilibrium situation.

The values of the constants for the assumed frozen flow can be found as follows. For a reacting mixture of perfect gases the thermal equation of state is given by equation (V 5.9) as

$$\frac{p}{\rho} = \mathfrak{n}\hat{R}T, \tag{5.1}$$

where $\mathfrak{n} = \sum_s \mathfrak{n}_s$ is the total number of moles per unit mass of gas and $\hat{R}$ is the universal gas constant (gas constant per mole). The equation of state for an imperfect gas can also be written formally [cf. equation (V 2.7)] as

$$\frac{p}{\rho} = ZR_0T, \tag{5.2}$$

where Z is the compressibility factor and R_0 is the ordinary gas constant (gas constant per unit mass) at a low-temperature reference state at which the gas is completely undissociated. Comparison of (5.1) and (5.2) shows that for a mixture of perfect gases $ZR_0 = \mathfrak{n}\hat{R}$. Since $R_0 = \mathfrak{n}_0\hat{R}$, where $\mathfrak{n}_0$ is the number of moles per unit mass at the reference state, this gives

$$Z = \frac{\mathfrak{n}}{\mathfrak{n}_0} = \frac{\mathcal{N}}{\mathcal{N}_0}, \tag{5.3}$$

where $\mathcal{N}$ denotes the total number of moles in the mixture. As already mentioned in Chapter V, Sec. 6, the compressibility factor for a mixture of perfect gases is thus equal to the number of moles in the mixture at the given conditions per mole at the low-temperature reference state. If the composition of the mixture is assumed frozen, the value of $\mathfrak{n}$ and hence of Z will be constant. In this case equations (5.1) and (5.2) are both of the usual perfect-gas form

$$\frac{p}{\rho} = R_f T,$$

where R_f is the ordinary gas constant for the frozen gas. The appropriate value of R for the frozen gas is thus given by

$$R_f = \mathfrak{n}\hat{R} = ZR_0. \tag{5.4}$$

Values of Z can be found in the various tables and charts for high-temperature gases, so that the value of R_f is readily obtained for any assumed condition of freezing.

To find the specific heats for the frozen gas, we assume that the mixture consists only of monatomic and diatomic gases, as is very nearly true for air at temperatures up to about 8000°K. If subscripts m and d denote the monatomic and diatomic species, respectively, we can then write

$$c_{v_f} = \mathfrak{n}_m \hat{c}_{v_m} + \mathfrak{n}_d \hat{c}_{v_d},$$

where c_{v_f} is the specific heat per unit mass of the mixture, $\mathfrak{n}_m$ and $\mathfrak{n}_d$ are the total number of moles per unit mass of mixture for the two classes of species, and $\hat{c}_v$ is the appropriate specific heat per mole of each class. We make no restriction as to how many different species there are in each class, so that $\mathfrak{n}_m$ and $\mathfrak{n}_d$ may each represent a sum of the values of $\mathfrak{n}_s$ for a number of monatomic and diatomic gases. From our work of Chapter IV we have $\hat{c}_{v_m} = \frac{3}{2}\hat{R}$ and $\hat{c}_{v_d} = A\hat{R}$, where $A = \frac{5}{2}$ if the vibrational energy is assumed to be frozen and $A = \frac{7}{2}$ if it is assumed to be in equilibrium at its fully excited value. The foregoing equation can thus be written

$$c_{v_f} = \hat{R}(\tfrac{3}{2}\mathfrak{n}_m + A\mathfrak{n}_d). \tag{5.5}$$

This equation can also be put in terms of Z as follows. We assume that the low-temperature state consists entirely of diatomic species, that is, $\mathfrak{n}_0 = \mathfrak{n}_{d_0}$. The number of moles of diatomic species that have dissociated at some high-temperature state is then $\mathfrak{n}_0 - \mathfrak{n}_d$, and since each such mole that dissociates contributes two moles of monatomic species we have $\mathfrak{n}_m = 2(\mathfrak{n}_0 - \mathfrak{n}_d)$. But we can also obviously write that $\mathfrak{n}_m = \mathfrak{n} - \mathfrak{n}_d$. Equating these two expressions for $\mathfrak{n}_m$ and solving for $\mathfrak{n}_d$, we find

$$\mathfrak{n}_d = 2\mathfrak{n}_0 - \mathfrak{n},$$

whence the relation $\mathfrak{n}_m = \mathfrak{n} - \mathfrak{n}_d$ gives

$$\mathfrak{n}_m = 2(\mathfrak{n} - \mathfrak{n}_0).$$

Replacing $\mathfrak{n}$ in these two expressions by $\mathfrak{n} = \mathfrak{n}_0 Z$ from equation (5.3) then leads to

$$\mathfrak{n}_d = \mathfrak{n}_0(2 - Z) \qquad \text{and} \qquad \mathfrak{n}_m = 2\mathfrak{n}_0(Z - 1),$$

and substitution of these expressions into (5.5) gives, since $\mathfrak{n}_0\hat{R} = R_0$,

$$c_{v_f} = R_0[(2A - 3) + (3 - A)Z].$$

Since for a perfect gas $c_p = c_v + R = c_v + ZR_0$, we also obtain

$$c_{p_f} = R_0[(2A - 3) + (4 - A)Z].$$

The value of γ_f is found from $\gamma_f = c_{p_f}/c_{v_f}$.

We can now write specific formulas for the two special cases as follows:

1. Chemical composition and vibrational energy both frozen.[5] In this case the value of A is $\frac{5}{2}$, and we obtain

$$c_{v_f} = R_0(2 + \tfrac{1}{2}Z), \qquad c_{p_f} = R_0(2 + \tfrac{3}{2}Z),$$

$$\gamma_f = \frac{4 + 3Z}{4 + Z}. \tag{5.6}$$

2. Chemical composition frozen, vibrational energy active at its fully excited equilibrium value. In this case A is $\frac{7}{2}$ and we find

$$c_{v_f} = R_0(4 - \tfrac{1}{2}Z), \qquad c_{p_f} = R_0(4 + \tfrac{1}{2}Z),$$

$$\gamma_f = \frac{8 + Z}{8 - Z}. \tag{5.7}$$

Note that for a purely monatomic mixture of gases we have, by equation (5.3), $Z = \mathfrak{n}_m/\mathfrak{n}_0 = 2$ and the foregoing equations for γ_f both give $\frac{5}{3}$. This is as it should be for purely monatomic gases. For a purely diatomic mixture we have $Z = 1$, and equation (5.6) gives $\gamma_f = \frac{7}{5}$ whereas equation (5.7) gives $\gamma_f = \frac{9}{7}$. These again are the correct values for a purely diatomic gas under the assumed conditions regarding the vibrational energy.

Exercise 5.1. In the equilibrium flow of an imperfect gas it is important to distinguish between the isentropic exponent and the ratio of specific heats, which are in general not one and the same thing. This can be seen from the following exercise.

(a) On the basis of the equations of equilibrium thermodynamics and the relations between partial derivatives, show that the speed of sound as defined by $a^2 \equiv (\partial p/\partial \rho)_s$ for a gas in equilibrium is given, with complete generality as

[5] This should not be taken to imply that the vibrational energy is zero. We merely assume that, whatever value it has, it plays no part in the energy interchange. The gas thus behaves *as if* the vibration of the diatomic species were unexcited.

regards the gas model, by

$$a^2 = \frac{c_p}{c_v}\left(\frac{\partial p}{\partial \rho}\right)_T, \tag{5.8}$$

where as usual $c_p = (\partial h/\partial T)_p$ and $c_v = (\partial e/\partial T)_\rho$.

(b) If the isentropic exponent γ_e is *defined* by $\gamma_e \equiv (\partial \ln p/\partial \ln \rho)_s$, show that a^2 is also given quite generally by

$$a^2 = \gamma_e \frac{p}{\rho}. \tag{5.9}$$

(c) Under what conditions with regard to the gas model will

$$\gamma_e = c_p/c_v = \gamma \text{ (common value)?}$$

For this situation to be realized is it necessary that the common value γ be a constant?

(The quantity γ_e is important in the equilibrium flow of an imperfect gas since it relates the speed of sound directly to the density and pressure. The ratio of specific heats c_p/c_v is of little use outside the dynamics of perfect gases.)

(d) With the aid of a Mollier diagram containing lines of constant a (e.g., Feldman, 1957), or of some suitable tabular data, use equations (5.8) and (5.9) to make a numerical evaluation of c_p/c_v and γ_e for air at some high-temperature condition ($T > 8000°$K). How do these values compare with the values of γ_f obtained at the same condition from equations (5.6) and (5.7)? In physical terms, what are the reasons for the differences of these various quantities relative to the value of $\gamma = \frac{7}{5}$ for air at ordinary room temperature?

Exercise 5.2. For the reservoir conditions after discharge as found in Chapter V, Exercise 6.2, calculate and plot a curve of Mach number M as a function of area ratio A/A^* for the nozzle of the hotshot tunnel under the following assumptions regarding the working medium:

(a) Air in complete thermochemical equilibrium.

(b) Air with the chemical composition and vibrational energy frozen at the conditions existing in the reservoir.

Include the subsonic portion of the curve, and carry the calculations as far as you can with the thermodynamic charts available (preferably the Mollier diagram of Feldman, 1957).

Exercise 5.3. Suppose that a pitot tube is placed in the flow of Exercise 5.2 at the point in the nozzle at which the static temperature is 3200°K.

(a) What is the value of the pressure measured by the pitot tube? Assume that the flow from the normal shock ahead of the tube to the stagnation point at the entrance to the tube is adiabatic and that the air is at all times in thermochemical equilibrium.

Ans. (as obtained with Mollier diagram of Feldman): 3300 psia.

(b) What is the temperature at the stagnation point? Compare this temperature with the temperature in the reservoir. How does the situation here differ qualitatively from what one finds in the flow of a perfect gas? What is the reason for this difference?

References

Ames Research Staff, 1953, Equations, Tables, and Charts for Compressible Flow. *NACA*, Rept. 1135.

Crutchfield, J. E., 1957, Normal Shock Tables for Air. *Arnold Engr. Devel. Center*, AEDC-TM-57-30.

Eggers, A. J., Jr., 1950, One-dimensional Flows of an Imperfect Diatomic Gas. *NACA*, Rept. 959.

Erickson, W. D., and H. S. Creekmore, 1960, A Study of Equilibrium Real-Gas Effects in Hypersonic Air Nozzles, including Charts of Thermodynamic Properties for Equilibrium Air. *NASA*, TN D-231.

Feldman, S., 1957, Hypersonic Gas Dynamic Charts for Equilibrium Air. *AVCO Res. Lab.*, Everett, Mass.

Hansen, C. F., and M. E. Hodge, 1961, Constant Entropy Properties for an Approximate Model of Equilibrium Air. *NASA*, TN D-352.

Hayes, W. D., and R. F. Probstein, 1959, *Hypersonic Flow Theory*, Academic.

Heims, S. P., 1958, Prandtl-Meyer Expansion of Chemically Reacting Gases in Local Chemical and Thermodynamic Equilibrium. *NACA*, TN 4230.

Hilsenrath, J., and C. W. Beckett, 1956, Tables of Thermodynamic Properties of Argon-Free Air to 15,000°K. *Arnold Engr. Devel. Center*, AEDC-TN-56-12.

Huber, P. W., 1958, Tables and Graphs of Normal-Shock Parameters at Hypersonic Mach Numbers and Selected Altitudes. *NACA*, TN 4352.

Jorgensen, L. H., and G. M. Baum, 1962, Charts for Equilibrium Flow Properties of Air in Hypervelocity Nozzles. *NASA*, TN D-1333.

Liepmann, H. W., and A. Roshko, 1957, *Elements of Gasdynamics*, Wiley.

Moeckel, W. E., 1957, Oblique-Shock Relations at Hypersonic Speeds for Air in Chemical Equilibrium. *NACA*, TN 3895.

Neel, C. A., and C. H. Lewis, 1964, Interpolations of Imperfect Air Thermodynamic Data I. At Constant Entropy. *Arnold Engr. Devel. Center*; AEDC-TDR-64-183.

Pridmore Brown, B. N., 1962, Conditions behind a Shock in Real-Gas Flow. *J. Aerospace Sci.*, vol. 29, no. 4, p. 484.

Treanor, C. E., 1960, A Graphical Solution for Normal Shock Waves in Real Gases. *J. Aerospace Sci.*, vol. 27, no. 2, p. 158.

Trimpi, R. L., and R. A. Jones, 1960, A Method of Solution with Tabulated Results for the Attached Oblique Shock-Wave System for Surfaces at Various Angles of Attack, Sweep, and Dihedral in an Equilibrium Real Gas including the Atmosphere. *NASA*, TR R-63.

Yoshikawa, K. K., and E. D. Katzen, 1961, Charts for Air-Flow Properties in Equilibrium and Frozen Flows in Hypervelocity Nozzles. *NASA*, TN D-693.

Chapter VII

Vibrational and Chemical Rate Processes

1 INTRODUCTION

In the preceding chapter we assumed that a moving element of fluid is in complete thermodynamic equilibrium at every instant as it moves through a continuously changing flow field. Actually, this assumption can never be realized exactly, since any readjustment of thermodynamic equilibrium to a change in environment requires a length of time. This is so because every redistribution of internal energy among the molecular states requires a number of molecular collisions and hence a certain characteristic time, depending on the conditions of temperature and density. The assumption of equilibrium demands that this characteristic time for readjustment by collisions be negligible compared with the time required for a fluid element to encounter a significant change in conditions as it moves through the flow field. In many cases this requirement is satisfied, and the assumption of complete thermodynamic equilibrium is a valid working approximation.

If, as can happen, however, the characteristic time for readjustment by collisions is of the same order as the characteristic time of the fluid flow, then the equilibrium assumption is not satisfied, and nonequilibrium effects must be taken into account. Such effects can occur in connection with any of the molecular processes—translation, rotation, vibration, chemical composition, etc. The adjustment of translation and rotation requires relatively few collisions, and the corresponding characteristic times are accordingly short. Nonequilibrium in these processes will therefore become important only when the characteristic flow time is very small, that is, when the gradients in the flow field are large. The non-equilibrium phenomena that then arise, which go by the familiar names of viscosity, thermal conductivity, and diffusion, do not depend for their existence on a high level of temperature. Their discussion from the molecular point of view in all but the most elementary way (cf. Chapter I) is difficult and will be deferred until Chapters IX and X. Other processes—for example, vibrational and chemical—are relatively slow to adjust. Nonequilibrium effects in these processes may therefore occur when the gradients in the flow are still small compared with those pertinent to translation and rotation. Since the processes in question are not active at low temperatures, the existence of a high temperature level is essential here. This kind of situation is found, for example, in the flow downstream of a strong shock wave or in the expansion through the nozzle of a hypervelocity wind tunnel. Vibrational and chemical nonequilibrium are in some respects simpler than translational and rotational. They are therefore appropriately discussed first.

The present chapter is concerned with establishing the basic differential equations that govern the vibrational and chemical rate processes in some special but practically important situations. As before, we establish these relations first for a closed stationary system. They can then be applied to flow problems by assuming that the rate processes in a nonuniform moving fluid are the same as in the closed stationary system. This will be done in Chapter VIII.

2 VIBRATIONAL RATE EQUATION

We consider first the case of vibrational nonequilibrium due to collisional processes.[1] To be specific we consider a system of harmonic oscillators, which have permissible energy states [see equation (IV 12.9a)]

$$\epsilon_i = (i + \tfrac{1}{2})h\nu \qquad \text{where } i = 0, 1, 2, \ldots \tag{2.1}$$

This system, which we may think of as made up of diatomic molecules, is assumed to be capable of collisional interchange of energy with the translational degrees of freedom of a heat bath having a *constant temperature T*. Such a heat bath can be provided, for example, by a large excess of inert gas containing a small concentration of the diatomic species. It can also be provided by the translational and rotational degrees of freedom of a *pure* diatomic species itself if only a small fraction of the molecules is excited. In either case, when the system of oscillators is in equilibrium

[1] For a more detailed development see Rubin and Shuler (1956). For an extended treatment of relaxation processes generally, see Herzfeld and Litovitz (1959).

with the heat bath, the distribution of oscillators over the energy states is given by the Boltzmann distribution [see equation (IV 6.12)]

$$N_i^* = N \frac{e^{-\epsilon_i/kT}}{\sum_i e^{-\epsilon_i/kT}}, \tag{2.2}$$

where N is now the total number of oscillators and N_i^* (written for convenience without the prime) is the number in the energy state ϵ_i at equilibrium. Individual oscillators are continually changing from one energy state to another upon interaction with the heat bath, but the overall effect at equilibrium is to maintain the distribution (2.2).

We now assume that the equilibrium distribution is disturbed by some external agency as, for example, by irradiation by a short-duration, high-intensity light. After the source of the disturbance is removed, the oscillators continue to change from one energy state to another upon collision with the heat bath. The overall effect, however, must be for the distribution numbers N_i to change with time in such a way as to return eventually to their equilibrium value. We wish here to study this nonequilibrium rate process, sometimes referred to as *vibrational relaxation.* In particular, we would like to obtain an equation telling us how the total energy contained in the system of oscillators approaches its equilibrium value.

We begin by establishing the differential equation that governs the rate of change of the nonequilibrium values of N_i as a function of time. To this end, we assume, consistent with detailed quantum-mechanical studies of transition probabilities during weak interaction, that the changes in energy of the oscillators upon collision take place only between adjacent states, that is, $\Delta i = \pm 1$. Let $k_{i,i\pm 1}$ denote the rate constant for transitions from one state to the next, that is, the fraction of oscillators in state i that make the transition to state $i \pm 1$ per unit time. The value of the k's, which need not concern us in detail at the moment, will depend on the following two factors:

1. The number of collisions per unit time between the oscillators and those molecules of the heat bath that possess sufficient energy to cause transition.[2] This will depend in turn on the temperature and density of the heat bath.
2. The probability that a single collision will result in a transition.

By means of the rate constants, we can write for the rate of change of the

[2] In line with our assumption of a small relative concentration of oscillators, we may neglect oscillator-oscillator collisions. The concentration of oscillators is then unimportant.

number of oscillators in any given state i

$$\frac{dN_i}{dt} = -k_{i,i+1}N_i + k_{i+1,i}N_{i+1} - k_{i,i-1}N_i + k_{i-1,i}N_{i-1}. \tag{2.3}$$

The first two terms represent the interchange of oscillators between the i and $i + 1$ states; the last two the interchange between the i and $i - 1$ states. An infinite number of equations (2.3) exists, one for each energy state.

The various rate constants that appear in equation (2.3) are not independent. In particular, in the special circumstance when the system is in equilibrium we must obviously have $dN_i/dt = 0$. Furthermore, by the principle of detailed balancing introduced in Chapter II, Sec. 4, the net interchange between any two adjacent states must individually be zero. This requirement, when applied to the i and $i - 1$ states, provides the condition

$$-k_{i,i-1}N_i^* + k_{i-1,i}N_{i-1}^* = 0.$$

Here the k's have the same value as when the system is out of equilibrium, since they depend only on the state of the heat bath, which is assumed constant. This leads, in view of equations (2.2) and (2.1), to the following relation between the k's:

$$k_{i-1,i} = k_{i,i-1}\frac{N_i^*}{N_{i-1}^*} = k_{i,i-1}\frac{e^{-\epsilon_i/kT}}{e^{-\epsilon_{i-1}/kT}} = k_{i,i-1}e^{-h\nu/kT}. \tag{2.4}$$

We see from this relation that $k_{i-1,i}$ is smaller than $k_{i,i-1}$. This is so because collisions that activate a molecule require more energy than collisions that deactivate a molecule and are therefore rarer. Since there are more molecules available to be activated than there are to be deactivated, however, the net interchange at equilibrium is zero.

A second relation between the k's is provided by the equation

$$k_{i,i-1} = ik_{1,0}, \tag{2.5}$$

which is obtained from the quantum-mechanical study of transition probabilities for a harmonic oscillator (see Pauling and Wilson, 1935, pp. 82 and 306). Equation (2.5) says, in effect, that the value of k for transitions from one state to the next state below is proportional to the quantum number of the upper state.

With the aid of the relations (2.4) and (2.5), the rate equation (2.3) can be written in terms of the single rate constant $k_{1,0}$. Putting (2.5) into (2.4), we write first

$$k_{i-1,i} = ik_{1,0}e^{-h\nu/kT}.$$

Replacing i by $i + 1$ in this relation and in relation (2.5), we also have

$$k_{i+1,i} = (i + 1)k_{1,0} \qquad \text{and} \qquad k_{i,i+1} = (i + 1)k_{1,0}e^{-h\nu/kT}.$$

By substituting the foregoing expressions into equation (2.3), we can write that equation as

$$\frac{dN_i}{dt} = k_{1,0}\{-iN_i + (i + 1)N_{i+1} + e^{-\Theta_v/T}[-(i + 1)N_i + iN_{i-1}]\}, \quad (2.6)$$

where we have again introduced the notation $\Theta_v \equiv h\nu/k$. Equation (2.6) was derived here for collisional exchange of energy between translation and vibration. An equation of the same form can be obtained for collisional vibration-vibration exchange (Montroll and Shuler, 1957) and for radiative exchange (Rubin and Shuler, 1957); only the multiplying factor in front of the braces is different. The infinite set of equations (2.6) can be solved directly for the N_i's as functions of t (see Montroll and Shuler, 1957), but we shall not go into that here.

Instead of the N_i's themselves, we are more concerned for gas-dynamic purposes with the total vibrational energy of the system of oscillators. This is given by the sum

$$E_v = \sum_{i=0}^{\infty} \epsilon_i N_i = h\nu \sum_{i=1}^{\infty} iN_i, \quad (2.7)$$

where the energy has been reckoned from the ground state in writing the last expression and the term for $i = 0$ is omitted since it then contributes nothing. Differentiation of (2.7) and substitution of dN_i/dt from (2.6) then gives

$$\frac{dE_v}{dt} = h\nu k_{1,0} \sum_{i=1}^{\infty} \{-i^2 N_i + i(i + 1)N_{i+1} + e^{-\Theta_v/T}[-i(i + 1)N_i + i^2 N_{i-1}]\}. \quad (2.8)$$

We note that each of the N_i's appears twice in the summation of the first two terms inside the braces, once when $i = s$ and once when $i = s - 1$. These two terms can be combined into a single sum as follows, where we begin by writing $i = s - 1$ in the second term[3]:

$$\sum_{i=1}^{\infty} \{-i^2 N_i + i(i + 1)N_{i+1}\} = -\sum_{i=1}^{\infty} i^2 N_i + \sum_{s=2}^{\infty} (s - 1)sN_s$$

$$= -\sum_{i=1}^{\infty} i^2 N_i + \sum_{s=2}^{\infty} s^2 N_s - \sum_{s=2}^{\infty} sN_s = -N_1 - \sum_{s=2}^{\infty} sN_s = -\sum_{i=0}^{\infty} iN_i.$$

By a similar procedure the last two terms can be combined into a single sum to obtain

$$\sum_{i=1}^{\infty} [-i(i + 1)N_i + i^2 N_{i-1}] = \sum_{i=0}^{\infty} (i + 1)N_i.$$

[3] This may be looked on as akin to changing the variable of integration in a definite integral.

With these two results equation (2.8) can be rewritten

$$\begin{aligned}\frac{dE_v}{dt} &= h\nu k_{1,0} \sum_{i=0}^{\infty} [-iN_i + e^{-\Theta_v/T}(i+1)N_i] \\ &= h\nu k_{1,0}\left[e^{-\Theta_v/T} \sum_{i=0}^{\infty} N_i - (1 - e^{-\Theta_v/T}) \sum_{i=0}^{\infty} iN_i\right].\end{aligned}$$

With the aid of equation (2.7) and the relation $N = \sum_{i=0}^{\infty} N_i$, this equation can further be rewritten

$$\begin{aligned}\frac{dE_v}{dt} &= h\nu k_{1,0}\left[e^{-\Theta_v/T}N - (1 - e^{-\Theta_v/T})\frac{E_v}{h\nu}\right] \\ &= k_{1,0}(1 - e^{-\Theta_v/T})\left[\frac{Nh\nu}{e^{\Theta_v/T} - 1} - E_v\right].\end{aligned}$$

Since $Nh\nu = mN(k/m)(h\nu/k) = mNR\Theta_v$, where m is the mass of an oscillator molecule, the first term inside the brackets can be recognized [see equation (IV 12.12)] as the vibrational energy that the system of oscillators would have if it were in equilibrium at the temperature T of the heat bath. Denoting this equilibrium energy by E_v^* and introducing the notation $\tau \equiv [k_{1,0}(1 - e^{-\Theta_v/T})]^{-1}$, we write finally

$$\boxed{\frac{dE_v}{dt} = \frac{E_v^* - E_v}{\tau}} . \tag{2.9}$$

This remarkably simple equation shows that the vibrational energy of the system of oscillators will always tend toward the equilibrium value, as we might expect. It shows further—and there is no obvious reason to expect this—that the rate at which it so tends at any instant is linearly proportional to the amount that it departs from equilibrium at that instant. It should be noted that in deriving this equation no assumption has been made as to how the oscillators are distributed over the energy states when they are out of equilibrium. In particular, no assumption of a Boltzmann distribution (at some temperature other than T) is necessary. It will also be noted that no direct assumption has been made regarding the size of the difference $(E_v^* - E_v)$. The linear result (2.9) is therefore valid even for large departures from equilibrium, at least insofar as the assumed harmonic oscillator is concerned. Since the harmonic oscillator itself is a valid approximation only for the lower vibrational levels, however (see Chapter IV, Sec. 12), the equation is in fact limited, to some unknown extent, to small departures from equilibrium. (For a discussion of anharmonic effects in vibrational nonequilibrium, see Bazley et al., 1958.) The quantity τ appearing in equation (2.9) has the dimensions of

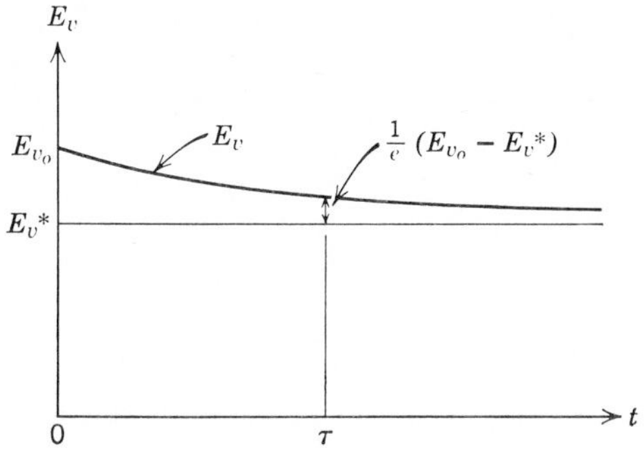

Fig. 1. Illustration of relaxation time.

time and is called the *relaxation time*. It is a function of the temperature and density (or pressure) of the heat bath and of the characteristics of the vibrating species.[4]

Since the conditions of the heat bath are constant, τ and E_v^* are constants and equation (2.9) can be solved at once. As can be verified by direct substitution, the solution is

$$E_v = Ce^{-t/\tau} + E_v^*,$$

where C is a constant. If $E_v = E_{v_0}$ at $t = 0$, then $C = E_{v_0} - E_v^*$, and the solution can be written

$$\frac{E_v - E_v^*}{E_{v_0} - E_v^*} = e^{-t/\tau}. \tag{2.10}$$

The relaxation time τ is thus seen to be the time required for the difference $E_v - E_v^*$ to fall to $1/e$ of its initial value (see Fig. 1). The smaller the value of τ, the faster the relaxation process.

The differential equation (2.9) was derived on the assumption that the system of oscillators is contained in a heat bath of constant state and that the number of oscillators is relatively small. The extent to which it is true in more complicated situations is a subject of current research. For practical purposes it is generally assumed that equation (2.9) is valid irrespective of the number of excited molecules and that it can be applied in a time-varying heat bath if we take E_v^* and τ as functions of the instantaneous state of the bath. For example, if T and p are the temperature

[4] Equation (2.9) was obtained, in effect, by taking the so-called "first moment" of the differential equation (2.6). This kind of procedure is used in the later discussion of translational nonequilibrium (Chapter IX) to obtain the macroscopic differential equations of fluid motion from the integro-differential Boltzmann equation, which governs the situation at the microscopic level in that case. For a general discussion of the relation between macroscopic and microscopic relaxation equations for vibrational nonequilibrium, see Shuler (1959).

and pressure of the heat bath, we set $E_v^* = mNR\Theta_v/(e^{\Theta_v/T} - 1)$ and $\tau = \tau(T, p)$ where $T = T(t)$ and $p = p(t)$, and write equation (2.9) as

$$\frac{dE_v}{dt} = \frac{E_v^*(T) - E_v}{\tau(T, p)}. \qquad (2.11)$$

Since T and p are now functions of t, the equation can no longer be integrated to obtain the solution (2.10), and τ loses its previous simple significance. It is now referred to as a *local relaxation time*, even though the process is no longer a relaxation (i.e., exponential) process in the strict sense. For the integration of (2.11) in any given situation, numerical methods must usually be used.

For application of equations (2.10) or (2.11) we require an explicit relation for $\tau = \tau(T, p)$. The theoretical derivation of this relation is a difficult problem, and the results depend on the approximations. The most commonly cited theory is the original one of Landau and Teller (1936) who obtained the result

$$\tau = \frac{K_1 T^{5/6} \exp{(K_2/T)^{1/3}}}{p(1 - e^{-\Theta_v/T})}, \qquad (2.12)$$

where K_1 and K_2 are positive constants that depend on the physical properties of the molecule.[5] At sufficiently low temperatures, the temperature variation in (2.12) is given largely by the exponential function in the numerator. We can then write to a good approximation

$$\tau = C \frac{\exp{(K_2/T)^{1/3}}}{p}, \qquad (2.13a)$$

or

$$\ln \tau = \left(\frac{K_2}{T}\right)^{1/3} - \ln p + \ln C, \qquad (2.13b)$$

where C is a constant. Thus, at a given pressure, τ decreases as T increases, and a plot of $\ln \tau$ versus $1/T^{1/3}$ would be expected to give approximately a straight line. For a given species of oscillator molecule, the value of C is different for different types of heat-bath molecule. This follows from the fact that the effectiveness of transfer of energy upon collision depends on the nature of the molecules involved. At high temperatures the factor $(1 - e^{-\Theta_v/T})$ becomes significant in equation (2.12), and simple results of the type (2.13) would not be expected to apply.

In the present state of the theory, the evaluation of τ must depend on experimental measurements interpreted in the light of the foregoing

[5] For alternative results see Widom (1957) and Parker (1959).

theoretical results. The basic results in this field insofar as the constituents of air are concerned are those of Blackman (1956), who made shock-tube studies of vibrational relaxation in pure oxygen and pure nitrogen. These results have since been confirmed by other investigators. Blackman, like most of the subsequent investigators, made use of the flow in the relaxation zone behind a normal shock wave. (For a description of this flow, see Chapter VIII, Sec. 9.) By assuming that the flow behind the shock wave provides a heat bath of constant T and p (which is in error by about 20%), he was able to deduce the values of τ from the density variation behind the wave as measured by an optical interferometer.[6] Similar work by Camac (1961) in oxygen-argon mixtures, but with ultraviolet-absorption rather than interferometric techniques, allows the separation of the effectiveness of O_2 and Ar as heat-bath molecules in oxygen relaxation. Similar data for nitric oxide have been obtained by Wray (1962b) in his work on mixtures of NO and Ar, again by ultraviolet techniques. Values of the constants C and K_2 deduced by fitting curves of the type (2.13b) to the available data are listed in Table A. We see that O_2-Ar collisions are only about $\frac{1}{7}$ as effective as O_2-O_2 collisions; NO-Ar collisions are less than $\frac{1}{100}$ as effective as NO-NO collisions. No such data are available

Table A Values of Constants in Equation (2.13a) *for* τ

Species	Heat-Bath Molecule	C, atm-microsec	K_2, °K	Approximate Range of T, °K
O_2	O_2	5.42×10^{-5}	2.95×10^{6}	800–3200
O_2	Ar	3.58×10^{-4}	2.95×10^{6}	1300–4300
N_2	N_2	7.12×10^{-3}	1.91×10^{6}	800–6000
NO	NO	4.86×10^{-3}	1.37×10^{5}	1500–3000
NO	Ar	6.16×10^{-1}	1.37×10^{5}	1500–4600

for N_2 with various heat-bath molecules at this writing. In the absence of data to the contrary, it is usually assumed that all heat-bath molecules have the same effectiveness in the vibrational relaxation of a given species. As the results of Table A indicate, however, this is probably not the case.

With the foregoing results we can estimate the number of molecular collisions required for the equilibration of vibrational energy. For example, if we consider pure O_2 at $p = 1$ atm and $T = 2000$°K, we find from equation (2.13a) and Table A that $\tau = 4.9 \times 10^{-6}$ sec. We know from the results of Chapter I, Sec. 6 that the frequency of collisions

[6] For a critical discussion of the methods used in the analysis of vibrational-relaxation data, see Blythe (1961).

experienced by a single molecule at $p = 1$ atm and $T = 273°K$ is $\Theta(273) = 10^{10}$/sec. From the results of Chapter I we also have

$$\Theta = \frac{\bar{C}}{\lambda} \sim \frac{\sqrt{T}}{1/\rho} \sim \frac{\sqrt{T}}{T/p} \sim \frac{p}{\sqrt{T}},$$

so that at the same pressure of 1 atm we find

$$\Theta(2000) = \sqrt{\frac{273}{2000}}\,\Theta(273) = 0.37 \times 10^{10} \text{ sec.}$$

Taking the product $\tau \times \Theta(2000)$ we see that a single molecule experiences 18,100 collisions in the time τ. At the given conditions, each molecule must thus undergo an average of about 18,000 molecular collisions to reduce the "out-of-equilibrium" vibrational energy to $1/e$ of its initial value. This is large compared with the number (less than 10) required for a similar effect on translational or rotational energy.

We have discussed here vibrational relaxation due only to inelastic collisions, which may be expected to be the primary mechanism at temperatures up to perhaps 8000°K. For a survey of other mechanisms that may be important at higher temperatures, see Bauer and Tsang (1963).

3 ENTROPY PRODUCTION BY VIBRATIONAL NONEQUILIBRIUM

Vibrational relaxation, being a nonequilibrium process, is irreversible. It is of interest to examine the resulting production of entropy.

For a discussion of entropy production, the system of harmonic oscillators must possess a definable entropy. This will be the case if we introduce the special assumption, not made in the work of the last section, that the oscillators have a Boltzmann distribution over their energy states. We can then speak of the system of oscillators as having an assignable *vibrational temperature* T_v, related to the vibrational energy by the equation [cf. equation (IV 12.12)]

$$E_v = \frac{mNR\Theta_v}{e^{\Theta_v/T_v} - 1}, \tag{3.1}$$

where m is the mass of an oscillator molecule and R is the corresponding gas constant. This equation, in effect, defines T_v. The Boltzmann distribution of oscillators is given correspondingly in terms of T_v by

$$N_i = N\frac{e^{-ih\nu/kT_v}}{Q_v}, \tag{3.2}$$

where the vibrational partition function Q_v is now [cf. equation (IV 12.11)]

$$Q_v = \sum_{i=0}^{\infty} e^{-ih\nu/kT_v} = \frac{1}{1 - e^{-\Theta_v/T_v}}. \tag{3.3}$$

Substitution of Q_v into the third of equations (IV 8.7) with $\partial(\ )/\partial T$ replaced by $\partial(\ )/\partial T_v$ leads directly to equation (3.1); the theoretical structure is thus self consistent. The entropy and other thermodynamic variables of the system of oscillators follow in similar fashion from the other of equations (IV 8.7). Note, however, that the pressure p_v of the oscillators is to be taken as zero, since Q_v is not a function of V. This is reasonable because the vibrational motion possesses no net translational momentum and hence is incapable of exerting a pressure.

Since the heat bath, which was assumed to be in equilibrium originally, and the system of oscillators now have assignable entropies, we can proceed to apply the ideas of irreversible thermodynamics. We consider the heat bath and the assembly of oscillators as together constituting a complete thermodynamic system. The heat bath and the oscillators then each constitute by themselves a separate (but coexisting) closed *sub*system, each with its appropriate temperature and with the oscillator subsystem contained within the heat bath (Fig. 2). We suppose that the heat bath can give up heat to the oscillators in infinitesimal amount dQ_1 and receive heat in amount dQ_0 from the external world, with corresponding negative amounts if the heat flows in the opposite direction. The heat bath thus constitutes the "surroundings" of the oscillators, and the oscillators and external world constitute the "surroundings" of the heat bath. For the complete system of heat bath plus oscillators, the surroundings are the external world only. The foregoing picture is easily modified, without change in the final results, to allow for heat flow directly from the external world to the oscillators as well. For collisional processes this mechanism is unlikely, however, in view of the short relaxation time of the translational heat bath as compared with that of the oscillators.

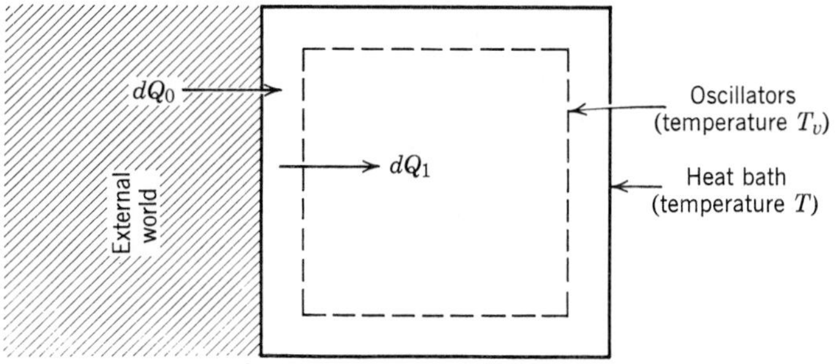

Fig. 2. Model for calculation of entropy production.

From the extensive property of entropy, the entropy S_c of the complete system of heat bath plus oscillators can be written as the sum of the entropy S_h of the heat bath and the entropy S_v of the oscillators:

$$S_c = S_h + S_v,$$

from which

$$dS_c = dS_h + dS_v. \tag{3.4}$$

From the statement of the second law of irreversible thermodynamics as given in Chapter III, Sec. 5, we can write for any closed system

$$dS = d_eS + d_iS = \frac{dQ}{T} + d_iS, \tag{3.5}$$

where T is the temperature at which the amount of heat dQ is received. We now analyze our assumed situation by applying equation (3.5) in turn to (a) the heat bath, (b) the oscillators, and (c) the complete system.

(a) For the heat bath, we have an amount of heat $dQ_h = dQ_0 - dQ_1$ received at the temperature T. Since the heat bath is internally in equilibrium, the production of entropy within the bath is zero, that is, $d_iS_h = 0$. Applying (3.5) to the heat bath, we therefore obtain

$$dS_h = \frac{dQ_0 - dQ_1}{T}. \tag{3.6}$$

(b) For the oscillators, we have an amount of heat $dQ_v = dQ_1$ received at the temperature T_v. Since the oscillators have a Boltzmann distribution, they are at all times internally at equilibrium at the instantaneous value of T_v. The production of entropy within the oscillators is therefore also zero, that is, $d_iS_v = 0$. Applying (3.5) to the subsystem of oscillators, we thus have

$$dS_v = \frac{dQ_1}{T_v}. \tag{3.7}$$

(c) For the complete system of heat bath plus oscillators, the amount of heat $dQ_c = dQ_0$ is received at the temperature T. Since the complete system is *not* in thermal equilibrium, however, the production of entropy in this case is not zero, that is, $d_iS_c \neq 0$. Applying (3.5) to the complete system we thus have

$$dS_c = \frac{dQ_0}{T} + d_iS_c. \tag{3.8}$$

By equation (3.4), however, we can also find dS_c by adding dS_h as given by (3.6) and dS_v as given by (3.7). We thus find alternatively

$$dS_c = \frac{dQ_0 - dQ_1}{T} + \frac{dQ_1}{T_v} = \frac{dQ_0}{T} + \left(\frac{1}{T_v} - \frac{1}{T}\right) dQ_1. \tag{3.9}$$

Comparing this with equation (3.8) we obtain for the entropy production in the complete system

$$d_iS_c = \left(\frac{1}{T_v} - \frac{1}{T}\right) dQ_1. \tag{3.10}$$

From the first law we can write in general that $dE = dQ - p\,dV$, and hence for the subsystem of oscillators, where $p_v = 0$, that $dE_v = dQ_v = dQ_1$. Equation (3.10) can therefore be written

$$\boxed{d_iS = \left(\frac{1}{T_v} - \frac{1}{T}\right) dE_v}\ , \tag{3.11}$$

where we have dropped the now superfluous subscript c. Since the two subsystems in our model are in internal equilibrium, this irreversible production of entropy must be thought of, within the context of our model, as arising entirely at the "interface" between the subsystems. The situation is analogous to that for chemical nonequilibrium as analyzed in Chapter III, Sec. 6.

Equation (3.11) shows that the entropy production can be zero under two conditions: (1) when the complete system is in equilibrium, which will be the case when $T = T_v$; or (2) when the system is not in equilibrium ($T \neq T_v$) but the vibrational energy is "frozen," that is, $dE_v = 0$. If neither of these conditions is satisfied, then d_iS will always be positive, for if $T > T_v$, then we know that $dE_v > 0$, whereas if $T < T_v$, then $dE_v < 0$. This is as it should be according to part (2) of the second law from irreversible thermodynamics (see Chapter III, Sec. 5).

Returning to equation (3.9) with $dQ_1 = dE_v$ and dropping the now superfluous subscripts c and 0, we can write for the complete system

$$dS = \frac{dQ}{T} + \left(\frac{1}{T_v} - \frac{1}{T}\right) dE_v.$$

With the aid of the first law applied to the system, this can be written

$$dS = \frac{1}{T}\,dE + \frac{p}{T}\,dV + \left(\frac{1}{T_v} - \frac{1}{T}\right) dE_v. \tag{3.12a}$$

If the heat bath is considered to be the translational and rotational degrees of freedom of the vibrating species itself, then this is the counterpart for a vibrating gas of the Gibbs equation (III 6.8) for a chemically reacting gas. In terms of specific quantities per unit mass of gas it becomes

$$ds = \frac{1}{T}\,de + \frac{p}{T}\,d\left(\frac{1}{\rho}\right) + \left(\frac{1}{T_v} - \frac{1}{T}\right) de_v. \tag{3.12b}$$

Exercise 3.1. (a) Obtain a formula for the *total* amount of entropy produced when a system of harmonic oscillators in contact with a constant-temperature heat bath relaxes from an initial nonequilibrium vibrational energy E_{v_0} to the final equilibrium energy E_v^*. This formula should give the total production of entropy as a function of E_{v_0} and E_v^* and any necessary physical constants of the system. It should not include any temperatures.

(b) Expand the total entropy production of part (a) in a power series in the quantity $\delta \equiv (E_{v_0} - E_v^*)/E_v^*$. Carry the expansion only as far as the first non-zero term in the series. What does this result demonstrate with regard to the total production of entropy?

4 CHEMICAL RATE EQUATIONS—GENERAL CONSIDERATIONS

The study of the rates of chemical reactions, commonly called *chemical kinetics*, is a complex and difficult science, both experimentally and theoretically. Although an enormous body of knowledge has been accumulated, the field is still in an incomplete state of development, and no generally satisfactory theory exists. We shall discuss here only a few of the ideas and results that are pertinent to the reactions that occur in gaseous air and shall take up only the simplest of the several theories, the so-called *collision theory*. We limit ourselves to homogeneous reactions, that is, to those that occur within the gas itself rather than at the solid or liquid surfaces that bound the gas. This is not because surface reactions are unimportant in practical gas dynamics—quite the contrary—but simply to keep the discussion within limits. We also exclude all photochemical reactions, that is, reactions that depend on radiation. Our discussion will depend rather heavily on that of Fowler and Guggenheim (1956), which the reader may consult for further details. More extended treatments of chemical kinetics in general, are given by Hinshelwood (1940), Laidler (1950), Benson (1960), and Frost and Pearson (1961).

If X_s denotes any given chemical species, the chemical equation describing the overall change from reactants to products may be written in the general form

$$\nu_1'X_1 + \nu_2'X_2 + \cdots \rightarrow \nu_1''X_1 + \nu_2''X_2 + \cdots, \tag{4.1a}$$

or

$$\sum_{s=1}^{l} \nu_s'X_s \rightarrow \sum_{s=1}^{l} \nu_s''X_s, \tag{4.1b}$$

where we for the time being consider the reaction proceeding in only the one direction as indicated. The reader will note that the convention for writing the chemical equation here is different from that used in Chapter III, Sec. 3. Here a given species X_s appears on both sides of the equation, and the stoichiometric coefficients ν'_s for reactants and ν''_s for products are given separate symbols and are all counted as positive. The earlier formulation was the more convenient for chemical equilibrium; the present has certain advantages in chemical kinetics. Both are in common use in the literature, and the student will have to become accustomed to them. In the present formulation, if a given chemical species X_s does not occur on one side or the other of the equation, then the corresponding value of ν'_s or ν''_s is zero.

The detailed mechanism of the actual reaction may or may not be represented by the overall reaction equation (4.1). For example, the reaction

$$H_2 + I_2 \rightarrow 2HI \tag{4.2}$$

for the formation of hydrogen iodide (cf. Chapter III, Sec. 7) does take place through a bimolecular collision involving a single hydrogen and a single iodine molecule. On the other hand, the reaction for the formation of hydrogen bromide, which is represented by the similar overall equation

$$H_2 + Br_2 \rightarrow 2HBr, \tag{4.3}$$

is known to actually take place via five elementary reaction processes as follows:

$$\begin{aligned} Br_2 + M &\rightarrow 2Br + M \\ Br + H_2 &\rightarrow HBr + H \\ H + Br_2 &\rightarrow HBr + Br \\ HBr + H &\rightarrow Br + H_2 \\ 2Br + M &\rightarrow Br_2 + M. \end{aligned} \tag{4.4}$$

In the first and last of these equations, the symbol M on both sides of the equation represents any molecule that may be present, for example, any one of the species H, Br, H_2, Br_2, or HBr. In contrast to the situation in equilibrium calculations, this so-called *catalytic molecule* must be included here. Although it does not itself undergo change in the given reaction, it plays an essential role, as will be explained later. In general, the detailed mechanism in a complex reaction cannot be predicted from the overall equation. It must be sorted out by careful experiments and theoretical reasoning.

The rates of chemical reactions are customarily studied in terms of the concentrations of the various species, denoted by $[X_s]$ and measured in, say, moles per unit volume. It is an experimentally observed fact that the rate of formation of any one of the *products* in many (but not all) reactions can be expressed by an equation of the form

$$\frac{d[X_s]}{dt} = \nu''_s k_c(T)[X_1]^{z_1}[X_2]^{z_2} \cdots, \tag{4.5a}$$

where the product on the right is taken over *all reactants*. The stoichiometric coefficient ν_s'' is included explicitly, since the change in the number of moles of a given species as a result of the reaction must be in proportion to its stoichiometric coefficient. This is apparent from equations (III 3.3) as applied to a unit volume; the remaining product on the right is then, in effect, an expression for $d\xi/dt$ in those equations. The so-called *rate constant* k_c, which appears in (4.5a) and which is associated with the given reaction, is a function only of the temperature. With the present convention for the stoichiometric coefficients, the rate of disappearance of any one of the *reactants* is given similarly by the equation

$$\frac{d[X_s]}{dt} = -\nu_s' k_c(T)[X_1]^{z_1}[X_2]^{z_2}\cdots, \tag{4.5b}$$

where a negative sign is included since the coefficient ν_s' is by definition positive. Letting $z = z_1 + z_2 + \cdots$, we say that the reaction to which equations (4.5) are applicable is of the zth order (z_1th order with respect to X_1, z_2th order with respect to X_2, etc.). It is evident that for a given reaction, the rate constant k_c must have the dimensions $(\text{moles/volume})^{1-z}$ $(\text{time})^{-1}$.

In general, the z_s's may be integers or fractions and may or may not be related to the stoichiometric coefficients in the chemical equation (4.1). When the overall reaction takes place in a single, simple step, as with the hydrogen-iodine reaction (4.2), then—as we shall see—the z_s's are equal to the respective stoichiometric coefficients ν_s'. The same is true of the rate expressions that can be written for the individual, elementary processes in a complex overall reaction such as (4.3). This is because the elementary processes themselves constitute simple, single-step reactions. For the overall reaction made up of a number of elementary processes, however, the z_s's in the overall rate expressions are not necessarily the same as the corresponding stoichiometric coefficients. Often one of the elementary steps will be slower than the rest and will thus determine the overall rate. The rates can then be described by expressions of the form (4.5), but the z_s's are not necessarily integers. If there is no single step much slower than the rest, it is not usually possible to describe the rates by a formula of the type (4.5). This is the case, for example, for the hydrogen-bromine reaction (4.3), for which the rate of formation of HBr follows the empirical relation

$$\frac{d[\mathrm{HBr}]}{dt} = \frac{k_a[\mathrm{H_2}][\mathrm{Br_2}]^{1/2}}{1 + k_b[\mathrm{HBr}]/[\mathrm{Br_2}]}, \tag{4.6}$$

where k_a and k_b are constants for a given temperature.

The task of theoretical chemical kinetics is to derive rate equations such as (4.5) and (4.6) and, in particular, to obtain expressions for the constants k. To do this in any given case requires some assumption regarding the detailed mechanism of the overall reaction, that is, some assumed breakdown of the reaction into its elementary processes. This will depend on past experience or on experimental evidence, often of spectroscopic nature. The decision is often a difficult one and must be checked at the end by comparing the derived theoretical rates against the measured values.

Once the reaction has been divided into its elementary processes, the rates of these processes must then be assessed. This breaks down in turn into two questions:

1. What conditions must be satisfied by the molecules if they are to react?
2. How frequently are these conditions satisfied?

The first of these is a problem in quantum theory that is usually too complex to be solved in practice. At present we can answer this question in most instances only by assuming certain physically reasonable conditions and at the end checking the derived rates against the measured values. With the answer to (1) assumed, usually in the form of some condition regarding the nature of the molecular collisions, the answer to (2) is calculated by means of equilibrium kinetic theory. This quasi-equilibrium procedure implies that the presence of the nonequilibrium reaction does not seriously disturb the distribution of molecular energy from what it would be if the gas were in equilibrium at the prevailing instantaneous state. The requirements for this assumption to be valid will be made more explicit further on.

Assuming an answer to question (1) implies an assumption regarding the mechanism whereby the elementary reaction takes place. This involves the concept of the *molecularity* of the reaction, that is, the number of molecules that come together in the course of the reaction. Various possibilities in this regard are as follows:

1. Bimolecular mechanism: Reaction occurs during the collision of two molecules. The dissociation reaction $AB + M \rightarrow A + B + M$ is a simple example of this.

2. Termolecular mechanism: Reaction occurs at a collision between three molecules. The recombination reaction $A + B + M \rightarrow AB + M$ is of this type.

3. Unimolecular mechanism: Reaction occurs spontaneously by decomposition of a single molecule that has previously been activated to a high energy level by collisions or otherwise. This is exemplified by the dissociation reaction $AB^* \rightarrow A + B$, where AB^* represents the activated

molecule. The unimolecular mechanism, which is the most difficult theoretically, is of little importance for gas-dynamic purposes, and we shall not be concerned with it.

Most gaseous reactions can be accounted for by one or more of the foregoing elementary mechanisms. Collisions involving four or more molecules are so infrequent as to be of no importance.

We now turn our attention specifically to the dissociation reaction $AB + M \rightarrow A + B + M$, which is of primary importance in the chemistry of air at high temperatures. This reaction takes place in a single step by a bimolecular mechanism. The simplest possible answer to question (1) regarding the conditions for the reaction would be that every collision between a molecule AB and a catalytic molecule M results in dissociation of the AB-molecule. The answer to question (2) regarding the frequency with which these conditions are satisfied is then given directly by the bimolecular collision rate Z_{CD} from equation (II 6.15b):

$$Z_{\mathrm{CD}} = \frac{n_{\mathrm{C}} n_{\mathrm{D}}}{\sigma} d_{\mathrm{CD}}^2 \left(\frac{8\pi kT}{m_{\mathrm{CD}}^*} \right)^{1/2}, \tag{4.7}$$

where we have altered the subscripts from the earlier equation to avoid confusion with the particular chemical species now being considered. Applying this to the present reaction and transforming to concentrations by means of the relation $n_{\mathrm{Y}} = \hat{N}[\mathrm{Y}]$, where $\hat{N}$ is Avogadro's number, we easily obtain a rate equation for [AB] having the form (4.5b). The expression for the rate constant so obtained will obviously depend on temperature as $T^{1/2}$. Unfortunately for this simple theory, the results are not borne out by experiment. First, the values of the rate constant calculated on this basis are usually about 10^8 times larger than those actually measured. Second, the actual dependence on T is much more pronounced than the factor $T^{1/2}$ would indicate. In fact, before any theory whatsoever was known, it was recognized on experimental grounds that for many reactions a plot of $\ln k_c$ versus $1/T$ gives an essentially straight line of negative slope, which leads to a relation of the form

$$\ln k_c = -\frac{\epsilon_a}{kT} + \text{const},$$

or

$$k_c = Ce^{-\epsilon_a/kT}, \tag{4.8a}$$

where C and ϵ_a are independent of T. This important equation is commonly called the *Arrhenius equation*, after the Swedish physicist-chemist Arrhenius who first (1889) gave it a theoretical interpretation. The constant ϵ_a is often referred to as the *Arrhenius activation energy*.

Precise measurements, usually very difficult to make, indicate that $\ln k_c$ versus $1/T$ is sometimes not exactly a straight line and that k_c is better fitted by an equation of the form

$$\boxed{k_c = CT^{\eta}e^{-\epsilon_0/kT}}, \tag{4.8b}$$

where C, η, and ϵ_0 are independent of T [and C and ϵ_0 are different from their counterparts in (4.8a)]. In any event, the exponential dependence in these empirical equations shows that something is obviously missing from the crude theory proposed earlier.

The fact that the rate at which molecules react is so much less than the rate at which they collide indicates that some condition is required for reaction beyond mere collision. It suggests, in particular, that a collision that results in reaction must involve considerably more energy than is available in just any collision. As we shall see, this condition is in accord with the exponential form of the equations (4.8). Furthermore, we might suppose that not even all collisions involving the required energy will result in reaction, but only those in which the relative orientation of the molecules at collision is such that the energy can be effective in altering the chemical bonds. Considerations of this kind lead us to attempt an expression for the rate of reaction in the form

$$\begin{pmatrix}\text{Rate of}\\ \text{reacting}\\ \text{collisions}\end{pmatrix} = \begin{pmatrix}\text{Rate of}\\ \text{collisions}\end{pmatrix} \times \begin{pmatrix}\text{Fraction of}\\ \text{collisions that}\\ \text{involve sufficient}\\ \text{energy}\end{pmatrix} \times \begin{pmatrix}\text{Fraction of sufficiently}\\ \text{energetic collisions}\\ \text{that actually result}\\ \text{in reaction}\end{pmatrix}. \tag{4.9}$$

The first factor on the right is known, for a bimolecular collision, from our previous work in equilibrium kinetic theory. The second factor, the *activation* or *energy factor*, will be evaluated in the next section, again from equilibrium theory. The third factor, the *steric factor* in chemical language, poses an insoluble problem in quantum theory and must be found empirically.

We can now state more explicitly the limitation imposed by the use of equilibrium theory to calculate rates in a system that is actually out of equilibrium. Since a reacting collision must involve at least one molecule with much more than ordinary energy, the reaction will tend to deplete the supply of these highly energetic molecules. But if we are to calculate the relative number of highly energetic molecules from equilibrium considerations, their number must not be seriously affected by this depletion process. This means that the molecular processes that tend to maintain the equilibrium (i.e., Boltzmann) distribution must operate much faster than the chemical processes that use up the highly energetic molecules. Experimental evidence indicates that this condition is fairly well satisfied for those chemical reactions that are slow enough to have been accessible to measurement.

The collision theory, which we have outlined above and which we shall formulate in detail in the following sections, actually leaves much to be desired. This is because it does not in fact come to grips with the basic question of how the molecules actually interact to break one set of chemical bonds and form another. As a result, the steric factor is left completely undetermined, as is also the activation energy that appears in the activation factor. The theory is nevertheless useful as a tool in the understanding and interpretation of experimental data. For an account of the more recent *theory of absolute reaction rates*, which goes more deeply into the underlying processes, see the references listed in the first paragraph of the section or the book by Glasstone, Laidler, and Eyring (1941).

Exercise 4.1. The decomposition of nitrogen dioxide (NO_2) is a second-order reaction with values of the rate constant k_c as follows (from experimental measurements):

T, °K:	592	603.2	627	651.5	656
k_c, cm^3/mole sec:	522	755	1700	4020	5030

By means of suitable plots, find the following:
(a) The Arrhenius activation energy ϵ_a;
(b) The energy ϵ_0 in the equation

$$k_c = CT^{1/2}e^{-\epsilon_0/kT}.$$

5 ENERGY INVOLVED IN COLLISIONS

Pursuing the foregoing ideas, we assume that a molecular collision, if it is to bring about a reaction, must involve an energy greater than a certain value ϵ_0. This turns out to be the activation energy in equation (4.8b) and must eventually be found empirically. To make the foregoing condition quantitative, we shall at first ignore the energy involved in the internal degrees of freedom of the molecules and consider only the kinetic energy of translation. This restriction is merely for simplicity and will be removed later. Not all the energy of translation, however, is of importance. First, the translational energy of the center of mass of the colliding molecules is irrelevant, and only the energy of their *relative* motion can be of importance. This is because two molecules that have identical velocities and hence no relative motion will never collide and hence can never react. Second—and this is open to some argument—we shall assume for the time being that only that part of the relative translational energy *in the direction of the line of centers* at the moment of impact is of consequence. The energy associated with motion perpendicular to the line of centers would correspond to a sideswiping and might therefore be expected to be ineffective in promoting a reaction. Under these assumptions the evaluation of the activation factor resolves itself into the determination of the fraction of collisions in which the relative translational energy along the line of centers at collision exceeds the specified value ϵ_0.

To solve this problem we return to our earlier discussion of molecular collisions in Chapter II, Sec. 6. In particular, we start with the expression (II 6.14) for the rate of bimolecular collisions having relative speed in the range g to $g + dg$ and angle between the relative velocity and the line of centers at impact in the range ψ to $\psi + d\psi$:

$$8\pi^2 n_C n_D \left(\frac{m^*_{CD}}{2\pi kT}\right)^{3/2} d^2_{CD} g^3 e^{-(m^*_{CD}/2kT)g^2} \sin\psi \cos\psi \, d\psi \, dg.$$

Dividing this by the total bimolecular collision rate, given by equation (II 6.15a) as

$$n_C n_D d^2_{CD} \left(\frac{8\pi kT}{m^*_{CD}}\right)^{1/2},$$

we obtain for the fraction of collisions within the aforementioned ranges[7]

$$4\pi^2 \left(\frac{m^*_{CD}}{2\pi kT}\right)^2 g^3 e^{-(m^*_{CD}/2kT)g^2} \sin\psi \cos\psi \, d\psi \, dg.$$

We now evaluate the fraction of collisions in which the component of g in the direction of the line of centers exceeds some value v_0, that is, such that $g \cos\psi \geqq v_0$ or $\cos\psi \geqq v_0/g$. This can be found by integration first with respect to $\cos\psi$ from 1 to v_0/g and then with respect to g from v_0 to ∞. The first integration provides the factor

$$\int_{\cos\psi=1}^{\cos\psi=v_0/g} \sin\psi \cos\psi \, d\psi = -\int_1^{v_0/g} \cos\psi \, d(\cos\psi) = \frac{1}{2}\left(1 - \frac{v_0^2}{g^2}\right).$$

Integrating with respect to g, we then write for the fraction of collisions in question

$$2\pi^2 \left(\frac{m^*_{CD}}{2\pi kT}\right)^2 \int_{v_0}^{\infty} g^3 \left(1 - \frac{v_0^2}{g^2}\right) e^{-(m^*_{CD}/2kT)g^2} \, dg.$$

[7] This last expression is correct as it stands for bimolecular collisions between both like and unlike molecules [cf. paragraph following equation (II 6.15a)]. In the case of the collision rate between like molecules, both of the foregoing expressions would have to be divided by 2, but this factor will cancel out in taking the ratio.

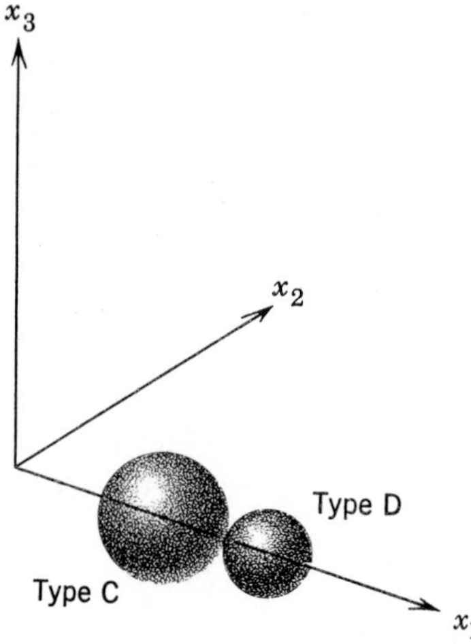

Fig. 3. Alignment of axes for calculation of ϵ_0.

The integral is easily evaluated, by means of the substitution $z = -(m^*_{\text{CD}}/2kT)g^2$ and use of standard integral forms, and has the value

$$\tfrac{1}{2}(2kT/m^*_{\text{CD}})^2 e^{-(m^*_{\text{CD}}/2kT)v_0^2}.$$

We thus obtain finally for the desired fraction of collisions

$$e^{-(m^*_{\text{CD}}/2kT)v_0^2}. \tag{5.1}$$

This reduces correctly to unity when $v_0 = 0$.

The expression $\frac{1}{2}m^*_{\text{CD}}v_0^2$ can be recognized as that part of the kinetic energy of the two molecules associated with their components of velocity taken along the line of centers and measured relative to the center of mass of the molecules. To show this, we can for present purposes align the x_1-axis with the line of centers at the instant of collision (Fig. 3). The kinetic energy in question is then

$$\epsilon_0 = \tfrac{1}{2}m_{\text{C}}(C_1 - W_1)^2 + \tfrac{1}{2}m_{\text{D}}(Z_1 - W_1)^2, \tag{5.2}$$

where C_1 and Z_1 are here the x_1-components of the velocity of the molecules of type C and D respectively and W_1 is the x_1-component of the velocity of the center of mass. But by equations (II 6.6), the values of $(C_1 - W_1)$ and $(Z_1 - W_1)$ corresponding to a relative speed v_0 along the line of centers (i.e., $g_1 \equiv v_0$) are

$$C_1 - W_1 = -\frac{m_{\text{D}}}{m_{\text{C}} + m_{\text{D}}}v_0, \qquad Z_1 - W_1 = \frac{m_{\text{C}}}{m_{\text{C}} + m_{\text{D}}}v_0.$$

We thus find, since $m^*_{\text{CD}} = m_{\text{C}}m_{\text{D}}/(m_{\text{C}} + m_{\text{D}})$,

$$\epsilon_0 = \tfrac{1}{2}m_{\text{C}}\left(\frac{m_{\text{D}}}{m_{\text{C}} + m_{\text{D}}}v_0\right)^2 + \tfrac{1}{2}m_{\text{D}}\left(\frac{m_{\text{C}}}{m_{\text{C}} + m_{\text{D}}}v_0\right)^2 = \tfrac{1}{2}m^*_{\text{CD}}v_0^2.$$

We can thus say from the expression (5.1) that, if $Z(\epsilon_0)$ denotes the frequency of collisions in which the relative translational energy along the line of centers exceeds some specified value $\epsilon_0 = \frac{1}{2}m^*_{\text{CD}}v_0^2$, then the fraction of such collisions compared with the total number is

$$\frac{Z(\epsilon_0)}{Z} = e^{-\epsilon_0/kT}. \tag{5.3}$$

This surprisingly simple expression is a special case of a more general result that covers types of energy other than translational and applies to collisions other than bimolecular. To demonstrate this we must recognize that all types of energy with which we deal can be written, when treated classically, in a similar quadratic form. For example, we have: (1) for kinetic energy of translation, $\frac{1}{2}mC^2$; (2) for kinetic energy of rotation, $\frac{1}{2}I\omega^2$; (3) for potential energy of vibration (as in a harmonic oscillator with spring constant k), $\frac{1}{2}kx^2$. All of these can be represented in general by the "square term" $\epsilon = \frac{1}{2}Az^2$, where A is a constant and z is some continuous variable. The corresponding momentum p in the generalized sense is Az, so that we can write alternatively $\epsilon = p^2/2A$. Let us now analyze the distribution of molecules with respect to the generalized, continuous energy ϵ. This can be done on the basis of the equilibrium, Boltzmann distribution of molecules over groups of quantum energy states, given by equation (IV 6.8) as

$$\frac{N_j^*}{N} = \frac{C_j e^{-\epsilon_j/kT}}{\sum_j C_j e^{-\epsilon_j/kT}}, \tag{5.4}$$

where N_j^* is the number of molecules in the C_j closely spaced states with essentially the same energy ϵ_j. It can be shown quite generally (see, e.g., Mayer and Mayer, 1940, pp. 42 and 126) that this can be used to represent the distribution in terms of continuous variables if these variables and the numbers C_j are appropriately chosen. Specifically, we must take as variables the momentum p and the corresponding generalized "position" coordinate q and set $C_j = dp\,dq/h$, where h is Planck's constant. The number N_j^* is then the number dN^* of molecules in the differential element $dp\,dq/h$. Equation (5.4) thus becomes, with the summation now appropriately replaced by an integral,

$$\frac{dN^*}{N} = \frac{e^{-\epsilon/kT}\,dp\,dq/h}{\int_0^\infty e^{-\epsilon/kT}dp\,dq/h} = \frac{e^{-\epsilon/kT}\,dp}{\int_0^\infty e^{-p^2/2AkT}\,dp} = \frac{e^{-\epsilon/kT}\,dp}{\sqrt{(\pi/2)AkT}}.$$

Since $p\,dp = A\,d\epsilon$ or $dp = A\,d\epsilon/\sqrt{2A\epsilon}$, this can also be written

$$\frac{dN^*}{N} = \frac{\epsilon^{-1/2}e^{-\epsilon/kT}\,d\epsilon}{\sqrt{\pi kT}}. \tag{5.5}$$

This gives the fraction of molecules for which the energy associated with a given square term lies in the range ϵ to $\epsilon + d\epsilon$. It thus gives the probability that the energy associated with the given square term lies in that range.

Let us now consider the joint contribution of two square terms 1 and 2. By means of the foregoing expression, we can write for the joint probability that two square terms have energies simultaneously in the ranges ϵ_1 to $\epsilon_1 + d\epsilon_1$ and ϵ_2 to $\epsilon_2 + d\epsilon_2$

$$\frac{1}{\pi kT}\,\epsilon_1^{-1/2} e^{-\epsilon_1/kT}\,d\epsilon_1\,\epsilon_2^{-1/2} e^{-\epsilon_2/kT}\,d\epsilon_2.$$

We now make the transformation $\epsilon_2 = \eta$ and $\epsilon_1 + \epsilon_2 = \epsilon$, from which $\epsilon_1 = \epsilon - \eta$. Using the Jacobian of the transformation, we obtain

$$d\epsilon_1\,d\epsilon_2 = \frac{\partial(\epsilon_1, \epsilon_2)}{\partial(\epsilon, \eta)}\,d\epsilon\,d\eta = \begin{vmatrix} \dfrac{\partial\epsilon_1}{\partial\epsilon} & \dfrac{\partial\epsilon_1}{\partial\eta} \\ \dfrac{\partial\epsilon_2}{\partial\epsilon} & \dfrac{\partial\epsilon_2}{\partial\eta} \end{vmatrix}\,d\epsilon\,d\eta = \begin{vmatrix} 1 & -1 \\ 0 & 1 \end{vmatrix}\,d\epsilon\,d\eta = d\epsilon\,d\eta,$$

so that the above expression transforms to

$$\frac{1}{\pi kT}(\epsilon - \eta)^{-1/2}\eta^{-1/2}e^{-\epsilon/kT}\,d\epsilon\,d\eta.$$

This represents the probability that the two square terms *together* have a total energy in the range ϵ to $\epsilon + d\epsilon$ while a particular one of them alone has energy in the range η to $\eta + d\eta$. By integrating with respect to η over all its possible values (i.e., from 0 to ϵ) we find the total probability that the two terms together have energy in the range ϵ to $\epsilon + d\epsilon$. This is

$$\frac{1}{\pi kT}e^{-\epsilon/kT}\,d\epsilon\int_0^{\epsilon}(\epsilon - \eta)^{-1/2}\eta^{-1/2}\,d\eta = \frac{1}{kT}e^{-\epsilon/kT}\,d\epsilon, \tag{5.6}$$

where the integral has been evaluated as π by means of the substitution $\eta = \epsilon \sin^2 \alpha$. We can now find the probability that the two square terms together have energy exceeding some value ϵ_0 by integrating this expression from ϵ_0 to ∞. The result is

$$\frac{1}{kT}\int_{\epsilon_0}^{\infty} e^{-\epsilon/kT}\,d\epsilon = e^{-\epsilon_0/kT}. \tag{5.7}$$

This result is identical to that obtained previously in equation (5.3). This is as it should be since, as seen from equation (5.2), the kinetic energy of relative motion along the line of centers is made up basically of two square terms.

By using the expression (5.5) repeatedly in the manner described in the last paragraph (see Hinshelwood, 1940, p. 39), it can be shown that the probability of $2s$ square terms together having energy in the range ϵ to $\epsilon + d\epsilon$ is

$$\frac{e^{-\epsilon/kT}\epsilon^{(s-1)}\,d\epsilon}{(s-1)!\,(kT)^s}.$$

This reduces to the previous result (5.6) when $s = 1$. The probability that the combined energy will exceed ϵ_0 is then

$$\frac{1}{(s-1)!\,(kT)^s}\int_{\epsilon_0}^{\infty} e^{-\epsilon/kT}\epsilon^{(s-1)}\,d\epsilon.$$

This can be integrated successively by parts to obtain

$$e^{-\epsilon_0/kT}\left[\frac{1}{(s-1)!}\left(\frac{\epsilon_0}{kT}\right)^{s-1} + \frac{1}{(s-2)!}\left(\frac{\epsilon_0}{kT}\right)^{s-2} + \cdots + 1\right].$$

The ratio of the first and second terms in this series is $(\epsilon_0/kT)/(s-1)$. If $\epsilon_0 \gg (s-1)kT$, as will usually be the case in situations of interest, the series may therefore be approximated by the first term. We thus obtain finally

$$\left.\begin{array}{l}\text{Probability that } 2s \text{ square}\\ \text{terms have combined}\\ \text{energy exceeding } \epsilon_0\end{array}\right\} \cong \frac{(\epsilon_0/kT)^{s-1}}{(s-1)!}\,e^{-\epsilon_0/kT}. \tag{5.8}$$

This reduces to the result of equations (5.3) and (5.7) when $s = 1$, in which case $\epsilon_0 \gg (s-1)kT = 0$, and the use of the first term in the series is exact.

We have represented the number of square terms by $2s$, where s is an integer, so that the total number of terms will be even. When the number of terms is odd, the development is more involved, and the simple result (5.8) must be modified (see Fowler and Guggenheim, 1956, p. 499). The assessment of the number of square terms that should be included, however, is not sufficiently exact to warrant this complication.

If in a collision between any number of molecules the energy involved can be described by $2s$ square terms, then equation (5.8) gives to a good approximation the fraction of such collisions with total energy greater than ϵ_0. There is nothing in the derivation that restricts the equation to any particular kind of energy. It thus provides the result that we need for the activation factor in the collision theory of reaction rates.

Exercise 5.1. Obtain an integrated expression for the fraction of collisions in which the *total* kinetic energy of relative translational motion exceeds some specified value ϵ_0. Compare your result with the equations preceding equation (5.8). From your comparison, how many square terms would you deduce that the total kinetic energy of relative motion must contain? What does this imply about the relative motion (cf. statements in next to last paragraph of Sec. 6)?

6 RATE EQUATION FOR DISSOCIATION-RECOMBINATION REACTIONS

We shall illustrate the ideas of the foregoing sections by applying them to the dissociation reaction

$$A_2 + M \xrightarrow{k_f} 2A + M. \tag{6.1}$$

Here the dissociation of an A_2-molecule comes about by collision with a second body M. This second body can be another molecule A_2, a dissociated atom A, or some entirely different species. The quantity k_f above the arrow in (6.1) is the rate constant of this so-called "forward" reaction, "forward" meaning that the reaction proceeds from left to right in the chemical equation. The value of k_f will differ depending on the nature of the particular body M under consideration.

To write a rate equation for the reaction (6.1), we begin by writing, in line with the ideas expressed in equation (4.9),

$$\left.\begin{array}{l}\text{Number of reacting}\\ \text{collisions per unit volume}\\ \text{per unit time}\end{array}\right\} = Z\frac{Z(\epsilon_0)}{Z}P,$$

where Z is the bimolecular collision rate, $Z(\epsilon_0)/Z$ is the fraction of collisions involving energy greater than some value ϵ_0 that is required to cause dissociation, and P is the steric factor. Since each reacting collision produces two A-atoms and removes one A_2-molecule, we can write for the rate of change of the molecular concentrations

$$\left(\frac{dn_A}{dt}\right)_f = 2P\frac{Z(\epsilon_0)}{Z}Z, \qquad \text{and} \qquad \left(\frac{dn_{A_2}}{dt}\right)_f = -P\frac{Z(\epsilon_0)}{Z}Z. \tag{6.2}$$

Either of these equations may be taken as the rate equation for the reaction.

The bimolecular collision rate Z is given by equation (4.7) and is, with obvious simplifications in notation,

$$Z = 2\frac{n_{A_2}n_M}{\sigma}\bar{d}^2\left(\frac{2\pi kT}{m^*}\right)^{1/2}, \tag{6.3}$$

where $\sigma = 1$ for $M \not\equiv A_2$ and $\sigma = 2$ for $M \equiv A_2$. The average diameter $\bar{d}$ and the reduced mass m^* are different for different catalytic bodies M. With reference to the fraction $Z(\epsilon_0)/Z$, it is reasonable to assume that the minimum energy ϵ_0 required for dissociation is equal to the dissociation energy D per molecule, known from spectroscopic data. Certainly, less energy would not be expected to cause dissociation.[8] With this assumption, the activation factor is given by equation (5.8) as

$$\frac{Z(\epsilon_0)}{Z} = \frac{Z(D)}{Z} = \left(\frac{D}{kT}\right)^{s-1} \frac{e^{-D/kT}}{(s-1)!}. \qquad (6.4)$$

The value of s is for the time being left unspecified. Putting these expressions into the first of equations (6.2), we can then write finally, in terms of molar concentrations $[Y] = n_Y/\hat{N}$,

$$\left(\frac{d[A]}{dt}\right)_f = 2\left[P\frac{2\hat{N}\bar{d}^2}{\sigma(s-1)!}\left(\frac{2\pi k}{m^*}\right)^{1/2} T^{1/2}\left(\frac{\Theta_d}{T}\right)^{s-1} e^{-\Theta_d/T}\right][A_2][M], \qquad (6.5)$$

where $\Theta_d \equiv D/k$ is the characteristic temperature for dissociation as in Chapter V, Sec. 2. This equation is of the form

$$\boxed{\left(\frac{d[A]}{dt}\right)_f = 2k_f[A_2][M]}, \qquad (6.6a)$$

where the rate constant k_f is identified with the complicated expression in brackets in (6.5). Following the same procedure with the second of equations (6.2), we can also obtain

$$\left(\frac{d[A_2]}{dt}\right)_f = -k_f[A_2][M]. \qquad (6.6b)$$

Equations (6.6) are of the same general form as equations (4.5). As in those equations, the stoichiometric coefficient of the species whose rate is being calculated appears as a factor on the right-hand side of the equation [e.g., 2 in (6.6a) and 1 in (6.6b)]. The exponents z_s in this elementary one-step reaction are seen to be equal to the corresponding stoichiometric coefficients ν_s' in the chemical equation (6.1). The order of the bimolecular reaction is thus 2. The expression for k_f in equation (6.5) has the form previously described in equation (4.8b), that is,

$$\boxed{k_f = C_f T^{\eta_f} e^{-\Theta_d/T}}, \qquad (6.7a)$$

where in this case

$$C_f = P\frac{2\hat{N}\bar{d}^2}{\sigma(s-1)!}\left(\frac{2\pi k}{m^*}\right)^{1/2}\Theta_d^{s-1} \quad \text{and} \quad \eta_f = \tfrac{3}{2} - s. \qquad (6.7b)$$

[8] For other than dissociation reactions such as (6.1), the value of the activation energy ϵ_0 is not so simply obtained and must come from reaction-kinetics experiments. Various rules have also been proposed for the rough calculation of ϵ_0 for certain types of reactions (see Benson, 1960, p. 316).

We now turn our attention to the recombination reaction

$$A_2 + M \xleftarrow[k_b]{} 2A + M, \tag{6.8}$$

where k_b is the rate constant for this so-called "backward" reaction. This reaction requires a three-body collision between two A-atoms and some third body M. The third body is necessary here to maintain the energy balance by carrying off at least part of the chemical energy released by the recombination. In the absence of such a third body, the recombination energy would merely be transferred to the one vibrational mode of the newly formed diatomic molecule and would thus remain to cause the molecule to dissociate again one-half period of vibration later. Recombination is thus just the reverse of dissociation, where the second body M is needed to supply part of the energy absorbed by the dissociation.

Expressions for the rate of the recombination reaction could be obtained by the same procedure as before, except that Z would now have to be the collision rate for termolecular instead of bimolecular collisions. It is simpler, however, to proceed via the already known conditions for chemical equilibrium as follows. For the combined forward and backward reactions, which will always occur simultaneously anyway, the net rate of change of [A] can be written

$$\frac{d[A]}{dt} = \left(\frac{d[A]}{dt}\right)_f + \left(\frac{d[A]}{dt}\right)_b = 2k_f[A_2][M] + \left(\frac{d[A]}{dt}\right)_b.$$

This equation is applicable at all times. In particular, it will apply at equilibrium when $d[A]/dt = 0$. Denoting equilibrium conditions by $[\ ]^*$, we thus obtain for the rate of the backward reaction at equilibrium

$$\left(\frac{d[A]}{dt}\right)_b^* = -2k_f[A_2]^*[M]^*.$$

By the principle of detailed balancing (Chapter II, Sec. 4), this same equation, with, of course, the appropriate value for k_f, must hold for each possible third body in an equilibrium mixture. By the law of mass action (III 9.2a), which must apply at equilibrium, we also have

$$\frac{[A]^{*2}}{[A_2]^*} = K_c(T), \tag{6.9}$$

where K_c is the equilibrium constant for the combined reactions (6.1) and (6.8). We can thus write, in terms of the reactants of the backward reaction,

$$\left(\frac{d[A]}{dt}\right)_b^* = -2\frac{k_f}{K_c}[A]^{*2}[M]^*.$$

We now introduce the common assumption that, for given values of the temperature and the reactant concentrations, the rate at which a reaction proceeds in a given direction is the same whether the system is or is not in equilibrium. This is not necessarily true since other simultaneous non-equilibrium processes, such as vibration, can in principle alter the rate of the reaction. The assumption is, however, the only one consistent with our present quasi-equilibrium approach, which considers the rate constant to be a function only of the temperature and independent of the non-equilibrium variables. It is also borne out by experiment in many cases. On the basis of this assumption, we remove the asterisks from the fore-going equation and write the rate of the recombination reaction in general as

$$\boxed{\left(\frac{d[\mathrm{A}]}{dt}\right)_b = -2k_b[\mathrm{A}]^2[\mathrm{M}]}\,, \tag{6.10}$$

where the backward rate constant k_b is related to k_f and K_c by

$$\boxed{K_c = \frac{k_f}{k_b}}\,. \tag{6.11}$$

This latter important relation is applicable not only to the present dissociation-recombination reaction, but to any simple, single-step reaction.

The recombination reaction is seen from equation (6.10) to be of third order. This is consistent with the necessity for a three-body collision in this elementary, single-step reaction. The equilibrium constant is known accurately from statistical mechanics and is of the form [cf. equation (6.9) and equation (IV 14.9)] $K_c = f(T)e^{-\Theta_d/T}$. As in equation (V 5.6), this can be approximated over a restricted range of temperatures by $K_c = C_c T^{\eta_c} e^{-\Theta_d/T}$. We thus find for the backward rate constant, from equations (6.11) and (6.7a),

$$\boxed{k_b = \frac{C_f}{C_c} T^{\eta_f - \eta_c} = C_b T^{\eta_b}}\,. \tag{6.12}$$

The exponential factor is thus absent from the backward rate constant; that is to say, the activation energy ϵ_0 of the recombination reaction is zero. The variation of k_b with T is thus much less than that of k_f. This is a result peculiar to recombination and certain other reactions. For reactions in general, both the forward and backward rate processes have a nonzero activation energy.

By combining equations (6.6a) and (6.10), a rate equation can be obtained for the combined dissociation-recombination reaction

$$A_2 + M \underset{k_b}{\overset{k_f}{\rightleftharpoons}} 2A + M. \tag{6.13}$$

This can be written, with the aid of equation (6.11),

$$\boxed{\frac{d[A]}{dt} = 2k_f[M]\left\{[A_2] - \frac{1}{K_c}[A]^2\right\}} \ . \tag{6.14}$$

When the system is in equilibrium (i.e., $d[A]/dt = 0$), the quantity in braces must be zero, and we reproduce, as we should, the equilibrium relation (6.9). If the system is out of equilibrium, for example in such a way that $[A]$ is greater than its equilibrium value, we see from equation (6.9) that $[A]^2/K_c > [A_2]$, and it follows from equation (6.14) that $d[A]/dt$ is negative. The system thus tends necessarily toward equilibrium.

Methods similar to the foregoing can be used to treat the reaction

$$AB + M \underset{k_b}{\overset{k_f}{\rightleftharpoons}} A + B + M. \tag{6.15}$$

This leads to the following equation analogous to (6.14):

$$\frac{d[A]}{dt} = k_f[M]\left\{[AB] - \frac{1}{K_c}[A][B]\right\}. \tag{6.16}$$

Here the initial factor of 2 that appears in (6.14) is missing, since the stoichiometric coefficient of A in equation (6.15) is 1 as compared with 2 in equation (6.13). The equation thus conforms with our previous convention for the definition of the rate constant.

If we are concerned with a mixture containing several possible catalytic bodies M, then equations of the type (6.14) or (6.16) must be summed over the various possible reactions. For example, in a mixture of dissociating-recombining A_2 and A, either of the two species may be the catalytic body. We must therefore take into account both of the reactions

$$A_2 + A \underset{k_{b,1}}{\overset{k_{f,1}}{\rightleftharpoons}} 2A + A \quad \text{and} \quad A_2 + A_2 \underset{k_{b,2}}{\overset{k_{f,2}}{\rightleftharpoons}} 2A + A_2. \tag{6.17}$$

Writing equation (6.14) for each of these reactions and adding, we obtain for the total change in [A]

$$\frac{d[A]}{dt} = 2\{k_{f,1}[A] + k_{f,2}[A_2]\}\left\{[A_2] - \frac{1}{K_c}[A]^2\right\}. \tag{6.18}$$

In the present state of knowledge, values of the rate constants k must be obtained experimentally. Collision theory is useful, despite its grossly oversimplified character, in helping to account for the proper form of these quantities. Because of the presence of the unknown steric factor P, however, complete numerical values cannot be obtained from the theory alone. The proper value for s, which is related to the number of square terms that contribute *effectively* to the energy of collision, is also open to doubt. Thus, in the final analysis, the values of C and η in equations (6.7a) or (6.12) must come from experiment.

To make a rough estimate of s, and hence of η_f [see equation (6.7b)], we can count up the number of square terms involved in a bimolecular collision. As we have seen, the relative translational energy along the line of centers provides two such terms. Whether or not the relative translation perpendicular to the line of centers should be taken into account is open to doubt. If it were, it would provide two more square terms. This is true because, to an observer moving with the center of mass, the motion of the molecules must appear coplanar (this will be shown in Chapter IX, Sec. 7) with the result that there are two transverse components of velocity to be taken into account. As regards vibration, there are two square terms for each vibrational mode, one for kinetic energy and one for potential energy. Thus, if we are concerned with diatomic molecules, there are four square terms present for vibration in a molecule-molecule collision and two in a molecule-atom collision. There is some doubt in the former case, however, as to whether the vibration of the catalytic molecule can be effective, so that the number of effective terms may be two in both instances. As regards rotation, there are again four square terms present in a molecule-molecule collision and two in a molecule-atom collision. It can be shown, however, that the requirements of conservation of angular momentum keep two of these from being effective in both cases. The number of effective square terms is thus 2 and 0 respectively. The entire situation can be summarized as follows:

	Mol.-Mol.	Mol.-Atom
Relative translation along line of centers	2	2
Relative translation perpendicular to line of centers	2 or 0	2 or 0
Vibration	4 or 2	2
Rotation	2	0
Total $= 2s$ =	10 to 6	6 to 4

We see that there is a wide choice possible for s, depending on what assumptions we are willing to make. The experimental evidence indicates that s is definitely larger than the value of unity that would be indicated if only the translational energy along the line of centers were considered, although how much larger is difficult to say. The problem is complicated by the fact that the overshadowing exponential variation with T in equation (6.7a) makes the experimental determination of η_f with any precision a difficult matter.

In all the foregoing we have ignored the coupling that must exist to some extent between the chemical reactions and other nonequilibrium processes, particularly vibration. Such phenomena, which may be expected to become more important as the temperature increases, have only recently begun to be investigated (see, for example, Wray, 1962c, Treanor and Marrone, 1962, Marrone and Treanor, 1963, and Heims, 1963). Their final understanding will probably require considerable experimental study of the detailed processes.

Exercise 6.1. Consider the dissociation reaction

$$O_2 + O_2 \xrightarrow{k_f} 2O + O_2$$

taking place in a mixture having a temperature of 6500°K. Other things being equal, what would be the numerical ratio between the values of the rate constant k_f calculated by means of collision theory on the basis of the following two alternative assumptions regarding the types of energy effective in the dissociation collisions: (a) The energy of relative translation along the line of centers at impact and the energy of vibration of both colliding molecules (but no other types of energy) are effective. (b) Only the energy of relative translation along the line of centers is effective. *Ans.:* 41 (approx.).

7 RATE EQUATION FOR COMPLEX MIXTURES

In a complex gas mixture, such as high-temperature air, a considerable number of chemical reactions can occur simultaneously. The rate equations applicable to such a mixture can be obtained as follows.

Consider a general gas mixture containing a total of l species X_s capable of undergoing y elementary chemical reactions. The chemical equation for the general elementary reaction r can be written, similarly to equation (4.1b), as

$$\sum_{s=1}^{l} \nu'_{s,r} X_s \underset{k_{b,r}}{\overset{k_{f,r}}{\rightleftharpoons}} \sum_{s=1}^{l} \nu''_{s,r} X_s, \tag{7.1}$$

where $\nu'_{s,r}$ and $\nu''_{s,r}$ are the stoichiometric coefficients appearing on the left and right in the reaction r and $k_{f,r}$ and $k_{b,r}$ are the forward and backward rate constants. On the basis of equations (4.5), the rate of change of the concentration of any species X_s as a result of this particular elementary reaction is

$$\left(\frac{d[X_s]}{dt}\right)_r = (\nu''_{s,r} - \nu'_{s,r})k_{f,r}\prod_{s=1}^{l}[X_s]^{\nu'_{s,r}} + (\nu'_{s,r} - \nu''_{s,r})k_{b,r}\prod_{s=1}^{l}[X_s]^{\nu''_{s,r}}. \quad (7.2)$$

By applying this at equilibrium and invoking the law of mass action plus the same assumptions that lead to equation (6.10), we can write again that

$$K_{c,r} = \frac{k_{f,r}}{k_{b,r}}. \quad (7.3)$$

Here $K_{c,r}$ is the equilibrium constant in the law of mass action (III 9.2a) applied to the reaction (7.1), namely,

$$\frac{\prod_{s=1}^{l}[X_s]^{*\nu''_{s,r}}}{\prod_{s=1}^{l}[X_s]^{*\nu'_{s,r}}} = K_{c,r}. \quad (7.4)$$

With equation (7.3), equation (7.2) can be written

$$\left(\frac{d[X_s]}{dt}\right)_r = (\nu''_{s,r} - \nu'_{s,r})k_{f,r}\left\{\prod_{s=1}^{l}[X_s]^{\nu'_{s,r}} - \frac{1}{K_{c,r}}\prod_{s=1}^{l}[X_s]^{\nu''_{s,r}}\right\}. \quad (7.5)$$

To find the total rate of change of $[X_s]$ we must sum equation (7.5) over all reactions r. We thus obtain finally

$$\boxed{\frac{d[X_s]}{dt} = \sum_{r=1}^{y}(\nu''_{s,r} - \nu'_{s,r})k_{f,r}\left\{\prod_{s=1}^{l}[X_s]^{\nu'_{s,r}} - \frac{1}{K_{c,r}}\prod_{s=1}^{l}[X_s]^{\nu''_{s,r}}\right\}}. \quad (7.6)$$

This is the general rate equation for a complex gas mixture. To make sure that he understands its application, the reader should satisfy himself that it reproduces equation (6.18) when applied formally to the simultaneous reactions (6.17).

8 HIGH-TEMPERATURE AIR

The kinetics of high-temperature air is a complex problem that is the subject of current research. Review articles appear from time to time in the literature, the most recent generally available one being that by Wray (1962a), who gives numerous references to the original sources. As we have seen in Chapter V, Sec. 6, if we limit ourselves to temperatures below about 8000°K, the only species present in sizeable amounts are O_2, N_2, O, N, and NO. Consideration of the numerous reactions possible between these species indicates that the following six are significant for nonequilibrium gas-dynamic calculations (Wray, 1962a, Lin and Teare, 1963):

$$
\begin{array}{lll}
r = 1: & O_2 + M \rightleftharpoons 2O + M & 5.12 \text{ ev} \\
2: & N_2 + M \rightleftharpoons 2N + M & 9.76 \text{ ev} \\
3: & NO + M \rightleftharpoons N + O + M & 6.49 \text{ ev} \\
4: & NO + O \rightleftharpoons O_2 + N & 1.37 \text{ ev} \\
5: & N_2 + O \rightleftharpoons NO + N & 3.27 \text{ ev} \\
6: & N_2 + O_2 \rightleftharpoons 2NO. & 1.90 \text{ ev}
\end{array}
$$

The reactions are all written so that the forward reaction is endothermic; the net amount of energy required to form the molecules on the right from the molecules on the left is given in electron-volts beside the equation. The first three reactions are the neutral-particle dissociation-recombination reactions used in the equilibrium considerations of Chapter V, Sec. 5; each one actually represents a number of possible reactions, one for each catalytic body M. The fourth and fifth reactions are rearrangement reactions involving NO; they provide nitrogen atoms at a much lower energy input than is required for the direct dissociation of the nitrogen molecule. The sixth reaction is the direct bimolecular reaction for the formation of NO. Only three of the six reactions in this list are linearly independent in the sense discussed in Chapter V, Sec. 5. In rate calculations, all the significant elementary reactions must be included. A linearly independent set is not sufficient, since it is the actual mechanism of reaction that is important, not just the overall balance.

If it is desired to include a first approximation to the ionization rate processes, the significant reaction is *not* the simple dissociation-recombination reaction used in Chapter V, Sec. 5:

$$NO + M \rightleftharpoons NO^+ + e + M. \qquad 9.25 \text{ ev}$$

In a noble gas such as argon, this type of reaction, with an electron serving as the catalytic body M, is indeed the important one (see Petschek and Byron, 1957). For nitric oxide, however, because of its diatomic nature, the first important reaction is the associative-ionization reaction

$$r = 7: \qquad N + O \rightleftharpoons NO^+ + e. \qquad 2.76 \text{ ev}$$

(For a discussion of this and the host of other ionization reactions that can occur, see Lin and Teare, 1963.) Here, because of the chemical energy released by the association of the N and O, the net energy that must be supplied is considerably less than the full ionization potential of 9.25 ev needed in the earlier reaction.

Experimental values for the constants in the equation for the forward rate constant

$$k_f = C_f T^{\eta_f} e^{-\epsilon_{0_f}/kT} \tag{8.1}$$

are given in Table B for reactions $r = 1$ through 7. These values were

Table B Values of Constants in Equation (8.1) *for* k_f

Reaction r	Catalytic Body	C_f	η_f	ϵ_{0_f}/k, °K	Approximate Range of T, °K	Source
1	Ar, (N), (NO)	3.6×10^{19}	-1.0	59,500	3300–7500	Camac and Vaughan (1961)
	(N_2)	4.8×10^{20}	-1.5	59,500	—	—
	O_2	1.9×10^{21}	-1.5	59,500	2800–5000	Byron (1959)
	O	6.4×10^{23}	-2.0	59,500	5000–7500	Camac and Vaughan (1961)
2	Ar, (O), (O_2), (NO)	1.9×10^{17}	-0.5	113,000	6000–9000	Byron (see Wray, 1962a)
	N_2	4.8×10^{17}	-0.5	113,000	6000–9000	Byron (see Wray, 1962a)
	N	4.1×10^{22}	-1.5	113,000	6000–9000	Byron (see Wray, 1962a)
3	Ar, (N_2), (O_2)	3.9×10^{20}	-1.5	75,500	4200–6600	Wray and Teare (1962)
	NO, (O), (N)	7.9×10^{21}	-1.5	75,500	3000–8000	Wray and Teare (1962)
4	—	3.2×10^{9}	$+1.0$	19,700	450–5000	Wray and Teare (1962)
5	—	7.0×10^{13}	0	38,000	2000–3000	Glick et al. (1957)
6	—	4.6×10^{24}	-2.5	64,600	1400–4300	Freedman and Daiber (1961)
7	—	6.5×10^{11}	0	31,900	4000–5000	Lin and Teare (1963)

selected by the authors from the available data; their source is shown in the table. The units in the table are such that k_f is in all cases in cm^3/mole sec. For the catalytic bodies listed without parentheses in reactions 1, 2, and 3, the values are experimental results taken from the stated sources; for the bodies in parentheses, they are merely estimates. The values given in this and other listings may at first appear considerably different. Such differences are due, in those cases where reliable data are available, primarily to differences in judgement regarding the choice of the value for η_f. In the range of temperature in which the data were obtained, the actual values of k_f calculated from the various listings will in most cases be in reasonable agreement. For catalytic bodies for which data are available, the rate constant for oxygen dissociation is known at the present writing to a factor of about 2. The rates for the remaining reactions are known with varying degrees of accuracy up to a factor of about 10. Since kinetic data are difficult to obtain, this sort of accuracy is not unusual. There is no discernable consistency to the observed variation of η_f; we can only say that the resulting value of s [see equation (6.7b)] is almost always greater than unity. In view of the tentative nature of some of the results in Table B, the reader should consult the latest sources and review articles before embarking on any serious calculations.

To find the backward rate constant k_b, we can use the foregoing data for k_f together with the relation $K_c = k_f/k_b$. For engineering purposes the equilibrium constant K_c can be obtained or deduced for the reactions of Table B from the data given in connection with equation (V 5.6).

9 SYMMETRICAL DIATOMIC GAS; IDEAL DISSOCIATING GAS

We shall be interested particularly in the symmetrical diatomic gas, whose rate equation has already been given in equation (6.18). It will be convenient to have this equation in terms of the degree of dissociation α introduced in Chapter V, Sec. 2. Since α is equal to the mass fraction of dissociated A-atoms [see equation (V 2.2)], we can write

$$[\mathrm{A}] = \frac{\rho\alpha}{\hat{M}_\mathrm{A}} \qquad \text{and} \qquad [\mathrm{A}_2] = \frac{\rho(1-\alpha)}{\hat{M}_{\mathrm{A}_2}} = \frac{\rho(1-\alpha)}{2\hat{M}_\mathrm{A}}, \tag{9.1}$$

where $\hat{M}_\mathrm{Y}$ is the molecular weight of the particular species Y. The chemical rate equations that we have discussed all give the rate of change of concentration [Y] as a result purely of chemical reactions. As they stand, therefore, they give $d[\mathrm{Y}]/dt$ for a closed system of fixed volume. This is because either changes in volume or the introduction of additional Y from the outside would cause a change in concentration even in the absence of chemical reactions. We accordingly regard ρ as a constant in substituting (9.1) into (6.18) and obtain for the rate equation in terms of α

$$\boxed{\frac{d\alpha}{dt} = \left(k_{f,1}\alpha + k_{f,2}\frac{1-\alpha}{2}\right)\frac{\rho}{\hat{M}_\mathrm{A}}\left[(1-\alpha) - \frac{2\rho}{\hat{M}_\mathrm{A}K_c}\alpha^2\right]}. \tag{9.2}$$

We see that $d\alpha/dt$ is a function of the three quantities T, ρ, and α, with T appearing through the equilibrium constant K_c and the rate constants k_f.

Equation (9.2) has been specialized by Freeman (1958) to a form appropriate for use with the ideal dissociating gas discussed in Chapter V, Sec. 3. To do this, we consider the equilibrium condition $d\alpha/dt = 0$, in which case equation (9.2) gives

$$\frac{\alpha^{*2}}{1-\alpha^*} = \frac{\hat{M}_A K_c}{2\rho}.$$

This may be compared with the law of mass action for the ideal dissociating gas [equation (V 3.2)]:

$$\frac{\alpha^{*2}}{1-\alpha^*} = \frac{\rho_d}{\rho} e^{-\Theta_d/T}.$$

We see that to make equation (9.2) reduce to the result appropriate to the ideal dissociating gas at equilibrium, we need only replace $2\rho/\hat{M}_A K_c$ by $(\rho/\rho_d)e^{\Theta_d/T}$. Freeman assumes further that the first parenthetical factor in (9.2) is for all practical purposes independent of α and can be written as a single rate constant of the form of equation (6.7a), that is,

$$\left(k_{f,1}\alpha + k_{f,2}\frac{1-\alpha}{2}\right) = C_f T^{\eta_f} e^{-\Theta_d/T}.$$

With these changes equation (9.2) can be approximated by

$$\boxed{\frac{d\alpha}{dt} = CT^{\eta}\rho\left[(1-\alpha)e^{-\Theta_d/T} - \frac{\rho}{\rho_d}\alpha^2\right]}, \tag{9.3}$$

where $C \equiv C_f/\hat{M}_A$ and $\eta \equiv \eta_f$ are constants. This rate equation is often used for approximate studies of the reacting flow of a symmetrical diatomic gas. The combination $(CT^{\eta}\rho)^{-1}$ has the dimensions of time and may be taken as a kind of characteristic time of the rate process.

The change in entropy associated with chemical reactions has already been discussed in general terms in Chapter III. For later work with the symmetrical diatomic gas it will be useful to have the Gibbs equation (III 6.8) in terms of α. Writing this equation for a unit mass of gas and applying it to the mixture of A and A_2, we have

$$ds = \frac{1}{T}\,de + \frac{p}{T}\,d\left(\frac{1}{\rho}\right) - \frac{1}{T}\left\{\hat{\mu}_A\,d\left(\frac{[A]}{\rho}\right) + \hat{\mu}_{A_2}\,d\left(\frac{[A_2]}{\rho}\right)\right\},$$

where $\hat{\mu}_Y$ is the chemical potential per mole of species Y. Substitution from equations (9.1) then gives

$$ds = \frac{1}{T}\,de + \frac{p}{T}\,d\left(\frac{1}{\rho}\right) + \frac{1}{T}(\mu_{A_2} - \mu_A)\,d\alpha, \tag{9.4}$$

where $\mu_Y = \hat{\mu}_Y/\hat{M}_Y$ is now the chemical potential per unit mass of species Y. When the gas mixture is in equilibrium, it follows from equation (III 7.2) that $2\hat{\mu}_A^* - \hat{\mu}_{A_2}^* = 0$. In terms of the combination of potentials per unit mass appearing in equation (9.4), this becomes $\mu_{A_2}^* - \mu_A^* = 0$. Equation (9.4) is general for any symmetrical diatomic gas. To apply it to the ideal dissociating gas it would be necessary to evaluate the chemical potentials appropriately for the constituents of that gas (cf. Chapter V, Sec. 3).

Exercise 9.1. (a) If α^* denotes the equilibrium degree of dissociation corresponding to arbitrary values of ρ and T, convert the rate equation (9.3) for the ideal dissociating gas to the equivalent pseudo-relaxation form

$$\frac{d\alpha}{dt} = F \times (\alpha^* - \alpha),$$

where F is a specific function of ρ, T, α, and α^* (but does not include ρ_d).

(b) Assuming that ρ and T are constant, obtain an expression relating α and t for an ideal dissociating gas that is initially out of equilibrium at some given value α_0 at $t = 0$.

(c) Assuming that the departures from equilibrium are *at all times* small, approximate the result of (b) to a form linear in these departures. Compare your result with the results for vibrational nonequilibrium in Sec. 2. In the light of this comparison, how is chemical nonequilibrium like vibrational nonequilibrium and how is it different?

Exercise 9.2. The dissociation-recombination of hydrogen iodide takes place according to the reaction

$$2\,\text{HI} \underset{k_b}{\overset{k_f}{\rightleftharpoons}} \text{H}_2 + \text{I}_2.$$

Both the forward and backward reactions take place by means of a bimolecular mechanism, no catalytic molecule being required. We consider a mixture of HI, H_2, and I_2 that is composed entirely of HI at sufficiently low temperatures. Assume that the mass and volume of the mixture are fixed. If α_{H_2} denotes the mass fraction of H_2 (i.e., the mass of H_2 per unit mass of mixture), derive the following chemical rate equation:

$$\frac{d\alpha_{H_2}}{dt} = \rho k_f \left\{ 4\hat{M}_{H_2} \left[\frac{1 - (1 + \hat{M}_{I_2}/\hat{M}_{H_2})\alpha_{H_2}}{\hat{M}_{I_2} + \hat{M}_{H_2}} \right]^2 - \frac{\alpha_{H_2}^2}{K_c \hat{M}_{H_2}} \right\},$$

where $\hat{M}_{I_2}$ and $\hat{M}_{H_2}$ are the molecular weights of I_2 and H_2.

10 GENERALIZED RATE EQUATION

We now take up some matters that apply to both the vibrational and chemical rate equations. If in the vibrational analysis we consider the heat bath to be the translational and rotational degrees of freedom of the vibrating species itself, the vibrational rate equation (2.11) applied to a unit mass of a single, vibrating gas is

$$\frac{de_v}{dt} = \frac{e_v^*(T) - e_v}{\tau(\rho, T)}. \tag{10.1}$$

Here we consider τ as a function of ρ and T in place of the previous p and T, the change being made by means of the thermal equation of state. As a representative chemical rate equation we take that for the symmetrical diatomic gas. In the form (6.18) this is

$$\frac{d[A]}{dt} = 2\{k_{f,1}(T)[A] + k_{f,2}(T)[A_2]\}\left\{[A_2] - \frac{1}{K_c(T)}[A]^2\right\}, \tag{10.2}$$

and in the form (9.2)

$$\frac{d\alpha}{dt} = \left[k_{f,1}(T)\alpha + k_{f,2}(T)\frac{1-\alpha}{2}\right]\frac{\rho}{\hat{M}_A}\left[(1-\alpha) - \frac{2\rho}{\hat{M}_A K_c(T)}\alpha^2\right]. \tag{10.3}$$

As has already been observed in connection with the chemical equations, these rate equations apply as derived to a closed system of fixed volume. They also assume a constant temperature T. By necessity they are customarily extended to open systems of varying volume and temperature by identifying ρ and T as the instantaneous values for the system. This procedure, which has already been mentioned in a more restricted sense in Sec. 2, implies that the instantaneous chemical or vibrational rate process is not altered by the presence of the additional rate processes necessary to change the mass, volume, or temperature. This is in the spirit of the entire quasi-equilibrium nature of the discussion leading to the rate equations; it is probably valid so long as the additional rate processes are not too rapid.

When the foregoing extension is made, changes in the quantity whose rate is being studied may come from changes in mass and volume as well as from the internal rate process. The quantities given by the foregoing rate equations should therefore be distinguished by $(de_v/dt)_{\text{vib}}$, $(d[A]/dt)_{\text{chem}}$, and $(d\alpha/dt)_{\text{chem}}$ to emphasize that they are now merely the contribution of the internal rate process noted in the subscript. Care must then be taken in calculating the total rate for any given quantity in any given system to include all other effects that may be pertinent. A simple instance of this is given in Chapter VIII, Sec. 2.

The rates of the internal processes as given by the foregoing equations depend only on the instantaneous state of the system. If q denotes a general nonequilibrium variable measured per unit mass [e_v in equation

(10.1), α in equation (10.3)], then equations (10.1) and (10.3) can both be written in the general form

$$\boxed{\left(\frac{dq}{dt}\right)_{\text{int proc}} = \omega(\rho, T, q)}\,. \tag{10.4}$$

Here ω denotes the complete function on the right, and the three variables ρ, T, q specify the state of the system. Rate equations for other than dissociation-recombination reactions can also be put in this form. By using this generalized equation we shall avoid the necessity for distinguishing between vibrational and chemical nonequilibrium (or in the latter case of specifying the particular chemical reaction) in much of our later theory.

11 LOCAL RELAXATION TIME; SMALL DEPARTURES FROM EQUILIBRIUM

For gas-dynamic purposes, it will be useful to reformulate equation (10.4) by generalizing the idea of a local relaxation time, which appears already in a special form in equation (10.1). To this end, we observe from equation (10.4) that the function ω has the dimensions of q divided by time. The partial derivative $(\partial\omega/\partial q)_{\rho,T}$ must therefore have the dimensions of $(\text{time})^{-1}$. We can thus define a local characteristic time τ of the rate process by

$$\boxed{\tau(\rho, T, q) \equiv -\frac{1}{(\partial\omega/\partial q)_{\rho,T}}}\,. \tag{11.1}$$

The negative sign has been introduced since τ should be positive and the value of the derivative is itself always negative.[9] The meaning of τ will be examined more precisely below. With the foregoing definition, the rate equation (10.4) can be rewritten

$$\boxed{\left(\frac{dq}{dt}\right)_{\text{int proc}} = \frac{\chi(\rho, T, q)}{\tau(\rho, T, q)}}\,, \tag{11.2}$$

[9] This can be demonstrated from the fact that in a stable system a change in q away from an equilibrium situation must give rise to a sign of dq/dt that tends to return the system to equilibrium.

where the function $\chi(\rho, T, q)$ is defined by

$$\chi(\rho, T, q) \equiv -\frac{\omega}{(\partial\omega/\partial q)_{\rho,T}}. \tag{11.3}$$

For the vibrational rate equation, the original equation (10.1) is already in the form (11.2) and the foregoing formalism was not really necessary. The reader can satisfy himself that when it is applied to equation (10.1) it merely reproduces that equation. The meaning of τ in that case is therefore clear.

To examine the meaning of τ in the general case, we must first generalize the notion of a local equilibrium condition, which appears also in a special form in (10.1). To do this we consider equation (11.2) and return temporarily to the original situation where ρ and T were constant. Under these conditions an equilibrium state can be thought to actually exist and corresponds to the condition $dq/dt = (dq/dt)_{\text{int proc}} = 0$. This defines, for all finite values of τ, a true equilibrium value of q—say q^*—given according to equation (11.2) by

$$\chi(\rho, T, q^*) = 0, \tag{11.4a}$$

from which we obtain

$$q^* = q^*(\rho, T). \tag{11.4b}$$

This expression for q^* must agree with that from equilibrium thermodynamics or statistical mechanics. In the completely time-dependent situation, when $\rho = \rho(t)$ and $T = T(t)$, it is useful to introduce a *local equilibrium value* $q^* = q^*(t)$. This is by definition to be calculated formally from equations (11.4), that is,

$$q^*(t) \equiv q^*[\rho(t), T(t)], \tag{11.5a}$$

or

$$\chi[\rho(t), T(t), q^*(t)] = 0. \tag{11.5b}$$

It is important to note that, so long as τ is non-zero, $q^*(t)$ cannot, in general, actually exist in the system over any interval of time as the result purely of internal processes. This follows from the fact that *if* it could, then q in both sides of equation (11.2) would be equal to q^* and the equation would lead, in view of equation (11.5b), to the conclusion that $dq^*/dt = (dq^*/dt)_{\text{int proc}} = 0$ or $q^* = \text{const.}$ But this is in contradiction to the requirement of the definition (11.5a) that q^* is a function of t.[10] The local equilibrium value must therefore be regarded, except in certain special situations, as purely fictitious. It is the value that q *would* have if

[10] This contradiction could be eliminated by taking $\tau = 0$, i.e., zero characteristic time for the rate process. This physically impossible but important singular case will be treated in a gas-dynamical context in the next chapter.

the system *were* in equilibrium at the local values of ρ and T. It has no relation to the value of q that actually exists at any time in a general nonequilibrium situation. The quantity $e_v^*(T)$ with $T = T(t)$ in equation (10.1) is an example of a local equilibrium value.

To see the meaning of τ we now consider nonequilibrium states that are a small departure from a known reference state that is independent of time. If $(\ )^{(r)}$ denotes the reference state and $(\ )'$ denotes the departures therefrom, we have

$$\rho = \rho^{(r)} + \rho',\ T = T^{(r)} + T',\ q = q^{(r)} + q',\ q^* = q^{*(r)} + q^{*\prime}.$$

For the reference state to be independent of time it must be an equilibrium state, that is, $q^{(r)} = q^{*(r)}$. Expanding the function χ of equation (11.2) about the reference state, we write to the first order in the departure quantities

$$\chi(\rho, T, q) = \cancel{\chi(\rho^{(r)}, T^{(r)}, q^{*(r)})} + \left(\frac{\partial \chi}{\partial \rho}\right)_{T,q}^{(r)} \rho' + \left(\frac{\partial \chi}{\partial T}\right)_{\rho,q}^{(r)} T' + \left(\frac{\partial \chi}{\partial q}\right)_{\rho,T}^{(r)} q', \tag{11.6}$$

where the leading term is zero by virtue of equation (11.4a). This expansion is exemplified in Fig. 4, where constant values of χ are to be imagined as represented by surfaces in a ρ, T, q-space. In the same manner, $\chi(\rho, T, q^*)$,

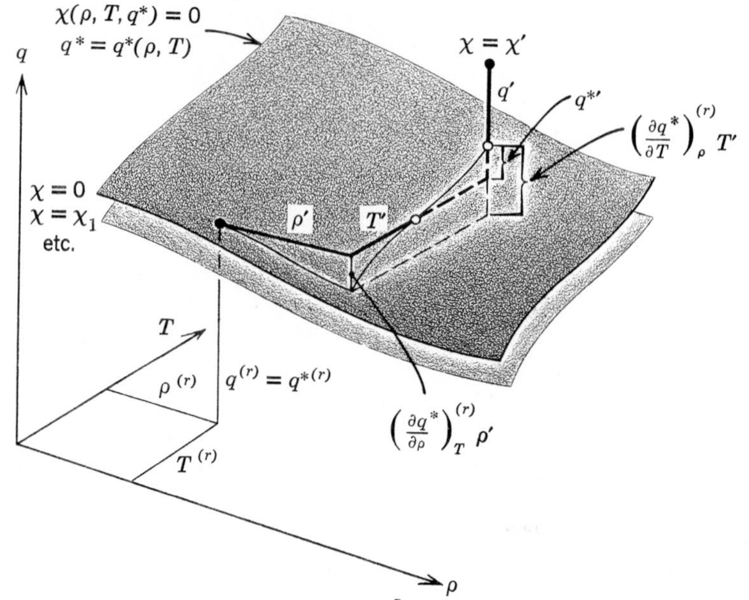

Fig. 4. Illustration of small departures from an equilibrium reference state.

which is always equal to zero by equation (11.5b), can be expanded to obtain

$$0 = \chi(\rho, T, q^*) = \chi(\rho^{(r)}, T^{(r)}, q^{*(r)}) + \left(\frac{\partial \chi}{\partial \rho}\right)_{T,q}^{(r)} \rho' + \left(\frac{\partial \chi}{\partial T}\right)_{\rho,q}^{(r)} T' + \left(\frac{\partial \chi}{\partial q}\right)_{\rho,T}^{(r)} q^{*\prime}. \tag{11.7}$$

This may be thought of (see Fig. 4) as giving the displacement $q^{*\prime}$ necessary to stay on the equilibrium surface $\chi(\rho, T, q^*) = 0$ for the same displacements ρ', T' appearing in equation (11.6). Subtracting (11.7) from (11.6) we thus obtain to the first-order

$$\chi(\rho, T, q) = -\left(\frac{\partial \chi}{\partial q}\right)_{\rho,T}^{(r)} (q^{*\prime} - q'). \tag{11.8}$$

To evaluate the derivative in this equation, we first rewrite equation (11.3) as $\chi(\partial\omega/\partial q)_{\rho,T} = -\omega$ and differentiate with respect to q to obtain

$$\left(\frac{\partial \chi}{\partial q}\right)_{\rho,T}\left(\frac{\partial \omega}{\partial q}\right)_{\rho,T} + \chi\left(\frac{\partial^2 \omega}{\partial q^2}\right)_{\rho,T} = -\left(\frac{\partial \omega}{\partial q}\right)_{\rho,T}.$$

Application of this relation at the reference state, where $\chi^{(r)} = \chi(\rho^{(r)}, T^{(r)}, q^{*(r)}) = 0$, then gives

$$\left(\frac{\partial \chi}{\partial q}\right)_{\rho,T}^{(r)} = -1.$$

Using equation (11.8) with this value for the derivative and writing $\tau = \tau^{(r)} + \tau' = \tau(\rho^{(r)}, T^{(r)}, q^{*(r)}) + \tau'$, we may approximate equation (11.2) by the linear equation

$$\boxed{\left(\frac{dq'}{dt}\right)_{\text{int proc}} = \frac{q^{*\prime}(\rho', T') - q'}{\tau(\rho^{(r)}, T^{(r)}, q^{*(r)})}}. \tag{11.9}$$

This is the general rate equation for small departures from an equilibrium reference state.

The meaning of τ as defined by equation (11.1) is now apparent. If ρ' and T' are zero, $q^{*\prime}(\rho', T')$ is zero and equation (11.9) can be integrated in the same manner as was done with equation (2.9). The quantity τ can thus be interpreted as the relaxation time for small departures of q from an equilibrium state at constant ρ and T. When the departures are large and ρ and T are not constant, it may be referred to with less precision as a local relaxation time.

Equation (11.9) will also be of direct use as the linearized form of equation (11.2). It shows, in terms of Fig. 4, that to a first order the rate depends only on the net displacement $(q' - q^{*\prime})$ from the equilibrium surface in the direction parallel to the q-axis. The quantity $q^{*\prime}$ in the equation can be found explicitly in terms of ρ', T' by solving equation (11.7). This gives

$$q^{*\prime}(\rho', T') = -\frac{(\partial\chi/\partial\rho)^{(r)}_{T,q}}{(\partial\chi/\partial q)^{(r)}_{\rho,T}}\rho' - \frac{(\partial\chi/\partial T)^{(r)}_{\rho,q}}{(\partial\chi/\partial q)^{(r)}_{\rho,T}}T'. \tag{11.10a}$$

The ratio of derivatives in the first term on the right can be written, from the usual relations for partial derivatives, as

$$-\frac{(\partial\chi/\partial\rho)^{(r)}_{T,q}}{(\partial\chi/\partial q)^{(r)}_{\rho,T}} = \left(\frac{\partial q}{\partial\rho}\right)^{(r)}_{T,\chi} = \left(\frac{\partial q^*}{\partial\rho}\right)^{(r)}_{T}.$$

The second equality follows from the fact that the reference state is an equilibrium state, which means that the middle derivative is taken at the constant value $\chi = 0$ and hence along the equilibrium surface $q = q^*$. A similar result applies for the ratio of derivatives in the second term. Equation (11.10a) thus becomes finally

$$\boxed{q^{*\prime}(\rho', T') = \left(\frac{\partial q^*}{\partial\rho}\right)^{(r)}_{T}\rho' + \left(\frac{\partial q^*}{\partial T}\right)^{(r)}_{\rho}T'}. \tag{11.10b}$$

This is the linearized form of the defining equation (11.5a) for $q^*[\rho(t), T(t)]$. Obviously, the value of $q^{*\prime}(\rho', T')$ depends, not only on ρ' and T' as recognized in the notation, but on $\rho^{(r)}$ and $T^{(r)}$ as well.

It is important to note that the foregoing kind of development is not unique. In line with our ideas of Chapter III, the system when out of vibrational or chemical equilibrium is still assumed to obey a thermal equation of state of the form $T = T(p, \rho, q)$—for example, equation (V 2.6) with the asterisk removed—and this equation can be used to rewrite (10.4) in the form

$$\left(\frac{dq}{dt}\right)_{\text{int proc}} = \omega(p, \rho, q). \tag{11.11}$$

Here p, ρ, q are taken as the primary state variables, and still other combinations would be possible. As before, equation (11.11) can be rewritten as

$$\left(\frac{dq}{dt}\right)_{\text{int proc}} = \frac{\bar{\chi}(p, \rho, q)}{\bar{\tau}(p, \rho, q)}, \tag{11.12}$$

where now

$$\bar{\tau}(p, \rho, q) \equiv -\frac{1}{(\partial\omega/\partial q)_{p,\rho}}, \tag{11.13a}$$

and

$$\bar{\chi}(p, \rho, q) \equiv -\frac{\omega}{(\partial\omega/\partial q)_{p,\rho}}. \tag{11.13b}$$

For any given state, the functions $\omega(\rho, T, q)$ and $\omega(p, \rho, q)$ in equations (10.4) and (11.11) are equal. The derivatives $(\partial\omega/\partial q)_{\rho,T}$ and $(\partial\omega/\partial q)_{p,\rho}$, however, are not. As a result the quantities $\bar{\tau}$ and $\bar{\chi}$ given by equations (11.13a) and (11.13b) are not in general equal to τ and χ given by equations (11.1) and (11.3). The relaxation time in a nonequilibrium situation is thus not unique but will differ depending on what is held constant in the defining derivative. The various relaxation times are connected through purely thermodynamic relations. Their numerical values are not greatly different, and in practical applications they may be interchanged without serious error. In analytical work, however, the different definitions must be kept distinct. Bars have therefore been added to the present notation as a reminder of the difference.

The local equilibrium values also differ depending on the choice of input variables. Consistent with our second development we can, for example, define a fictitious equilibrium value of q to be calculated by substituting the local nonequilibrium values of $p(t)$ and $\rho(t)$ formally into the equilibrium relation

$$q^* = \bar{q}^*(p, \rho) \tag{11.14}$$

to obtain

$$\bar{q}^*(t) \equiv \bar{q}^*[p(t), \rho(t)]. \tag{11.15}$$

This is in contrast to the previous value (11.5a) that was based on the equilibrium relation (11.4b):

$$q^* = q^*(\rho, T). \tag{11.16}$$

Since they derive from purely equilibrium considerations, the functions on the right in (11.14) and (11.16) are related to each other through the *equilibrium* equation of state $T = T^*(p, \rho)$, that is,

$$q^*(\rho, T) = q^*[\rho, T^*(p, \rho)] = \bar{q}^*(p, \rho).$$

This relation would be satisfied by a system actually in equilibrium. For a system out of equilibrium, however, T, p, and ρ are connected through the *nonequilibrium* equation of state $T = T[p(t), \rho(t), q(t)]$, and we have

$$\begin{aligned} q^*[\rho(t), T(t)] &= q^*\{\rho(t), T[p(t), \rho(t), q(t)]\} \\ &= \bar{\bar{q}}^*[p(t), \rho(t), q(t)] \neq \bar{q}^*[p(t), \rho(t)]. \end{aligned}$$

This says that the value of $q^*[\rho(t), T(t)]$, when re-expressed in terms of $p(t)$ and $\rho(t)$, depends also on $q(t)$. Its value for particular instantaneous values of p and ρ will thus vary depending on q; it is therefore not equal to $\bar{q}^*[p(t), \rho(t)]$, which is fixed purely by the instantaneous values of p and ρ. The fictitious local equilibrium value, like the relaxation time, is therefore not unique. It is best to look on it in a nonequilibrium situation as merely a convenient, shorthand notation for a certain formally defined function of ρ and T, or p and ρ, or whatever variables are chosen.

Following the same procedures as before we can also approximate equation (11.12) for small departures from an equilibrium reference state. We obtain

$$\left(\frac{dq'}{dt}\right)_{\text{int proc}} = \frac{\bar{q}^{*\prime}(p', \rho') - q'}{\bar{\tau}(p^{(r)}, \rho^{(r)}, q^{*(r)})}, \tag{11.17}$$

where

$$\bar{q}^{*\prime}(p', \rho') = \left(\frac{\partial \bar{q}^*}{\partial p}\right)_\rho^{(r)} p' + \left(\frac{\partial \bar{q}^*}{\partial \rho}\right)_p^{(r)} \rho'. \tag{11.18}$$

These are analogous to equations (11.9) and (11.10b). The differences between $\bar{q}^{*\prime}$ and $q^{*\prime}$ and between $\bar{\tau}$ and τ are such that the right-hand sides of (11.17) and (11.9) are equal within the accuracy of the linear approximation. This can be proven formally by manipulation of partial-derivative relations based on the definitions of the various quantities involved.

The formalism of this section can, of course, also be carried out using other combinations of variables, such as p, T, q; h, ρ, q; etc.

Exercise 11.1. The words "in general" were included in the sentence following equation (11.5b) in recognition of the existence of a special circumstance in which the sentence is not true. What is this circumstance? That is, if τ is nonzero, how can density and temperature be changed with time in a special way such that q remains equal to $q^*(t)$ indefinitely, provided it is so initially? What relation must connect $\rho(t)$ and $T(t)$ for this exceptional situation to exist? Interpret this situation in terms of the equilibrium surface of Fig. 4.

Exercise 11.2. Show that the vibrational rate equation (2.11) can be written, for small departures from an equilibrium reference state, in the form

$$\frac{dT_v'}{dt} = \frac{T' - T_v'}{\tau(\rho^{(r)}, T^{(r)})},$$

where T_v' is the departure in the vibrational temperature T_v defined in Sec. 3. This equation is sometimes used as the vibrational rate equation. Unlike equations (2.9) and (2.11), however, it is not valid for large departures from equilibrium, even for the harmonic oscillator.

Exercise 11.3. (a) Obtain an expression for $\tau(\rho, T, \alpha)$ for (1) the general rate equation (10.3) for the symmetrical diatomic gas, and (2) the specialized rate equation (9.3) for the ideal dissociating gas.

(b) What is the relationship between the result for the ideal dissociating gas when applied at an equilibrium reference state and the result previously obtained in Exercise 9.1(c)?

References

Bauer, S. H., and S. C. Tsang, 1963, Mechanisms for Vibrational Relaxation at High Temperatures. *Phys. Fluids*, vol. 6, no. 2, p. 182.

Bazley, N. W., E. W. Montroll, R. J. Rubin, and K. E. Shuler, 1958, Studies in Non-equilibrium Rate Processes. III. The Vibrational Relaxation of a System of Anharmonic Oscillators. *J. Chem. Phys.*, vol. 28, no. 4, p. 700. (For errata, see also vol. 29, no. 5, p. 1185.)

Benson, S. W., 1960, *The Foundations of Chemical Kinetics*, McGraw-Hill.

Blackman, V. H., 1956, Vibrational Relaxation in Oxygen and Nitrogen. *J. Fluid Mech.*, vol. 1, pt. 1, p. 61.

Blythe, P. A., 1961, Comparison of Exact and Approximate Methods for Analysing Vibrational Relaxation Regions. *J. Fluid Mech.*, vol. 10, pt. 1, p. 33.

Byron, S. R., 1959, Measurement of the Rate of Dissociation of Oxygen. *J. Chem. Phys.*, vol. 30, no. 6, p. 1380.

Camac, M., 1961, O_2 Vibration Relaxation in Oxygen-Argon Mixtures. *J. Chem. Phys.*, vol. 34, no. 2, p. 448.

Camac, M., and A. Vaughan, 1961, O_2 Dissociation Rates in O_2-Ar Mixtures. *J. Chem. Phys.*, vol. 34, no. 2, p. 460.

Fowler, R., and E. A. Guggenheim, 1956, *Statistical Thermodynamics*, Cambridge University Press.

Freedman, E., and J. W. Daiber, 1961, Decomposition Rate of Nitric Oxide between 3000 and 4300°K. *J. Chem. Phys.*, vol. 34, no. 4, p. 1271.

Freeman, N. C., 1958, Non-equilibrium Flow of an Ideal Dissociating Gas. *J. Fluid Mech.*, vol. 4, pt. 4, p. 407.

Frost, A. A., and R. G. Pearson, 1961, *Kinetics and Mechanism*, 2nd ed., Wiley.

Glasstone, S., K. J. Laidler, and H. Eyring, 1941, *The Theory of Rate Processes*, McGraw-Hill.

Glick, H. S., J. J. Klein, and W. Squire, 1957, Single-Pulse Shock Tube Studies of the Kinetics of the Reaction $N_2 + O_2 \rightleftharpoons 2NO$ between 2000–3000°K. *J. Chem. Phys.*, vol. 27, no. 4, p. 850.

Heims, S. P., 1963, Moment Equations for Vibrational Relaxation Coupled with Dissociation. *J. Chem. Phys.*, vol. 38, no. 3, p. 603.

Herzfeld, K. F., and T. A. Litovitz, 1959, *Absorption and Dispersion of Ultrasonic Waves*, Academic.

Hinshelwood, C. N., 1940, *The Kinetics of Chemical Change*, Oxford University Press.

Laidler, K. J., 1950, *Chemical Kinetics*, McGraw-Hill.

Landau, L., and E. Teller, 1936, Zur Theorie der Schalldispersion. *Physik Z. Sowjetunion*, b. 10, h. 1, p. 34.

Lin, S-C., and J. D. Teare, 1963, Rate of Ionization behind Shock Waves in Air. II. Theoretical Interpretations. *Phys. Fluids*, vol. 6, no. 3, p. 355.

Marrone, P. V., and C. E. Treanor, 1963, Chemical Relaxation with Preferential Dissociation from Excited Vibrational Levels. *Phys. Fluids*, vol. 6, no. 3, p. 1215.

Mayer, J. E., and M. G. Mayer, 1940, *Statistical Mechanics*, Wiley.

Montroll, E. W., and K. E. Shuler, 1957, Studies in Nonequilibrium Rate Processes. I. The Relaxation of a System of Harmonic Oscillators. *J. Chem. Phys.*, vol. 26, no. 3, p. 454.

Parker, J. G., 1959, Rotational and Vibrational Relaxation in Diatomic Gases. *Phys. Fluids*, vol. 2, no. 4, p. 449.

Pauling, L., and E. B. Wilson, 1935, *Introduction to Quantum Mechanics*, McGraw-Hill.

Petschek, H., and S. Byron, 1957, Approach to Equilibrium Ionization behind Strong Shock Waves in Argon. *Ann. Phys.*, vol. 1, no. 3, p. 270.

Rubin, R. J., and K. E. Shuler, 1956, Relaxation of Vibrational Nonequilibrium Distributions. I. Collisional Relaxation of a System of Harmonic Oscillators. *J. Chem. Phys.*, vol. 25, no. 1, p. 59.

Rubin, R. J., and K. E. Shuler, 1957, On the Relaxation of Vibrational Nonequilibrium Distributions. III. The Effect of Radiative Transitions on the Relaxation Behavior. *J. Chem. Phys.*, vol. 26, no. 1, p. 137.

Shuler, K. E., 1959, Relaxation Processes in Multistate Systems. *Phys. Fluids*, vol. 2, no. 4, p. 442.

Treanor, C. E., and P. V. Marrone, 1962, Effect of Dissociation on the Rate of Vibrational Relaxation. *Phys. Fluids*, vol. 5, no. 9, p. 1022.

Widom, B., 1957, Inelastic Molecular Collisions with a Maxwellian Interaction Energy. *J. Chem. Phys.*, vol. 27, no. 4, p. 940.

Wray, K. L., 1962a, Chemical Kinetics of High Temperature Air; in *Hypersonic Flow Research*, ed. by F. R. Riddell, Academic.

Wray, K. L., 1962b, Shock-Tube Study of the Vibrational Relaxation of Nitric Oxide. *J. Chem. Phys.*, vol. 36, no. 10, p. 2597.

Wray, K. L., 1962c, Shock-Tube Study of the Coupling of the O_2-Ar Rates of Dissociation and Vibrational Relaxation. *J. Chem. Phys*,. vol. 37, no. 6, p. 1254.

Wray, K. L., and J. D. Teare, 1962, Shock-Tube Study of the Kinetics of Nitric Oxide at High Temperatures. *J. Chem. Phys.*, vol. 36, no. 10, p. 2582.

Chapter VIII

Flow with Vibrational or Chemical Nonequilibrium

1 INTRODUCTION

With the necessary rate equations in hand, we can take up the study of flow with vibrational or chemical nonequilibrium. To this end we utilize the assumption, anticipated in the preceding chapter, that the internal rate processes in a moving fluid at a given instantaneous state are the same as those that occur at the same state in a closed system of fixed volume and temperature. As we shall see, the resulting theory, in addition to providing useful results for the nonequilibrium flow itself, sheds new light on the conditions underlying the classical equilibrium flow. In particular, we shall find that the equilibrium flow is, with certain exceptions, the limiting case of nonequilibrium flow when the characteristic time of the rate process tends to zero.

We shall assume throughout this chapter that the flow is everywhere in instantaneous translational and rotational equilibrium, or, in other words, that all effects of viscosity, heat conduction, and diffusion are negligible. For simplicity, we shall allow for only one nonequilibrium process, that is, vibrational or chemical nonequilibrium, but not both. An extension to multiple processes is possible but greatly complicates the details. To keep the treatment within the desired length, we presume that the reader is acquainted with the governing equations and fundamental results for three-dimensional classical flow as presented in the standard textbooks on gas dynamics (see, e.g., Liepmann and Roshko, 1957, to which we shall make frequent reference). An alternative treatment of some of the problems discussed in this and later chapters is found in the book by Clarke and McChesney (1964).

The neglect of translational and rotational nonequilibrium is consistent with our philosophy, mentioned in the Preface, of considering only one kind of nonequilibrium at a time. The nonequilibrium effects of translation and rotation are considered in Chapter IX and X, there with the assumption of no vibration or chemical reaction. We shall make no attempt to combine the various phenomena, as one must, for example, in the important problems of combustion or chemically reacting boundary layers. For the former the reader may consult the book by Williams (1965) and for the latter the articles by Lees (1958) and Moore (1964).

2 BASIC NONLINEAR EQUATIONS

We first set down the basic equations under the assumptions just outlined. We take as spatial variables the Cartesian coordinates x_1, x_2, x_3 and allow for variations with both time and position. For brevity, we adopt the repeated-subscript notation. With this notation, the repetition of a letter subscript within a given term indicates that this term is to be thought of as written three times with the repeated subscript having the successive values 1, 2, 3 and the sum then taken of these three terms. Thus, for example, the term $u_j u_j$ is shorthand notation for the sum $u_1^2 + u_2^2 + u_3^2$. The substantial derivative, that is, the derivative following a fluid element, is then written

$$\frac{D(\ )}{Dt} \equiv \frac{\partial(\ )}{\partial t} + u_1 \frac{\partial(\ )}{\partial x_1} + u_2 \frac{\partial(\ )}{\partial x_2} + u_3 \frac{\partial(\ )}{\partial x_3} = \frac{\partial(\ )}{\partial t} + u_j \frac{\partial(\ )}{\partial x_j}, \quad (2.1)$$

where u_i is the flow-velocity vector with components u_1, u_2, u_3.

We first set down the usual macroscopic conservation equations for nonviscous, nonconducting, nondiffusing flow. (For a detailed derivation in the same notation used here, see Liepmann and Roshko, 1957, Chapter 7.) These equations are derived in most books on the assumption of complete thermodynamic equilibrium. The derivations are equally valid for flows with vibrational and chemical nonequilibrium, however, provided the thermodynamic variables are given their extended definition appropriate to the nonequilibrium situation (cf. Chapter III, Sec. 4). A derivation of the same equations from the microscopic (i.e., kinetic-theory) point of view will be given in Chapter IX, Sec. 5.

The continuity equation for the fluid, as obtained by applying the law of conservation of mass to the flow through a fixed volume element, is

$$\frac{\partial \rho}{\partial t} + \frac{\partial}{\partial x_j}(\rho u_j) = 0. \quad (2.2a)$$

This can also be written with the aid of (2.1) as

$$\frac{D\rho}{Dt} + \rho \frac{\partial u_j}{\partial x_j} = 0. \tag{2.2b}$$

Euler's equations of motion, which can be obtained by applying Newton's law to an element of unit mass moving with the fluid, are, in the absence of body forces,

$$\rho \frac{Du_i}{Dt} + \frac{\partial p}{\partial x_i} = 0. \tag{2.3}$$

This is a vector equation with three scalar components, one for each of the values of i. Under the present assumptions, the energy equation can be obtained most readily by applying the first law of thermodynamics to an adiabatic element of unit mass moving with the fluid. Taking the first law in the form (III 4.8), we thus have

$$\rho \frac{Dh}{Dt} - \frac{Dp}{Dt} = 0. \tag{2.4a}$$

An alternative form can be obtained by multiplying equation (2.3) through by u_i (which has the effect at the same time of summing the equation over the three coordinate directions) and using the result to replace Dp/Dt in (2.4a). This gives

$$\rho \frac{D}{Dt}(h + \tfrac{1}{2}u^2) - \frac{\partial p}{\partial t} = 0. \tag{2.4b}$$

The conservation equations (2.2), (2.3), and (2.4) provide five scalar equations for the six unknowns ρ, u_i, p, and h. In the study of equilibrium flow in Chapter VI, the set was completed by the addition of an equation of state in the form $h = h^*(p, \rho)$, where an asterisk is now added to distinguish the equilibrium-flow situation. For nonequilibrium flow, however, we must introduce a third variable to specify the vibrational or chemical state of the gas. If, as in Chapter VII, Sec. 10, we give this nonequilibrium variable the general symbol q, the state equation is then of the form

$$h = h(p, \rho, q). \tag{2.5}$$

The equilibrium-flow situation corresponds to setting q everywhere identically equal to its equilibrium value $q^* = q^*(p, \rho)$, in which case equation (2.5) becomes

$$h = h[p, \rho, q^*(p, \rho)] = h^*(p, \rho). \tag{2.6}$$

The introduction of the additional variable q requires an additional equation. This is provided by the rate equation for the nonequilibrium process. To write this equation for a moving fluid, we follow a procedure similar to that commonly used to obtain the continuity equation (2.2a). We take as our system of interest a fixed infinitesimal element of volume (Fig. 1) and consider the rate of change inside that volume of a quantity Q, which is the extensive counterpart of the intensive variable q. Here, however, the net rate of change of Q in the volume is influenced not only by the inflow of Q into the volume but by the nonequilibrium production of

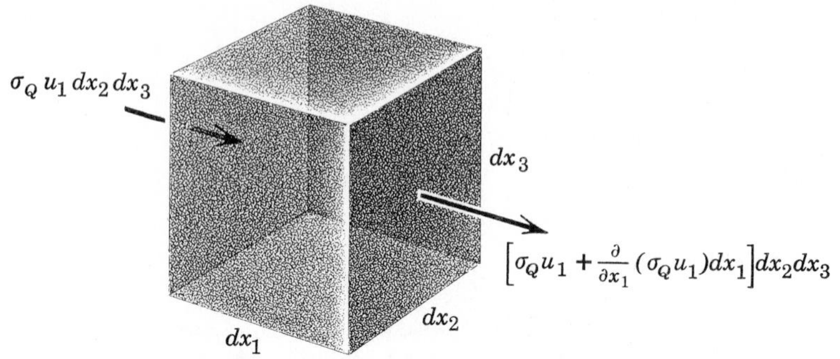

Fig. 1. Fixed infinitesimal element of volume.

Q inside the volume as well. Assuming that the nonequilibrium process is the same as in a static system, we can separate the effect into two parts and write

$$\frac{\partial \sigma_Q}{\partial t} = \left(\frac{\partial \sigma_Q}{\partial t}\right)_{\text{net inflow}} + \left(\frac{\partial \sigma_Q}{\partial t}\right)_{\text{int proc}} \tag{2.7}$$

Here $\sigma_Q = \rho q$ is the amount of Q per unit volume inside the element and $(\partial \sigma_Q / \partial t)_{\text{int proc}}$ is the internal rate of production of Q per unit volume obtained from consideration of the processes in a closed system of fixed volume. As usual the inflow term can be evaluated by considering the net inflow for a pair of faces of the volume element (Fig. 1) and then summing over the three coordinate directions. We thus obtain

$$\left(\frac{\partial \sigma_Q}{\partial t}\right)_{\text{net inflow}} = -\frac{\partial}{\partial x_j}(\sigma_Q u_j). \tag{2.8}$$

This assumes, when applied to chemical nonequilibrium, that diffusion is negligible, so that the mean velocity of each species is identical with the mass velocity of the fluid as a whole. If diffusion is not negligible, the velocity in (2.8) must be the mean velocity of the species identified with Q, and the difference between this and the mass velocity u_i of the fluid as a whole then becomes significant (see Chapter IX, Sec. 10). For zero diffusion, substitution of (2.8) into (2.7) gives

$$\frac{\partial \sigma_Q}{\partial t} + \frac{\partial}{\partial x_j}(\sigma_Q u_j) = \left(\frac{\partial \sigma_Q}{\partial t}\right)_{\text{int proc}}, \tag{2.9a}$$

which has an obvious resemblance to equation (2.2a). In view of the equality $\sigma_Q = \rho q$, this equation can also be written

$$\frac{\partial}{\partial t}(\rho q) + \frac{\partial}{\partial x_j}(\rho q u_j) = \left[\frac{\partial(\rho q)}{\partial t}\right]_{\text{int proc}} = \rho\left(\frac{\partial q}{\partial t}\right)_{\text{int proc}}. \tag{2.9b}$$

Since the derivative on the right is evaluated for a closed system of fixed volume (see Sec. 1), we may take ρ inside or outside the derivative as we wish. Taking it outside and using equation (VII 11.12) to find $(\partial q/\partial t)_{\text{int proc}}$, we obtain

$$\frac{\partial}{\partial t}(\rho q) + \frac{\partial}{\partial x_j}(\rho q u_j) = \rho \frac{\chi(p, \rho, q)}{\tau(p, \rho, q)}. \tag{2.10}$$

We have chosen to use p, ρ, q in specifying the right-hand side of (2.10) because these appear as the natural variables in the earlier conservation equations. Since we shall use these variables exclusively from now on, we have dropped the bars from χ and τ. Equation (2.10) is one form of the general rate equation for the moving fluid. Along with equations (2.2) through (2.5), it provides seven equations for the seven unknowns ρ, u_i, p, h, and q.

If σ_Q in equation (2.9a) is taken to be the number of moles of a species Y per unit volume, we have $\sigma_Q = [\mathrm{Y}] = \rho_{\mathrm{Y}}/\hat{M}_{\mathrm{Y}}$, where ρ_{Y} is the mass density of species Y and $\hat{M}_{\mathrm{Y}}$ is its molecular weight. Equation (2.9a) can then be specialized to

$$\frac{\partial \rho_{\mathrm{Y}}}{\partial t} + \frac{\partial}{\partial x_j}(\rho_{\mathrm{Y}} u_j) = \hat{M}_{\mathrm{Y}}\left(\frac{\partial[\mathrm{Y}]}{\partial t}\right)_{\text{chem}}. \tag{2.11}$$

This so-called *continuity equation for species Y* is sometimes used as the rate equation in chemically reacting flow.

Returning to the general equation (2.10), we can put this equation in simpler form by rewriting the left-hand side as follows:

$$\rho \frac{\partial q}{\partial t} + q \frac{\partial \rho}{\partial t} + \rho u_j \frac{\partial q}{\partial x_j} + q \frac{\partial}{\partial x_j}(\rho u_j) = \rho \frac{\chi}{\tau}.$$

By the continuity equation (2.2a), the second and fourth terms on the left combine to give zero. We thus obtain the rate equation for the moving fluid in the form

$$\frac{Dq}{Dt} = \frac{\chi(p, \rho, q)}{\tau(p, \rho, q)}. \tag{2.12}$$

This equation is formally similar to the rate equation (VII 11.12) for the internal process alone. It shows that (provided we neglect diffusion in the case of chemical nonequilibrium) the change in Q in an element of fixed mass moving with the mass velocity of the fluid is due entirely to the internal rate process. We shall find it convenient to use the rate equation in the form (2.12) in our subsequent work.

In any given problem the foregoing equations must be supplemented by appropriate boundary conditions. Insofar as the velocity or pressure are concerned, these are the same as the well-known conditions from classical theory. The new feature here is that we must now add appropriate conditions on the additional variable q. In the nonequilibrium flow behind a shock wave, for example, we impose the condition that at a given location on the wave the value of q immediately behind the wave is the same as that immediately ahead of the wave. The reasons for this will be explained in Sec. 9. Insofar as solid surfaces are concerned, in problems of chemical nonequilibrium they are assumed to be noncatalytic, that is, their presence has no direct influence on the chemical reaction. To satisfy any other chemical condition at the surface is, in fact, impossible unless diffusion is taken into account. The situation is analogous to that of the classical no-slip condition on tangential velocity at a solid surface, which can be satisfied only if viscosity is included. In problems of vibrational nonequilibrium the surfaces are assumed, consistent with our assumption of adiabatic flow, to undergo no exchange of energy with the vibrational mode of the gas. The variable q at the surface is thus in either case free to take on whatever value is dictated by the other conditions of the problem.

Although the entropy is not essential in the above set of equations, it is sometimes of interest to examine the entropy production in nonequilibrium flow. To do this we note that both equation (VII 3.12b) for the vibrating gas and equation (VII 9.4) for the dissociating gas can be put in the common form

$$T\,ds = de + p\,d\left(\frac{1}{\rho}\right) + \zeta\,dq,$$

where $\zeta = (T/T_v - 1)$ for vibration ($q = e_v$) and $\zeta = (\mu_{A_2} - \mu_A)$ for dissociation ($q = \alpha$). The quantity ζ is a function of the state of the gas and can be written $\zeta = \zeta(p, \rho, q)$. From our previous work, its value is zero when the gas is in equilibrium, that is, $\zeta(p, \rho, q^*) = 0$. In terms of the enthalpy $h \equiv e + p/\rho$, the foregoing relation becomes

$$T\,ds = dh - \frac{1}{\rho}\,dp + \zeta\,dq. \qquad (2.13)$$

Applying this to the changes following a fluid element, we obtain, in view of the energy equation (2.4a),

$$\frac{Ds}{Dt} = \frac{\zeta}{T}\frac{Dq}{Dt}. \tag{2.14}$$

This equation gives the rate of entropy production following a fluid element, which in general is not zero in a nonequilibrium flow.

It should be noted that for steady flow (i.e., $\partial(\ )/\partial t = 0$) the energy equation in the form (2.4b) can be integrated at once to obtain

$$h + \tfrac{1}{2}u^2 = \text{constant along a streamline}, \tag{2.15}$$

where the combination on the left is the usual total enthalpy $h_t \equiv h + u^2/2$. The presence of vibrational or chemical nonequilibrium thus does not alter the result, well known from equilibrium-flow theory, that in steady flow the total enthalpy is constant along a streamline.

The foregoing formulation is not valid in general for flows with ionization nonequilibrium. In such flows, because of the large mass ratio between the electrons and the molecules and ions, very many collisions are required to distribute the translational energy between the electron gas and the molecule-ion gas (see Petschek and Byron, 1957). Consequently it cannot always be assumed that the translational energy of the electrons corresponds to the same temperature as does that of the heavy particles. It is necessary, therefore, to introduce at least two nonequilibrium variables, the electron temperature and the degree of ionization. An additional equation is then provided by an energy balance on the electrons (see, e.g., Appleton and Bray, 1964).

Exercise 2.1. Note that the rate equation for the moving fluid can also be written in the form

$$\frac{D(\rho q)}{Dt} = \rho\frac{\chi}{\tau} + q\frac{D\rho}{Dt}.$$

What is the physical meaning of the terms in this equation?

3 EQUILIBRIUM AND FROZEN FLOW

It is of interest to examine the behavior of the foregoing equations, in particular the rate equation (2.12), when the relaxation time τ of the nonequilibrium process is vanishingly small or infinitely large.

We first examine the limit as $\tau \to 0$. In general, we do not expect to find infinite values of Dq/Dt in a nonequilibrium flow, and it is reasonable to assume that this will remain true in the limit (except, as we shall see later, at certain singular instants or singular surfaces, lines, or points). On the basis of this assumption that Dq/Dt remains finite, it follows from equation (2.12) that $\chi \to 0$ as $\tau \to 0$. But we have seen [cf. equation (VII 11.4a)] that $\chi(p, \rho, q) = 0$ defines the equilibrium value $q^*(p, \rho)$, which was used as in equation (2.6) to complete the set of equations for equilibrium flow. We thus see that the limit $\tau \to 0$ leads to the equilibrium flow discussed in Chapter VI. This is in line with the intuitive idea that an infinitely fast rate process will lead to the instantaneous establishment of local equilibrium as a fluid element moves through the flow. In Secs. 6 and 8 we shall see specific examples of how the equilibrium flow appears as the natural limit of the nonequilibrium flow as $\tau \to 0$. We also see from the foregoing that when τ is identically zero the differential equation (2.12) is replaced by the algebraic equation $\chi(p, \rho, q^*) = 0$. Thus the order of the system of differential equations drops discontinuously by one at the limit of equilibrium flow. This behavior, which is characteristic of so-called "singular perturbation problems," will appear again later.

It is important to note clearly that equilibrium *flow* is essentially different from simple *equilibrium* at fixed conditions in a closed system. In both cases we have $\chi = 0$. In equilibrium flow, however, we have $Dq^*/Dt \neq 0$ and $\tau = 0$, whereas in simple equilibrium we have $dq/dt = 0$ for all values of τ. That is to say, equilibrium flow involves changes in q (e.g., species production) whereas equilibrium in a static system does not.

Actually, of course, the limit of equilibrium flow (i.e., $\tau = 0$) can never be physically realized. All internal rate processes require molecular collisions, and these in turn require a certain amount of time. The departure from equilibrium, in fact, provides the "driving force" for the internal changes, and without it no changes would take place. The name "equilibrium flow" is thus in reality a contradiction in terms. If, however, the characteristic time of the rate process is very small compared with the time intervals of interest in the flow, then the flow will behave essentially *as if* it were in local instantaneous equilibrium. Under these conditions the assumption of equilibrium flow, as made in Chapter VI, is a useful working approximation.

The reasoning of our earlier paragraph regarding the equilibrium-flow limit has an important exception. For the equilibrium flow to be approached uniformly everywhere in the flow field as $\tau \to 0$, the initial and boundary conditions on q must in the limit satisfy the equilibrium relation $q = q^*(p, \rho)$. There are situations, however, where this is not so. For example, as mentioned earlier, the value of q immediately behind a shock wave is by requirement equal to the value immediately ahead of the wave. As a result the value of $\chi(p, \rho, q)$ at such a boundary is fixed at some nonzero value. It follows from the rate equation (2.12) that in this situation

$Dq/Dt \to \infty$ as $\tau \to 0$, and this contradicts our original assumption that Dq/Dt remains everywhere finite. Under these conditions the flow cannot approach the equilibrium flow uniformly at all points in the field. Adjacent to the boundaries there will be, even as we approach the limit, narrow regions of nonequilibrium flow embedded in the otherwise nearly equilibrium field. A simple example of this appears later in Sec. 9. This sort of nonuniform behavior is again a characteristic of singular perturbation problems. It is a reflection of the drop in the order of the system of differential equations precisely at the limit, with a resulting inability of the equilibrium-flow equations to satisfy the boundary condition on q. The situation is analogous to that encountered in viscous-flow theory with regard to the condition on the tangential velocity at a solid surface as the viscosity coefficient tends to zero. (For a general discussion of singular perturbation problems in fluid mechanics, see Van Dyke, 1964.)

The limit $\tau \to \infty$ is both different and similar. Here the rate equation (2.12) shows that $Dq/Dt \to 0$, irrespective of the value of χ. Thus in the limit of an infinitely slow rate process the value of q for a fluid element remains constant. This is the special case of frozen flow discussed previously in Chapter VI, Sec. 5. The frozen-flow limit also is singular in that the differential rate equation reduces in the limit to $Dq/Dt = 0$, which integrates immediately to the algebraic equation $q =$ constant. The order of the system of differential equations thus again drops discontinuously by one at the limit. In this instance the singular behavior is reflected in the flow field by a nonuniform approach to frozen flow in the vicinity of positions at which the velocity goes to zero, as for example, a stagnation point or an infinite reservoir. At such positions the *local* characteristic time of the flow (or residence time of the fluid element) tends to infinity, and its ratio to the relaxation time can remain finite locally even when $\tau \to \infty$. This situation is mathematically less obvious than the nonuniformity at the boundary in the equilibrium-flow limit. It is evident in the calculated results for the nonequilibrium flow over blunt bodies (see, e.g., Conti, 1964).

When a gas that is capable of vibrating or reacting at high temperatures is at sufficiently low temperature, then the characteristic time τ is very large and the equilibrium value of q is essentially constant independent of p and ρ. Under these conditions nonequilibrium effects never arise in any flow that starts from an equilibrium condition, and such a flow is at the same time both an equilibrium and a frozen flow. In terms of the foregoing limits, we have here $\tau \to \infty$ and $\chi \to 0$ simultaneously. This is the situation, for example, in air at room temperatures.

For both equilibrium flow and frozen flow, the entropy production as given by equation (2.14) is zero. This follows from the fact that $Dq/Dt = 0$

when the flow is frozen and $\zeta = \zeta(p, \rho, q^*) = 0$ when the flow is everywhere in local equilibrium. For any intermediate case ($Dq/Dt \neq 0$ and $\zeta \neq 0$), Ds/Dt must be greater than zero in accord with the second law of thermodynamics.

4 ACOUSTIC EQUATIONS

In the study of nonequilibrium flow, one of the first questions that arises concerns the speed of sound, that is, the speed of propagation of a small disturbance. This question arises immediately if we set out to calculate the speed of sound from the familiar formula $a^2 = (\partial p/\partial \rho)_s$. In the present case, the thermochemical state is defined by the three variables p, ρ, q, so that we have for the entropy $s = s(p, \rho, q)$ or, inversely, $p = p(\rho, s, q)$. We thus see that the derivative $(\partial p/\partial \rho)_s$ is not unique. It can, for example, be taken with q fixed, as indicated by

$$\left(\frac{\partial p}{\partial \rho}\right)_{s,q}, \tag{4.1a}$$

or with q held equal to its equilibrium value $q = q^*(\rho, s)$, as indicated by

$$\left(\frac{\partial p}{\partial \rho}\right)_{s,q=q^*}. \tag{4.1b}$$

There is also an unlimited number of other possibilities. The question thus arises as to which of the possible derivatives, properly speaking, gives *the* speed of sound in a nonequilibrium situation. As it turns out, the problem is somewhat complicated, and both the foregoing derivatives play an important role.

To study this question we proceed on the basis of the linearized, or "acoustic," equivalent of the equations of the preceding section. These equations will also be of use to us later in providing convenient access to linearized flow theory through a suitable transformation. We concern ourselves, as usual in acoustic theory, with small disturbances propagating into a uniform fluid initially at rest at conditions p_0, ρ_0, q_0, h_0. The undisturbed fluid is assumed to be in equilibrium, that is, $q_0 = q_0^*$. With disturbance quantities denoted by primes, the state quantities within the disturbance are $p = p_0 + p'$, $\rho = \rho_0 + \rho'$, etc. Since the undisturbed fluid is at rest, the velocity components are given directly by the disturbance quantities, that is, $u_i = u_i'$.

In the foregoing situation, the convective term in the expression (2.1) for the substantial derivative becomes $u_j \partial(\ )/\partial x_j = u_j' \partial(\ )'/\partial x_j$, which is of second order in the disturbance quantities. The linearized form of the substantial derivative is therefore simply $D(\ )/Dt = \partial(\ )/\partial t$. The conservation equations (2.2b), (2.3), and (2.4a) thus become after linearization

$$\frac{\partial \rho'}{\partial t} + \rho_0 \frac{\partial u_j'}{\partial x_j} = 0, \tag{4.2}$$

$$\rho_0 \frac{\partial u_i'}{\partial t} + \frac{\partial p'}{\partial x_i} = 0, \tag{4.3}$$

$$\rho_0 \frac{\partial h'}{\partial t} - \frac{\partial p'}{\partial t} = 0. \tag{4.4}$$

The linearized form of the state equation (2.5) is obtained by expanding h in a Taylor's series about the conditions in the undisturbed fluid and retaining only the first-order terms. This gives

$$h' = h_{p_0} p' + h_{\rho_0} \rho' + h_{q_0} q', \tag{4.5}$$

where we have for simplicity adopted the subscript notation for differentiation. The first subscript thus indicates differentiation with respect to the noted variable, the other two of the three variables p, ρ, q remaining fixed, for example, $h_{p_0} \equiv [(\partial h/\partial p)_{\rho,q}]_0$. We will understand, unless stated otherwise, that all thermodynamic quantities are taken as functions of these three variables. The linearized form for the right-hand side of the rate equation (2.12) has been given in Chapter VII, Sec. 11. Using equation (VII 11.17) we thus write the linear counterpart of equation (2.12) as

$$\frac{\partial q'}{\partial t} = \frac{q^{*\prime} - q'}{\tau_0}, \tag{4.6}$$

where the undisturbed fluid has been taken as the equilibrium reference state. Here we have simplified the notation again by understanding, consistent with the foregoing, that $q^{*\prime} \equiv \bar{q}^{*\prime}(p', \rho')$ and $\tau_0 \equiv \bar{\tau}(p_0, \rho_0, q_0)$. The disturbance quantity $q^{*\prime} = q^* - q_0^* = q^* - q_0$ is related to p' and ρ' by equation (VII 11.18), which in subscript notation is

$$q^{*\prime} = \overset{*}{q}_{p_0} p' + \overset{*}{q}_{\rho_0} \rho'. \tag{4.7}$$

Equations (4.2) through (4.7) constitute eight constant-coefficient linear equations for the eight unknowns ρ', u_i', p', h', q', and $q^{*\prime}$.

We shall now show that the foregoing equations can be combined into a single equation for a potential function. To do this we first differentiate the rate equation (4.6) with respect to t and write the result as

$$\tau_0 \frac{\partial}{\partial t}\left(\frac{\partial q'}{\partial t}\right) + \frac{\partial}{\partial t}(q' - q^{*\prime}) = 0. \tag{4.8}$$

We now consider the derivative $\partial q'/\partial t$ in the first term of this equation. By inverting the state equation (4.5) to find q', differentiating the resulting equation with respect to t, and then eliminating $\partial \rho'/\partial t$ and $\partial h'/\partial t$ by means of equations (4.2) and (4.4), we find

$$\frac{\partial q'}{\partial t} = \frac{h_{\rho 0}}{h_{q0}}\left(-\frac{h_{p0} - 1/\rho_0}{h_{\rho 0}}\frac{\partial p'}{\partial t} + \rho_0 \frac{\partial u_j'}{\partial x_j}\right). \tag{4.9}$$

Now let us consider the second term in (4.8). By differentiating equation (4.7) with respect to t and using equation (4.2) to eliminate $\partial \rho'/\partial t$, we obtain

$$\frac{\partial q^{*\prime}}{\partial t} = q_{p0}^* \frac{\partial p'}{\partial t} - \rho_0 q_{\rho 0}^* \frac{\partial u_j'}{\partial x_j}.$$

Subtracting this result from equation (4.9), we have for the second term in (4.8), after some simple algebraic manipulation,

$$\frac{\partial}{\partial t}(q' - q^{*\prime}) = \frac{h_{\rho 0} + h_{q0}q_{\rho 0}^*}{h_{q0}}\left(-\frac{h_{p0} + h_{q0}q_{p0}^* - 1/\rho_0}{h_{\rho 0} + h_{q0}q_{\rho 0}^*}\frac{\partial p'}{\partial t} + \rho_0 \frac{\partial u_j'}{\partial x_j}\right). \tag{4.10}$$

The expressions (4.9) and (4.10) could be substituted into (4.8) to obtain a single equation in terms of p' and u_i'.

The coefficients of $\partial p'/\partial t$ in (4.9) and (4.10) are related to the partial derivatives (4.1) mentioned at the outset of the section. To see this we utilize the differential state relationship (2.13):

$$T\,ds = dh - \frac{1}{\rho}\,dp + \zeta\,dq.$$

By means of the equation of state (2.5) in the differential form $dh = h_p\,dp + h_\rho\,d\rho + h_q\,dq$, this can also be written

$$T\,ds = \left(h_p - \frac{1}{\rho}\right)dp + h_\rho\,d\rho + (h_q + \zeta)\,dq. \tag{4.11}$$

This is purely a differential state relation for $s = s(p, \rho, q)$ and is in no way restricted to the flow situation under discussion. By taking $ds = 0$ and $dq = 0$, we find at once

$$\boxed{\left(\frac{\partial p}{\partial \rho}\right)_{s,q} = -\frac{h_\rho}{h_p - 1/\rho} \equiv a_f^2}. \tag{4.12}$$

We give this the symbol a_f^2 for reasons that will appear later. Returning to equation (4.11), we can also take $ds = 0$ and $q = q^*(p, \rho)$. The latter implies that $dq = dq^* = q_p^*\,dp + q_\rho^*\,d\rho$ and $\zeta = \zeta(p, \rho, q^*) = 0$, so that we obtain

$$\boxed{\left(\frac{\partial p}{\partial \rho}\right)_{s,q=q^*} = -\frac{h_\rho + h_q q_\rho^*}{h_p + h_q q_p^* - 1/\rho} \equiv a_e^2}. \tag{4.13}$$

We give this the symbol a_e^2. The coefficients of $\partial p'/\partial t$ in (4.9) and (4.10) are thus each the inverse of one of the derivatives (4.1), evaluated at the conditions of the undisturbed fluid.

A potential function φ can be introduced through the linearized Euler equations (4.3), which were not used in obtaining (4.9) and (4.10). It can be seen by direct substitution that these equations are satisfied if we define $\varphi = \varphi(x_i, t)$ such that

$$p' = -\rho_0 \frac{\partial \varphi}{\partial t}, \tag{4.14a}$$

$$u_i' = \frac{\partial \varphi}{\partial x_i}. \tag{4.14b}$$

The governing equation for φ is then found by substituting these expressions into (4.9) and (4.10) and putting the results into the differentiated rate equation (4.8). Using the notation (4.12) and (4.13), we thus find the following as the single equation for the one unknown φ:

$$\boxed{\tau_0^+ \frac{\partial}{\partial t}\left(\frac{1}{a_{f_0}^2}\frac{\partial^2 \varphi}{\partial t^2} - \frac{\partial^2 \varphi}{\partial x_j\,\partial x_j}\right) + \left(\frac{1}{a_{e_0}^2}\frac{\partial^2 \varphi}{\partial t^2} - \frac{\partial^2 \varphi}{\partial x_j\,\partial x_j}\right) = 0}, \tag{4.15}$$

where τ_0^+ is a new relaxation time given by

$$\tau_0^+ \equiv \frac{\tau_0 h_{\rho_0}}{h_{\rho_0} + h_{q_0} q_{\rho_0}^*}.$$

The linear, third-order equation (4.15) is the counterpart for non-equilibrium flow of the classical wave equation

$$\frac{1}{a_0^2}\frac{\partial^2 \varphi}{\partial t^2} - \frac{\partial^2 \varphi}{\partial x_j\,\partial x_j} = 0. \tag{4.16}$$

This classical second-order equation is derived, in effect, on the assumption that h is a function only of p and ρ, which we know to be true [cf. equations (2.5) and (2.6)] for either frozen or equilibrium flow. The general solution of the classical equation in the one-dimensional case ($\partial^2\varphi/\partial x_2^2 = \partial^2\varphi/\partial x_3^2 = 0$) is[1]

$$\varphi = f(x_1 - a_0 t) + g(x_1 + a_0 t), \tag{4.17}$$

where f and g are arbitrary functions. These functions represent plane waves traveling in the positive and negative x_1-directions respectively without change of form and with a unique speed $a_0 = (\partial p/\partial \rho)^{1/2}_{s0}$. This speed is fixed by the conditions in the undisturbed gas. Similar results hold for cylindrical and spherical waves, except that the form of these waves changes as a result of their geometrical spreading.

Certain properties of nonequilibrium acoustic propagation emerge at once from the differential equation (4.15). This equation is a combination of two classical wave operators, one based on a_{e_0} and one based on a_{f_0}. In the limit as $\tau_0^+ \to 0$—that is, for equilibrium flow—the equation reduces to the classical equation (4.16) with $a_0 = a_{e_0}$ (provided the derivatives remain finite as assumed earlier). It follows from the foregoing remarks concerning the classical equation that the speed of propagation of an acoustic wave is then given by a_{e_0}. The quantity $a_e \equiv (\partial p/\partial \rho)^{1/2}_{s,q=q^*}$ is therefore known as the *equilibrium speed of sound.* In the limit as $\tau_0^+ \to \infty$ —that is, for frozen flow—only the derivative of the first wave operator remains in the equation (4.15). This can be integrated at once to obtain

$$\frac{1}{a_{f_0}^2}\frac{\partial^2\varphi}{\partial t^2} - \frac{\partial^2\varphi}{\partial x_j\,\partial x_j} = F(x_i),$$

where $F(x_i)$ is a function only of position. Since the disturbances at some time must have been zero throughout the fluid, we set $F(x_i) = 0$. The equation thus reduces in this limit to the classical equation (4.16) with $a_0 = a_{f_0}$. In this limit the speed of an acoustic wave is therefore given by a_{f_0}. The quantity $a_f \equiv (\partial p/\partial \rho)^{1/2}_{s,q}$ is known accordingly as the *frozen speed of sound.* The foregoing results will appear formally as the limits in the solution for harmonic plane waves in Sec. 6. It will also develop that for values of τ_0^+ intermediate between the two limits the classical results do not hold, that is, there is no unique speed of propagation that depends only on the conditions in the undisturbed gas.

Consistent with the discussion in the preceding section, the governing equation (4.15) is seen to drop discontinuously from third to second order at both limits.

[1] We assume that the reader is familiar with the classical wave equation and the interpretation of its solution, as given for example by Liepmann and Roshko (1957), pp. 65–69.

5 FROZEN AND EQUILIBRIUM SPEEDS OF SOUND

The frozen and equilibrium speeds of sound defined by equations (4.12) and (4.13) are thermodynamic state variables that have a meaning quite apart from any particular flow problem. It must be noted, however, that the equilibrium speed $a_e \equiv (\partial p/\partial \rho)^{1/2}_{s,q=q^*}$ is defined only in an actual equilibrium state. This is true because the partial derivative can be taken with q held equal to q^* only at a state at which q is equal to q^* to begin with. This fact was crucial in setting $dq = dq^*$ and $\zeta = 0$ in equation (4.11) and thus obtaining equation (4.13). It is therefore implicit in (4.13) that the derivatives h_p, h_ρ, h_q are evaluated at an equilibrium state $p, \rho, q^*(p, \rho)$ and therefore that

$$a_e = a_e[p, \rho, q^*(p, \rho)] = a_e^*(p, \rho). \tag{5.1}$$

The frozen speed of sound $a_f \equiv (\partial p/\partial \rho)^{1/2}_{s,q}$, on the other hand, can be evaluated for arbitrary values of q and is thus defined whether the fluid is in an equilibrium state or not. We may therefore write

$$a_f = a_f(p, \rho, q), \tag{5.2}$$

where q may have any arbitrary, nonequilibrium value.

The calculation of a_f and a_e for a particular gas model may be rather tedious because of the necessity for evaluating the various derivatives in equations (4.12) and (4.13) (see, for example, Exercise 5.1). Typical final results are shown in Figs. 2 and 3 for a mixture of dissociating oxygen

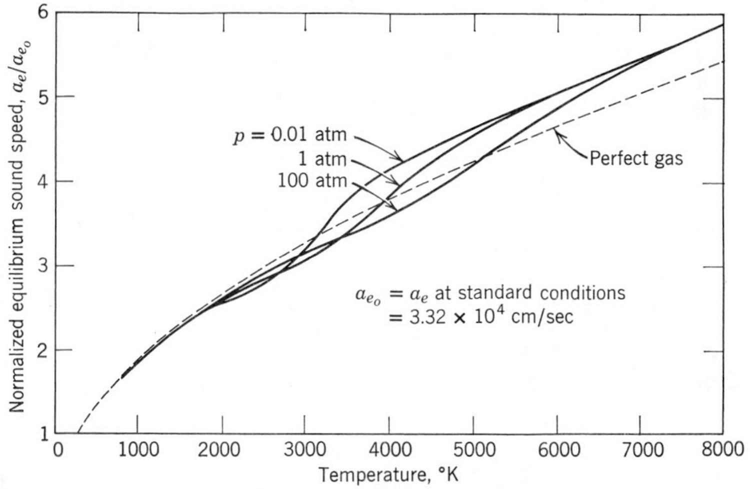

Fig. 2. Equilibrium sound speed for mixture of reacting O_2 and O and inert N_2 (after Der, 1963).

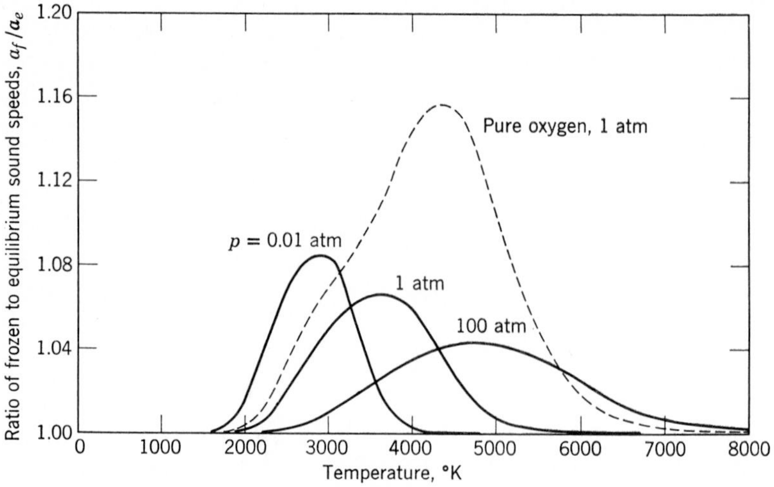

Fig. 3. Ratio of frozen to equilibrium sound speeds for mixture of reacting O_2 and O and inert N_2 (after Der, 1963).

and inert nitrogen in percentages by mass of 23.3 and 76.7 (see Der, 1963). This may be taken as a rough approximation to air at temperatures up to about 5000°K (see Chapter V, Fig. 4). Figure 2 shows a_e as a function of temperature for three representative pressures. Also shown for comparison is the value calculated from the perfect-gas formula $a = (\gamma RT)^{1/2}$, with γ and R held constant at the values appropriate to the mixture at standard conditions ($p = 1$ atm, $T = 273$°K). The departures from the perfect-gas curve are a measure of the effects of oxygen dissociation and of caloric imperfections due to the vibration of the O_2 and N_2, which are assumed to behave as harmonic oscillators. The ratio a_f/a_e is shown in Fig. 3. Also included is a curve for pure dissociating oxygen to show the effect of the dilution by the inert nitrogen.

The result, observed in Fig. 3, that a_f is greater than a_e is a general one. It can be demonstrated, quite independently of the gas model, from general thermodynamic considerations based on the assumption of thermodynamic stability, that is, of the tendency of the disturbed gas to return to an equilibrium state (see De Groot and Mazur, 1962, pp. 323–325). In most practical cases, the difference between a_f and a_e is not large.

The result that $a_f = a_e$ at low temperatures is due to the fact, mentioned near the end of Sec. 3, that at such temperatures q^* is for all practical purposes independent of p and ρ, so that $q_p^* = q_\rho^* = 0$. The equality of a_f and a_e then follows from equations (4.12) and (4.13). This result is valid for both chemical reactions and molecular vibration. The result of Fig. 3 that a_f is again equal to a_e at sufficiently high temperatures, however, is peculiar to the chemical case and depends on the choice of the gas model. It is due in the present instance to the fact that at high temperatures the oxygen is completely dissociated so that q^* is once more essentially independent of p and ρ. This is not true for molecular vibration, however. Once the vibration is fully excited, for example, we have $q^* = e_v^* = RT = p/\rho$, which is not independent of p and ρ. For vibration we thus have $a_f = a_e$ only at low temperatures.

Results for the speeds of sound in more complicated models of air are available in the literature. For typical values for a case similar to the seven-species model discussed in Chapter V, Sec. 5, see Gross and Eisen (1959).

Exercise 5.1. (a) Derive the following expressions for the frozen and equilibrium speeds of sound for the ideal dissociating gas:

$$a_f^2 = R_{A_2}T\frac{(4+\alpha)(1+\alpha)}{3},$$

$$a_e^2 = R_{A_2}T\frac{\alpha^*(1-\alpha^{*2})(1+2T/\Theta_d)+(8+3\alpha^*-\alpha^{*3})(T/\Theta_d)^2}{\alpha^*(1-\alpha^*)+3(2-\alpha^*)(T/\Theta_d)^2}.$$

(b) Using these and any other necessary expressions and employing the numerical constants appropriate to oxygen, calculate and plot the ratio a_f/a_e as a function of T for a gas in thermodynamic equilibrium at 1 atm pressure.

6 PROPAGATION OF PLANE ACOUSTIC WAVES

For a detailed examination of the propagation of acoustic disturbances in a nonequilibrium situation, we shall study the case of plane waves, as, for example, in a duct of constant area. If we adopt the subscript notation for partial differentiation, equation (4.15) for $\varphi(x, t)$ is in this case

$$\tau_0^+\left(\frac{1}{a_{f_0}^2}\varphi_{tt}-\varphi_{xx}\right)_t+\frac{1}{a_{e_0}^2}\varphi_{tt}-\varphi_{xx}=0, \tag{6.1}$$

where x is now the single space variable.

A simple general solution to this equation, analogous to (4.17) for the classical equation (4.16), is not possible. We restrict ourselves therefore to the special case of harmonic disturbances propagating to infinity in the positive x-direction. These could be caused, for example, by a piston with mean position at $x = 0$ (see Fig. 4) and executing the harmonic

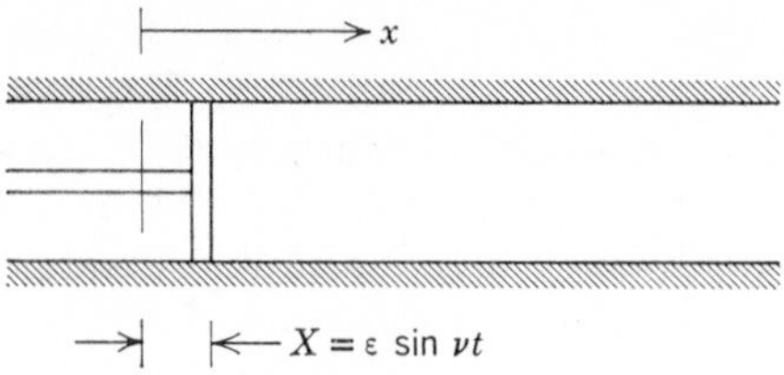

Fig. 4. Harmonically oscillating piston.

motion $X = \varepsilon \sin \nu t$, where ν is the radian frequency. For small amplitudes of oscillation ε, the appropriate boundary condition, which expresses the fact that the fluid at the face of the piston must move with the velocity of the piston, is

$$u'(0, t) = \frac{dX}{dt} = \nu\varepsilon \cos \nu t. \tag{6.2}$$

Here the boundary condition has been transferred from $x = X$ to $x = 0$, as is permissible in linear theory (see Liepmann and Roshko, 1957, p. 207). Since the boundary condition is on $u' = \varphi_x$ rather than φ, it is convenient to work with a differential equation for u'. This can be found by differentiating equation (6.1) with respect to x to obtain

$$\tau_0^+\left(\frac{1}{a_{f_0}^2} u'_{tt} - u'_{xx}\right)_t + \frac{1}{a_{e_0}^2} u'_{tt} - u'_{xx} = 0. \tag{6.3}$$

We therefore seek a solution to (6.3) subject to the boundary condition (6.2).

To simplify the equations, we first introduce the dimensionless variables

$$\bar{x} \equiv \nu x / a_{f_0}, \qquad \bar{t} \equiv \nu t, \qquad \bar{u}' \equiv u'/\nu\varepsilon, \tag{6.4}$$

and the notation

$$k \equiv \nu\tau_0^+, \qquad b \equiv a_{f_0}^2/a_{e_0}^2 > 1, \tag{6.5}$$

where the inequality in the second expression follows from the fact that $a_{f_0} > a_{e_0}$. The differential equation (6.3) then becomes

$$k(\bar{u}'_{\bar{t}\bar{t}} - \bar{u}'_{\bar{x}\bar{x}})_{\bar{t}} + b\bar{u}'_{\bar{t}\bar{t}} - \bar{u}'_{\bar{x}\bar{x}} = 0, \tag{6.6}$$

and the boundary condition (6.2) is

$$\bar{u}'(0, \bar{t}) = \cos \bar{t}. \tag{6.7}$$

The quantity k introduced by the first of equations (6.5) is the dimensionless ratio of the characteristic time τ_0^+ of the rate process to the characteristic time $1/\nu$ of the piston oscillation.

It is clear from the nature of the boundary condition (6.7) that at any point along the duct the motion of the fluid will be periodic with the same frequency as that of the piston. Furthermore, since the problem is linear, this periodic motion, like that of the piston, will be harmonic. This suggests a solution of the form

$$\bar{u}'(\bar{x}, \bar{t}) = \mathrm{Re}[f(\bar{x})e^{i\bar{t}}], \tag{6.8}$$

where $f(\bar{x})$ is complex, and we take the real part in order eventually to satisfy the boundary condition (6.7). Putting this assumed solution into (6.6) gives the following complex, ordinary differential equation for f:

$$f'' + \frac{b + ik}{1 + ik} f = 0, \tag{6.9}$$

where a prime indicates differentiation with respect to $\bar{x}$. A solution of this equation is given formally by

$$f = Ce^{c\bar{x}}, \tag{6.10}$$

where C and c are complex constants. The value of c is given by the following equation obtained by substituting (6.10) into (6.9):

$$c^2 = -\frac{b + ik}{1 + ik}. \tag{6.11}$$

This equation has two complex roots with equal positive and negative values. We can represent these roots in terms of real quantities δ and λ by

$$c = \pm(\delta + i\lambda), \tag{6.12}$$

where δ is by definition positive. The general solution of equation (6.9) can then be written

$$f = C^+e^{-(\delta+i\lambda)\bar{x}} + C^-e^{(\delta+i\lambda)\bar{x}}, \tag{6.13}$$

where C^+ and C^- are constants. The notation for these constants is oppositely related to the sign of the corresponding root c for reasons that will appear later.

To find δ and λ we substitute the expression (6.12) into (6.11) and rewrite the resulting equation as

$$\begin{aligned}(\delta + i\lambda)^2 &= \delta^2 - \lambda^2 + i2\delta\lambda \\ &= -\frac{b + ik}{1 + ik} \times \frac{1 - ik}{1 - ik} = -\frac{b + k^2 - ik(b - 1)}{1 + k^2},\end{aligned}$$

from which we obtain the two simultaneous equations

$$\delta^2 - \lambda^2 = -\frac{b + k^2}{1 + k^2} \equiv P, \tag{6.14a}$$

$$2\delta\lambda = \frac{k(b - 1)}{1 + k^2} \equiv Q. \tag{6.14b}$$

Since δ is by definition positive and b is greater than 1, it follows from the second of these equations that λ is also positive. The formal solution of equations (6.14) in terms of P and Q is

$$\delta = \pm[\tfrac{1}{2}(P \pm \sqrt{P^2 + Q^2})]^{1/2}, \qquad \lambda = \pm[\tfrac{1}{2}(-P \pm \sqrt{P^2 + Q^2})]^{1/2}.$$

Since $|\sqrt{P^2 + Q^2}| > |P|$, it is necessary to take the plus sign with the inner radical in order to have δ and λ real. The outer sign must in both cases be taken as plus since δ and λ are both positive. The final expressions for δ and λ, obtained by substituting for P and Q from (6.14), are then

$$\left.\begin{matrix}\delta\\ \lambda\end{matrix}\right\} = \left\{\frac{1}{2(1 + k^2)}\left[\mp(b + k^2) + \sqrt{(1 + k^2)(b^2 + k^2)}\right]\right\}^{1/2}, \tag{6.15}$$

where the upper sign goes with δ and the lower sign with λ.

The general solution for $\bar{u}'$ is obtained by putting (6.13) into (6.8) and is

$$\bar{u}'(\bar{x}, \bar{t}) = \operatorname{Re}\{C^+ \exp[-\delta\bar{x} + i(\bar{t} - \lambda\bar{x})] + C^- \exp[\delta\bar{x} + i(\bar{t} + \lambda\bar{x})]\} \tag{6.16a}$$

$$= C^+ e^{-\delta\bar{x}} \cos(\bar{t} - \lambda\bar{x}) + C^- e^{\delta\bar{x}} \cos(\bar{t} + \lambda\bar{x}). \tag{6.16b}$$

In the second form it is understood that C^+ and C^- are now real constants. The C^+- and C^--terms in the solution (6.16) represent damped harmonic waves traveling in the positive and negative $\bar{x}$-directions respectively. The notation for the constants was chosen with this in mind. To solve the problem posed in Fig. 4, we retain the first term and set $C^- = 0$. The boundary condition (6.7) then gives $C^+ = 1$. We thus find for the final solution in the original dimensional variables

$$\boxed{u'(x, t) = \nu\varepsilon \exp\left[-\left(\frac{\nu\delta}{a_{f_0}}\right)x\right] \cos\left[\nu\left(t - \frac{\lambda}{a_{f_0}}x\right)\right]}, \tag{6.17}$$

where δ and λ are given by equation (6.15). The speed of propagation V of the disturbance is thus

$$V = \frac{a_{f_0}}{\lambda}, \tag{6.18}$$

while the damping per unit length of x increases exponentially with $\nu\delta/a_{f_0}$. Both λ and δ, and hence the speed and the damping, depend on the relaxation time and the frequency through the dimensionless quantity k, and on the ratio of the frozen and equilibrium speeds of sound through the dimensionless ratio b.

Equations (6.15) and (6.17) show explicitly how the nonequilibrium results tend to the classical findings in the limits of equilibrium and frozen flow. The equilibrium-flow case corresponds to $\tau_0^+ \to 0$ and hence $k \to 0$, in which case (6.15) shows that $\delta \to 0$ and $\lambda \to \sqrt{b} = a_{f_0}/a_{e_0}$. In this limit, therefore, the wave propagates without change of form and at the equilibrium speed of sound a_{e_0}. The frozen-flow case corresponds to $\tau_0^+ \to \infty$ and hence $k \to \infty$, in which case we obtain $\delta \to 0$ and $\lambda \to 1$. Thus the wave propagates again without change of form but at the frozen speed of sound a_{f_0}.

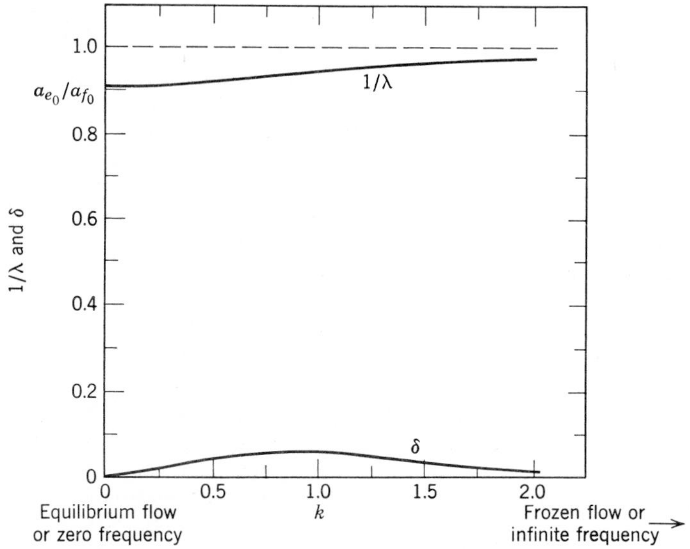

Fig. 5. Variation of δ and $1/\lambda$ with k for $a_{f_0}/a_{e_0} = \frac{11}{10}$.

For intermediate values of k, the values of δ and $1/\lambda$ for a representative fixed value of a_{f_0}/a_{e_0} vary as shown in Fig. 5. As we go from equilibrium flow to frozen flow, the wave speed increases monotonically from a_{e_0} to the somewhat larger value a_{f_0}. The damping increases from zero to a maximum and then decreases back again toward zero.

In the discussion thus far, changes in k have been regarded as corresponding to changes in the relaxation time, that is, to changes in τ_0^+ for a fixed, finite frequency ν. Since $k = \nu\tau_0^+$, the same variation in k can also be obtained by varying ν for a fixed τ_0^+. Regarded from this point of view, the wave speed and damping are seen to be frequency dependent. The limits $k \to 0$ and $k \to \infty$ now correspond to $\nu \to 0$ and $\nu \to \infty$. It follows at once that waves of vanishingly small frequency travel at the speed a_{e_0}, whereas those of infinitely high frequency travel at the speed a_{f_0}. For this reason, a_{e_0} and a_{f_0} are known also as the "low-frequency" and "high-frequency" speeds of sound. The damping, which depends on the product $\nu\delta/a_{f_0}$, now requires careful consideration. In the low-frequency limit, both ν and δ tend to 0, with the result that the waves are undamped. In the high-frequency limit, ν tends to ∞ whereas δ tends to 0, so that a detailed investigation is required. To examine this point, we return to equation (6.15) and approximate δ^2 to the first order in $1/k^2$ as follows, where use is made of the approximation $\sqrt{1+x} \cong 1 + x/2$, valid for $x \ll 1$:

$$\delta^2 = \frac{1}{2(1 + 1/k^2)}\left[-\left(1 + \frac{b}{k^2}\right) + \sqrt{\left(1 + \frac{1}{k^2}\right)\left(1 + \frac{b^2}{k^2}\right)}\right]$$

$$\cong \frac{1}{2(1 + 1/k^2)}\left[-1 - \frac{b}{k^2} + \sqrt{1 + \frac{b^2 + 1}{k^2}}\right]$$

$$\cong \frac{1}{2(1 + 1/k^2)} \times \frac{b^2 - 2b + 1}{2k^2} \cong \left(\frac{b - 1}{2k}\right)^2.$$

From this we obtain in the limit as $\nu \to \infty$

$$\frac{\nu\delta}{a_{f_0}} \to \frac{\nu}{a_{f_0}} \times \frac{b - 1}{2k} = \frac{1}{2a_{f_0}\tau_0^+}\left(\frac{a_{f_0}^2}{a_{e_0}^2} - 1\right), \tag{6.19}$$

which shows the existence of damping at very high frequencies. Other things being equal, low-frequency waves will thus persist over greater distances than do high-frequency waves. This is a reflection of the fact that in the first case, a fluid element remains close to equilibrium, and the rate of entropy production is essentially zero. In the second case the changes occur so rapidly that deviations from equilibrium are significant, and the rate of entropy increase is relatively large.

It is also of interest to examine the pressure disturbance p'. To do this, we first find φ by integrating the following relation, which we obtain by transforming variables in equation (4.14b) and substituting from equation (6.16a) with $C^+ = 1$ and $C^- = 0$:

$$\varphi_{\bar{x}} = \frac{a_{f_0}}{\nu}\varphi_x = \varepsilon a_{f_0}\bar{u}' = \varepsilon a_{f_0}\,\mathrm{Re}\,\{\exp\,[-\delta\bar{x} + i(\bar{t} - \lambda\bar{x})]\}.$$

Carrying out the integration with respect to $\bar{x}$ gives

$$\varphi = \varepsilon a_{f_0}\,\mathrm{Re}\left\{-\frac{\delta - i\lambda}{\delta^2 + \lambda^2}\exp\,[-\delta\bar{x} + i(\bar{t} - \lambda\bar{x})] + f(\bar{t})\right\}.$$

The function of integration $f(\bar{t})$ can be evaluated from the fact that φ, like $\bar{u}'$, must be a periodic function of $\bar{t}$. But a periodic function of $\bar{t}$ alone (i.e., unmultiplied by a function of $\bar{x}$) cannot satisfy the differential equation (6.1), hence we must take $f(\bar{t}) = 0$. Equation (4.14a) then gives p' as

$$p' = -\rho_0 \varphi_t = -\nu \rho_0 \varphi_{\bar{t}} = \nu \varepsilon \rho_0 a_{f_0} \operatorname{Re} \left\{ \frac{\lambda + i\delta}{\delta^2 + \lambda^2} \exp\left[-\delta \bar{x} + i(\bar{t} - \lambda \bar{x})\right] \right\}$$

$$= \nu \varepsilon \rho_0 a_{f_0} \operatorname{Re} \left\{ \frac{1}{\sqrt{\delta^2 + \lambda^2}} \exp\left[-\delta \bar{x} + i(\bar{t} + \Delta \bar{t} - \lambda \bar{x})\right] \right\}.$$

where $\Delta \bar{t} \equiv \tan^{-1}(\delta/\lambda)$. Reverting to the dimensional variables, we thus find finally

$$p' = \frac{\nu \varepsilon \rho_0 a_{f_0}}{\sqrt{\delta^2 + \lambda^2}} e^{-(\nu\delta/a_{f_0})x} \cos\left[\nu\left(t + t_p - \frac{\lambda}{a_{f_0}} x\right)\right], \qquad (6.20)$$

where

$$t_p = \frac{1}{\nu} \tan^{-1} \frac{\delta}{\lambda}.$$

The pressure disturbance thus also propagates as a damped harmonic wave with speed a_{f_0}/λ. Comparison of equations (6.17) and (6.20) shows that, in general, the pressure and velocity disturbances are out of phase, the pressure at a given station x leading the velocity in time by the amount t_p. Since $\delta \to 0$ as k approaches both 0 and ∞, this phase shift disappears in both the equilibrium-flow and frozen-flow (or low-frequency and high-frequency) limits.

The foregoing results can be used to obtain the disturbance resulting from any arbitrary piston motion by analysis of the disturbance in terms of its Fourier components. Since the high-frequency components travel faster and are more damped than the low-frequency ones, the disturbance in the present theory must change form as it propagates. This is in contrast to the situation in classical acoustic theory, where any small plane disturbance propagates without distortion. In the language of acoustics, the distortion due to the frequency-dependent velocity is called *dispersion* and that due to the damping is called *absorption*. It should be noted that the wave distortion that occurs here as the result of dispersion and absorption is purely a linear, small-disturbance effect associated with the nonequilibrium process. It is in no way related to the well-known distortion that is found in a large-amplitude wave in classical theory as a result of nonlinear effects (see, e.g., Liepmann and Roshko, 1957, pp. 74–77).

The propagation of a discrete acoustic wave has been studied by Chu (1957). Here the piston is assumed to start impulsively from rest at $t = 0$ and move with a constant velocity u_p'. The analysis involves Laplace-transform methods and is too lengthy to reproduce here. The results, which are instructive, are as shown qualitatively in Fig. 6. At the initial instant the disturbance is a step wave. As would be anticipated from the preceding results, this step begins at once to distort, the initial signal running out ahead at the frozen speed a_{f_0} and the rest of the wave following along at a somewhat lesser speed. The initial signal remains a step. Like the high-frequency harmonic waves, it is highly damped; the distributed part of the wave less so. At a sufficiently large time t the wave thus appears as a step disturbance of reduced size traveling at the speed a_{f_0}, followed at a distance proportional to t by the main part of the wave traveling at more or less the speed a_{e_0}. At a time $t \gg 2\tau_0^+/(b - 1)$ [cf. equation (6.19)], the front of the wave is practically negligible, and the

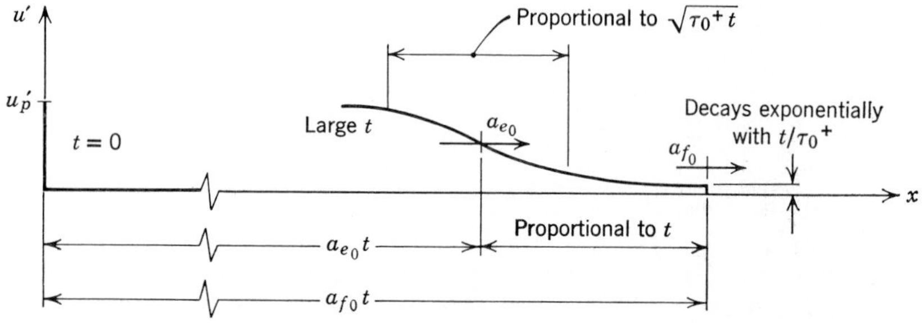

Fig. 6. Propagation of discrete acoustic wave due to impulsive piston motion.

disturbance appears as a somewhat diffuse wave having a width proportional to $\sqrt{\tau_0^+ t}$ and traveling at essentially the equilibrium speed of sound a_{e_0}. If τ_0^+ is very small, this condition arises very close to the piston. In the limit $\tau_0^+ \to 0$ (equilibrium flow), the distortion disappears, and the disturbance travels purely as a step wave at the equilibrium speed a_{e_0}. In the limit $\tau_0^+ \to \infty$ (frozen flow), it travels as a step wave at the frozen speed a_{f_0}.

It must be emphasized that the wave *front* travels at the frozen speed of sound a_{f_0} *irrespective of the value of the relaxation time*, unless it is identically zero. Physically, this is because the wave is propagating into a medium that has no way of knowing of the approaching disturbance. Thus the fluid has no opportunity to alter its state during the passage of the initial signal, and that signal must therefore travel at the speed appropriate to the frozen internal state of the gas.

Equations such as (6.1), which are made up of several wave operators with different signal speeds, are characteristic of acoustic problems involving dissipative effects. We shall encounter another example, in this case a fifth-order equation, in our discussion of radiative gas dynamics in

Chapter XII. A general equation of this type of arbitrary order has been studied in a fundamental way by Whitham (1959), with results that are as exemplified by the discrete wave discussed above. In general, he finds that for small time the wave motion defined by the highest-order terms dominates; for very large time the lowest-order terms ultimately describe the main disturbance. For the wave motion defined by terms of a *given* order, the presence of lower-order terms produces an exponential damping, while the presence of higher-order terms produces a diffusion.

As in classical theory, the inclusion of nonlinear effects will to some extent modify the results discussed here. These matters are the subject of current research. A treatment of the nonlinear effects due to fluid convection in the expansion wave from an impulsively retracting piston has been given by Jones (1964).

Exercise 6.1. Obtain an expression, analogous to (6.20), for the density disturbance ρ' in a harmonic wave. How does the phase shift of the density disturbance compare with that of the pressure disturbance, both measured relative to the velocity?

Exercise 6.2. Develop a formula for the work done on the gas by the piston of Fig. 4 (per unit piston area) during one cycle of oscillation. Show by means of a suitable qualitative graph how the work done by the piston varies with the nonequilibrium parameter k for a fixed frequency ν. What would you say happens, within the context of acoustic theory, to the energy given to the gas by the work done by the piston?

Exercise 6.3. Consider the exponential piston motion $X = \varepsilon e^{\kappa t}$, where the piston started from $x = 0$ at $t = -\infty$. The quantities ε and κ are positive real constants. Derive the following expression for the resulting disturbance velocity according to the nonequilibrium acoustic theory of this section:

$$u'(x, t) = \kappa\varepsilon \exp\left[\kappa\left(t - \sqrt{\frac{b+k}{1+k}}\frac{x}{a_{f_0}}\right)\right],$$

where in this case $k \equiv \kappa\tau_0^+$. Discuss the nature of this solution and its behavior in the equilibrium- and frozen-flow limits.

7 EQUATION FOR SMALL DEPARTURES FROM A UNIFORM FREE STREAM

In gas dynamics we often want to study the flow around an object placed in a uniform free stream. A great part of classical aerodynamics, in particular, is concerned with the disturbances caused by slender objects under conditions of equilibrium or frozen flow. Here we shall examine the same problem in the nonequilibrium case.

A linearized equation governing the steady, nonequilibrium flow about a slender object in a uniform parallel stream can be obtained by returning to the basic equations of Sec. 2 and proceeding suitably therefrom. This procedure is followed, for example, by Vincenti (1959). Here we shall obtain the desired equation from the acoustic equation of Sec. 4.

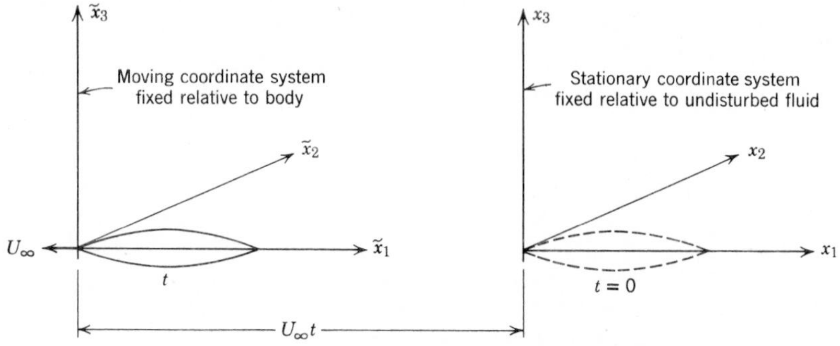

Fig. 7. Fixed and moving coordinate systems.

Consider an object flying at a constant speed U_∞ parallel to the x_1-axis of a Cartesian coordinate system x_1, x_2, x_3 fixed relative to undisturbed fluid (Fig. 7). The state of the undisturbed fluid is denoted by subscript ∞. If the disturbances caused by the object are small (and U_∞ is not too large or too close to the speeds of sound), we might hope to analyze the flow by means of the acoustic equation (4.15), which will here be written

$$\tau_\infty^+ \frac{\partial}{\partial t}\left(\frac{1}{a_{f\infty}^2}\frac{\partial^2}{\partial t^2} - \frac{\partial^2}{\partial x_j\,\partial x_j}\right)\varphi + \left(\frac{1}{a_{e\infty}^2}\frac{\partial^2}{\partial t^2} - \frac{\partial^2}{\partial x_j\,\partial x_j}\right)\varphi = 0. \qquad (7.1)$$

In the foregoing situation it is often convenient to transform to a coordinate system $\tilde{x}_1$, $\tilde{x}_2$, $\tilde{x}_3$ fixed relative to the moving body. To an observer moving with this coordinate system the body will appear to be immersed in a uniform parallel stream with undisturbed velocity U_∞ in the direction opposite to that of the original motion of the body. If the coordinate systems are taken to coincide at some arbitrary time $t = 0$, the situation at time t will be as shown in the figure. The transformation from one coordinate system to the other is given by

$$\tilde{x}_1 = x_1 + U_\infty t, \qquad \tilde{x}_2 = x_2, \qquad \tilde{x}_3 = x_3, \qquad \tilde{t} = t. \qquad (7.2)$$

If the disturbance field in the new coordinate system is time dependent, as would be the case with a time-dependent attitude or shape of the body, it will be described by a disturbance potential $\tilde{\varphi}(\tilde{x}_i, \tilde{t})$. Since the disturbance field itself is not changed by the coordinate transformation, we have

$$\tilde{\varphi}(\tilde{x}_i, \tilde{t}) = \varphi(x_i, t). \tag{7.3}$$

Now, if z is any one of the original independent variables, we have in general

$$\frac{\partial(\ )}{\partial z} = \frac{\partial(\ )}{\partial \tilde{t}}\frac{\partial \tilde{t}}{\partial z} + \frac{\partial(\ )}{\partial \tilde{x}_j}\frac{\partial \tilde{x}_j}{\partial z}.$$

In particular we find

$$\frac{\partial(\ )}{\partial x_i} = \frac{\partial(\ )}{\partial \tilde{x}_i}, \qquad \frac{\partial(\ )}{\partial t} = \frac{\partial(\ )}{\partial \tilde{t}} + U_\infty \frac{\partial(\ )}{\partial \tilde{x}_1}. \tag{7.4}$$

Applying these expressions in equation (7.1), we obtain for the potential equation in the new coordinate system

$$\tau_\infty^+\left(\frac{\partial}{\partial \tilde{t}} + U_\infty \frac{\partial}{\partial \tilde{x}_1}\right)\left[\frac{1}{a_{f\infty}^2}\left(\frac{\partial}{\partial \tilde{t}} + U_\infty \frac{\partial}{\partial \tilde{x}_1}\right)^2 - \frac{\partial^2}{\partial \tilde{x}_j \partial \tilde{x}_j}\right]\tilde{\varphi}$$
$$+ \left[\frac{1}{a_{e\infty}^2}\left(\frac{\partial}{\partial \tilde{t}} + U_\infty \frac{\partial}{\partial \tilde{x}_1}\right)^2 - \frac{\partial^2}{\partial \tilde{x}_j \partial \tilde{x}_j}\right]\tilde{\varphi} = 0. \tag{7.5}$$

When the attitude and shape of the body are independent of time, the disturbance field in the new coordinate system is also independent of time, that is, $\partial(\ )/\partial \tilde{t} = 0$. If we define frozen and equilibrium Mach numbers according to

$$M_{f\infty} \equiv \frac{U_\infty}{a_{f\infty}}, \qquad M_{e\infty} \equiv \frac{U_\infty}{a_{e\infty}}, \tag{7.6}$$

equation (7.5) then reduces to

$$\boxed{\tau_\infty^+ U_\infty \frac{\partial}{\partial x_1}\left[(M_{f\infty}^2 - 1)\frac{\partial^2 \varphi}{\partial x_1^2} - \frac{\partial^2 \varphi}{\partial x_2^2} - \frac{\partial^2 \varphi}{\partial x_3^2}\right] + \left[(M_{e\infty}^2 - 1)\frac{\partial^2 \varphi}{\partial x_1^2} - \frac{\partial^2 \varphi}{\partial x_2^2} - \frac{\partial^2 \varphi}{\partial x_3^2}\right] = 0}, \tag{7.7}$$

where we have discarded the now superfluous tilde notation. The product $\tau_\infty^+ U_\infty$ that appears here defines a *relaxation length* for the rate process in the present situation. The velocity components in the new coordinate system are given, in view of equations (4.14b) and (7.2) through (7.4), by

$$u_1 = U_\infty + u_1', \qquad u_2 = u_2', \qquad u_3 = u_3', \quad \text{where} \quad u_i' = \partial\varphi/\partial x_i, \tag{7.8}$$

where the tilde notation has again been dropped. The disturbance pressure in steady flow is given, in view of equations (4.14a) and (7.4), by

$$p' = -\rho_\infty \frac{\partial\varphi}{\partial t} = -\rho_\infty U_\infty \frac{\partial\tilde{\varphi}}{\partial\tilde{x}_1} = -\rho_\infty U_\infty u_1'. \tag{7.9}$$

This agrees with the usual result obtained by linearizing the Euler equations (2.3) for the case of steady flow.[2]

Equation (7.7) is the governing equation for steady nonequilibrium flows that are a small departure from a uniform parallel stream. It has an obvious relationship to the classical Prandtl-Glauert equation (see Liepmann and Roshko, 1957, p. 205):

$$(M_\infty^2 - 1)\frac{\partial^2\varphi}{\partial x_1^2} - \frac{\partial^2\varphi}{\partial x_2^2} - \frac{\partial^2\varphi}{\partial x_3^2} = 0, \tag{7.10}$$

where $M_\infty \equiv U_\infty/a_\infty$ is the free-stream Mach number. This equation, like the classical wave equation (4.16), is valid under the assumption that h is a function only of p and ρ. If we set $\tau_\infty^+ = 0$ in equation (7.7), it reduces, as we would expect, to the Prandtl-Glauert equation for equilibrium flow. In the limit $\tau_\infty^+ \to \infty$, it goes over into the Prandtl-Glauert equation for frozen flow.

The coefficients of $\partial^2\varphi/\partial x_1^2$ in equation (7.7) are related to the regions of influence of an infinitesimal disturbance in supersonic flow. As we have seen, in nonequilibrium flow the front of a discrete acoustic disturbance travels in general with the frozen speed $a_{f\infty}$ relative to the fluid into which it propagates. It follows from the well-known construction of Fig. 8(*a*) that when $U_\infty > a_{f\infty}$ the influence of an infinitesimal disturbance is confined to the interior of a cone with semi-apex angle

$$\mu_{f\infty} = \tan^{-1}\frac{a_{f\infty}t}{\sqrt{(U_\infty t)^2 - (a_{f\infty}t)^2}} = \tan^{-1}\frac{1}{\sqrt{M_{f\infty}^2 - 1}} \tag{7.11}$$

(see Liepmann and Roshko, 1957, p. 228). This result is true irrespective of the value of the relaxation time τ_∞^+, so long as it is not identically zero. In two-dimensional or axially symmetric flow, the up- and down-going directions that make the angle $\mu_{f\infty}$ with the x_1-axis are the *characteristic*

[2] Equation (7.9) is applicable only for two-dimensional flow or for so-called planar systems in three-dimensional flow. For a consistent approximation for axially symmetric bodies it is necessary to retain certain second-order terms (cf. Liepmann and Roshko, 1957, p. 224).

directions of the differential equation (7.7), which are defined by the third-order terms when $\tau^+_\infty \neq 0$.[3] In the situation where τ^+_∞ is identically zero, the front of a discrete disturbance travels at the equilibrium speed $a_{e\infty}$, and the region of influence is given by the angle

$$\mu_{e\infty} = \tan^{-1} \frac{1}{\sqrt{M^2_{e\infty} - 1}}. \tag{7.12}$$

The directions defined by this angle are the characteristic directions of the

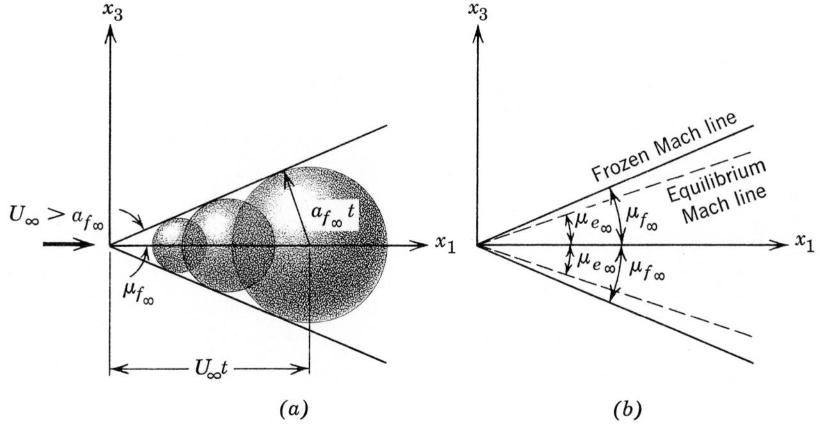

Fig. 8. Regions of influence in nonequilibrium and equilibrium flow.

second-order, Prandtl-Glauert equation that appears discontinuously from equation (7.7) when $\tau^+_\infty = 0$. By analogy to the classical theory, lines laid off according to equations (7.11) and (7.12) are known as *frozen* and *equilibrium Mach lines*, and $\mu_{f\infty}$ and $\mu_{e\infty}$ are the frozen and equilibrium Mach angles. Since $a_{f\infty} > a_{e\infty}$, it follows that $M_{f\infty} < M_{e\infty}$ and hence that $\mu_{f\infty} > \mu_{e\infty}$. The frozen Mach line thus stands at a steeper inclination to the flow than does the equilibrium Mach line [see Fig. 8(*b*)]. The discontinuous change in the direction of the characteristics when $\tau^+_\infty = 0$ is another manifestation of the singular nature of the equilibrium-flow limit.

The characteristic directions defined by equations (7.11) and (7.12) are real only if the sign of the quantity under the radical is positive. When

[3] For a discussion of the theory of characteristics of partial differential equations, which would take us too far afield, see Courant and Hilbert (1962), Chapter V, or Courant and Friedrichs (1948), Chapter II. According to this theory, the third-order equation (7.7) has an additional characteristic direction given by the lines parallel to the x_1-axis. This is the counterpart in the linear approximation of the fact that the streamlines are also characteristics, as will be discussed for the nonlinear equations in Sec. 14.

this is the case the differential equation is said in mathematical terminology to be *hyperbolic*; when real characteristics do not exist it is said to be *elliptic*. For all values of τ_∞^+ other than zero, the differential equation (7.7) thus changes from elliptic to hyperbolic at the stream speed defined by $M_{f_\infty} = U_\infty/a_{f_\infty} = 1$. When τ_∞^+ is identically zero the change takes place at the somewhat lower speed defined by $M_{e_\infty} = U_\infty/a_{e_\infty} = 1$.[4]

For later use we shall also need the steady-flow equivalent of equation (4.9). Introducing a_{f_0} and φ into equation (4.9) by means of equations (4.12) and (4.14) and transforming to steady flow by means of the equations of this section, we obtain finally

$$U_\infty \frac{\partial q'}{\partial x_1} = -\frac{\rho_\infty h_{p\infty}}{h_{q\infty}}\left[(M_{f\infty}^2 - 1)\frac{\partial^2 \varphi}{\partial x_1^2} - \frac{\partial^2 \varphi}{\partial x_2^2} - \frac{\partial^2 \varphi}{\partial x_3^2}\right]. \tag{7.13}$$

Before leaving this section, it should be pointed out that the differential equation (7.7) is not suitable for the study of the flow over slender objects at high speeds and high altitudes, which is a problem of primary importance in nonequilibrium aerodynamics. In this problem the undisturbed flow is at a low temperature so that τ_∞^+ is essentially ∞. Reactions and hence nonequilibrium processes come into play only because of the large temperature disturbances that accompany the passage of even a slender object at hypersonic speeds. In equation (7.7) on the other hand, the undisturbed stream, although in equilibrium, is assumed in effect to be already at a high enough temperature so that τ_∞^+ is finite and the flow is capable of undergoing a reaction when even a small disturbance passes. The linear equation (7.7) is therefore incapable of handling the hypersonic flight problem, which is essentially nonlinear. It is useful nevertheless in illustrating some of the basic properties of nonequilibrium flow. It may also be of use in studying slender objects immersed in an already reactive flow, as, for example, a vane in a rocket exhaust or a disturbance in a supersonic nozzle flow.

8 FLOW OVER A WAVY WALL

The flow over a wavy wall offers a simple example of the solution of equation (7.7). The solution of this problem for the Prandtl-Glauert equation (7.10) was first given by Ackeret (1928) and is reproduced in any

[4] The original acoustic equation (4.15) could similarly be discussed in terms of its characteristic directions. In that case, however, the characteristics are always real, and the equation is always hyperbolic.

standard textbook on gas dynamics. It is instructive to reexamine the problem for nonequilibrium flow.

The solution follows lines similar to those of Sec. 6 for plane harmonic waves. To begin we write equation (7.7) in the two-dimensional, subscript form

$$\tau_{\infty}^{+} U_{\infty}[(M_{f\infty}^{2} - 1)\varphi_{xx} - \varphi_{yy}]_{x} + (M_{e\infty}^{2} - 1)\varphi_{xx} - \varphi_{yy} = 0, \quad (8.1)$$

where we have changed to x and y for the independent variables. The corresponding disturbance velocities are denoted by u' and v'. We concern ourselves (see Fig. 9) with the flow in the semi-infinite space above the

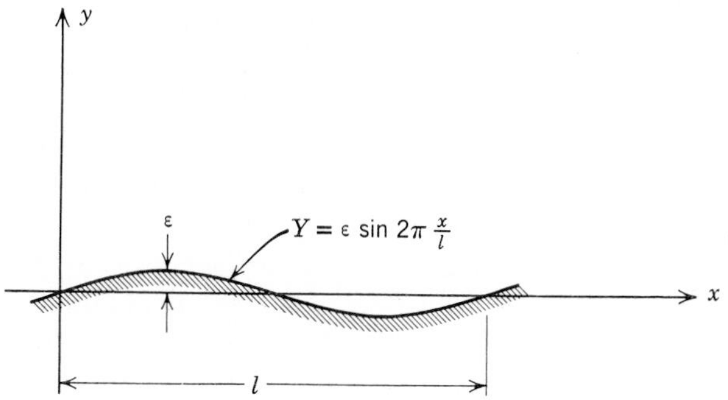

Fig. 9. Sinusoidally wavy wall.

infinite sinusoidal wall $Y = \varepsilon \sin 2\pi(x/l)$. The linearized boundary condition expressing the fact that the flow must be tangential to the wall is then

$$\frac{v'(x, 0)}{U_{\infty}} = \frac{dY}{dx} = 2\pi \frac{\varepsilon}{l} \cos\left(2\pi \frac{x}{l}\right). \quad (8.2)$$

Here we have again transferred the boundary condition to the axis $y = 0$, as is permissible in the linear approximation. Since the boundary condition is on $v' = \varphi_y$, we shall replace equation (8.1) with the following equation obtained by differentiating (8.1) with respect to y:

$$\tau_{\infty}^{+} U_{\infty}[(M_{f\infty}^{2} - 1)v'_{xx} - v'_{yy}]_{x} + (M_{e\infty}^{2} - 1)v'_{xx} - v'_{yy} = 0. \quad (8.3)$$

In addition to (8.2), we impose the usual condition that the disturbances remain finite as $y \to \infty$.

The solution proceeds much as in Sec. 6. To simplify the equation, we introduce the dimensionless variables

$$\bar{x} \equiv 2\pi \frac{x}{l}, \qquad \bar{y} \equiv 2\pi \frac{y}{l}, \qquad \bar{v}' \equiv \frac{v'}{2\pi(\varepsilon/l)U_{\infty}}, \quad (8.4)$$

and the notation

$$K \equiv \frac{2\pi\tau_\infty^+ U_\infty}{l}, \qquad a \equiv M_{f_\infty}^2 - 1, \qquad b \equiv M_{e_\infty}^2 - 1. \tag{8.5}$$

The differential equation (8.3) then becomes

$$K(a\bar{v}'_{\bar{x}\bar{x}} - \bar{v}'_{\bar{y}\bar{y}})_{\bar{x}} + b\bar{v}'_{\bar{x}\bar{x}} - \bar{v}'_{\bar{y}\bar{y}} = 0, \tag{8.6}$$

and the boundary condition (8.2) is

$$\bar{v}'(\bar{x}, 0) = \cos \bar{x}. \tag{8.7}$$

The dimensionless quantity K is proportional to the ratio of the relaxation length $\tau_\infty^+ U_\infty$ of the rate process to the characteristic length l of the boundary. Equation (8.6) is seen to contain one more constant than does the corresponding equation (6.6) of Sec. 6. One of these could have been formally eliminated by including a factor $\sqrt{a}$ in the definition of $\bar{y}$. Since the value of a can be negative as well as positive, however, this introduces the complication of complex quantities into the transformation. It is simpler to avoid these complications by retaining the additional constant.

Following reasoning similar to that in Sec. 6, we again attempt a solution in the form

$$\bar{v}'(\bar{x}, \bar{y}) = \text{Re}\,[f(\bar{y})e^{i\bar{x}}]. \tag{8.8}$$

Proceeding as before we find for the function $f(\bar{y})$

$$f = C^+e^{-(\Delta+i\Lambda)\bar{y}} + C^-e^{(\Delta+i\Lambda)\bar{y}}, \tag{8.9}$$

where Δ and Λ are real quantities, with Δ positive by definition, and C^+ and C^- are constants. The values of Δ and Λ are given by the simultaneous equations

$$\Delta^2 - \Lambda^2 = -\frac{b + aK^2}{1 + K^2} \equiv P, \tag{8.10a}$$

$$2\Delta\Lambda = \frac{K(b - a)}{1 + K^2} \equiv Q. \tag{8.10b}$$

Since $M_{e_\infty} > M_{f_\infty}$, we have $b > a$. It therefore follows from equation (8.10b) that Λ, like Δ, is positive. The formal solution of equations (8.10) is given once more by (see Sec. 6)

$$\Delta = \pm[\tfrac{1}{2}(P \pm \sqrt{P^2 + Q^2})]^{1/2}, \qquad \Lambda = \pm[\tfrac{1}{2}(-P \pm \sqrt{P^2 + Q^2})]^{1/2}. \tag{8.11}$$

It is again necessary to take the plus sign with the inner radical in order to have Δ and Λ real. The outer signs must also be taken as plus in order to have Δ and Λ positive. The final expressions, put in terms of K, M_{e_∞}, and M_{f_∞} by means of equations (8.10), are then

$$\left.\begin{array}{c}\Delta\\\Lambda\end{array}\right\} = \left\{\frac{1}{2(1 + K^2)}\left[\mp (M_{e_\infty}^2 - 1) \mp K^2(M_{f_\infty}^2 - 1)\right.\right.$$
$$\left.\left. + \sqrt{(1 + K^2)(|M_{e_\infty}^2 - 1|^2 + K^2\,|M_{f_\infty}^2 - 1|^2)}\right]\right\}^{\frac{1}{2}}. \quad (8.12)$$

Absolute-value signs have been introduced under the radical to emphasize the fact that, as a result of the squaring, the sign of the quantities $(M_{e_\infty}^2 - 1)$ and $(M_{f_\infty}^2 - 1)$ is here of no importance. The same is not true, however, for these quantities outside the radical.

The general solution for $\bar{v}'$, obtained by putting (8.9) into (8.8), is

$$\bar{v}'(\bar{x}, \bar{y}) = \text{Re}\ \{C^+ \exp\ [-\Delta\bar{y} + i(\bar{x} - \Lambda\bar{y})] + C^- \exp\ [\Delta\bar{y} + i(\bar{x} + \Lambda\bar{y})]\}$$
$$\hspace{11cm} (8.13\text{a})$$

$$= C^+e^{-\Delta\bar{y}} \cos\ (\bar{x} - \Lambda\bar{y}) + C^-e^{\Delta\bar{y}} \cos\ (\bar{x} + \Lambda\bar{y}), \quad (8.13\text{b})$$

where it is understood in the second form that C^+ and C^- are real. The condition that the disturbances remain finite as $\bar{y} \to \infty$ requires that $C^- = 0$, and the boundary condition (8.7) gives $C^+ = 1$. After transformation back to the original variables, the solution for v' is then complete. The value of u' can be found by a procedure similar to that used to obtain p' in Sec. 6. We first integrate $\varphi_{\bar{y}} = (l/2\pi)\varphi_y = \varepsilon U_\infty \bar{v}'$ with respect to $\bar{y}$ to obtain φ and then differentiate with respect to $\bar{x}$ to find $u' = \varphi_x = (2\pi/l)\varphi_{\bar{x}}$. The final results for u' and v' are thus, in the original dimensional variables,

$$\frac{u'}{U_\infty} = \frac{2\pi(\varepsilon/l)}{\sqrt{\Delta^2 + \Lambda^2}}\ e^{-2\pi\Delta(y/l)} \sin\left[2\pi\left(\frac{x - \Lambda y}{l} - \frac{x_p}{l}\right)\right], \quad (8.14\text{a})$$

$$\frac{v'}{U_\infty} = 2\pi\frac{\varepsilon}{l}\ e^{-2\pi\Delta(y/l)} \cos\left[2\pi\frac{x - \Lambda y}{l}\right], \quad (8.14\text{b})$$

where

$$\frac{x_p}{l} = \frac{1}{2\pi} \tan^{-1}\frac{\Lambda}{\Delta}. \quad (8.15)$$

The nature of the flow is apparent from these results. Under all conditions the disturbance velocities decay exponentially with y along the lines $x - \Lambda y = $ constant, which are straight lines with slope (measured from the vertical) of

$$\frac{dx}{dy} = \Lambda. \quad (8.16)$$

The rate of decay is proportional to the value of Δ. Along the straight lines (8.16) the vertical disturbance velocity is in phase with the corresponding slope of the wall. The horizontal disturbance velocity, however, is out of phase with the wall. At a given value of y, it lags behind the corresponding ordinate of the wall by the horizontal distance x_p given by equation (8.15).

In the limits of equilibrium flow ($K = 0$) and frozen flow ($K = \infty$), equation (8.12) gives the following simple results for Δ and Λ:

	$K = 0$ Equilibrium flow		$K = \infty$ Frozen flow	
	Subsonic	Supersonic	Subsonic	Supersonic
	$M_{e\infty} < 1$	$M_{e\infty} > 1$	$M_{f\infty} < 1$	$M_{f\infty} > 1$
Δ	$\sqrt{1 - M_{e\infty}^2}$	0	$\sqrt{1 - M_{f\infty}^2}$	0
Λ	0	$\sqrt{M_{e\infty}^2 - 1}$	0	$\sqrt{M_{f\infty}^2 - 1}$.

In both limiting cases, therefore, the disturbance velocities are given by equations of the following form:

For subsonic flow,

$$\frac{u'}{U_\infty} = \frac{2\pi(\varepsilon/l)}{\sqrt{1 - M_\infty^2}} \exp\left(-2\pi\sqrt{1 - M_\infty^2}\,\frac{y}{l}\right) \sin\left(2\pi\frac{x}{l}\right),$$

$$\frac{v'}{U_\infty} = 2\pi\frac{\varepsilon}{l} \exp\left(-2\pi\sqrt{1 - M_\infty^2}\,\frac{y}{l}\right) \cos\left(2\pi\frac{x}{l}\right);$$

For supersonic flow,

$$\frac{u'}{U_\infty} = -\frac{2\pi(\varepsilon/l)}{\sqrt{M_\infty^2 - 1}} \cos\left(2\pi\frac{x - \sqrt{M_\infty^2 - 1}\,y}{l}\right),$$

$$\frac{v'}{U_\infty} = 2\pi\frac{\varepsilon}{l} \cos\left(2\pi\frac{x - \sqrt{M_\infty^2 - 1}\,y}{l}\right).$$

These equations have the same form as the classical results obtained by Ackeret on the basis of the Prandtl-Glauert equation (see, e.g., Liepmann and Roshko, 1957, pp. 208–214). The decay of the velocity field is zero for supersonic flow. For subsonic flow, decay takes place along lines whose rotation from the vertical is zero. The only difference between the results for equilibrium and frozen flow is in the speed of sound on which the Mach number is based.

The variation of Δ and Λ for an intermediate value of K is shown in Fig. 10, together with curves for the two limiting cases. The results are plotted as functions of $M_{e\infty}$ for the representative value $a_{f\infty}/a_{e\infty} = \frac{11}{10}$. It follows that $M_{f\infty} = (a_{e\infty}/a_{f\infty})M_{e\infty} = \frac{10}{11}M_{e\infty}$. The value $K = 1$ gives very nearly the maximum value of the variables in the regions in which the variation with K is not monotonic.

Figure 10 shows that the presence of the nonequilibrium process removes certain of the qualitative differences that distinguish subsonic

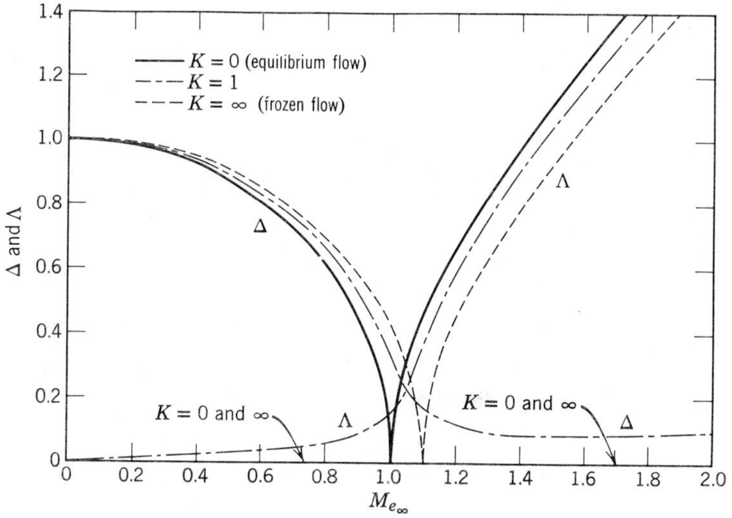

Fig. 10. Values of Δ and Λ as functions of $M_{e\infty}$ for $a_{f\infty}/a_{e\infty} = \frac{11}{10}$.

flow from supersonic flow in the two limiting cases. (The change from one type of flow to the other, in fact, is no longer precisely defined.) The exponential decay of the velocity field, as measured by the value of Δ, now persists throughout the Mach-number range. As before, it decreases rapidly in the near-sonic region. It never disappears entirely, however, as it does for both equilibrium and frozen flow at supersonic speed. The rearward rotation of the lines of exponential decay, which is measured by the value of Λ, is now zero for incompressible flow only ($M_{e\infty} = 0$). As in the two limiting cases, it grows rapidly in the near-sonic region. It is evident to a slight extent, however, even at lower speeds.

To study the drag on the wall we can examine the horizontal force d exerted on one wavelength and over a unit width of wall perpendicular to the direction of flow. This can be evaluated by integrating the x-component of the force exerted by the disturbance pressure p' according to the formula

$$d = \int_0^l p'(x, 0) \frac{dY}{dx} \, dx.$$

In terms of the drag coefficient c_d per unit width of wall, this gives, in view of equation (7.9),

$$c_d \equiv \frac{d}{\frac{1}{2}\rho_\infty U_\infty^2 l} = -2 \int_0^1 \frac{u'(x, 0)}{U_\infty} \frac{dY}{dx} \, d\left(\frac{x}{l}\right).$$

Substituting from equations (8.2) and (8.14a) and carrying out the integration, we find

$$c_d = 4\pi^2 \left(\frac{\varepsilon}{l}\right)^2 \frac{\Lambda}{\Delta^2 + \Lambda^2}. \tag{8.17}$$

The factor $\Lambda/(\Delta^2 + \Lambda^2)$ is plotted in Fig. 11. In the two limits we see the classical Ackeret result of zero drag at subsonic speeds, a discontinuous

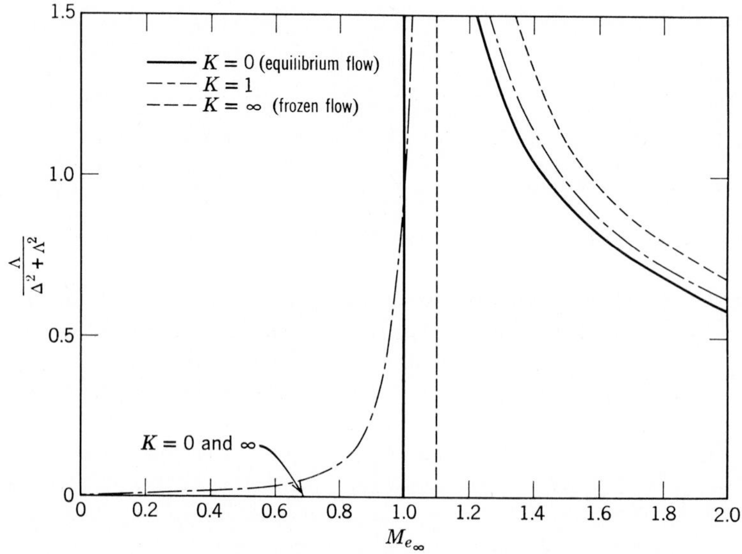

Fig. 11. Drag parameter as a function of M_{e_∞} for $a_{f_\infty}/a_{e_\infty} = \frac{11}{10}$.

jump to infinite drag at the sonic speed, and a decreasing drag coefficient in the supersonic range. As would be expected from the previous discussion, nonequilibrium effects eliminate the discontinuous behavior at transonic speeds. The maximum in the near-sonic range for $K = 1$ is still large, so much so as to be well outside the figure. It is, nevertheless, not infinite. The non-zero drag that is now evident for $M_{e_\infty} < 1$ is due to the presence of a mechanism for entropy increase even at subsonic speeds [cf. equation (2.14)]. At supersonic speeds there are now two sources of entropy rise. One of these is provided by shock waves, whose action is well known from classical gas dynamics. Such waves do not appear explicitly in the linearized solution for the wavy wall, but they would have to be present in the flow some distance above the wall in any nonlinear treatment. Their influence appears in the drag even in the linear theory. The second mechanism is again the nonequilibrium process, which acts in the regions between shock waves. The combined effect is evidently to cause the drag for $M_{e\infty} > a_{f\infty}/a_{e\infty}$ to lie between the values calculated for equilibrium and frozen flow.

Although nonequilibrium effects eliminate the discontinuous behavior that obviously invalidates the Prandtl-Glauert theory at transonic speeds, it should not be inferred that the linearized nonequilibrium results are reliable at such speeds. To obtain an accurate nonequilibrium transonic theory it is still necessary to include the essentially nonlinear character of the transonic range (see Ryhming, 1963).

Exercise 8.1. On the basis of the results (8.14), develop an expression for the disturbance q' in the nonequilibrium parameter at the surface of the wavy wall.

9 LINEARIZED FLOW BEHIND A NORMAL SHOCK WAVE

In terms of the discussion of Sec. 3, the wavy wall is an example in which the flow tends to the corresponding equilibrium or frozen flow uniformly in all parts of the flow field as τ tends to its limiting values. A simple example in which this is *not* so is given by the relaxation zone behind a normal shock wave.

When the stream ahead of the wave is in equilibrium, the velocity profile through a standing normal shock in a vibrating and reacting gas is, for a considerable range of conditions, as shown in Fig. 12(*a*). Since molecular translation and rotation equilibrate in fewer collisions than do vibration or chemical composition, the profile divides into two overlapping but distinguishable regions. In the first of these, which has typically a thickness of a few mean free paths, the velocity drops precipitately as translational and rotational adjustment take place. In this region the dissipative mechanisms of viscosity and thermal conductivity are controlling, as will be discussed in Chapter X. Following this region is a much longer region where the vibration and chemical composition come more slowly into equilibrium. Here the velocity continues to fall toward its final downstream value, but at a more gradual rate. If conditions are such that the vibration comes into equilibrium much faster than the chemical composition, this region may itself appear as two subregions. Whatever the situation, the final condition (subscript b) is one of complete equilibrium, and we can write that $h_b = h^*(p_b, \rho_b)$. When the upstream condition (subscript a) is an equilibrium condition also, then the shock wave separates two regions of uniform equilibrium flow, and the overall change from u_a to u_b can be calculated from the equilibrium-flow theory

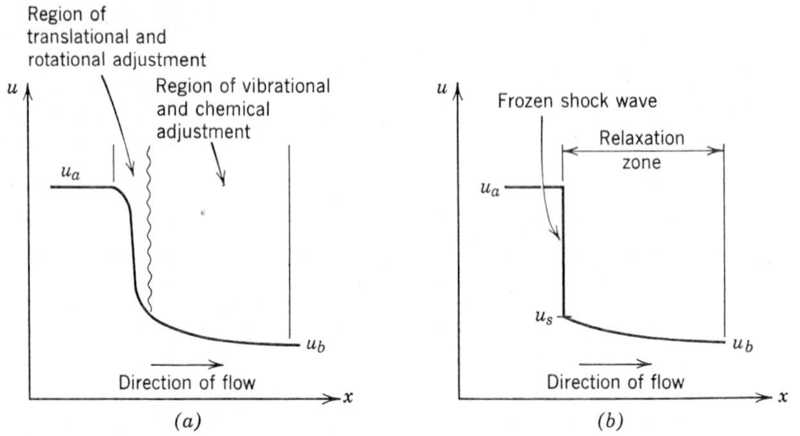

Fig. 12. Velocity distribution through a normal shock wave. (a) True situation; (b) Idealized model.

of Chapter VI, Sec. 2. For the shock wave to exist, the flow ahead of the wave must, as usual, be supersonic. Since only equilibrium conditions are in question both ahead of and behind the wave, the appropriate speed of sound to use here is the equilibrium one. This is, in fact, the only sound speed that enters naturally into the equilibrium-flow theory. A condition for the existence of a wave such as we have described is therefore that $u_a > a_{e_a}$.

For many purposes the true wave can be replaced by an idealized model as shown in Fig. 12(b). Here the regions are completely separated, and the adjustment to equilibrium is assumed to take place in two distinct stages. In the first of these the translation and rotation are assumed to adjust to a new equilibrium value instantaneously. If, as here, we allow for only one nonequilibrium process and this is the vibrational one, the vibrational energy is assumed frozen during this change. If chemistry is the pertinent nonequilibrium process, the vibration is also taken to adjust instantaneously in the first stage, and the chemical composition is taken to be frozen. In either case, if condition a is one of equilibrium, the velocity across this *frozen shock wave* drops discontinuously to a value $u_s > u_b$. In the second stage the vibration (or chemical composition) is assumed to relax gradually to an equilibrium value while the translation and rotation (and possibly vibration) continue in local equilibrium. In this *relaxation zone* the velocity decreases continuously from u_s to u_b. By our assumption, the nonequilibrium variable on the downstream side of the frozen shock wave has the known value $q_s = q_a$. The corresponding values of p_s, ρ_s, and u_s can be found from the usual conservation equations for a normal shock wave (see Chapter VI, Sec. 2) with $h_s = h(p_s, \rho_s, q_a)$. These values and the known value $q_s = q_a$ then provide the initial conditions for the nonequilibrium flow in the relaxation zone. Because the flow ahead of the frozen shock wave must be supersonic, the condition for the application of this flow model is that $u_a > a_{f_a}$. The appropriate speed of sound here is the frozen one, since the flow through the shock wave is frozen. Since $a_{f_a} > a_{e_a}$, there will be a narrow range of upstream conditions for which a wave of the general type described in the preceding paragraph may exist but where the idealized model of this paragraph is not applicable. The situation under these conditions is discussed in Sec. 12.

The reader will have noticed a difference of terminology in the foregoing paragraphs. From one point of view, the entire disturbance region from u_a to u_b may be regarded as a shock wave that is embedded in an otherwise equilibrium flow. The whole idealized model of Fig. 12(*b*) is then sometimes referred to as a "partly dispersed" wave. Alternatively, the discontinuous jump in the idealized model may be referred to as the shock wave and the following region as a relaxation zone. We shall generally prefer the latter terminology, although we shall not hesitate to use the former when it is appropriate. The student of the literature must become accustomed to both.

If conditions are such that the difference between u_s and u_b is small, the relaxation zone may be treated with the linearized equations of Sec. 7. The appropriate reference condition in this case is the equilibrium state far downstream. Under these conditions the governing equation (7.7), applied to one-dimensional flow and written in terms of $u' = \partial\varphi/\partial x$, is

$$\frac{d}{dx}\left(\frac{du'}{dx}\right) + \frac{1}{\overset{+}{\tau}_b u_b} \times \frac{\beta_e^2}{\beta_f^2}\frac{du'}{dx} = 0, \tag{9.1}$$

where $\beta_e^2 \equiv M_{e_b}^2 - 1$ and $\beta_f^2 \equiv M_{f_b}^2 - 1$. This can be integrated at once to obtain

$$\frac{du'}{dx} = C \exp\left(-\frac{\beta_e^2}{\beta_f^2} \times \frac{x}{\overset{+}{\tau}_b u_b}\right). \tag{9.2}$$

The constant C can be evaluated with the aid of equation (7.13). Applying this equation at the point s, we obtain the following condition on du'/dx:

$$\left(\frac{du'}{dx}\right)_s = -\frac{h_{q_b}}{\rho_b h_{\rho_b}\beta_f^2} u_b \left(\frac{dq'}{dx}\right)_s = -\frac{h_{q_b}(q_s^{*\prime} - q_s')}{\tau_b \rho_b h_{\rho_b}\beta_f^2}, \tag{9.3}$$

where the linearized, steady-flow rate equation $u_b(dq'/dx) = (q^{*\prime} - q')/\tau_b$ [cf. equations (4.6) and (7.4)] has been used in the second equality. Equation (9.3) shows clearly that the velocity gradient at the downstream side of the frozen shock wave depends on the condition on q at that point. If the origin of x is taken at the frozen shock wave and (9.3) is used to evaluate C, equation (9.2) can be written

$$du' = -\frac{h_{q_b}(q_s^* - q_s)}{\tau_b \rho_b h_{\rho_b}\beta_f^2} \exp\left(-\frac{\beta_e^2}{\beta_f^2} \times \frac{x}{\tau_b^+ u_b}\right) dx,$$

where the identity $q_s^{*\prime} - q_s' = (q_s^* - q_b^*) - (q_s - q_b) = q_s^* - q_s$ has also been employed. The solution can now be obtained by integrating on the left from u' to 0 and on the right from x to $+\infty$. We thus obtain finally, with the aid of the equation for τ_b^+ following equation (4.15),

$$\boxed{u - u_b = \frac{u_b h_{q_b}(q_s^* - q_s)}{\rho_b(h_{\rho_b} + h_{q_b} q_{\rho_b}^*)\beta_e^2} \exp\left(-\frac{\beta_e^2}{\beta_f^2} \times \frac{x}{\tau_b^+ u_b}\right)}. \tag{9.4}$$

Of the various quantities in the coefficient on the right, h_q is generally positive,[5] $(h_\rho + h_q q_\rho^*)$ is generally negative, and $\beta_e^2 \equiv M_{e_b}^2 - 1$ is in this instance negative. The velocity difference $(u - u_b)$ will thus have the sign of $(q_s^* - q_s)$.

The resemblance of the present results to the vibrational relaxation process of equation (VII 2.10) is evident. If we apply equation (9.4) at $x = 0$, we see that the complicated multiplier of the exponential function on the right must equal $u_s - u_b$ (see also Exercise 9.1). As the flow proceeds downstream, this velocity difference relaxes with a relaxation length given by $\tau_b^+ u_b \beta_f^2/\beta_e^2$. Here, however, the temperature, density, and other flow quantities change as well. The variation of u with x for different values of τ_b^+, all other parameters being fixed, is shown qualitatively in Fig. 13. In this figure the stream ahead of the shock has been assumed in equilibrium as in our original discussion. As a result of the increase in temperature and density through the frozen shock wave, we then have $q_s^* > q_s = q_a$ and the velocity u_s is greater than u_b as in Fig. 12(*b*). We see that as $\tau_b^+ \to 0$, the solution does not everywhere approach the

[5] Except possibly for a dissociating gas at very high temperatures.

equilibrium-flow result, which in this case would be a uniform velocity u_b downstream of the shock. There always remains a small relaxation zone, which in the limit squeezes down to a spike coincident with the downstream side of the frozen shock wave. This situation is in contrast to that for the wavy wall, where the solution approaches the equilibrium-flow result uniformly as $\tau_\infty^+ \to 0$. This difference is an example of the situation discussed in Sec. 3. It is the result of the existence in the present case of the boundary condition on q at the downstream side of the frozen shock wave. This condition imposes the requirement that $q_s = q_a \neq q_s^*$, and it follows from equation (9.3) that as $\tau_b \to 0$, $(du'/dx)_s$ must approach

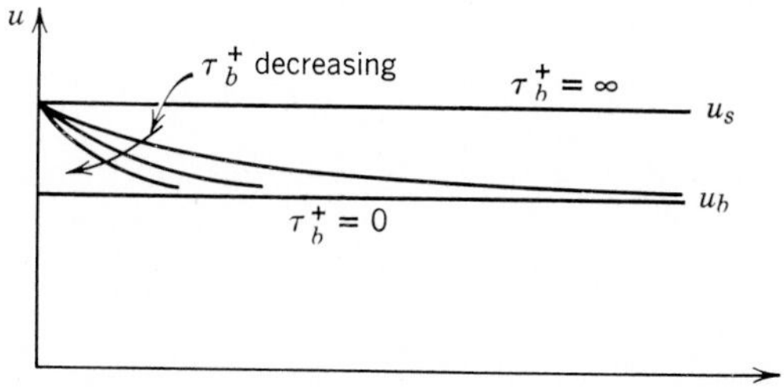

Fig. 13. Velocity distribution in the relaxation zone behind a normal shock wave according to linearized theory.

∞ rather than 0. For the wavy wall there is no fixed condition on q anywhere in the flow, and the entire field is free to approach the equilibrium-flow solution.

In the preceding paragraph—although *not* in the analysis leading to equation (9.4)—we have considered that the stream ahead of the shock wave is in equilibrium, that is, that $q_a = q_a^*$. Since we always have $q_a^* < q_s^*$, it follows in this situation that $q_s = q_a < q_s^*$. Thus, if we speak specifically in terms of a dissociation-recombination reaction, the flow in the relaxation zone is a dissociating one. It also follows, as we have seen from equation (9.4), that the velocity then decreases with x. It may sometimes happen, however, that the flow ahead of the frozen shock wave is *out* of equilibrium, usually in the direction of overdissociation, that is, $q_a > q_a^*$. This is the case, for example, in the nonequilibrium nozzle flow that will be discussed in Sec. 13. If the overdissociation ahead of the wave is sufficient, the flow behind the frozen shock, although heated and compressed, may still be overdissociated, that is, $q_s = q_a > q_s^*$. In this case the flow in the relaxation zone involves recombination, and the velocity given by equation (9.4) increases with x. Under special conditions it can happen that the nonequilibrium degree of dissociation ahead of the wave is precisely equal to the equilibrium value in the heated and compressed flow behind the shock, that is, $q_s = q_a = q_s^*$. Under these conditions the flow behind the frozen shock wave neither dissociates nor recombines, and the velocity given by equation (9.4) is uniform. Thus the compression due to the shock puts the fluid precisely into equilibrium and nothing further need happen.

Here, as throughout the chapter, we have considered flow with only a single nonequilibrium process. For a basic discussion of the complications introduced by multiple, coupled reactions in steady flow in the linearized one-dimensional case, see Napolitano (1964).

Exercise 9.1. By setting $x = 0$ in equation (9.4) we see that the velocity u_s immediately behind the frozen shock wave must be given by

$$u_s - u_b = \frac{u_b h_{q_b}(q_s^* - q_s)}{\rho_b(h_{\rho_b} + h_{q_b} q_{\rho_b}^*)\beta_e^2}.$$

Verify this relationship directly by applying the conservation equations of Chapter VI, Sec. 2 from station s to station b and introducing the linearizing assumption that the changes in all quantities from one station to the other are small.

10 EQUATIONS FOR STEADY QUASI-ONE-DIMENSIONAL FLOW

In Secs. 4 through 8 we discussed the linearized theory of nonequilibrium flow in more than one independent variable. In this section we return to the nonlinear situation of Secs. 2 and 3 but restrict ourselves to one independent variable, that is, to steady quasi-one-dimensional flow in a channel of varying area. In the nonlinear case little can be done analytically and recourse must be had to numerical calculations.

The steady quasi-one-dimensional equivalents of the nonlinear conservation equations of Sec. 2 have the following well-known form (see, e.g., Liepmann and Roshko, 1957, Chap. 2):

Mass: $$\frac{d\rho}{\rho} + \frac{du}{u} + \frac{dA}{A} = 0, \tag{10.1}$$

Momentum: $$\rho u\, du + dp = 0, \tag{10.2}$$

Energy: $$u\, du + dh = 0. \tag{10.3}$$

Equations (10.2) and (10.3) can be obtained by specializing equations (2.3) and (2.4b) to steady one-dimensional flow. Equation (10.1) cannot be obtained directly from equation (2.2), but is found by differentiating the continuity relation $\rho u A$ = constant. As in Sec. 2 the set is completed by adding the equation of state

$$h = h(p, \rho, q), \tag{10.4}$$

and a rate equation of the form

$$u\frac{dq}{dx} = \frac{\chi(p, \rho, q)}{\tau(p, \rho, q)}. \tag{10.5}$$

Here the substantial derivative Dq/Dt of equation (2.12) has been replaced by the special form $u\,dq/dx$ applicable to steady one-dimensional flow. If $A(x)$ is a known function of position along the channel, these equations constitute five equations for the five unknowns ρ, u, p, h, and q. The same arguments as before apply concerning entropy production, total enthalpy, and the limits of equilibrium and frozen flow.

To obtain the equivalent of the usual stream-tube relation we replace dp in equation (10.2) by $(dp/d\rho)\,d\rho$ and then eliminate $d\rho$ by means of equation (10.1). This gives

$$\frac{du}{u} = \frac{dA/A}{\dfrac{u^2}{dp/d\rho} - 1}. \tag{10.6}$$

It follows as usual that accelerating or decelerating flow through a throat (i.e., $du \neq 0$ at $dA = 0$) can exist only if

$$u^2_{\text{throat}} = \left(\frac{dp}{d\rho}\right)_{\text{throat}}. \tag{10.7}$$

This is the same as the usual result from classical gas dynamics except that here the entropy s in general is not a constant in the flow. As a result, $dp/d\rho$ cannot be related to the speed of sound (either equilibrium or frozen), and is thus no longer a function purely of the local thermodynamic state of the gas. The condition (10.7) therefore loses its previous utility and importance. It is, in particular, no longer useful for a direct evaluation of conditions at the throat when reservoir conditions are given (cf. Chapter VI, Sec. 3). Such conditions can now be found only by a complete integration of the equations of motion from the reservoir (or other initial point) to the throat.

For carrying out a solution (e.g., by numerical integration), equation (10.6) is not particularly useful in view of the appearance of the flow derivative $dp/d\rho$ in the denominator. A more suitable equation can be found by eliminating dh from equation (10.3) by means of the state equation in the differential form $dh = h_p\,dp + h_\rho\,d\rho + h_q\,dq$ and then substituting for dp and $d\rho$ from equations (10.2) and (10.1). We thus obtain finally

$$\boxed{\frac{du}{u} = \frac{\dfrac{dA}{A} - \dfrac{h_q}{\rho h_\rho}\,dq}{u^2/a_f^2 - 1}}, \tag{10.8}$$

where, as in equation (4.12), $a_f = a_f(p, \rho, q)$ is given by

$$a_f^2 = -\frac{h_\rho}{h_p - 1/\rho}. \tag{10.9}$$

With a little algebraic manipulation it can be seen that equation (10.8) is essentially the quasi-one-dimensional nonlinear equivalent of equation (7.13), the combination of terms $(\partial^2\varphi/\partial x_2^2 + \partial^2\varphi/\partial x_3^2)$ in that equation being replaced here by $(u/A)\,dA/dx$. If $q =$ constant, equation (10.8) reduces to the classical streamtube equation applied to frozen flow, that is, to equation (10.6) with $dp/d\rho = a_f^2 = (\partial p/\partial\rho)_{s,q}$. The behavior in the equilibrium-flow limit is not as immediately obvious as in the linear equations (4.15) and (7.7), where we considered small disturbances from an equilibrium state and the equilibrium speed of sound appeared directly. Here we must set $q = q^*$ and hence $dq = dq^* = q_p^*\,dp + q_\rho^*\,d\rho$ in the right-hand side of equation (10.8). If we eliminate dp and $d\rho$ again by means of (10.2) and (10.1) and solve once more for du/u, we now find

$$\frac{du}{u} = \frac{dA/A}{-u^2\dfrac{h_p + h_q q_p^* - 1/\rho}{h_\rho + h_q q_\rho^*} - 1} = \frac{dA/A}{u^2/a_e^2 - 1}.$$

In the second form on the right we have introduced the equilibrium speed of sound given by equation (4.13). Thus in this limit equation (10.8) goes over, as we might expect, into the classical streamtube equation (10.6) applied to equilibrium flow. In either limiting case, a change in velocity can be caused only by a change in streamtube area.

When the flow is neither frozen nor in equilibrium, equation (10.8) shows that a change in velocity can occur even at constant area. This was the case, for example, in the relaxation zone behind a normal shock wave as discussed in the preceding section. In this situation it follows from equation (10.6) that the flow must be such that $u^2 = dp/d\rho$ at *all* points in the streamtube, although as pointed out previously, $dp/d\rho$ can be equated neither to a_f^2 nor a_e^2.

In supercritical (i.e., continuously accelerating) nozzle flow, the numerator and denominator of equation (10.8) must go to zero simultaneously at some point in the nozzle. In general, this does not occur precisely at the throat, as it does, for example, in frozen flow. The direction of displacement of this point relative to the throat can be seen by considering the sign of the quantities in the numerator of equation (10.8). In general, h_ρ is negative and, except perhaps at extremely high temperatures, h_q is positive. In an accelerating flow, dq is negative as a result of the falling temperature. It follows that the combination $(h_q/\rho h_\rho)\, dq$ in equation (10.8) is positive and hence that the numerator can go to zero only where dA is positive. Since the denominator must go to zero simultaneously, we thus see that in a supercritical nonequilibrium nozzle flow the point at which $u = a_f$ must lie somewhat downstream of the throat. The amount of the displacement depends on the degree to which the flow departs from frozen flow. It reaches a maximum in the limit of equilibrium flow, in which case the condition $u = a_e < a_f$ must then occur at the throat.

For actual calculations, dq in equation (10.8) is given by the rate equation (10.5). Substituting this equation into (10.8) and reverting to the original notation $\omega = \chi/\tau$ from Chapter VII, Sec. 11, we can write

$$\frac{du}{dx} = \frac{\dfrac{u}{A}\dfrac{dA}{dx} - \dfrac{h_q}{\rho h_\rho}\,\omega}{u^2/a_f^2 - 1}\,. \tag{10.10}$$

This is a convenient form for the numerical integration of a nonequilibrium channel flow when $A(x)$ is given. For any such integration, a specific gas model must be assumed.

Exercise 10.1. Consider the steady, adiabatic, quasi-one-dimensional flow of a gas in vibrational nonequilibrium, but with no other nonequilibrium processes. The variation with x of the local equilibrium vibrational energy e_v^* is as follows:

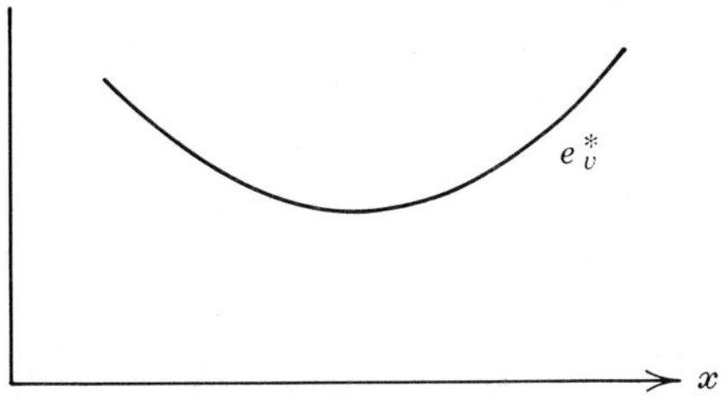

On a similar plot, draw the corresponding variation of the actual vibrational energy e_v, assuming that the flow started from an equilibrium situation at some distance in the negative x-direction and never goes far out of equilibrium, that is, e_v, although visibly different from e_v^*, is always close to it. Also draw a qualitatively correct curve showing the variation with x of the specific entropy of a fluid element. These curves will have certain essential and mutually related features. What are they?

11 NONLINEAR FLOW BEHIND A NORMAL SHOCK WAVE

For flow at constant area, that is, for truly one-dimensional flow, equation (10.10) simplifies to

$$\frac{du}{dx} = \frac{-h_q\omega}{\rho h_\rho(u^2/a_f^2 - 1)}. \tag{11.1}$$

An equation such as this can be used to study the nonlinear flow in the relaxation zone behind a normal shock wave, discussed already in the linear approximation in Sec. 9. This has been done by Freeman (1958) for the ideal dissociating gas. In this case the nonequilibrium variable q is taken as α, the degree of dissociation. The state and rate expressions are given by equation (V 3.6), with the asterisks removed, and equation (VII 9.3):

$$h = R_{A_2}[(4 + \alpha)T + \alpha\Theta_d], \tag{11.2}$$

$$\omega = CT^\eta\rho\left\{(1 - \alpha)e^{-\Theta_d/T} - \frac{\rho}{\rho_d}\alpha^2\right\}, \tag{11.3}$$

where T can be expressed in terms of p, ρ, and α by means of the thermal equation of state (V 3.3), again without the asterisk:

$$\frac{p}{\rho} = (1 + \alpha)R_{A_2}T. \tag{11.4}$$

Typical results obtained by Freeman by numerical integration of an equation equivalent to (11.1) are shown in Fig. 14. Freeman assumes that the shock wave is sufficiently strong that the pressure and enthalpy in the stream ahead of it are negligible (cf. Chapter VI, Sec. 2). The results can then be presented in generalized form by considering the dependent variables to be functions of the dimensionless distance $\bar{x} = x\rho_a C/R_{A_2}^\eta u_a^{1-2\eta}$ for various values of the dimensionless ratios $\bar{\rho} = \rho_a/\rho_d$ and $\bar{\mu} = (u_a^2/2)/R_{A_2}\Theta_d$. These are a measure of the density and specific kinetic energy respectively of the flow ahead of the shock.[6] The free stream is taken to be in equilibrium at $\alpha_a \simeq 0$. The results shown are for $\eta = 0$; the values of $\bar{\rho}$ are typical of the part of the atmosphere at which the density is from a tenth to a hundredth of its sea-level value.

[6] The characteristic time $(Cu_a^{2\eta}\rho_a/R_{A_2}^\eta)^{-1}$ appears here only in the dimensionless independent variable. This is possible because the only given length in the problem is the relaxation length $u_a(Cu_a^{2\eta}\rho_a/R_{A_2}^\eta)^{-1}$. A similar procedure could have been followed in Fig. 13, but we chose not to do so in order to illustrate the dependence on τ_b^+.

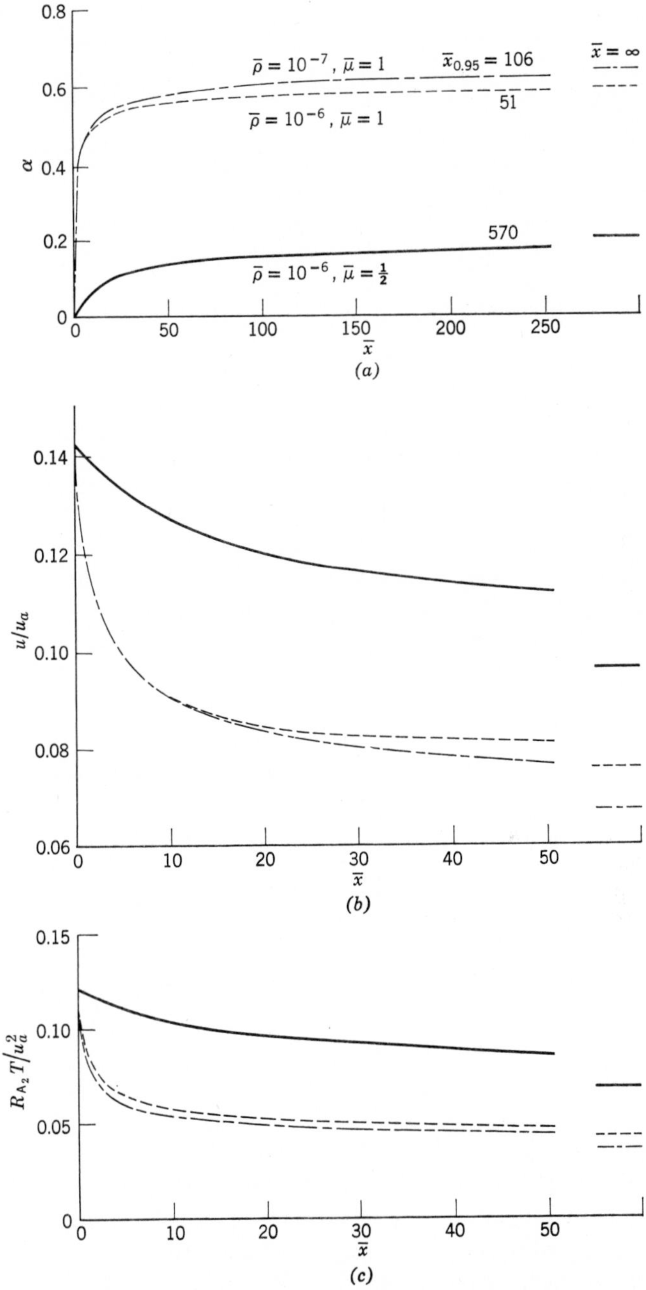

Fig. 14. Flow variables in the relaxation zone behind a normal shock wave; $\eta = 0$ (after Freeman, 1958). (*a*) Degree of dissociation; (*b*) Velocity; (*c*) Temperature.

All curves show a rapid variation near the frozen shock and a much slower variation near equilibrium. As might be anticipated, the overall effects of dissociation are increased by increasing the kinetic energy or decreasing the density of the free stream. A measure of the distance required to reach equilibrium is given by the distance required for α to reach 95 percent of its final value. This distance, denoted by $\bar{x}_{0.95}$, is included in Fig. 14(*a*). A decrease in the kinetic energy of the free stream from the dissociation energy ($\bar{\mu} = 1$) to half that value ($\bar{\mu} = \frac{1}{2}$) is seen to cause an increase in $\bar{x}_{0.95}$ by a factor of about 10. The decrease in temperature in the relaxation zone [Fig. 14(*c*)] may be thought of as due to the "heat-sink" effect of the dissociation process, that is, as dissociation proceeds energy is withdrawn from the translational, rotational, and vibrational degrees of freedom and the temperature drops accordingly. The variation in pressure through the relaxation zone (not shown in Fig. 14) is small, amounting at most to about 10 percent for the cases considered.

The literature on nonequilibrium effects behind steady shock waves is large and continually growing. Representative references are as follows: vibrational nonequilibrium, Johannesen et al. (1962); ionization nonequilibrium, Lin and Teare (1963); chemical nonequilibrium of complex mixtures, Duff and Davidson (1959); coupled vibrational and chemical nonequilibrium, Marrone and Treanor (1963). A general review article with useful references to the Russian literature is that of Losev and Osipov (1962). For a discussion of the effects of vibrational and chemical nonequilibrium on the propagation of *unsteady* shock waves, see Spence (1961).

12 FULLY DISPERSED SHOCK WAVE

The question remains (cf. Sec. 9) as to what happens when $a_{e_a} < u_a < a_{f_a}$, so that a shock wave of some kind can exist but the partly dispersed model of Fig. 12(*b*) is not applicable. The answer to this question is implicit in equation (11.1).

In general, u may be expected to decrease through a shock wave while the temperature and hence a_f increase. Since we specify that $u_a < a_{f_a}$ to begin with, it follows that $u < a_f$ everywhere within the wave. Furthermore, as we have mentioned previously, h_q is positive and h_ρ is negative; and, in a shock wave, $\omega = u\, dq/dx$ must be positive. It follows that the value of du/dx given by equation (11.1) will be everywhere negative and finite. We can therefore imagine constructing step by step a solution of equation (11.1) in which u decreases *continuously* from u_a to u_b. This would not be so if we had started with $u_a > a_{f_a}$. In this case we would need to have $u = a_f$ at some point within the wave, which would imply an infinite value of du/dx. It turns out that equation (11.1) alone cannot then provide a solution. The only way we can overcome the difficulty is to insert a discontinuity.

We thus see that when $a_{e_a} < u_a < a_{f_a}$ a continuous shock profile can exist *purely as a result of the dissipative effects of vibrational or chemical nonequilibrium*. Such a profile is sometimes known as a "fully dispersed" shock wave. Since a_f is only slightly greater than a_e, such fully dispersed waves are necessarily weak. For stronger waves, that is, $u_a > a_{f_a}$, the vibrational and chemical processes are insufficient to give rise to a continuous profile, and the dissipative effects of viscosity and thermal conductivity must come into play. If the details of the profile are not important, these effects can be represented by a discontinuous, frozen shock, as in the partly dispersed model of Sec. 9.

An analytical solution for the fully dispersed wave in the case of vibrational nonequilibrium has been given by Lighthill (1956) and subsequently modified by Griffith (1961). The existence of such a wave was confirmed experimentally by Griffith and Kenny (1957) using interferometric, shock-tube measurements of wave propagation in carbon dioxide.

13 NOZZLE FLOW

For the flow in a nozzle or other channel of varying area, the full differential equations of Sec. 10 must be integrated. In the special situations of equilibrium and frozen flow, the basic conservation equations, when written so as to involve the entropy, integrate immediately into a system of purely algebraic equations (cf. Chapter VI, Sec. 3). As a result, the conditions at a point can be expressed purely as a function of the initial conditions and of the area ratio at that point [see equation (VI 3.12)]. The solution at a given location is thus independent of the detailed past history of the flow—this, in fact, is the physical meaning of the appearance of the purely algebraic equations. In nonequilibrium flow the conservation equations cannot be integrated independently of the details of the flow situation. As a result, conditions at a point now depend not only on the area ratio at that point, but on the entire history of the flow between the initial point and the point in question.

A simple solution that illustrates this fact can be found by considering a uniform flow, originally in equilibrium, entering an area change starting at a point 0 (Fig. 15). If the area change is small, the equations may be linearized about the equilibrium conditions at 0. A governing equation analogous to (7.7) can then be obtained by starting from the linearized form of the basic equations (10.1) through (10.5) and combining them in a manner similar to that of Sec. 4. It is simpler, however, to take advantage of the equivalence that was noted immediately following equations (10.8) and (10.9). With this equivalence the quasi-one-dimensional counterpart of the linearized equation (7.7) can be written down at once as follows, where

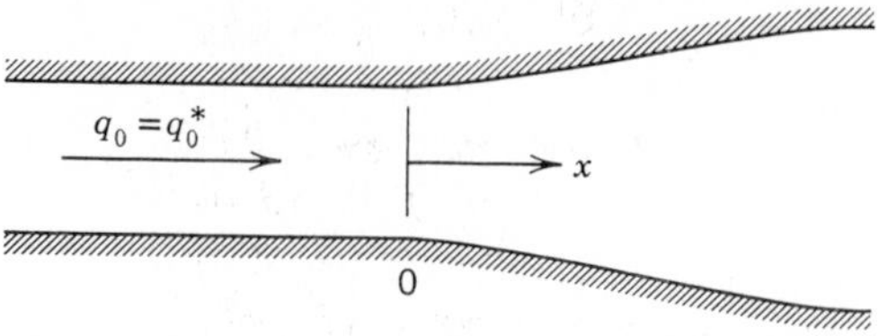

Fig. 15. Uniform equilibrium flow entering an area change.

$u' = u - u_0 = \partial\varphi/\partial x$, $A' = A - A_0$, and reference conditions are now those at 0:

$$\tau_0^+ u_0 \frac{d}{dx}\left(\beta_f^2 \frac{du'}{dx} - \frac{u_0}{A_0}\frac{dA'}{dx}\right) + \beta_e^2 \frac{du'}{dx} - \frac{u_0}{A_0}\frac{dA'}{dx} = 0. \tag{13.1}$$

As in Sec. 9, we have introduced the notation $\beta_e^2 \equiv M_{e_0}^2 - 1$ and $\beta_f^2 \equiv M_{f_0}^2 - 1$. (Note that when $dA'/dx = 0$, equation (13.1) reduces to equation (9.1), which we have already used to analyze the linearized flow behind a normal shock wave.) The one-dimensional counterpart of equation (7.13) is similarly

$$u_0 \frac{dq'}{dx} = -\frac{\rho_0 h_{\rho_0}}{h_{q_0}}\left(\beta_f^2 \frac{du'}{dx} - \frac{u_0}{A_0}\frac{dA'}{dx}\right). \tag{13.2}$$

Equation (13.1) can immediately be integrated once from 0 to x. Since $q_0 = q_0^*$, it follows from the rate equation [e.g., equation (10.5)] that $(dq'/dx)_0 = 0$. Hence, by equation (13.2), the combination in parentheses in equation (13.1) is zero at the lower limit of integration. Since we also have $u_0' = 0$ and $A_0' = 0$, the integration of (13.1) thus gives

$$\tau_0^+ u_0 \left(\beta_f^2 \frac{du'}{dx} - \frac{u_0}{A_0}\frac{dA'}{dx}\right) + \beta_e^2 u' - \frac{u_0}{A_0} A' = 0, \tag{13.3}$$

which can be written after rearrangement as

$$\frac{du'}{dx} + \frac{\beta_e^2}{\beta_f^2} \times \frac{1}{\tau_0^+ u_0} u' = \frac{u_0}{\beta_f^2 A_0}\left(\frac{dA'}{dx} + \frac{A'}{\tau_0^+ u_0}\right). \tag{13.4}$$

This equation is of the type $dy/dx + Py = Q$, which has the general solution (see, e.g., Coddington, 1961, p. 43)

$$y = \exp\left(-\int_0^x P\,dz\right)\left[C + \int_0^x Q \exp\left(\int_0^\xi P\,dz\right) d\xi\right], \tag{13.5}$$

where C is a constant. Applying this to equation (13.4) and setting $C = 0$ to satisfy the condition that $u' = u_0' = 0$ at $x = 0$, we obtain

$$u' = \frac{u_0}{\beta_f^2 A_0}\int_0^x \left(\frac{dA'}{d\xi} + \frac{A'}{\tau_0^+ u_0}\right) \exp\left(-\frac{\beta_e^2}{\beta_f^2} \times \frac{x-\xi}{\tau_0^+ u_0}\right) d\xi.$$

The derivative $dA'/d\xi$ can be removed by integrating the first term in the integral once by parts. We thus obtain finally

$$u'(x) = \frac{u_0}{\beta_f^2 A_0}\left[A'(x) - \frac{\beta_e^2 - \beta_f^2}{\beta_e^2}\int_0^x A'(\xi) \exp\left(-\frac{\beta_e^2}{\beta_f^2} \times \frac{x-\xi}{\tau_0^+ u_0}\right)\frac{\beta_e^2}{\beta_f^2}\frac{d\xi}{\tau_0^+ u_0}\right]. \tag{13.6}$$

This result shows explicitly that the velocity at a point x depends not only on the net change of area, as given by $A'(x)$, but on the area variation $A'(\xi)$ over the entire interval of integration. In the limit $\tau_0^+ \to \infty$ (frozen flow) the integral term is zero and we obtain simply

$$u'(x) = \frac{u_0}{\beta_f^2 A_0} A'(x). \tag{13.7a}$$

This is the classical result that the change in velocity depends only on the net area change. In the limit $\tau_0^+ \to 0$ (equilibrium flow) the integrand is indeterminate of the form 0/0, and we must proceed more carefully. We note that for very small values of τ_0^+, the value of the exponential function in the integrand dies off very rapidly as $x - \xi$ changes from zero. The major contribution by far to the integral thus comes from the region very close to $\xi = x$, in which we may consider $A'(\xi)$ to have essentially the fixed value $A'(x)$. We thus write for very small values of τ_0^+

$$\int_0^x A'(\xi) \exp\left(-\frac{\beta_e^2}{\beta_f^2} \times \frac{x-\xi}{\tau_0^+ u_0}\right)\frac{\beta_e^2}{\beta_f^2}\frac{d\xi}{\tau_0^+ u_0}$$
$$\cong A'(x)\int_0^x \exp\left(-\frac{\beta_e^2}{\beta_f^2} \times \frac{x-\xi}{\tau_0^+ u_0}\right)\frac{\beta_e^2}{\beta_f^2}\frac{d\xi}{\tau_0^+ u_0}.$$

It can be shown quite rigorously, in fact, that the difference between these two expressions vanishes as $\tau_0^+ \to 0$. Carrying out the integration on the right, we obtain

$$\int_0^x A'(\xi) \exp\left(-\frac{\beta_e^2}{\beta_f^2} \times \frac{x-\xi}{\tau_0^+ u_0}\right)\frac{\beta_e^2}{\beta_f^2}\frac{d\xi}{\tau_0^+ u_0} \cong A'(x)\left[1 - \exp\left(-\frac{\beta_e^2}{\beta_f^2}\frac{x}{\tau_0^+ u_0}\right)\right].$$

It follows that in the limit as $\tau_0^+ \to 0$, the value of the integral is simply $A'(x)$. Putting this result into (13.6), we obtain for the equilibrium-flow limit

$$u'(x) = \frac{u_0}{\beta_f^2 A_0}\left(1 - \frac{\beta_e^2 - \beta_f^2}{\beta_e^2}\right)A'(x) = \frac{u_0}{\beta_e^2 A_0} A'(x). \qquad (13.7b)$$

Thus we recover again the classical result that the change in velocity depends only on the net change in area, except that here β_f^2 is replaced by β_e^2. Only in the two limits, however, does this simple result hold true.

Because of the dependence of the nonequilibrium flow on its entire past history, it is not possible to find the mass flow in the supercritical case by proceeding at once to the throat, as can be done in equilibrium or frozen flow [see equation (VI 3.7) et seq.]. Instead one must perform repeated numerical integrations downstream from the reservoir on the basis of assumed values of the mass flow. This is done until a value is found that allows smooth passage through the singular point of equation (10.10) at the point where $u = a_f$. Since the location of this point is unknown and must itself come out of the solution, the problem is difficult. (For a systematic approach, see Emanuel, 1964.)

Integration of equations equivalent to (10.10) have been carried through by Bray (1959) for the ideal dissociating gas. The appropriate thermochemical equations have already been set down in Sec. 11. The nozzle was taken to have the area distribution

$$A = A_{\text{throat}} + \pi(\tan^2 \theta)x^2,$$

where x is measured from the throat. For large values of x the nozzle, if axisymmetric, approaches a cone with half angle θ. Typical results are shown in Fig. 16. These results are for a dimensionless reservoir temperature and pressure of $T_0/\Theta_d = 0.1$ and $p_0/R_{A_2}\rho_d\Theta_d = 5 \times 10^{-6}$, which correspond to 5900°K and 115 atm for oxygen and 11,300°K and 215 atm for nitrogen. The accompanying degree of dissociation is $\alpha_0 = 0.690$. Results are given for several values of a dimensionless rate parameter

$$\Phi = \frac{CT_0^\eta \rho_d}{2\sqrt{\pi}\tan\theta}\sqrt{\frac{A_{\text{throat}}}{R_{A_2}\Theta_d}},$$

which contains the characteristics of the nozzle as well as the chemical properties of the gas.[7] All results shown are for $\eta = 0$.

[7] Here the characteristic time $(CT_0^\eta \rho_d)^{-1}$ must appear in a parameter of the problem and cannot be absorbed into the independent variable, which has been taken for convenience as the area ratio A/A_{throat} instead of x. This is because a characteristic geometrical length $\sqrt{A_{\text{throat}}}$ appears in the statement of the problem.

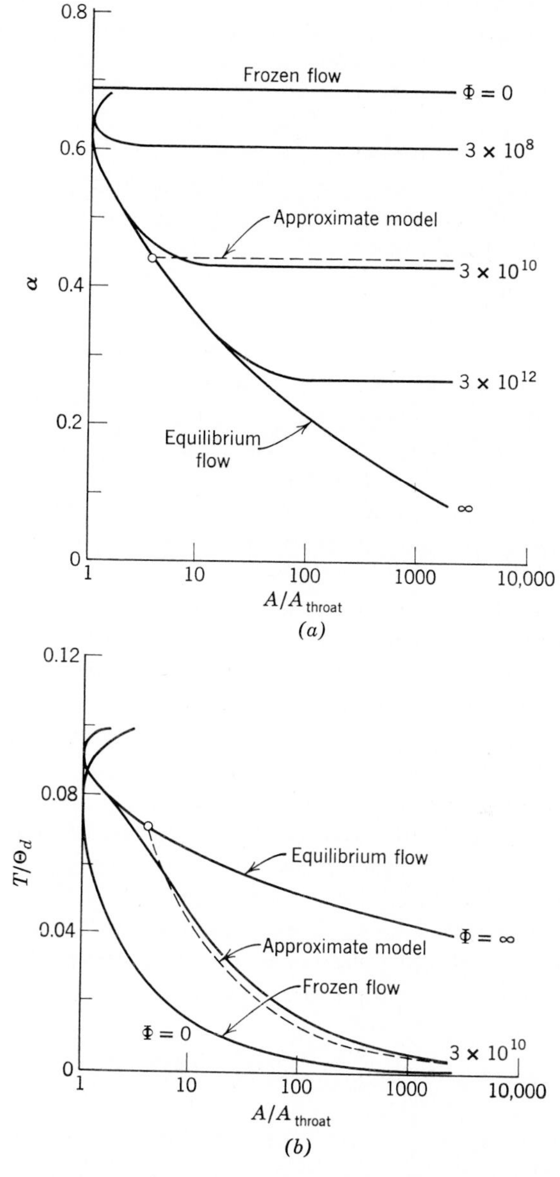

Fig. 16. Flow variables in hyperbolic nozzle; $T_0/\Theta_d = 0.1$, $p_0/R_{A_2}\rho_d\Theta_d = 5 \times 10^{-6}$, $\eta = 0$ (after Bray, 1959). (*a*) Degree of dissociation; (*b*) Temperature.

For a given value of Φ between 0 (frozen flow) and ∞ (equilibrium flow), the flow starts out by following very closely to the equilibrium-flow curve. As the fluid expands down the nozzle, however, α begins to lag noticeably behind its equilibrium-flow value and then very quickly approaches a constant [Fig. 16(*a*)]. Once this happens the flow is effectively frozen at a value of α intermediate between the reservoir value α_0 and the continually falling equilibrium-flow value. This is consistent with the form of the rate expression (11.3). Because of the falling temperature, this expression is dominated in the downstream part of the nozzle by the recombination term, with the result that

$$u \frac{d\alpha}{dx} = \omega \cong -CT^{\eta} \frac{\rho^2}{\rho_d} \alpha^2.$$

Thus, for any finite value of C, the density will eventually become sufficiently low as a result of the expansion that for all practical purposes $d\alpha/dx = 0$. Once the freezing occurs, the temperature drops very quickly from the equilibrium-flow value toward the value for frozen flow [Fig. 16(*b*)]. This is because in the equilibrium flow the exothermic recombination process transfers energy to the mechanical degrees of freedom of the molecules and thus helps to keep the temperature up. With the disappearance of the recombination process, the temperature therefore falls. Bray's results also show that the nonequilibrium effects are relatively small on ρ and u, but large on a_f and hence on M_f. For a given set of reservoir conditions, the nonequilibrium effects are delayed and decreased in magnitude if C and hence Φ are increased, as in, say, changing from oxygen to nitrogen. For a given value of C they are decreased by an increase in ρ_0 at fixed T_0 and increased by an increase in T_0 at fixed ρ_0.

The suddenness of the freezing process in Fig. 16(*a*) suggests an approximate model in which the freezing is assumed to take place instantaneously from the equilibrium flow at an appropriate point in the nozzle. On the basis of certain qualitative arguments, Bray proposes the following criterion for estimating the position of this point:

$$-\left(u \frac{d\alpha}{dx}\right)_e = KC[T^{\eta}\rho(1 - \alpha)e^{-\Theta_d/T}]_e.$$

Here K is a constant of order unity, and the subscript e denotes that the various quantities are evaluated for the equilibrium flow. This criterion is equivalent to saying that at the freezing point the value of $-D\alpha/Dt$ is of the same order of magnitude as the rate of dissociation [cf. equation (11.3)]. Results obtained from this relationship with $K = 1$ are included in Fig. 16(*a*) for $\Phi = 3 \times 10^{10}$. The variation in temperature that follows is shown in Fig. 16(*b*). Other approximate methods for finding the frozen value of the nonequilibrium variable have been proposed in the literature. These methods are discussed in the light of a more rigorous examination of the freezing process by Blythe (1964).

The essential correctness of the theoretical results has been confirmed by Wegener (1959) in nozzle experiments on the dissociation–recombination

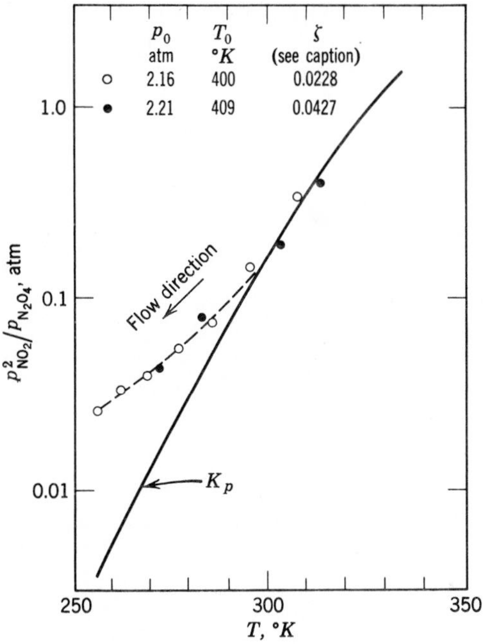

Fig. 17. Experimental results for nozzle flow; ζ = mole fraction of NO_2 in reservoir (after Wegener, 1959).

of nitrogen tetroxide. The reacting gas was carried in inert nitrogen at mole fractions of less than 0.05, so that the predominant reaction was

$$N_2O_4 + N_2 \rightleftharpoons 2NO_2 + N_2.$$

This reaction has the advantage that its kinetic and equilibrium properties are well known and that it is completed in the moderate temperature range from 200 to 400°K. It also has the convenient property that, whereas the N_2O_4 and N_2 are highly transparent to blue light, the NO_2 is absorbent, and its concentration can be measured by photometric means. The mixture has the disadvantage that it is toxic and corrosive, so that care must be taken in its handling. Results are shown in terms of partial pressures in

Fig. 17, which is taken from Wegener's paper. If the flow were in equilibrium the ordinate $p_{NO_2}^2/p_{N_2O_4}$ would be equal to the equilibrium constant K_p [see equation (III 9.1a)]. A considerable departure from equilibrium is seen to occur in the downstream part of the nozzle. It can be shown that the results shortly after the point of departure correspond to frozen flow, in qualitative agreement with Bray's theoretical findings.

Typical additional work on nonequilibrium nozzle flow has been reported for vibrational nonequilibrium by Erickson (1963), Stollery, Smith, and Park (1964), and Hurle, Russo, and Hall (1964); for ionization nonequilibrium by Eschenroeder (1962), Rosner (1962), and Bray (1964); and for complex gas mixtures involving multiple nonequilibrium processes by Penner (1955), Eschenroeder, Boyer, and Hall (1962), Emanuel and Vincenti (1962), and Emanuel (1963).

14 METHOD OF CHARACTERISTICS

As in classical gas dynamics, a useful method of calculation for certain nonlinear problems in nonequilibrium flow is given by the method of

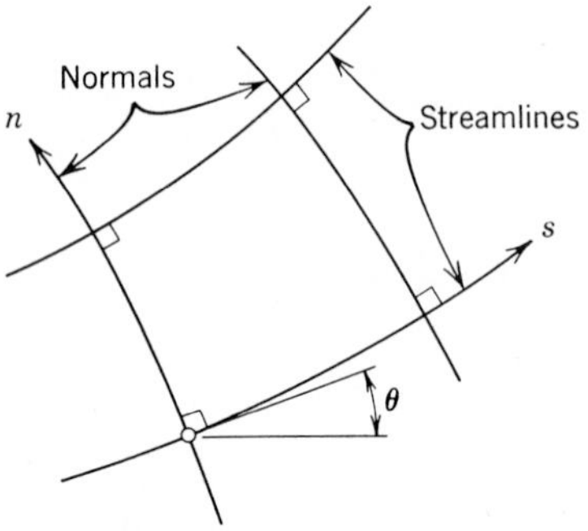

Fig. 18. Local natural coordinate system.

characteristics. This method takes advantage of the properties of the characteristic directions of hyperbolic differential equations, already mentioned in connection with the linear equation (7.7). The following brief treatment assumes that the reader is acquainted with similar discussions for a perfect gas (see, for example, Liepmann and Roshko, 1957, Chapter 12 or Meyer, 1953). We limit ourselves to the case of steady, two-dimensional supersonic flow, which is sufficient to illustrate the complications introduced by the nonequilibrium process. The extension to axially symmetric steady flow or to unsteady one-dimensional flow is then obvious from the classical treatments.

We start with the steady two-dimensional equivalent of the nonlinear conservation equations of Sec. 2. It is convenient here to use as independent variables the local, natural coordinates s and n, where s is the distance along a streamline and n the distance along its normal trajectory (Fig. 18). The momentum equation in the s-direction and the energy equation can be obtained at once by specializing equations (2.3) and (2.4b) in an x_1, x_2 coordinate system oriented to coincide locally with the s, n system. The continuity equation and the momentum equation in the n-direction are best set up directly in the s, n coordinates (see, e.g., Rotty, 1962, Chapter 10). If θ is the angle of inclination of the streamline relative to some arbitrary direction, the equations are as follows:

$$\text{Mass:} \qquad \frac{1}{\rho}\frac{\partial \rho}{\partial s} + \frac{1}{u}\frac{\partial u}{\partial s} + \frac{\partial \theta}{\partial n} = 0, \tag{14.1}$$

$$s\text{-momentum:} \qquad \rho u \frac{\partial u}{\partial s} + \frac{\partial p}{\partial s} = 0, \tag{14.2}$$

$$n\text{-momentum:} \qquad \rho u^2 \frac{\partial \theta}{\partial s} + \frac{\partial p}{\partial n} = 0, \tag{14.3}$$

$$\text{Energy:} \qquad u \frac{\partial u}{\partial s} + \frac{\partial h}{\partial s} = 0. \tag{14.4}$$

Again the set is completed by adding the equation of state

$$h = h(p, \rho, q) \tag{14.5}$$

and the rate equation, written here in the form

$$u \frac{\partial q}{\partial s} = \omega(p, \rho, q). \tag{14.6}$$

This is a set of six equations for the six unknowns ρ, u, θ, p, h, and q. The energy equation (14.4) can, of course, be integrated at once to obtain

$$\tfrac{1}{2}u^2 + h = \text{constant on a streamline.} \tag{14.7}$$

Thus the minimum number of first-order differential equations in the set is four.

A useful combination of the foregoing equations can be arrived at by first differentiating the state equation (14.5) with respect to s and replacing $\partial h/\partial s$ and $\partial q/\partial s$ in the result by means of (14.4) and (14.6). We then use the resulting equation to replace $\partial \rho/\partial s$ in the continuity equation (14.1). If we finally eliminate $\partial u/\partial s$ from the result in favor of $\partial p/\partial s$ by means of (14.2), we obtain

$$\left(\frac{u^2}{a_f^2} - 1\right)\frac{\partial p}{\partial s} + \rho u^2 \frac{\partial \theta}{\partial n} - \frac{h_q}{h_\rho} u\omega = 0, \tag{14.8}$$

where a_f is given again by equation (10.9). In equations (14.2), (14.3), (14.6), and (14.8) we have four first-order differential equations with only the four unknowns u, p, θ, and q appearing in the derivatives. Equation (14.8) will be recognized as the two-dimensional analog of equation (10.10) with p used in place of u.

The characteristic directions of a system of partial differential equations can be defined in a number of ways, all of which are essentially equivalent (see Courant and Hilbert, 1962, p. 408). A useful definition for present purposes is that they are the special directions along which the system of partial differential equations reduces to a system of ordinary differential equations. The resulting ordinary differential equations, which then form a convenient basis for numerical calculations, are known as the *compatibility relations*. Of the present equations, (14.2), (14.4), and (14.6) are already, in effect, ordinary differential equations, since they contain derivatives with respect only to s. It follows that the streamline is a triply covered characteristic direction, and equations (14.2), (14.4), and (14.6) supply as they stand the corresponding compatibility relations. As pointed out, (14.4) can be integrated at once to obtain the algebraic relation (14.7).

The remaining equations (14.3) and (14.8) can be used to obtain two more characteristic directions as follows. Let us form a linear combination of these two equations by multiplying (14.3) by a factor λ and adding the result to (14.8). This gives

$$\left[\left(\frac{u^2}{a_f^2} - 1\right)\frac{\partial}{\partial s} + \lambda\frac{\partial}{\partial n}\right]p + \rho u^2\left[\lambda\frac{\partial}{\partial s} + \frac{\partial}{\partial n}\right]\theta - \frac{h_q}{h_p}u\omega = 0. \quad (14.9)$$

Now, the first expression in brackets is a directional derivative in a direction whose slope relative to the s-direction is

$$\left(\frac{dn}{ds}\right)_1 = \frac{\lambda}{u^2/a_f^2 - 1}.$$

Similarly, the expression in the second set of brackets is a directional derivative in the direction

$$\left(\frac{dn}{ds}\right)_2 = \frac{1}{\lambda}.$$

The equation will involve derivatives in only one direction, and so be effectively an ordinary differential equation, if these two slopes are the same. This is the case if we take $\lambda = \pm\sqrt{u^2/a_f^2 - 1}$. The common slope thus has the two possible values

$$\frac{dn}{ds} = \pm\frac{1}{\sqrt{u^2/a_f^2 - 1}} = \pm\frac{a_f}{\sqrt{u^2 - a_f^2}}. \quad (14.10)$$

The directions so defined are thus recognized as characteristic directions. Equation (14.9) then becomes

$$\sqrt{u^2/a_f^2 - 1}\left[\sqrt{u^2/a_f^2 - 1}\,\frac{\partial}{\partial s} \pm \frac{\partial}{\partial n}\right]p \pm \rho u^2\left[\sqrt{u^2/a_f^2 - 1}\,\frac{\partial}{\partial s} \pm \frac{\partial}{\partial n}\right]\theta - \frac{h_q}{h_\rho}\,u\omega = 0.$$

These are the compatibility relations in the directions (14.10). They can be put in convenient form by multiplying through by a_f/u and by noting that (cf. Fig. 19)

$$\frac{a_f}{u}\left[\sqrt{u^2/a_f^2 - 1}\,\frac{\partial}{\partial s} \pm \frac{\partial}{\partial n}\right] = \frac{\sqrt{u^2 - a_f^2}}{u}\frac{\partial}{\partial s} \pm \frac{a_f}{u}\frac{\partial}{\partial n} = \begin{cases}\dfrac{d}{d\eta} \\ \dfrac{d}{d\xi}\end{cases},$$

where $d\eta$ and $d\xi$ are differential elements of length along the left- and right-running characteristics, respectively.

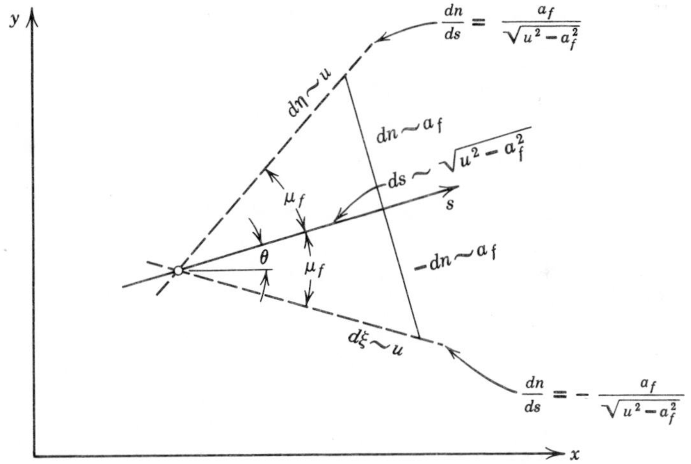

Fig. 19. Left- and right-running characteristics.

If θ is measured relative to the x-axis of an x, y coordinate system, the five compatibility relations can finally be written as follows:

$$\left.\begin{aligned}\rho u\,du + dp &= 0\\ u\,dq - \omega\,ds &= 0\\ \tfrac{1}{2}u^2 + h &= \text{const}\end{aligned}\right\} \tag{14.11a}$$

$$\text{along}\quad \frac{dy}{dx} = \tan\theta, \tag{14.11b}$$

and

$$\sqrt{u^2/a_f^2 - 1}\, dp \pm \rho u^2\, d\theta - \frac{h_q}{h_\rho} a_f \omega \begin{Bmatrix} d\eta \\ d\xi \end{Bmatrix} = 0 \qquad (14.12a)$$

$$\text{along } \frac{dy}{dx} = \tan(\theta \pm \mu_f). \qquad (14.12b)$$

Here $\mu_f \equiv \tan^{-1}(1/\sqrt{u^2/a_f^2 - 1})$, and the upper and lower signs go with $d\eta$ and $d\xi$ respectively. The characteristics (14.12b) are recognized as the frozen Mach lines already introduced in Sec. 7, and μ_f is the frozen Mach angle. For frozen flow ($\omega = 0$), equations (14.12a) reduce at once to a particularly simple form of the compatibility relations for such flow (see Hayes and Probstein, 1959, Chapter VII).

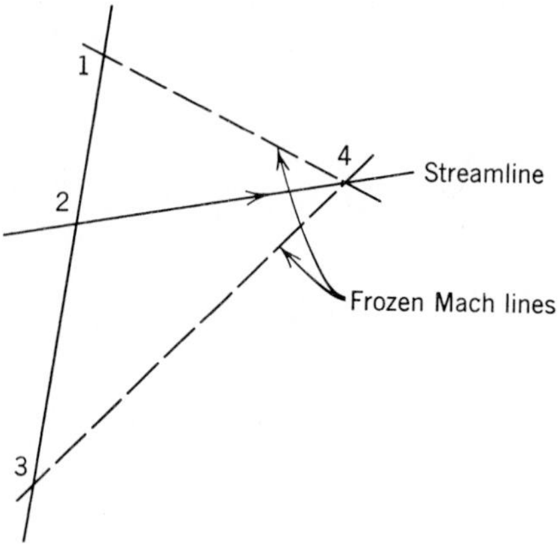

Fig. 20. Construction for finite-difference calculation.

Equations (14.11) and (14.12) form a convenient basis for the step by step calculation of nonequilibrium flow when $u > a_f$. Written in finite-difference form (cf. Liepmann and Roshko, 1957, pp. 293–296), they provide five simultaneous equations for the calculation of p, θ, u, h, and q at a point 4 when the data are known at points 1, 2, and 3 on a non-characteristic line (Fig. 20). The details of a typical application are given by Der (1963).

The compatibility relations are given in different form by different writers (see, for example, Wood and Kirkwood, 1957, and Chu, 1957). In particular, the entropy is often introduced in place of one of the dependent variables used here. The entropy loses much of its usefulness in nonequilibrium flow, however, since it no longer has the property, so convenient in classical flow, of being constant along streamlines. Thermodynamic relations involving entropy are also more complicated and tedious to use than state relations of the form (14.5). The use of entropy as a variable may therefore be inadvisable in nonequilibrium flow.

15 SUPERSONIC FLOW OVER A CONCAVE CORNER

We conclude this chapter with a description of the supersonic nonequilibrium flow over concave and convex corners in the two-dimensional case [Fig. 21(*a*) and (*b*)]. In the corresponding classical flow, there is no characteristic length in the statement of the problem, and the flow fields are correspondingly simple. They consist in both cases of two uniform parallel flows, one upstream and one downstream, separated for a concave corner by a straight oblique shock wave and for a convex corner by a

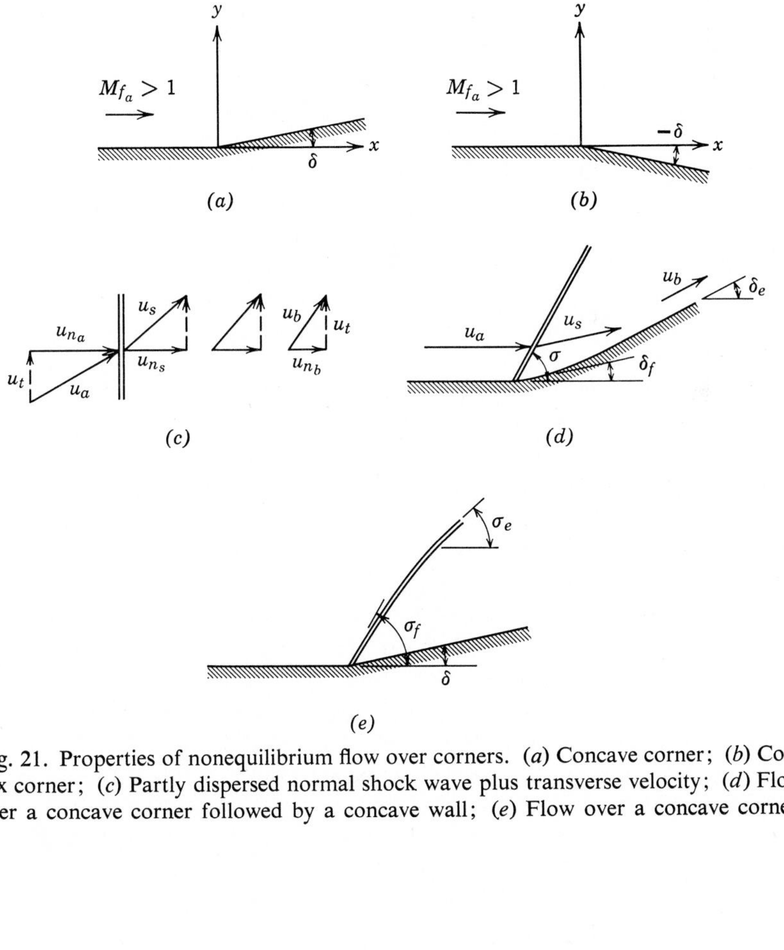

Fig. 21. Properties of nonequilibrium flow over corners. (*a*) Concave corner; (*b*) Convex corner; (*c*) Partly dispersed normal shock wave plus transverse velocity; (*d*) Flow over a concave corner followed by a concave wall; (*e*) Flow over a concave corner.

Prandtl-Meyer fan. In the nonequilibrium flow a characteristic relaxation length enters through the fluid properties, and the problem is no longer a simple one.

To gain some insight into the flow over a concave corner, we can return to our results for the partly dispersed normal shock wave [Fig. 12(b)] and superpose a uniform transverse velocity u_t parallel to the wave. This is a well-known device for converting a normal shock wave into an oblique wave (see Liepmann and Roshko, 1957, p. 85). It leads in the classical case to a uniform resultant velocity both ahead of and behind the wave. In the nonequilibrium case the result is as shown in Fig. 21(c), where the velocity component u_n normal to the wave is identical with the velocity u of Secs. 9 and 11. Here, if we take the flow ahead of the wave to be in equilibrium, the normal component decreases with distance behind the wave (see Fig. 14), and the resultant velocity is rotated accordingly toward the wave. If we reorient the velocity ahead of the wave in a more conventional manner [Fig. 21(*d*)], we thus obtain a flow that can be interpreted as that over a sharp concave corner followed by a concave wall. The oblique shock wave from the corner is straight at the angle σ, and the resultant velocity is constant along lines parallel to the wave. To assess the turning angle we remember that the initial part of the partly dispersed wave is a frozen-flow normal shock. It follows that the turning angle at the corner must be the angle δ_f appropriate to a frozen-flow oblique shock at the wave angle σ. The *entire* partly dispersed wave, however, may be looked on as a normal shock in an equilibrium flow. This means that the turning angle far downstream must be the angle δ_e appropriate to an oblique shock in equilibrium flow at the same wave angle σ.

We can now imagine that the simple concave corner of Fig. 21(*a*) is obtained by straightening out the concave wall of Fig. 21(*d*). This may be expected to give rise to expansion waves propagating out from the wall. As in classical flow over a convex body, these would be expected to interact with the oblique wave causing it to weaken and bend back as it proceeds out from the corner. Since the interaction cannot yet be effective precisely at the corner, the wave angle there is unaltered from what it was in Fig. 21(*d*). In a notation more suitable to the present situation, it is the angle σ_f appropriate to a frozen-flow oblique wave at the corner turning angle δ. To assess conditions far from the corner, we can regard the entire flow from the point of view of an observer removed to a distance large compared with the relaxation length. At this distance the nonequilibrium effects might be expected to seem insignificant and the flow field to appear essentially as an equilibrium one. At large distances from the corner, therefore, the wave angle would be expected to approach the angle σ_e appropriate to an oblique shock wave in equilibrium flow at the same turning angle δ. Consistent with the foregoing, the conditions on the wall immediately behind the shock wave will be those for frozen flow behind a corner of angle δ; far downstream they would be expected to approach the values for equilibrium flow over the same corner. For reasons that we shall see presently, however, this last conclusion is only partially correct.

The problem of the concave corner in the linearized theory has been solved approximately by Moore and Gibson (1960) and exactly by Clarke (1960). Clarke's solution proceeds by Laplace-transform methods applied to the linear equation (7.7). He finds that the initial disturbance, which in the linear theory lies along the frozen Mach line through the corner, decays exponentially with distance from the corner. This is the best that the linear theory can do to simulate a shock wave of decreasing angle. For the pressure $p(x)$ on the wall, Clarke finds

$$\frac{p(x) - p_a}{\frac{1}{2}\rho_a u_a^2} = \frac{2\delta}{\beta_f}\left[\exp\left(-\frac{\beta_e^2/\beta_f^2 + 1}{2} \times \frac{x}{\overset{+}{\tau}_a u_a}\right) I_0\left(\frac{\beta_e^2/\beta_f^2 - 1}{2} \times \frac{x}{\overset{+}{\tau}_a u_a}\right)\right.$$
$$\left. + \int_0^{x/\overset{+}{\tau}_a u_a} \exp\left(-\frac{\beta_e^2/\beta_f^2 + 1}{2} \times \frac{\xi}{\overset{+}{\tau}_a u_a}\right) I_0\left(\frac{\beta_e^2/\beta_f^2 - 1}{2} \times \frac{\xi}{\overset{+}{\tau}_a u_a}\right) d\left(\frac{\xi}{\overset{+}{\tau}_a u_a}\right)\right], \tag{15.1}$$

where $\beta_e^2 \equiv M_{e_a}^2 - 1$, $\beta_f^2 \equiv M_{f_a}^2 - 1$, I_0 is the modified Bessel function of the first kind of order 0, and x is measured downstream from the corner. At $x = 0$ the bracketed quantity is 1, and we have $[p(0) - p_a]/\frac{1}{2}\rho_a u_a^2 = 2\delta/\beta_f$. This agrees with the classical result from the Prandtl-Glauert equation applied to frozen flow (see Liepmann and Roshko, 1957, p. 216). As $x \to \infty$ the bracketed quantity can be shown to approach β_f/β_e, and we obtain $[p(\infty) - p_a]/\frac{1}{2}\rho_a u_a^2 = 2\delta/\beta_e$. This again is the classical result, now applied to equilibrium flow. The linear theory thus agrees with our intuitive ideas regarding conditions at small and large distances. For values of x between 0 and ∞ the pressure given by (15.1) decreases monotonically with x.

To solve the nonlinear problem it is necessary to use a numerical method of characteristics similar to that discussed in Sec. 14. This has been done for the ideal dissociating gas by Capiaux and Washington (1963). Results from their work are reproduced in Fig. 22 for the case specified by the data in part (*a*) of the figure. These results correspond to an equilibrium free stream with $\alpha_a \simeq 0$. The results are plotted as functions of the dimensionless distance $\bar{x} = x\rho_a C/R_{A_2}^{\eta} u_a^{1-2\eta}$, where x is measured parallel to the initial stream. This is the same dimensionless variable used previously in Fig. 14. In Figs. 22(*b*) and (*c*) results are given along several streamlines, identified by their initial vertical location $\bar{y}_a = y_a\rho_a C/R_{A_2}^{\eta} u_a^{1-2\eta}$ in the undisturbed stream. The calculations here were based on a value of -2.5 for the exponent η in the rate expression (11.3).

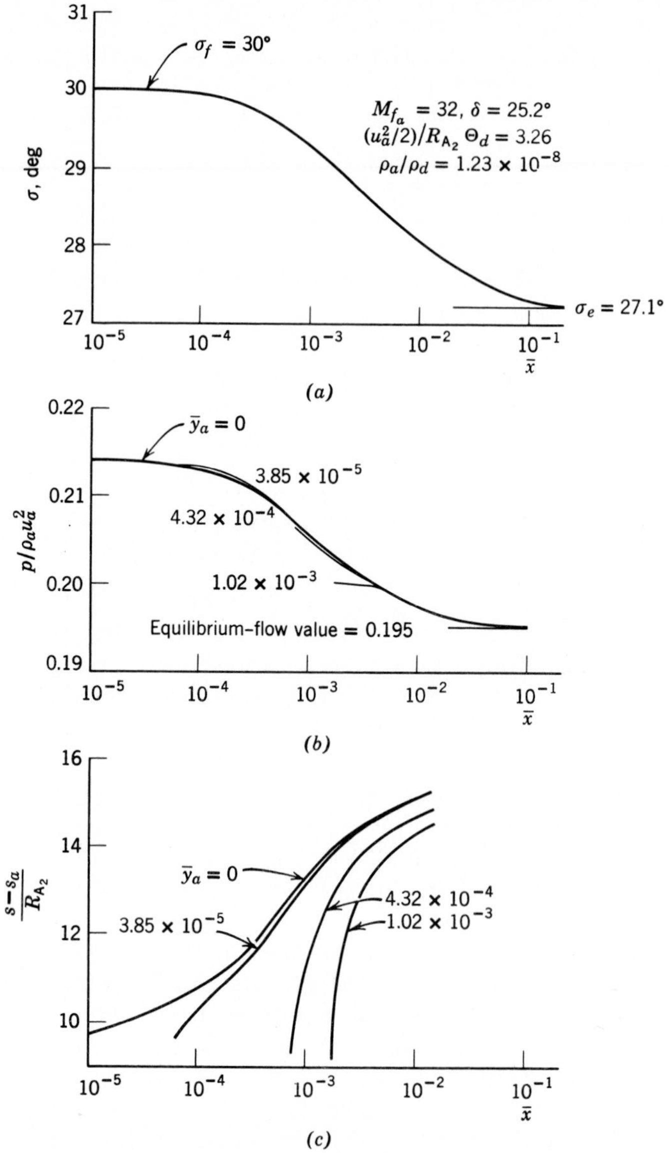

Fig. 22. Results for nonequilibrium flow over a concave corner; $\eta = -2.5$ (after Capiaux and Washington, 1963). (*a*) Shock-wave angle; (*b*) Pressure; (*c*) Entropy.

Figure 22(*a*) illustrates how the shock wave bends back as we proceed along the wave. Figure 22(*b*) shows that, except close to the shock wave itself, the pressure on all streamlines has a uniform value at any given station $\bar{x}$ and that this approaches the value for completely equilibrium flow as $\bar{x}$ becomes large. This again is consistent with our earlier ideas that an observer at a large distance would see the flow as essentially a uniform, equilibrium one. The entropy of Fig. 22(*a*), however, shows no sign of approaching a uniform value as $\bar{x}$ increases, and this situation continues no matter how far downstream the calculations are carried. The entropy thus fails to approach the uniform, equilibrium-flow solution at large distances, and other state variables show a similar behavior. This does not mean that the flow on each streamline does not itself approach an equilibrium state—obviously it must if we go far enough downstream. It does mean, however, that the eventual equilibrium state on each streamline is different and is, in particular, not the same as the uniform state for completely equilibrium flow. Our earlier ideas are thus seen to be oversimplified and in fact correct only as regards the pressure.

An explanation that accounts for the observed phenomena can be had by assuming that the flow at large distances behind the shock is parallel to the wall. We further assume that at large distances above the wall the partly dispersed wave formed by the shock and the subsequent relaxation zone is straight, which means that the overall changes across it are the same as for an oblique shock in equilibrium flow. Taken together, these two assumptions imply that at large distances behind the shock *and* large distances above the wall, the pressure is the same as that calculated on a wedge of angle δ in an equilibrium flow. Furthermore, since the curvature of the parallel flow assumed to exist far downstream is zero, this same pressure will prevail all the way in to the wall. This is in conformity with the numerical findings. No such argument can be made for other quantities, however. The entropy, on the contrary, should be expected to undergo a different increase on different streamlines, both from the different wave angles at the shock and the different history in the nonequilibrium flow behind the shock. The approach observed here to the equilibrium-flow value in the pressure but not in the other quantities is a reflection of the singular-perturbation nature of nonequilibrium flow in the equilibrium-flow limit. The precise mathematical reasons for this have been studied by Lee (1963), who also gives an analytical treatment of the corner problem valid in certain situations (see also Lee, 1964). Numerical calculations for vibrational nonequilibrium by Sedney, South, and Gerber (1964) also lead to similar conclusions.

The foregoing discussion and example are for a situation in which the initial flow is in equilibrium at a low temperature and the flow downstream of the shock is therefore one of dissociation. If the initial stream is frozen out of equilibrium, as discussed previously at the end of Sec. 9, the downstream flow may involve either dissociation or recombination, and the shock wave may be curved either backward or forward. If conditions are such that the shock wave puts the flow precisely *into* equilibrium, the flow behind the wave is uniform, as was the case in Sec. 9, and the wave itself is straight. This problem has been analyzed in the linear approximation by Vincenti (1962). In certain unusual circumstances, especially at very high speeds, the wave may be curved forward or have a reflex shape, even when the initial flow is in equilibrium. This is discussed in detail by Capiaux and Washington (1963).

Exercise 15.1. Consider the linearized, two-dimensional potential equation (8.1), and assume that the flow is everywhere supersonic.

(a) Find a solution of this equation of the form $\varphi = f(x - \lambda y)$, where λ is a given positive real constant and f is a specific function that you are to find.

(b) Show that this solution with $x > \lambda y$ can be used to obtain a linearized representation of the flow over a curved wall supporting a straight oblique shock wave, as depicted in Fig. 21(*d*). For this purpose consider, similar to the procedure in Sec. 9, that the linearization is made with respect to the equilibrium condition at infinity downstream (condition *b*). In addition, take the origin of the x, y coordinate system at the concave corner with the x-axis parallel to the wall at downstream infinity. The constant λ will then be related to the direction of the oblique shock. Find, in particular, an expression for the pressure in the entire flow field behind the shock in terms of the angles σ, δ_f, and δ_e, the conditions at infinity downstream, and the independent variables x and y.

16 SUPERSONIC FLOW OVER A CONVEX CORNER

The nonequilibrium flow over a convex corner takes place, as in the classical case, through an expansion fan centered on the corner (Fig. 23). Here, however, the dependence of the flow speed u on the local flow direction θ is not unique, as in a Prandtl-Meyer fan, but differs from one streamline to the next. Near the corner the transit time through the fan is small compared with the relaxation time, with the result that the nonequilibrium variable q undergoes little change. At the corner itself the transit time is zero, and q remains constant at the initial value q_a. The dependence of velocity on flow direction right at the corner is therefore identical to that for frozen Prandtl-Meyer flow. The relaxation of q to a new equilibrium value then takes place entirely along the wall following the corner. On streamlines far from the corner the situation is different. Here the transit time through the fan is large compared with the relaxation time, and the values of q can approach those for local equilibrium. At infinity the relationship between velocity and flow direction is that for an equilibrium Prandtl-Meyer flow, and all the change of q to a final equilibrium value takes place within the fan. Because of these differences, the Mach lines from the corner are not straight as in Prandtl-Meyer flow, except for the initial Mach line, which is everywhere at the frozen Mach angle μ_{f_a} of the undisturbed stream. Similarly, the flow following the expansion fan is not uniform but varies from point to point. As a result, a simple analytical treatment, such as that for Prandtl-Meyer flow, is not possible.

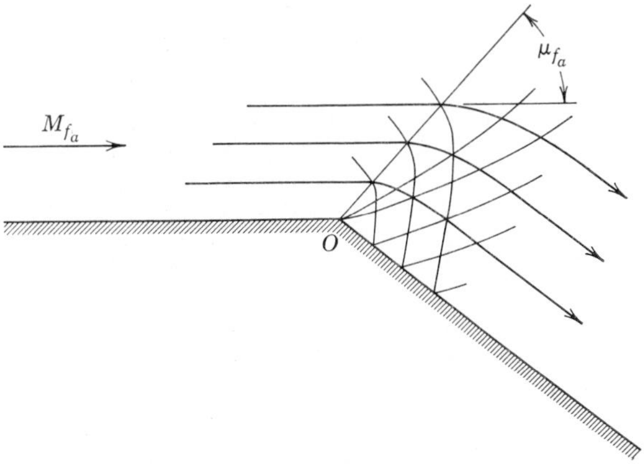

Fig. 23. Flow over a convex corner.

To the linear approximation, Clarke's solution (15.1) serves equally well for the convex and concave corner. It is only necessary that here δ be counted as negative. Again the pressure on the wall immediately behind the corner is that for linearized frozen flow; far downstream it approaches the value for linearized equilibrium flow. In between, the pressure now increases monotonically with distance along the wall.

Results for the nonlinear flow for an ideal dissociating gas have been obtained by Appleton (1963) by a numerical method of characteristics. Typical results for quantities on the wall downstream of the corner are shown in Fig. 24 for three values of the overall deflection angle. The value of η in the rate expression (11.3) was taken as 0; the data for the equilibrium initial stream are listed in part (*b*) of the figure. The independent variable (in the present case of $\eta = 0$) is the dimensionless distance $\bar{x} \equiv xC\rho_d/\sqrt{R_{A_2}\Theta_d}$ measured parallel to the undisturbed stream. The pressure [Fig. 24(*a*)] is seen to rise downstream from the corner, but not monotonically as in the linear theory. Instead it overshoots the final value, which it then approaches from above. As with the concave corner,

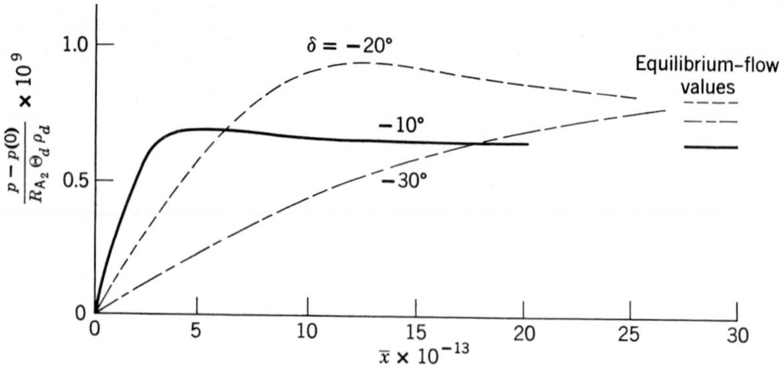

(*a*)

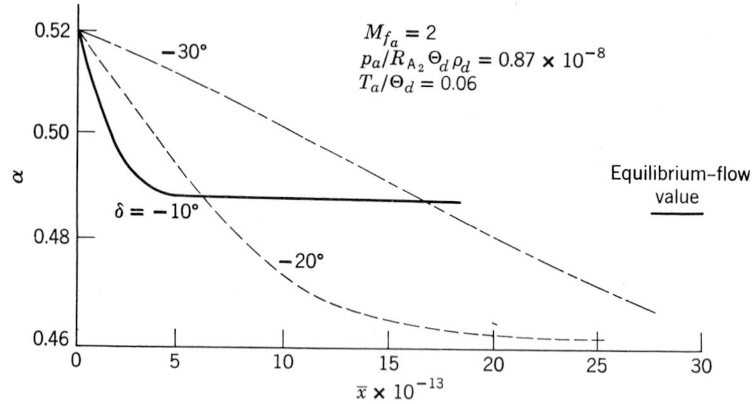

(*b*)

Fig. 24. Flow variables on wall downstream of a convex corner; $\eta = 0$ (after Appleton, 1963). (*a*) Pressure measured relative to value immediately behind corner; (*b*) Degree of dissociation.

the final value (at least for $\delta = -10°$ where the results extend sufficiently far to leave no doubt) agrees with that for completely equilibrium flow. The degree of dissociation α falls downstream of the corner, but apparently does not approach the equilibrium-flow value. Thus the equilibrium state that must eventually be reached on each streamline is again not the same from one streamline to another. As with the concave corner, physical arguments can be made that the pressure should approach the uniform equilibrium-flow value far downstream but that other quantities should not.

Exercise 16.1. What will be the behavior of the flow in the limit as the relaxation time of the fluid tends to zero for (*a*) a concave corner and (*b*) a convex corner? What would the behavior be if the convex corner were not sharp but slightly rounded?

References

Ackeret, J., 1928, Über Luftkräfte bei sehr grossen Geschwindigkeiten insbesondere bei ebenen Strömungen. *Helv. Phys. Acta.* vol. 1, p. 301.

Appleton, J. P., 1963, Structure of a Prandtl-Meyer Expansion in an Ideal Dissociating Gas. *Phys. Fluids*, vol. 6, no. 8, p. 1057.

Appleton, J. P., and K. N. C. Bray, 1964, The Conservation Equations for a Non-equilibrium Plasma. *J. Fluid Mech.*, vol. 20, pt. 4, p. 659.

Blythe, P. A., 1964, Asymptotic Solutions in Non-equilibrium Nozzle Flow. *J. Fluid Mech.*, vol. 20, pt. 2, p. 243.

Bray, K. N. C., 1959, Atomic Recombination in a Hypersonic Wind-Tunnel Nozzle. *J. Fluid Mech.*, vol. 6, pt. 1, p. 1.

Bray, K. N. C., 1964, Electron-Ion Recombination in Argon Flowing through a Supersonic Nozzle; in *The High Temperature Aspects of Hypersonic Flow*, ed. by W. C. Nelson, Macmillan.

Capiaux, R., and M. Washington, 1963, Nonequilibrium Flow past a Wedge. *AIAA Jour.*, vol. 1, no. 3, p. 650.

Chu, B. T., 1957, Wave Propagation and the Method of Characteristics in Reacting Gas Mixtures with Applications to Hypersonic Flow. *Wright Air Development Center*, TN-57-213.

Clarke, J. F., 1960, The Linearized Flow of a Dissociating Gas. *J. Fluid Mech.*, vol. 7, pt. 4, p. 577.

Clarke, J. F., and M. McChesney, 1964, *The Dynamics of Real Gases*, Butterworths.

Coddington, E. A., 1961, *Ordinary Differential Equations*, Prentice-Hall.

Conti, R. J., 1964, Stagnation Equilibrium Layer in Nonequilibrium Blunt-Body Flows. *AIAA Jour.*, vol. 2, no. 11, p. 2044.

Courant, R., and K. O. Friedrichs, 1948, *Supersonic Flow and Shock Waves*, Interscience.

Courant, R., and D. Hilbert, 1962, *Methods of Mathematical Physics*, vol. II, Interscience.

De Groot, S. R., and P. Mazur, 1962, *Non-equilibrium Thermodynamics*, Interscience.

Der, J. J., 1963, Theoretical Studies of Supersonic Two-Dimensional and Axisymmetric Nonequilibrium Flow, including Calculations of Flow through a Nozzle. *NASA*, TR R-164.

Duff, R. E., and N. Davidson, 1959, Calculations of Reaction Profiles behind Steady State Shock Waves. II. The Dissociation of Air. *J. Chem. Phys.*, vol. 31, no. 4, p. 1018.

Emanuel, G., 1963, Problems Underlying the Numerical Integration of the Chemical and Vibrational Rate Equations in a Near-Equilibrium Flow. *Arnold Engr. Devel. Center*, AEDC-TDR-63-82.

Emanuel, G., 1964, A General Method for Numerical Integration through a Saddle-Point Singularity, with Application to One-Dimensional Nonequilibrium Nozzle Flow. *Arnold Engr. Devel. Center*, AEDC-TDR-64-29.

Emanuel, G., and W. G. Vincenti, 1962, Method for Calculation of the One-Dimensional Nonequilibrium Flow of a General Gas Mixture through a Hypersonic Nozzle. *Arnold Engr. Devel. Center*, AEDC-TDR-62-131.

Erickson, W. D., 1963, Vibrational-Nonequilibrium Flow of Nitrogen in Hypersonic Nozzles. *NASA*, TN-D-1810.

Eschenroeder, A. Q., 1962, Ionization Nonequilibrium in Expanding Flows. *ARS Jour.*, vol. 32, no. 2, p. 196.

Eschenroeder, A. Q., D. W. Boyer, and J. G. Hall, 1962, Nonequilibrium Expansions of Air with Coupled Chemical Reactions. *Phys. Fluids*, vol. 5, no. 5, p. 615.

Freeman, N. C., 1958, Non-equilibrium Flow of an Ideal Dissociating Gas. *J. Fluid Mech.*, vol. 4, pt. 4, p. 407.

Griffith, W. C., and A. Kenny, 1957, On Fully-Dispersed Shock Waves in Carbon Dioxide. *J. Fluid Mech.*, vol. 3, pt. 3, p. 286.

Griffith, W. C., 1961, Vibrational Relaxation Times; in *Fundamental Data Obtained from Shock-Tube Experiments*, ed. by, A. Ferri, Pergamon.

Gross, R. A., and C. L. Eisen, 1959, On the Speed of Sound in Air. *Phys. Fluids*, vol. 2, no. 3, p. 276.

Hayes, W. D., and R. F. Probstein, 1959, *Hypersonic Flow Theory*, Academic.

Hurle, I. R., A. L. Russo, and J. G. Hall, 1964, Spectroscopic Studies of Vibrational Nonequilibrium in Supersonic Nozzle Flows. *J. Chem. Phys.*, vol. 40, no. 8, p. 2076.

Johannesen, N. H., H. K. Zienkiewicz, P. A. Blythe, and J. H. Gerrard, 1962, Experimental and Theoretical Analysis of Vibrational Relaxation Regions in Carbon Dioxide. *J. Fluid Mech.*, vol. 13, pt. 2, p. 213.

Jones, J. G., 1964, On the Near-Equilibrium and Near-Frozen Regions in an Expansion Wave in a Relaxing Gas. *J. Fluid Mech.*, vol. 19, pt. 1, p. 81.

Lee, R. S., 1963, *A Study of Supersonic Nonequilibrium Flow over a Wedge*, Ph.D. Dissertation, Stanford University (available from University Microfilms, Ann Arbor, Mich.).

Lee, R. S., 1964, A Unified Analysis of Supersonic Nonequilibrium Flow over a Wedge. I. Vibrational Nonequilibrium. *AIAA Jour.*, vol. 2, no. 4, p. 637.

Lees, L., 1958, Convective Heat Transfer with Mass Addition and Chemical Reactions; in *Combustion and Propulsion, Third AGARD Colloquium*, ed. by M. W. Thring et al., Pergamon.

Liepmann, H. W., and A. Roshko, 1957, *Elements of Gasdynamics*, Wiley.

Lighthill, M. J., 1956, Viscosity Effect in Sound Waves of Finite Amplitude; in *Surveys in Mechanics*, ed. by G. K. Batchelor and R. M. Davies, Cambridge University Press.

Lin, S. C., and J. D. Teare, 1963, Rate of Ionization behind Shock Waves in Air. II. Theoretical Interpretations. *Phys. Fluids*, vol. 6, no. 3, p. 355.

Losev, S. A., and A. I. Osipov, 1962, The Study of Nonequilibrium Phenomena in Shock Waves. *Soviet Phys. Usp.*, vol. 4, no. 4, p. 525.

Marrone, P. V., and C. E. Treanor, 1963, Chemical Relaxation with Preferential Dissociation from Excited Vibrational Levels. *Phys. Fluids*, vol. 6, no. 9, p. 1215.

Meyer, R. E., 1953, The Method of Characteristics; in *Modern Developments in Fluid Dynamics, High Speed Flow*, vol. I, ed. by L. Howarth, Oxford University Press.

Moore, F. K., 1964, Hypersonic Boundary Layer Theory; in *Theory of Laminar Flows*, ed. by F. K. Moore, Princeton University Press.

Moore, F. K., and W. E. Gibson, 1960, Propagation of Weak Disturbances in a Gas subject to Relaxation Effects. *J. Aerospace Sci.*, vol. 27, no. 2, p. 117.

Napolitano, L. G., 1964, Chemical Relaxation for Channel Flows of Doubly Reacting Mixtures; in *Supersonic Flow, Chemical Processes and Radiative Transfer*, ed. by D. B. Olfe and V. Zakkay, Macmillan.

Penner, S. S., 1955, *Chemical Reactions in Flow Systems*, Butterworths.

Petschek, H., and S. Byron, 1957, Approach to Equilibrium Ionization behind Strong Shock Waves in Argon. *Ann. Phys.*, vol. 1, no. 3, p. 270.

Rosner, D. E., 1962, Estimation of Electrical Conductivity at Rocket Nozzle Exit Sections. *ARS Jour.*, vol. 32, no. 10, p. 1602.

Rotty, R. M., 1962, *Introduction to Gas Dynamics*, Wiley.

Ryhming, I. L., 1963, Non-equilibrium Flow inside a Wavy Cylinder. *J. Fluid Mech.*, vol. 17, pt. 4, p. 551.

Sedney, R., J. C. South, and N. Gerber, 1964, Characteristic Calculation of Non-Equilibrium Flows; in *The High Temperature Aspects of Hypersonic Flow*, ed. by W. C. Nelson, Macmillan.

Spence, D. A., 1961, Unsteady Shock Propagation in a Relaxing Gas. *Proc. Roy. Soc.*, ser. A, vol. 264, p. 221.

Stollery, J. L., J. E. Smith, and C. Park, 1964, The Effects of Vibrational Relaxation on Hypersonic Nozzle Flows; in *The High Temperatures Aspects of Hypersonic Flow*, ed. by W. C. Nelson, Macmillan.

Van Dyke, M. D., 1964, *Perturbation Methods in Fluid Mechanics*, Academic.

Vincenti, W. G., 1959, Non-equilibrium Flow over a Wavy Wall. *J. Fluid Mech.*, vol. 6, pt. 4, p. 481.

Vincenti, W. G., 1962, Linearized Flow over a Wedge in a Nonequilibrium Oncoming Stream. *J. Mécan.*, vol. I, no. 2, p. 193.

Wegener, P. P., 1959, Supersonic Nozzle Flow with a Reacting Gas Mixture. *Phys. Fluids*, vol. 2, no. 3, p. 264.

Whitham, G. B., 1959, Some Comments on Wave Propagation and Shock Wave Structure with Application to Magnetohydrodynamics. *Comm. Pure and Appl. Math.*, vol. XII, no. 1, p. 113.

Williams, F. A., 1965, *Combustion Theory*, Addison-Wesley.

Wood, W. W., and J. G. Kirkwood, 1957, Hydrodynamics of a Reacting and Relaxing Fluid. *J. Appl. Phys.*, vol. 28, no. 4, p. 395.

Chapter IX

Nonequilibrium Kinetic Theory

1 INTRODUCTION

In Chapter VII we observed that the adjustment of translational and rotational nonequilibrium in a gas flow requires relatively few collisions as compared with that of vibrational and chemical nonequilibrium. Accordingly, we were able to discuss the latter effects separately, considering translation and rotation to be in local equilibrium. When gradients in the flow field are large, however, as for example in a boundary layer or shock wave, translational nonequilibrium results in the phenomena known as viscous stress and heat conduction, and the Maxwellian distribution function ceases to give an adequate description of the translational molecular velocities. Rotational nonequilibrium, through the effect of the so-called bulk viscosity, is also of quantitative importance, although it does not alter the qualitative description in as significant a way. The molecular study of these situations, involving from the thermodynamic point of view, mechanical and thermal nonequilibrium, requires the application of new concepts and methods of analysis, which will be the subject of this and the following chapter.

As before we shall for the most part treat separately the major effect under consideration, translational nonequilibrium. Hence, *unless explicitly stated to the contrary*, we shall focus our attention on a monatomic gas composed of a single chemical species and ignore internal atomic structure. Gases with internal degrees of freedom will be included, however, in our discussion of the conservation equations in Sec. 2, and a brief description will be given in Chapter X regarding the calculation of their transport properties. Rotational nonequilibrium will be discussed in Chapter X, Sec. 8. Certain important characteristics of gas mixtures will be described in Sec. 10 and in Chapter X, Sec. 7, but for a detailed treatment of such phenomena as diffusion the reader will have to consult the cited references. The interesting kinetic theory of ionized gases is beyond the scope of this book, although an understanding of the theory of neutral gases provides an appropriate background for such a study. In spite of the restrictions that we shall place on our discussion, it is hoped that the reader will nevertheless gain some appreciation of the fact that the application of kinetic theory in practical situations is not limited to simple monatomic gases.

These chapters, which comprise an introduction to advanced kinetic theory, are intended to serve two principal functions. The first is to provide a molecular interpretation of macroscopic gas dynamics, including an indication of the methods whereby transport properties can be evaluated from molecular considerations. The second is to develop concepts and methods useful in the study of physical situations that may be outside the range of validity of conventional gas dynamics, as are encountered in the interior of strong shock waves and in the study of rarefied-gas dynamics. The present chapter is concerned with the framework of kinetic theory including the general conservation equations the derivation of the *Boltzmann equation* for the distribution function, and the analysis of molecular collisions. In Chapter X several methods of evaluating the distribution function are discussed together with certain applications, with an emphasis on the *Chapman-Enskog expansion* leading to the Navier-Stokes equations.[1]

2 THE CONSERVATION EQUATIONS OF GAS DYNAMICS

Before examining in detail the concepts and consequences of advanced kinetic theory, we can gain in understanding and insight from a molecular formulation of mass, momentum, and energy conservation in a gas. In addition to providing general results for later use, this will enable us to extend in more precise terms the molecular interpretation of thermodynamic properties, viscous stress, and heat flux given in Chapter I.

Let us consider a volume of arbitrary dimensions fixed in space and enclosed by an imaginary surface through which a gas is flowing. We can

[1] For brevity we use the term Navier-Stokes equations to refer to the set of partial differential equations, including the energy equation, that describe the motion of a viscous, heat-conducting fluid, rather than in the precise sense of only the momentum equations. These equations can be derived by considering the fluid as a continuous medium (continuum) in which the viscous stress is proportional to the rate of strain and the heat flux is proportional to the temperature gradient. A prior knowledge of the Navier-Stokes equations is not necessary to an understanding of these chapters and is not presumed.

then say the following:

1. The time rate of change of mass within the volume must equal the net inward rate at which mass crosses the surface.
2. The time rate of change of the total momentum of the molecules within the volume (a vector quantity) must equal the net inward rate at which molecular momentum crosses the surface, plus the net force of external origin acting on the molecules within the volume.
3. The time rate of change of the total energy of the molecules within the volume must equal the net inward rate at which energy crosses the surface, plus the net rate of increase in energy resulting from the action of external forces.

To formulate the conservation relations quantitively, we shall need expressions for the molecular densities (quantities per unit volume) and fluxes (quantities per unit area per unit time) of mass, momentum, and energy. In writing these expressions we depart, for clarity, from our previous convention and now interpret quantities written in the subscript notation in terms of *components*. Thus, we now use the symbol c_i to represent the components c_1, c_2, c_3 of the molecular velocity vector. The principal motivation for this distinction arises in connection with the molecular momentum flux (2.6b). We shall revert later in this section to the interpretation of a quantity such as c_i as a vector.

For a gas composed of a single chemical species, the mass density is the product of the mass m per molecule and the number n of molecules per unit volume:

$$\rho = mn. \qquad \text{(mass density)} \tag{2.1}$$

The momentum density is obtained by summing the momentum per molecule over all molecules in a unit volume, that is, by integrating over velocity space the momentum per molecule times the product of the number density and the distribution function.[2] The component of momentum per molecule in the i-direction is mc_i. The i-momentum density is then

$$\int_{-\infty}^{\infty}\int_{-\infty}^{\infty}\int_{-\infty}^{\infty} mc_i\, nf\, dc_1\, dc_2\, dc_3 = \rho\int_{-\infty}^{\infty}\int_{-\infty}^{\infty}\int_{-\infty}^{\infty} c_i f\, dc_1\, dc_2\, dc_3.$$

This can be written more compactly with the aid of the general expression for the mean value $\bar{Q}$ of an arbitrary function $Q(c_i)$ of the components of

[2] We remind the reader that our definition of the distribution function is such that $\int_{-\infty}^{\infty}\int_{-\infty}^{\infty}\int_{-\infty}^{\infty} f\, dc_1\, dc_2\, dc_3 = 1$ [see equation (II 2.7)]. Other authors (e.g., Chapman and Cowling, 1952) use the alternate normalization $\int_{-\infty}^{\infty}\int_{-\infty}^{\infty}\int_{-\infty}^{\infty} f\, dc_1\, dc_2\, dc_3 = n$. Our f is therefore their f/n.

molecular velocity, defined by equation (II 2.8):

$$\bar{Q} \equiv \int_{-\infty}^{\infty}\int_{-\infty}^{\infty}\int_{-\infty}^{\infty} Q(c_i) f \, dc_1 \, dc_2 \, dc_3.$$

An integral of this form is often referred to as a moment of the distribution function. Since we are describing a general rather than an equilibrium situation, the distribution function must here be interpreted as a function of position and time as well as molecular velocity, namely,

$$f = f(c_i, x_i, t)$$

and hence

$$\bar{Q} = \bar{Q}(x_i, t).$$

With this understood, the i-momentum density is then equal to the product of the local mass density and the component in the i-direction of the local mean velocity:

$$\rho \bar{c}_i. \qquad \text{(momentum density)} \tag{2.2}$$

The translational energy per unit volume can similarly be written

$$\int_{-\infty}^{\infty}\int_{-\infty}^{\infty}\int_{-\infty}^{\infty} \tfrac{1}{2} m(c_1^2 + c_2^2 + c_3^2) \, nf \, dc_1 \, dc_2 \, dc_3 = \tfrac{1}{2}\rho\overline{(c_1^2 + c_2^2 + c_3^2)}. \tag{2.3}$$

It is convenient here to define a *thermal* or *peculiar* velocity, with components C_i, measured with respect to the mean velocity, that is, the velocity seen by an observer moving with the mean velocity of the gas. Thus we write

$$C_i \equiv c_i - \bar{c}_i. \tag{2.4}$$

The symbol C_i was used for the molecular velocity in Chapters I and II, since there we were concerned primarily with the random motion of the molecules. The right-hand side of equation (2.3) for the energy density can now be expanded and simplified by means of the definition (2.4) and certain properties of mean quantities that can be deduced from their definition as averages over the distribution function. For example, we have

$$\bar{C}_i \equiv \int_{-\infty}^{\infty}\int_{-\infty}^{\infty}\int_{-\infty}^{\infty} (c_i - \bar{c}_i) f \, dc_1 \, dc_2 \, dc_3 = \bar{c}_i - \bar{c}_i = 0.$$

We can also write

$$\overline{c_1^2 + c_2^2 + c_3^2} \equiv \int_{-\infty}^{\infty}\int_{-\infty}^{\infty}\int_{-\infty}^{\infty} (c_1^2 + c_2^2 + c_3^2) f \, dc_1 \, dc_2 \, dc_3 = \overline{c_1^2} + \overline{c_2^2} + \overline{c_3^2},$$

where, with $\bar{C}_i = 0$,

$$\overline{c_1^2} = \overline{(\bar{c}_1 + C_1)^2} = \bar{c}_1^2 + 2\bar{c}_1\bar{C}_1 + \overline{C_1^2} = \bar{c}_1^2 + \overline{C_1^2}, \text{ etc.}$$

If we denote the absolute magnitude of a vector quantity by the omission of the subscript, we have

$$\bar{c}^2 \equiv \bar{c}_1^2 + \bar{c}_2^2 + \bar{c}_3^2,$$

$$\overline{C^2} \equiv \overline{C_1^2} + \overline{C_2^2} + \overline{C_3^2},$$

and the energy density (2.3) becomes

$$\tfrac{1}{2}\rho\bar{c}^2 + \tfrac{1}{2}\rho\overline{C^2}. \qquad \text{(density of translational energy)} \tag{2.5}$$

The molecular fluxes of mass, i-momentum, and translational energy across a plane that is perpendicular to the j-direction can be calculated by means of the general expression, derived in Chapter II, Sec. 3, for the flux across such a plane of molecules in the velocity range between c_1, c_2, c_3 and $c_1 + dc_1$, $c_2 + dc_2$, $c_3 + dc_3$:[3]

$$c_j nf\, dc_1\, dc_2\, dc_3.$$

The fluxes of mass, i-momentum, and translational energy are thus

$$\int_{-\infty}^{\infty}\int_{-\infty}^{\infty}\int_{-\infty}^{\infty} mc_j nf\, dc_1\, dc_2\, dc_3 = \rho\bar{c}_j, \qquad \text{(mass flux)} \tag{2.6a}$$

$$\int_{-\infty}^{\infty}\int_{-\infty}^{\infty}\int_{-\infty}^{\infty} mc_i c_j nf\, dc_1\, dc_2\, dc_3 = \rho\overline{c_i c_j}, \qquad \text{(flux of } i\text{-momentum)} \tag{2.6b}$$

and

$$\int_{-\infty}^{\infty}\int_{-\infty}^{\infty}\int_{-\infty}^{\infty} \tfrac{1}{2}mc^2 c_j nf\, dc_1\, dc_2\, dc_3 = \tfrac{1}{2}\rho\overline{c^2 c_j}. \qquad \text{(flux of translational energy)} \tag{2.6c}$$

It will be convenient for later use to rewrite the last two expressions by expansion in terms of the components C_i of the thermal velocity. With the definition (2.4) for C_i, the right-hand side of equation (2.6b) becomes

$$\rho\overline{c_i c_j} = \rho\overline{(\bar{c}_i + C_i)(\bar{c}_j + C_j)} = \rho(\bar{c}_i\bar{c}_j + \bar{c}_i\overline{C_j} + \bar{c}_j\overline{C_i} + \overline{C_i C_j})$$
$$= \rho\bar{c}_i\bar{c}_j + \rho\overline{C_i C_j},$$

and thus the momentum flux is

$$\rho\bar{c}_i\bar{c}_j + \rho\overline{C_i C_j}. \qquad \text{(flux of } i\text{-momentum)} \tag{2.7}$$

[3] This expression follows from equation (II 3.1), which is expressed in terms of the velocity components C_i. There the mean velocity (and hence the distinction between C_i and c_i) is taken to be zero. To convert to the present situation we replace C_i by c_i.

In the expression (2.6c) for the energy flux, we have

$$\begin{aligned}\overline{c^2c_j} &= \overline{[(\bar{c}_1 + C_1)^2 + (\bar{c}_2 + C_2)^2 + (\bar{c}_3 + C_3)^2][\bar{c}_j + C_j]} \\ &= \bar{c}_j\bar{c}^2 + \bar{c}_j\overline{C^2} + 2(\bar{c}_1\overline{C_jC_1} + \bar{c}_2\overline{C_jC_2} + \bar{c}_3\overline{C_jC_3}) + \overline{C_jC^2},\end{aligned}$$

where zero terms such as $2\bar{c}_j\bar{c}_1\overline{C_1}$ have been omitted (see Exercise 2.1). With the repeated-subscript notation for summation, the expression (2.6c) then becomes

$$\rho\bar{c}_j(\tfrac{1}{2}\bar{c}^2 + \tfrac{1}{2}\overline{C^2}) + \tfrac{1}{2}\rho\overline{C_jC^2} + \rho\overline{C_jC_k}\bar{c}_k. \qquad \text{(flux of translational energy)} \tag{2.8}$$

With the expressions that we have obtained for the densities and fluxes of mass, momentum, and energy, the conservation equations in differential form can now be derived by considering an elementary volume of fixed dimensions dx_1, dx_2, dx_3. The rate of change with respect to time of the mass within this volume is, from equation (2.1),

$$\frac{\partial}{\partial t}(\rho\, dx_1\, dx_2\, dx_3) = \frac{\partial \rho}{\partial t}\, dx_1\, dx_2\, dx_3.$$

From Fig. 1, which is based on equation (2.6a) and shows the mass flow across the six faces of the volume, it can be seen that the net *inward* mass flow across the two faces that are perpendicular to the x_1-direction is

$$-\frac{\partial}{\partial x_1}(\rho\bar{c}_1)\, dx_1\, dx_2\, dx_3.$$

Conservation of mass requires that the time derivative of the mass within the volume be equal to the sum of this expression and two similar quantities for the net mass flow across the faces perpendicular to the x_2- and x_3-directions, as follows:

$$\begin{aligned}\frac{\partial \rho}{\partial t}\, dx_1\, dx_2\, dx_3 = &-\frac{\partial}{\partial x_1}(\rho\bar{c}_1)\, dx_1\, dx_2\, dx_3 - \frac{\partial}{\partial x_2}(\rho\bar{c}_2)\, dx_2\, dx_1\, dx_3 \\ &-\frac{\partial}{\partial x_3}(\rho\bar{c}_3)\, dx_3\, dx_1\, dx_2.\end{aligned}$$

Dividing by $dx_1\, dx_2\, dx_3$ and using the repeated-subscript summation notation, we obtain the familiar continuity equation

$$\frac{\partial \rho}{\partial t} + \frac{\partial}{\partial x_j}(\rho\bar{c}_j) = 0. \tag{2.9}$$

By derivations similar to that of the continuity equation, the momentum and energy conservation equations in the absence of external forces can be

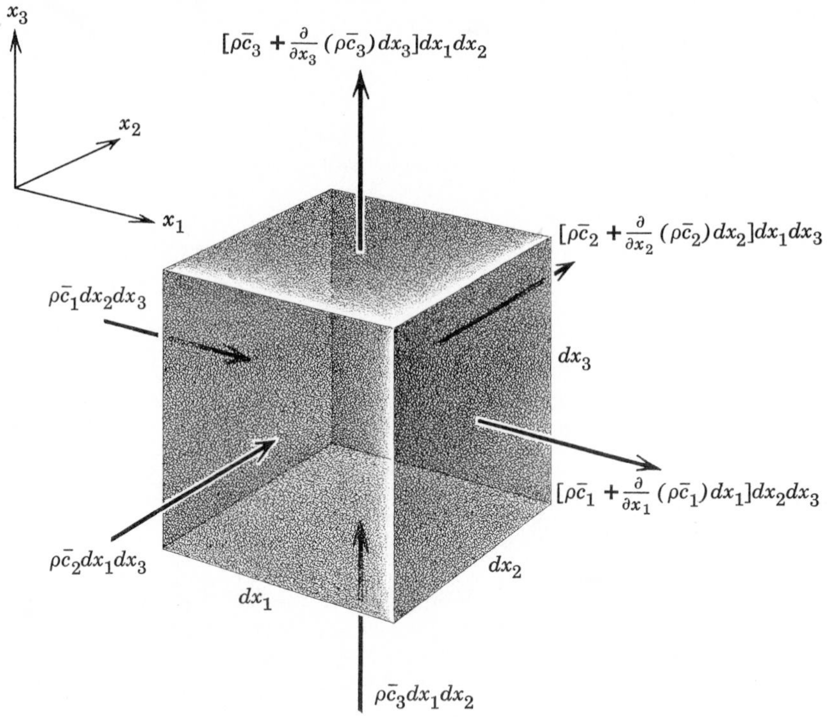

Fig. 1. Mass flow across the six surfaces of dV_x.

stated: *the time derivative of the pertinent density plus the divergence of the corresponding flux is equal to zero.* In the subscript notation the divergence of a quantity containing an unsummed subscript j is obtained by multiplication of the quantity by $\partial/\partial x_j$, as can be seen from the continuity equation (2.9). If there exists an external force with components F_i per unit mass, which in general may depend on the molecular velocity, an additional term $nm\bar{F}_i = \rho\bar{F}_i$ must be added to the momentum equation. Using the momentum density and flux relations (2.2) and (2.7), we then have for the momentum equation

$$\frac{\partial}{\partial t}(\rho\bar{c}_i) + \frac{\partial}{\partial x_j}[\rho\bar{c}_i\bar{c}_j + \rho\overline{C_iC_j}] = \rho\bar{F}_i. \tag{2.10}$$

The reader will notice that it has not been necessary to consider separately "surface forces" such as pressure. As we shall see shortly, these "forces" are in fact included in our calculation of the molecular momentum flux.

The rate of increase of the energy of a molecule resulting from the action of the external force is mF_jc_j; when we sum over all molecules the net rate is $\rho\overline{F_jc_j}$. Incorporating the energy density and flux from the expressions (2.5) and (2.8), we thus obtain the energy equation

$$\frac{\partial}{\partial t}[\tfrac{1}{2}\rho\bar{c}^2 + \tfrac{1}{2}\rho\overline{C^2}] + \frac{\partial}{\partial x_j}[\bar{c}_j(\tfrac{1}{2}\rho\bar{c}^2 + \tfrac{1}{2}\rho\overline{C^2}) + \tfrac{1}{2}\rho\overline{C_jC^2} + \rho\overline{C_jC_k}\bar{c}_k] = \rho\overline{F_jc_j}. \tag{2.11}$$

Here it should be noted that the repeated-subscript notation is used twice in the last term on the left, so that this term represents the sum of nine terms.

Implicit in our derivation of the conservation equations is the assumption that the state of the gas may be meaningfully described by certain mean quantities such as density and mean velocity. These quantities are often interpreted as molecular averages taken over a volume in physical space or a time interval that is sufficiently large that the average includes a great many molecules, yet small enough that the quantities of interest vary smoothly with position and time. Since there are about 3×10^{19} molecules per cubic centimeter at standard conditions, this restriction is amply satisfied in many practical situations. On the other hand, it is important to note that these mean quantities and the distribution function itself cannot be applied literally to a point in space and time. This observation leads to two interpretations of the distribution function: one, as previously stated, in terms of an *actual* density in velocity space defined only for volumes in physical space and time intervals that are "not too small or not too large," and the other as a probability or expected value. The latter interpretation refers to averages taken (in thought) over a great many repeated experiments or identical systems (such an *ensemble* of systems was mentioned in Chapter IV, Sec. 8). Although we do not wish to go into the more subtle points here, the definition of the distribution function as an *expected* value both clarifies its interpretation and extends its applicability to situations for which the time and volume averages suggested earlier are no longer appropriate (see Grad, 1958, Sec. 9). Such an interpretation will be implied in the material that follows.

The conservation equations (2.9), (2.10), and (2.11) can be put in more conventional form by relating the mean quantities appearing therein to the familiar terms of gas dynamics. The mass density and flux and hence the continuity equation [equations (2.1), (2.6a), and (2.9)], as well as the momentum density [equation (2.2)], depend as they stand only on ρ and $\bar{c}_i$ as in conventional gas dynamics. (The symbol u_i was used for the gas-dynamic velocity in the conservation equations in Chapter VIII and corresponds to $\bar{c}_i$ here.) The energy density [equation (2.5)] can be rewritten in terms of the energy of translational thermal motion per unit mass e_{tr} as

$$e_{tr} = \tfrac{1}{2}\overline{C^2}, \tag{2.12}$$

which in turn can be related to the temperature. Here there are two alternatives: to identify the translational thermal energy per molecule with a temperature defined in thermodynamics, or to define independently a translational kinetic temperature

$$\boxed{\tfrac{3}{2}kT \equiv \tfrac{1}{2}m\overline{C^2}} \tag{2.13}$$

that agrees with the thermodynamic temperature under conditions of mutual applicability. The former approach has been taken in previous chapters (see Chapter I, Sec. 3 and Chapter IV, Sec. 1); we shall here adopt the latter interpretation to avoid questions of validity in nonequilibrium situations such as in the interior of a strong shock wave. Recalling that the gas constant per unit mass R is related to Boltzmann's constant k, the gas constant per molecule, by

$$R = \frac{k}{m},$$

we then have as in Chapter I

$$e_{tr} = \frac{3}{2}\frac{k}{m}T = \tfrac{3}{2}RT. \tag{2.14}$$

The term $\rho\overline{C_iC_j}$ appearing in the momentum equation (2.10) is the flux of i-momentum across a plane that is perpendicular to the j-direction as a result of the thermal motions of the molecules. This momentum flux is that which would be seen by an observer moving with the mean velocity of the gas. In a nonequilibrium situation the pressure is defined as the average over the three directions of the normal components of the thermal momentum flux:

$$p \equiv \tfrac{1}{3}\rho[\overline{C_1^2} + \overline{C_2^2} + \overline{C_3^2}] = \tfrac{1}{3}\rho\overline{C^2}. \tag{2.15}$$

Thus, from the definition of temperature (2.13), we have the perfect-gas relation

$$\boxed{p = nkT = \rho RT}, \tag{2.16}$$

in agreement with the familiar phenomenological relation of gas dynamics and with our equilibrium calculations of Chapter II. The viscous stress τ_{ij} is defined as the negative of the thermal momentum flux with the pressure subtracted from the normal components, that is,

$$\boxed{\tau_{ij} \equiv -[\rho\overline{C_iC_j} - p\delta_{ij}]}\,, \tag{2.17}$$

where the Kronecker delta δ_{ij} is such that

$$\begin{aligned} \delta_{ij} &= 1 \quad \text{if} \quad i = j, \\ \delta_{ij} &= 0 \quad \text{if} \quad i \neq j. \end{aligned}$$

With these definitions the gradient of the thermal momentum flux can be written

$$\frac{\partial}{\partial x_j}(\rho\overline{C_iC_j}) = \frac{\partial p}{\partial x_i} - \frac{\partial \tau_{ij}}{\partial x_j}.$$

We continue to apply the notation of gas dynamics to the conservation equations by noting that the energy equation (2.11) contains terms in $\rho\overline{C^2}$ and $\rho\overline{C_iC_j}$ (written there $\rho\overline{C_jC_k}$). When these terms are replaced by e_{tr} and τ_{ij} from equations (2.12) and (2.17), the enthalpy

$$h_{\text{tr}} \equiv e_{\text{tr}} + \frac{p}{\rho} = \tfrac{5}{2}RT \tag{2.18}$$

arises naturally on the left-hand side of the energy equation as a consequence of the molecular energy flux. The remaining term to be rewritten in the energy equation is $\frac{1}{2}\rho\overline{C_jC^2}$, the only energy flux that is non-zero when viewed by an observer moving with the mean velocity of the gas. This term can hence be identified with the *heat flux* q_i, and written[4]

$$\boxed{q_i \equiv \tfrac{1}{2}\rho\overline{C_iC^2}}\,. \tag{2.19}$$

Arguments similar to those used in the calculation of the pressure at a wall in Chapter II show that, in addition to describing the momentum and energy flux in the interior of a gas, p, τ_{ij}, and q_i as defined here are appropriate to the momentum and energy interaction between a gas and a solid surface, with the restriction that the mean velocity of the gas at the wall be zero.

In setting up the conservation equations, we have interpreted an equation such as (2.19) as representing three equations for each of three components of C_i and q_i. Since a knowledge of the three components of a vector completely specifies the vector, we can now revert to our previous

[4] With the summation convention, quantities such as τ_{ij} and q_i often appear in our equations in equivalent forms that are written with other subscripts (e.g., τ_{jk} and q_j). When we make a general reference to such quantities we shall use the notation τ_{ij} and q_i.

notation introduced in Chapter II, interpreting equation (2.19) as a vector equation and the quantities q_i, C_i, and $\overline{C_iC^2}$ as vectors. We shall use this convention in the material that follows, so that for example the momentum equation (2.10) will be considered a vector equation. Thus any quantity with a single unsummed subscript (e.g., $\bar{c}_i$) is a vector, whereas a quantity with no unsummed subscripts (e.g., $\overline{C^2} = \overline{C_jC_j}$) is a scalar. There remain, however, certain quantities such as $\rho\overline{C_iC_j}$ and τ_{ij} with two unsummed subscripts. These quantities represent the nine components, obtained by taking all possible values of the subscripts i and j, of the momentum flux and viscous stress. They have a mathematical as well as a physical significance, and as such they are termed second-order tensors (vectors being first-order tensors). The nine components are often represented as arranged in a three-by-three matrix, the column denoted by the subscript i and the row by j. Our use of tensors will be largely a matter of convenience, with the hope of gaining clarity through a simple and concise notation. Application of the straightforward rules of subscript notation makes a detailed discussion of tensor products and tensor calculus unnecessary for our purposes; for a mathematical treatment of tensors and their properties the reader is referred to Jeffreys (1957) and Chapman and Cowling (1952).

Before we rewrite the conservation equations in the notation of gas dynamics, it is also useful to consider the modifications required for a gas with internal molecular structure. A review of our development will show that only those quantities involving energy are affected. Although the definition of translational kinetic temperature is unchanged, we must include, as in Chapter IV, the energy associated with the internal structure of the molecule in our expressions for the energy per unit mass e and the enthalpy h:

$$e = e_{\text{tr}} + e_{\text{int}} = \tfrac{3}{2}RT + e_{\text{int}}, \tag{2.20a}$$

$$h \equiv e + \frac{p}{\rho} = \frac{5}{2}RT + e_{\text{int}}. \tag{2.20b}$$

If the internal molecular energy is distributed over the various internal energy states according to the Boltzmann distribution, the methods of Chapter IV can be used to calculate e_{int}. In addition if ϵ_{int} denotes the internal energy of a single molecule, the heat-flux vector q_i must now be written[5]

$$q_i = \tfrac{1}{2}\rho\overline{C_iC^2} + n\overline{C_i\epsilon_{\text{int}}}. \tag{2.21}$$

[5] The second term on the right in equation (2.21) is most easily interpreted directly as a sum of the internal energy transported across a plane per unit area per unit time by the thermal translational motions of the molecules. The interpretation as a moment of the distribution function requires an extension of our use of the distribution function.

In general the internal molecular energy can also be involved in an energy exchange with an external force field or by radiation. We shall not include these effects here.

When we combine the results of this section, the mass, momentum, and energy conservation equations (2.9), (2.10), and (2.11) become

$$\frac{\partial \rho}{\partial t} + \frac{\partial}{\partial x_j}(\rho \bar{c}_j) = 0 \tag{2.22}$$

$$\frac{\partial}{\partial t}(\rho \bar{c}_i) + \frac{\partial}{\partial x_j}(\rho \bar{c}_i \bar{c}_j) = -\frac{\partial p}{\partial x_i} + \frac{\partial \tau_{ij}}{\partial x_j} + \rho \bar{F}_i \tag{2.23a}$$

$$\frac{\partial}{\partial t}(\rho e + \tfrac{1}{2}\rho \bar{c}^2) + \frac{\partial}{\partial x_j}[\rho \bar{c}_j(h + \tfrac{1}{2}\bar{c}^2)] = \frac{\partial}{\partial x_j}(\tau_{jk}\bar{c}_k - q_j) + \rho \overline{F_j c_j} \,. \tag{2.24a}$$

These equations frequently appear in other forms, obtained from the above by subtracting the continuity equation, multiplied by the mean velocity, from the momentum equation and by similar manipulations involving the energy equation (see Exercise 2.3). In particular the momentum and energy equations in the forms

$$\rho \frac{D\bar{c}_i}{Dt} = -\frac{\partial p}{\partial x_i} + \frac{\partial \tau_{ij}}{\partial x_j} + \rho \bar{F}_i \tag{2.23b}$$

$$\rho \frac{D}{Dt}(h + \tfrac{1}{2}\bar{c}^2) = \frac{\partial p}{\partial t} + \frac{\partial}{\partial x_j}(\tau_{jk}\bar{c}_k - q_j) + \rho \overline{F_j c_j} \,, \tag{2.24b}$$

where

$$\frac{D}{Dt} \equiv \frac{\partial}{\partial t} + \bar{c}_j \frac{\partial}{\partial x_j},$$

reduce to equations (VIII 2.3) and (VIII 2.4b) when τ_{ij}, q_i, and F_i are zero.

At this point the reader would do well to examine the various terms that appear in the foregoing conservation equations and review their molecular and macroscopic interpretations. Consideration should be given to such matters as to whether a given term arises from the mean or the thermal motion of the gas, and the role of external and "surface" forces.

With our molecular analysis we have obtained the general macroscopic expressions of the fundamental conservation principles that govern the motion of gases. Even for a monatomic gas in the absence of external forces, however, these five equations [the momentum equation (2.23a) or (2.23b) is counted three times] and the equations of state (2.14), (2.16), and (2.18) are not sufficient to determine the five quantities usually of greatest interest: ρ, $\bar{c}_i$, and T (or e). This results from the appearance in the conservation equations of the additional moments τ_{ij} and q_i of the distribution function. A major task of kinetic theory is to establish methods by which either the distribution function itself or a relationship between the moments τ_{ij} and q_i and the quantities ρ, $\bar{c}_i$, and T can be obtained.

Exercise 2.1. Verify the expression (2.8) for the flux of translational energy by performing in detail the expansion of $\frac{1}{2}\rho\overline{c^2 c_j}$ in terms of C_j and $\bar{c}_j$.

Exercise 2.2. Use the reasoning leading to Eucken's relation in Chapter I, Sec. 5 to explain why, in equation (2.21) for the heat flux in a gas with internal molecular energy, $n\overline{C_i \epsilon_{\text{int}}}$ need not be zero even though $\overline{C_i}$ is by definition zero. Find a simple mean-free-path expression for $n\overline{C_i \epsilon_{\text{int}}}$.

Exercise 2.3. From the conservation equations (2.22), (2.23a), and (2.24a) obtain: (a) momentum and energy equations in terms of the momentum-flux tensor (also called the pressure tensor) $p_{ij} \equiv p\delta_{ij} - \tau_{ij}$ rather than the viscous-stress tensor τ_{ij}, (b) the momentum and energy equations (2.23b) and (2.24b), and (c) an energy equation in the form of equation (VIII 2.4a), but including viscous and heat-flux terms. Assume in part (c) that the external force is independent of molecular velocity.

Exercise 2.4. For a monatomic gas in the absence of external forces, show that with the use of the equations of state there are thirteen dependent variables in the conservation equations (2.22), (2.23a), and (2.24a). Count separately each component of a vector or tensor but note that $\tau_{12} = \tau_{21}$, etc., and that the sum $\tau_{jj} = 0$.

3 THE BOLTZMANN EQUATION

Since the conservation equations derived in the previous section are incomplete by themselves, the application of kinetic theory to problems of translational nonequilibrium requires additional information concerning the moments τ_{ij} and q_i of the distribution function. In the continuum formulation leading to the Navier-Stokes equations, the relations between viscous stress and velocity gradient and heat flux and temperature gradient are taken as assumptions that can be verified by experiment. It would seem, however, that if the nature of the molecular-collision process were known the motion of the gas could be completely specified without such *a priori* assumptions. With this reasoning in mind it might be expected at first that the logical approach to kinetic theory would be to follow the dynamical trajectories of individual molecules from given initial conditions. Such a procedure is seldom if ever feasible, except in a certain sense in problems of free-molecule flow, where intermolecular collisions are unimportant in some region of the gas. Still another method is the physically motivated mean-free-path theory discussed in Chapter I. Although this method is helpful in obtaining a qualitative understanding of translational nonequilibrium, it has not proved suitable either for accurate calculations or for extension to general situations. In our discussion we shall give up any hope of following individual particles and shall concentrate instead on the distribution function and on average or expected values of the quantities of interest. In contrast to the limitations inherent in the mean-free-path theory, an accurate knowledge of such average quantities could be realized generally if we could develop methods of obtaining the distribution function itself.

With this goal in mind we shall formulate an equation that describes the rate of change, with respect to position and time, of the distribution function. This equation, known as the *Boltzmann equation*, can be derived by a procedure that is analogous to that used in obtaining the conservation equations. Since we are now concerned with the number of molecules in a given velocity range, rather than a "conserved" quantity, we expect a source term to appear in the Boltzmann equation.

If we fix our attention on those molecules (the class c_i) with a velocity differing by only a small amount from the vector c_i, the expected number of such molecules simultaneously in the volume element $dV_x \equiv dx_1\, dx_2\, dx_3$ in physical space and the volume element $dV_c \equiv dc_1\, dc_2\, dc_3$ in velocity space is $nf(c_i)\, dV_x\, dV_c$. Here we have written $f(c_i)$ rather than $f(c_i, x_i, t)$ for brevity and to emphasize our concentration upon molecules of class c_i. Any rate of increase in the number of molecules of class c_i in these (fixed) volume elements, as denoted by

$$\frac{\partial}{\partial t}\,[nf(c_i)]\, dV_x\, dV_c, \tag{3.1a}$$

must result from either a "convection" of molecules across the surfaces of dV_x and dV_c or from intermolecular collisions within dV_x (the source term).

We first consider convection across the surfaces of dV_x. The flux of molecules in the velocity range dV_c across a surface of dV_x that is perpendicular to the j-direction is $c_j nf(c_i)\, dV_c$. By the reasoning used in deriving the continuity equation (2.9), the net inward flux of such molecules across the six surfaces of dV_x is

$$-\frac{\partial}{\partial x_1}[c_1 nf(c_i)]\,dV_c\,dx_1\,dx_2\,dx_3 - \frac{\partial}{\partial x_2}[c_2 nf(c_i)]\,dV_c\,dx_2\,dx_1\,dx_3$$
$$-\frac{\partial}{\partial x_3}[c_3 nf(c_i)]\,dV_c\,dx_3\,dx_1\,dx_2$$
$$= -c_j\frac{\partial}{\partial x_j}[nf(c_i)]\,dV_c\,dV_x. \quad (3.1b)$$

On the right-hand side of this expression we have used the fact that the position and velocity coordinates are independent, so that c_j can be taken outside the differentiation with respect to x_j. It is important to note the difference between this concept and the familiar idea from gas dynamics, which remains true in the present context, that the mean velocity vector $\bar{c}_i$ is a function of position and time. Thus we have $c_i \neq c_i(x_i, t)$ but $\bar{c}_i = \bar{c}_i(x_i, t)$.

We next consider convection across the surfaces of dV_c. An external force mF_i per molecule will cause an acceleration F_i of the molecules in dV_x. This alters the number of molecules of class c_i and will be seen as a drift of the otherwise stationary molecules through the volume element dV_c in velocity space. The time rate of change of "position" in velocity space (the drift "velocity") is the acceleration F_i. Since the density in dV_c of molecules under consideration is $(n\,dV_x)f(c_i)$, the flux of molecules across a surface of dV_c that is perpendicular to c_j is then $F_j nf(c_i)\,dV_x$. Hence the net inward flux across the six surfaces of dV_c is[6]

$$-\frac{\partial}{\partial c_j}[F_j nf(c_i)]\,dV_x\,dV_c. \quad (3.1c)$$

Finally we denote by

$$\left\{\frac{\partial}{\partial t}[nf(c_i)]\right\}_{\text{coll}} dV_x\,dV_c \quad (3.1d)$$

the rate of increase of the number of molecules of class c_i resulting from collisions. If we now equate the time derivative (3.1a) to the sum of the convection and source terms (3.1b, c, d) and divide all terms by $dV_x\,dV_c$, we obtain the Boltzmann equation

$$\frac{\partial}{\partial t}[nf(c_i)] + c_j\frac{\partial}{\partial x_j}[nf(c_i)] + \frac{\partial}{\partial c_j}[F_j nf(c_i)] = \left\{\frac{\partial}{\partial t}[nf(c_i)]\right\}_{\text{coll}}. \quad (3.2a)$$

The collision term that appears on the right-hand side of the Boltzmann equation was evaluated in Chapter II for elastic-sphere molecules.

[6] The number density n could be removed from within the differentiation in this expression. This form is used here for symmetry with the other terms.

Applying that result [equation (II 4.5)] in the present situation, we have[7]

$$\left\{\frac{\partial}{\partial t}[nf(c_i)]\right\}_{\text{coll}} = \int_{-\infty}^{\infty}\int_0^{2\pi}\int_0^{\pi/2} n^2[f(c_i')f(\zeta_i')$$
$$- f(c_i)f(\zeta_i)]\, gd^2 \sin\psi \cos\psi \, d\psi \, d\varepsilon \, dV_\zeta. \quad (3.2b)$$

We have used here the variables c_i and ζ_i in place of C_i and Z_i, so that the relative speed g is given by $|\zeta_i - c_i|$. The collision integral (3.2b) represents the effect on the distribution function of two competing processes: collisions between molecules with velocities c_i and ζ_i which deplete the number of molecules of class c_i, and inverse collisions between molecules with velocities c_i' and ζ_i' which replenish the number of such molecules. It will be remembered that for a fixed direction of the line of centers at impact (see Chapter II, Figs. 5 and 7) the velocities c_i' and ζ_i' before an inverse, replenishing collision are equal to the velocities after a depleting collision (the converse being true as well). The velocities c_i' and ζ_i' can be determined from the dynamics of a binary collision as functions of c_i, ζ_i, ψ, and ε, where ψ and ε are the two angles that specify the direction of the line of centers. (This relation will be worked out in detail in Sec. 7, and the statements regarding the collision process given without proof here and in Chapter II will be verified.)

At this point the reader would do well to examine the Boltzmann equation (3.2a, b) in detail to assure that he understands the meaning of each term. For example, it should be apparent that each term including the collision integral is in general a function of c_i, x_i, and t. A useful exercise would be to determine, in thought, how it could be established whether or not a trial distribution function actually satisfies the Boltzmann equation, with particular reference to how the collision integral (3.2b) would be evaluated given an explicit form of the distribution function. The reader might also confirm that if the collision integral were written as the difference of two integrals, $f(c_i)$ could be taken outside the integrals in the term representing the frequency of depleting collisions, whereas such a simplification would not be possible with the term arising from the replenishing collisions.

We shall be interested later in representing the collision process in a more general way than by assuming that the molecules behave as rigid elastic spheres. There are several ways of doing this, depending on which geometric variables are used to specify the orientation of the molecules in a collision. For example, if we define the *impact parameter* b as the distance of closest approach of the centers of two molecules if there were

[7] The appearance of c_i and ζ_i as arguments of f in $f(c_i)f(\zeta_i)$ does not imply use of the summation convention; the distribution function itself is a scalar, although its argument can be considered a vector.

no collision (see Fig. 2), then for elastic spheres we have $b = d \sin \psi$, $db = d \cos \psi \, d\psi$, and $d^2 \sin \psi \cos \psi \, d\psi \, d\varepsilon = b \, db \, d\varepsilon$. A straightforward extension of the arguments leading to the calculation of the collision frequency in Chapter II will show that $b \, db \, d\varepsilon$ remains a valid representation of the *differential cross-section for collision* if the molecules are no longer taken as elastic spheres but are considered point centers of force.

For generality at this point, we shall rewrite the collision integral (3.2b) in terms of such a differential cross-section, which will be denoted by dP_c

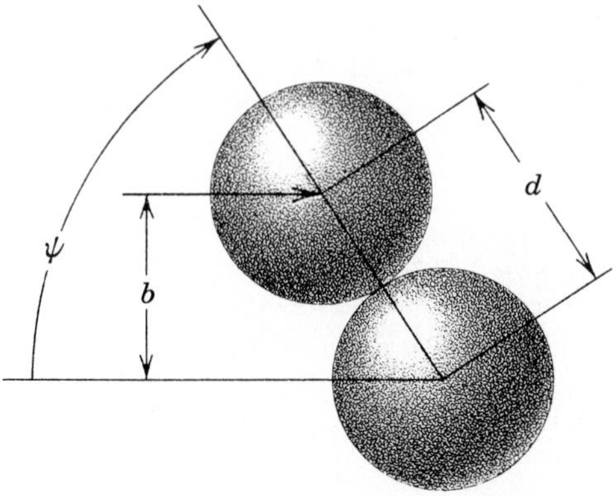

Fig. 2. The impact parameter b, elastic-sphere molecules.

and will involve two geometric variables such as ψ and ε or b and ε. The exact form of dP_c for use in our later calculations is established in Sec. 7. For the reader who prefers a definite representation, dP_c can be considered an abbreviated notation for $b \, db \, d\varepsilon$. Replacing $d^2 \sin \psi \cos \psi \, d\psi \, d\varepsilon$ by dP_c in the collision integral and dropping the argument c_i of $f(c_i)$ on the left for compactness, we then can write the Boltzmann equation (3.2a, b) in the form

$$\boxed{\begin{aligned}&\frac{\partial}{\partial t}(nf) + c_j \frac{\partial}{\partial x_j}(nf) + \frac{\partial}{\partial c_j}(F_j nf) \\ &\qquad = \int_{-\infty}^{\infty} \int_{dP_c} n^2 [f(c_i')f(\zeta_i') - f(c_i)f(\zeta_i)] g \, dP_c \, dV_\zeta\end{aligned}} \quad . \tag{3.3}$$

At this point a note of caution is probably warranted. Although the derivation of the Boltzmann equation given here satisfactorily represents the physical situation, a more general derivation based on the *Liouville equation* better illustrates the limitations of the Boltzmann equation. These limitations are often overlooked, and there may be a tendency to regard the Boltzmann equation as the last word in the microscopic description of gases. The more basic Liouville equation is a continuity equation for the *N-particle distribution function* in a six-N dimensional phase space that represents the position and momenta of a system of N particles, and is an exact consequence of the laws of mechanics. The Boltzmann equation for the one-particle distribution function $f(c_i)$ is obtained by integrating the Liouville equation over all six-N coordinates except for x_i and mc_i corresponding to a single particle (see Hirschfelder, Curtiss, and Bird, 1964, Sec. 7.1c or Grad, 1958, for such a derivation).

In obtaining the final form of the Boltzmann equation from the Liouville equation certain restrictions must be made, restrictions that have been implied in our derivation as well. In particular we have assumed a molecular representation of a gas in agreement with the order-of-magnitude estimates given in Chapter I, Sec. 6. That is, we have assumed that the range of intermolecular forces is small compared with the molecular spacing which in turn is small compared with the average distance a molecule travels between collisions. This representation implies a straight-line trajectory through relatively empty space interrupted occasionally by abrupt collisions.[8] It is a valid approximation for an electrically neutral gas when the density is not too high. Thus we were able to consider only binary (two-body) collisions in our evaluation of the collision integral, and intermolecular forces did not appear in the formulation of the conservation equations. (In view of the latter simplification, the model that we are considering corresponds to a thermally perfect gas.) Since successive collisions of a single molecule are widely separated relative to the molecular spacing, we assumed no correlation between the initial velocities c_i and ζ_i of the two molecules in a collision (molecular chaos). In addition to these conditions, we also assumed implicitly that the distribution function does not vary appreciably over a distance of the order of the range of the intermolecular force or in a time comparable to the duration of a collision. If this were not so the functions $f(c_i')$ and $f(\zeta_i')$ would have to be evaluated at a different position or time than $f(c_i)$ and $f(\zeta_i)$.

Using the Boltzmann equation derived in this section, we can now turn our attention to a rigorous derivation of the Maxwellian distribution function first obtained in Chapter II.

[8] As a result of the straight-line molecular trajectories, the Cartesian coordinate system is particularly appropriate for kinetic theory. The left-hand side of the Boltzmann equation takes on a different form in other coordinate systems (see Lees, 1959, or Lees and Liu, 1962).

4 EQUILIBRIUM AND ENTROPY

An exact solution to the Boltzmann equation can be obtained for the important special case of a gas in equilibrium. If we restrict our attention to situations without external force fields, at equilibrium all position and time derivatives must be zero. The left-hand side of the Boltzmann equation (3.2) is then zero, and the equilibrium distribution function is the solution of

$$\left[\frac{\partial}{\partial t}(nf)\right]_{\text{coll}} = \int_{-\infty}^{\infty}\int_{dP_c} n^2[f(c_i')f(\zeta_i') - f(c_i)f(\zeta_i)]g\,dP_c\,dV_\zeta = 0. \tag{4.1}$$

This equation represents the balance at equilibrium between the total number of depleting collisions, which reduce the number of molecules with the velocity c_i, and the total number of replenishing collisions, which add to the number of such molecules. The solution can be obtained without further consideration of the details of intermolecular collisions. As we saw in Chapter II, Sec. 4, a *sufficient* condition for equilibrium is for the integrand in equation (4.1) to be always zero. Then we have

$$f(c_i')f(\zeta_i') = f(c_i)f(\zeta_i), \tag{4.2}$$

and *each* depleting collision is exactly balanced by its inverse, replenishing collision (detailed balancing).

Referring to the analysis of Chapter II, Sec. 5, we find that the most general solution of equation (4.2), consistent with the definition (2.13) of temperature and the normalization of the distribution function [equation (II 2.7)] is the Maxwellian distribution function f_0 given by

$$\boxed{f_0 = \left(\frac{m}{2\pi kT}\right)^{3/2} e^{-mC^2/2kT} = \left(\frac{m}{2\pi kT}\right)^{3/2} \exp\left[-\frac{m}{2kT}(|c_i - \bar{c}_i|)^2\right]} \tag{4.3}$$

The first form of f_0 in equation (4.3) was given by equation (II 5.8). In Chapter II we were concerned only with the random molecular motion. As we have previously noted, in such a situation wherein the mean velocity is zero the distinction between C_i and c_i vanishes. Here we have written the second form for f_0 to indicate explicitly for later use that this form, which is obtained from the first with the definition $C_i \equiv c_i - \bar{c}_i$, is an equivalent expression of the solution of equations (4.1) and (4.2). This form corresponds to the general solution (II 5.2). For a gas in equilibrium, $\bar{c}_i$ in equation (4.3), if not zero, must be regarded as resulting from a uniform translation of the gas as a whole relative to the observer, that is, from a Galilean transformation of the coordinate system. Our present purpose is to show that equation (4.2) is also a *necessary* condition for equilibrium, and hence that at equilibrium the Maxwellian distribution (4.3) is the most general solution of the Boltzmann equation.

Now, if the collision integral (4.1) is zero, it is necessary that all of its *moments* must also be zero. In particular we shall examine the moment involving the function $1 + \ln(Anf)$, where A is a constant as yet unspecified; the physical motivation for this choice will become apparent later in this section. If we multiply equation (4.1) by $1 + \ln(Anf)$ and integrate over all velocities c_i, then by the foregoing reasoning we have as a necessary condition for equilibrium

$$\int_{-\infty}^{\infty} [1 + \ln(Anf)] \left[\frac{\partial}{\partial t}(nf)\right]_{\text{coll}} dV_c$$
$$= \int_{-\infty}^{\infty}\int_{-\infty}^{\infty}\int_{dP_c} [1 + \ln(Anf)]$$
$$\times\, n^2[f(c_i')f(\zeta_i') - f(c_i)f(\zeta_i)]g\, dP_c\, dV_\zeta\, dV_c = 0. \quad (4.4)$$

Thus to prove that equation (4.2) is a necessary condition for equilibrium we need only show that (4.2) is necessary in order that the single moment (4.4) be zero.

At this point it is convenient for later use to proceed indirectly by discussing several general transformations involving moments of the collision integral. Let $Q(c_i)$ be any function of the molecular velocity c_i. Then it can be shown that

$$\int_{-\infty}^{\infty}\int_{-\infty}^{\infty}\int_{dP_c} Q(c_i)n^2[f(c_i')f(\zeta_i') - f(c_i)f(\zeta_i)]g\, dP_c\, dV_\zeta\, dV_c \quad (4.5\text{a})$$
$$= -\tfrac{1}{4}\int_{-\infty}^{\infty}\int_{-\infty}^{\infty}\int_{dP_c} (\delta_c Q + \delta_\zeta Q)$$
$$\times\, n^2[f(c_i')f(\zeta_i') - f(c_i)f(\zeta_i)]g\, dP_c\, dV_\zeta\, dV_c \quad (4.5\text{b})$$
$$= \tfrac{1}{2}\int_{-\infty}^{\infty}\int_{-\infty}^{\infty}\int_{dP_c} (\delta_c Q + \delta_\zeta Q)n^2 f(c_i)f(\zeta_i)g\, dP_c\, dV_\zeta\, dV_c \quad (4.5\text{c})$$
$$= \int_{-\infty}^{\infty}\int_{-\infty}^{\infty}\int_{dP_c} \delta_c Q\, n^2 f(c_i)f(\zeta_i)g\, dP_c\, dV_\zeta\, dV_c, \quad (4.5\text{d})$$

where

$$\delta_c Q \equiv Q(c_i') - Q(c_i)$$

and

$$\delta_\zeta Q \equiv Q(\zeta_i') - Q(\zeta_i)$$

are the changes in Q evaluated for each of the two molecules in a collision.

The equality of the various integrals (4.5) follows from a careful consideration of their physical meaning. We first note that a simple change in notation for the velocities of the two molecules in a collision by the interchange of c_i and ζ_i (and hence of c_i' and ζ_i'), cannot affect the value of the integral (4.5a). Hence we have

$$\int_{-\infty}^{\infty}\int_{-\infty}^{\infty}\int_{dP_c} Q(c_i)n^2[f(c_i')f(\zeta_i') - f(c_i)f(\zeta_i)]g\, dP_c\, dV_\zeta\, dV_c$$
$$= \int_{-\infty}^{\infty}\int_{-\infty}^{\infty}\int_{dP_c} Q(\zeta_i)n^2[f(\zeta_i')f(c_i') - f(\zeta_i)f(c_i)]g\, dP_c\, dV_c\, dV_\zeta$$
$$= \tfrac{1}{2}\int_{-\infty}^{\infty}\int_{-\infty}^{\infty}\int_{dP_c} [Q(c_i) + Q(\zeta_i)] \times n^2[f(c_i')f(\zeta_i') - f(c_i)f(\zeta_i)]g\, dP_c\, dV_\zeta\, dV_c. \quad (4.6)$$

Here the last expression follows from the fact that in general two quantities that are equal to each other are each equal to one-half their sum. Similarly if $Q(c_i, c_i')$ is a function of both c_i and c_i', we obtain

$$\int_{-\infty}^{\infty}\int_{-\infty}^{\infty}\int_{dP_c} Q(c_i, c_i')n^2[f(c_i')f(\zeta_i') - f(c_i)f(\zeta_i)]g\, dP_c\, dV_\zeta\, dV_c$$
$$= \int_{-\infty}^{\infty}\int_{-\infty}^{\infty}\int_{dP_c} Q(\zeta_i, \zeta_i')n^2[f(\zeta_i')f(c_i') - f(\zeta_i)f(c_i)]g\, dP_c\, dV_c\, dV_\zeta$$
$$= \tfrac{1}{2}\int_{-\infty}^{\infty}\int_{-\infty}^{\infty}\int_{dP_c} [Q(c_i, c_i') + Q(\zeta_i, \zeta_i')] \times n^2[f(c_i')f(\zeta_i') - f(c_i)f(\zeta_i)]g\, dP_c\, dV_\zeta\, dV_c. \quad (4.7)$$

The next point is somewhat more difficult. The integral

$$\int_{-\infty}^{\infty}\int_{-\infty}^{\infty}\int_{dP_c} n^2 f(c_i)f(\zeta_i)g\, dP_c\, dV_\zeta\, dV_c,$$

which was used in Chapter II, Sec. 6, represents the total number of collisions per unit volume per unit time. This integral is unchanged by a change from unprimed to primed velocities in our method of counting collisions, so that

$$\int_{-\infty}^{\infty}\int_{-\infty}^{\infty}\int_{dP_c} n^2 f(c_i)f(\zeta_i)g\, dP_c\, dV_\zeta\, dV_c = \int_{-\infty}^{\infty}\int_{-\infty}^{\infty}\int_{dP_c} n^2 f(c_i')f(\zeta_i')g'\, dP_c\, dV_{\zeta'}\, dV_{c'}.$$

That is to say, since every collision is the inverse of some other collision, the total number of collisions can be calculated either by summing over all original collisions or by summing over all inverse collisions. (This does not imply that the collision integral (4.1) is always zero, since the latter is not summed over all velocities c_i.) Similarly it follows that

$$\int_{-\infty}^{\infty}\int_{-\infty}^{\infty}\int_{dP_c} Q(c_i)n^2 f(c_i)f(\zeta_i)g\,dP_c\,dV_\zeta\,dV_c$$
$$= \int_{-\infty}^{\infty}\int_{-\infty}^{\infty}\int_{dP_c} Q(c_i')n^2 f(c_i')f(\zeta_i')g'\,dP_c\,dV_{\zeta'}\,dV_{c'}. \quad (4.8)$$

Remembering that the velocities after an inverse collision are equal to the velocities before the original collision (which means that the inverse of an inverse collision is the original collision), we can also write

$$\int_{-\infty}^{\infty}\int_{-\infty}^{\infty}\int_{dP_c} Q(c_i')n^2 f(c_i)f(\zeta_i)g\,dP_c\,dV_\zeta\,dV_c$$
$$= \int_{-\infty}^{\infty}\int_{-\infty}^{\infty}\int_{dP_c} Q(c_i)n^2 f(c_i')f(\zeta_i')g'\,dP_c\,dV_{\zeta'}\,dV_{c'}. \quad (4.9)$$

This equation can be interpreted physically as the equality of two equivalent expressions for the sum over all collisions of the value after collision of the quantity Q. In equation (4.9) c_i on the right-hand side is to be regarded as a function of c_i', ζ_i', and the two geometric variables appearing in dP_c, whereas the converse is true on the left. Again each of the equal integrals in equations (4.8) and (4.9) is equal to one-half their sum. Since, as we shall verify in Sec. 7, $g = g'$ and $dV_\zeta\,dV_c = dV_{\zeta'}\,dV_{c'}$, by direct combination using equations (4.8) and (4.9) we then have

$$\int_{-\infty}^{\infty}\int_{-\infty}^{\infty}\int_{dP_c} Q(c_i)n^2[f(c_i')f(\zeta_i') - f(c_i)f(\zeta_i)]g\,dP_c\,dV_\zeta\,dV_c$$
$$= \tfrac{1}{2}\int_{-\infty}^{\infty}\int_{-\infty}^{\infty}\int_{dP_c} [Q(c_i) - Q(c_i')]$$
$$\times\, n^2[f(c_i')f(\zeta_i') - f(c_i)f(\zeta_i)]g\,dP_c\,dV_\zeta\,dV_c. \quad (4.10)$$

Now by considering $\frac{1}{2}[Q(c_i) - Q(c_i')] = -\frac{1}{2}\,\delta_c Q$ in equation (4.10) as a special case of $Q(c_i, c_i')$ in equation (4.7), we obtain the first of the desired transformations (4.5):

$$\int_{-\infty}^{\infty}\int_{-\infty}^{\infty}\int_{dP_c} Q(c_i)n^2[f(c_i')f(\zeta_i') - f(c_i)f(\zeta_i)]g\,dP_c\,dV_\zeta\,dV_c$$
$$= -\tfrac{1}{4}\int_{-\infty}^{\infty}\int_{-\infty}^{\infty}\int_{dP_c} (\delta_c Q + \delta_\zeta Q)$$
$$\times\, n^2[f(c_i')f(\zeta_i') - f(c_i)f(\zeta_i)]g\,dP_c\,dV_\zeta\,dV_c. \quad (4.11)$$

The remaining two equalities in (4.5) can be established by a straightforward extension of these arguments and are considered in Exercise 4.3.

Returning now to our consideration of the equilibrium distribution function, we can apply the transformation (4.11) to the moment (4.4) of the collision integral, remembering that a necessary condition for equilibrium is for the integral (4.4) to be zero. With $Q(c_i) = 1 + \ln (Anf)$ in (4.11), equation (4.4) becomes

$$-\tfrac{1}{4}\int_{-\infty}^{\infty}\int_{-\infty}^{\infty}\int_{dP_c}\{\ln [Anf(c_i')] - \ln [Anf(c_i)] + \ln [Anf(\zeta_i')] - \ln [Anf(\zeta_i)]\} \times n^2[f(c_i')f(\zeta_i') - f(c_i)f(\zeta_i)]g\, dP_c\, dV_\zeta\, dV_c = 0 \quad (4.12a)$$

or

$$-\tfrac{1}{4}\int_{-\infty}^{\infty}\int_{-\infty}^{\infty}\int_{dP_c}\ln\left[\frac{f(c_i')f(\zeta_i')}{f(c_i)f(\zeta_i)}\right] \times n^2[f(c_i')f(\zeta_i') - f(c_i)f(\zeta_i)]g\, dP_c\, dV_\zeta\, dV_c = 0. \quad (4.12b)$$

Examination of this expression shows that the logarithmic factor is positive or negative depending upon whether $f(c_i')f(\zeta_i')$ is greater or less than $f(c_i)f(\zeta_i)$. The two factors involving f in the integrand must therefore be of the same algebraic sign; hence the integrand itself must be either always positive or zero. Since the integrand (and thus the integral) is zero only when $f(c_i')f(\zeta_i') = f(c_i)f(\zeta_i)$, we are led to the conclusion that $f(c_i')f(\zeta_i') = f(c_i)f(\zeta_i)$ is a necessary as well as a sufficient condition for the equilibrium solution of the Boltzmann equation (4.1). We have thus proved that the equilibrium distribution function is necessarily Maxwellian.

In our discussion of equilibrium we have assumed a one-to-one correspondence between equilibrium and a state of the gas that is uniform in space and time. It is also interesting to consider the approach to equilibrium. In particular, we shall examine the idealized situation of the approach to equilibrium with time, starting from an initially nonequilibrium state, of a gas that is isolated from its environment and spatially uniform. This situation of a spatially uniform nonequilibrium gas, which we adopt for simplicity, is from a strict point of view physically unrealistic. For an extension of the arguments to a spatially nonuniform gas, see Chapman and Cowling (1952), Sec. 4.13.

From thermodynamics we know that an important property of the gas in the description of a nonequilibrium process is the entropy; we further expect that the entropy of the gas will increase with time, reaching a maximum value at equilibrium. If we wish then to investigate this entropy change on a microscopic basis, we can extend the methods of statistical mechanics discussed in Chapter IV. It is quite important, however, to distinguish between the approach of that chapter and our present requirements. In Chapter IV we were interested in a microscopic evaluation of *equilibrium* thermodynamic properties, and we considered these properties defined within the context of macroscopic thermodynamics. Here we are concerned with the *approach* to equilibrium and so require an interpretation of entropy that is appropriate to nonequilibrium circumstances. As with temperature and pressure, we shall now adopt a definition of entropy that is valid whether or not the gas is in equilibrium, and that agrees with the thermodynamic definition in their range of mutual applicability. A thorough discussion of such distinctions is given by Grad (1961) in his article, "The Many Faces of Entropy."

In equations (IV 7.6) and (IV 4.6), we found that the entropy of a system of N particles can be written

$$S = k \ln \Omega = k \ln \left[\sum_{N_j} W(N_j) \right], \tag{4.13}$$

where $W(N_j)$ is the number of possible arrangements or microstates associated with the macrostate described by the distribution numbers N_j. The number Ω of microstates available to the system is then the sum of $W(N_j)$ over all possible distributions N_j consistent with a fixed number of particles N and energy E. For purely mathematical convenience in our evaluation of S, we replaced $\sum_{N_j} W(N_j)$ by $W_{\max} = W(N_j^*)$, the number of microstates associated with the most probable distribution N_j^* [equation (IV 5.1)].

In our present nonequilibrium context we shall write in place of (4.13)

$$S = k \ln W(N_j), \tag{4.14}$$

where N_j pertains to the expected distribution at a given time. It should be noted that equation (4.14) is not an approximation to (4.13) but a reinterpretation of entropy appropriate to our new purposes. As the gas approaches equilibrium, $N_j \to N_j^*$ and therefore the entropy that we shall calculate will approach the value obtained in Chapter IV.

To perform this calculation of the entropy we note that the gas molecules are distributed widely over the permissible energy states so that we can apply the Boltzmann limit in evaluating $\ln W(N_j)$. (If the situation were otherwise our classical treatment of kinetic theory would be invalid.) Then, using equation (4.14) and equation (IV 6.2), we have

$$S = k \ln W(N_j) = k \sum_j N_j \left(\ln \frac{C_j}{N_j} + 1 \right), \tag{4.15}$$

where C_j is the number of quantum states associated with the jth subdivision of such states.[9] If we associate each subdivision with a volume

[9] Here the notation is that of Chapter IV and summations are indicated explicity rather than by repeated subscripts.

element dV_c in velocity space, N_j and C_j are given by

$$N_j = Nf(c_i)\,dV_c,$$

and

$$C_j = \left(\frac{m^3V}{h^3}\right) dV_c,$$

where the expression for C_j follows from equation (IV 9.10) and the transformation

$$\epsilon^{1/2}\,d\epsilon = \frac{1}{2\pi}\left(\frac{m}{2}\right)^{3/2}(4\pi c^2\,dc) = \frac{1}{2\pi}\left(\frac{m}{2}\right)^{3/2} dV_c.$$

Our expression for the entropy then becomes

$$\begin{aligned} S &= k\int_{-\infty}^{\infty} Nf(c_i)\left[\ln\left(\frac{m^3V}{h^3Nf(c_i)}\right) + 1\right] dV_c \\ &= kN - kV\int_{-\infty}^{\infty} \ln\left(\frac{h^3}{m^3}nf\right) nf\,dV_c. \end{aligned} \tag{4.16}$$

(The reader may wish to compare our discussion with a brief and largely self-contained development of this result, in somewhat different notation, given by Landau and Lifshitz, 1958, §40.)

We can now apply equation (4.16) to our examination of the change in entropy with time in a spatially uniform gas in the absence of external forces. Differentiating (4.16) with respect to time, we obtain

$$\frac{dS}{dt} = -kV\int_{-\infty}^{\infty}\left[1 + \ln\left(\frac{h^3}{m^3}nf\right)\right]\frac{d}{dt}(nf)\,dV_c.$$

The Boltzmann equation (3.2a) under these conditions is simply

$$\frac{\partial}{\partial t}(nf) = \frac{d}{dt}(nf) = \left[\frac{\partial}{\partial t}(nf)\right]_{\text{coll}}.$$

We can then write

$$\frac{dS}{dt} = -kV\int_{-\infty}^{\infty}\left[1 + \ln\left(\frac{h^3}{m^3}nf\right)\right]\left[\frac{\partial(nf)}{\partial t}\right]_{\text{coll}} dV_c. \tag{4.17}$$

The right-hand side of this expression for the entropy change can be evaluated by means of the analysis of the first part of this section. Comparing the integral in equation (4.17) with the moment of the collision integral for $Q = 1 + \ln(Anf)$ in equation (4.4), we see that they are identical if we let $A = (h/m)^3$. Thus by applying the transformation (4.11), as we did in obtaining equations (4.12), we find that the rate of change of entropy with respect to time is given by

$$\frac{dS}{dt} = \frac{kV}{4}\int_{-\infty}^{\infty}\int_{-\infty}^{\infty}\int_{dP_c} \ln\left[\frac{f(c_i')f(\zeta_i')}{f(c_i)f(\zeta_i)}\right] \times n^2[f(c_i')f(\zeta_i') - f(c_i)f(\zeta_i)]g\, dP_c\, dV_\zeta\, dV_c. \quad (4.18)$$

As before, the integrand in this expression is always positive unless $f(c_i')f(\zeta_i') = f(c_i)f(\zeta_i)$, in which case the integrand is zero. Hence we conclude that

$$\frac{dS}{dt} \geq 0, \quad (4.19a)$$

and

$$\frac{dS}{dt} = 0 \quad \text{if and only if} \quad f(c_i')f(\zeta_i') = f(c_i)f(\zeta_i). \quad (4.19b)$$

Since we have shown that the condition $f(c_i')f(\zeta_i') = f(c_i)f(\zeta_i)$ leads to $f = f_0$, the entropy of an isolated system must continually increase until at equilibrium it reaches a maximum and the distribution function is Maxwellian. These arguments were first made by Boltzmann and are frequently called the *Boltzmann H-theorem.* The symbol H is then used for the quantity $\int_{-\infty}^{\infty} (\ln nf)\, dV_c$, although Boltzmann himself used E in his original paper.

If we evaluate the entropy at equilibrium when $f = f_0$, we find by direct substitution of equation (4.3) for f_0 into equation (4.16) for S that S is given by

$$S = kN\left\{\ln\left[\frac{1}{n}\left(\frac{2\pi mkT}{h^2}\right)^{3/2}\right] + \frac{5}{2}\right\}, \quad (4.20)$$

in complete agreement, as we would expect, with the Sackur-Tetrode equation for the translational entropy of a perfect gas [equation (IV 9.5a)].

In the foregoing discussion we have been able to find a relationship and agreement between the disciplines of kinetic theory, statistical mechanics, and thermodynamics. There remains, however, one point that requires further clarification. Nowhere, it would seem, within our formulation of kinetic theory have we introduced on the molecular level an irreversible mechanism. We would expect, for example, that our analysis would be appropriate to gas composed literally of perfectly elastic spheres. Yet the results indicate an irreversible approach to equilibrium [equations (4.19)]. The question then arises as to how microscopically reversible phenomena can lead to macroscopic irreversibility. This apparent paradox has been the subject of much discussion in the period since the publication of Boltzmann's H-theorem. For a time the fundamental foundations of kinetic theory and statistical mechanics were seriously questioned. Even today the arguments used to justify and interpret irreversibility are involved and closely reasoned, and there still exist differences of opinion, at least with regard to emphasis.

For clarity we shall illustrate our discussion with an example of one aspect of the paradox. If we consider the familiar insulated container—divided initially by a partition into two parts, one at a perfect vacuum and the other containing a gas—then removal of the partition will result after a time in a new equilibrium state with a redistribution of the gas throughout the container and an increase in entropy. Now if we were able to stop the motion at some time after removing the partition and reverse the velocities of the gas molecules, the motion would be reversed. The gas would then move back toward its initial configuration, with a consequent decrease in entropy. (We assume that molecular collisions with the walls of the container are exactly reversed and for clarity consider only times prior to reattainment of the initial configuration.) Thus certain molecular motions are possible for which the entropy of an isolated system decreases rather than increases, in apparent contradiction to our previous conclusions. It should be noted here that we have momentarily taken the entropy as a definite function of the actual molecular velocities rather than as an expectation or a thermodynamic property. We shall have more to say about this distinction shortly.

There are three points that aid substantially in the clarification of the situation. First, the principle of irreversible increase in entropy is more than amply verified by experiment. In our present example there can for practical purposes be no question as to the outcome of removing the partition. Second, our analysis leading to the H-theorem [equations (4.19)], although simplified physically, is mathematically exact and follows directly from the Boltzmann equation, so that no contradictions arise from purely mathematical approximation. Thus any criticism of the H-theorem implies a criticism of the Boltzmann equation itself. Third, and this is the key point, the microscopic derivation of macroscopic irreversibility is a theorem of *statistical mechanics* rather than of *particle dynamics*. We have renounced from the beginning any hope of following the exact individual molecular trajectories from precise initial conditions; the Boltzmann equation and the H-theorem must be regarded as giving a description of the *expected* motion of the gas. Thus, as we go more deeply into the subject, it is appropriate to consider the kinetic theory of nonuniform gases as one aspect of the broader field of nonequilibrium statistical mechanics.

The implications of the third point require further discussion. Although "reverse" molecular motions as in our example are possible, it can be shown that the probability of their occurrence is so small that they may be neglected, except when we are explicitly considering fluctuation phenomena such as Brownian motion. (Such effects can themselves also be successfully understood by means of statistical mechanics.) The principle of increase of entropy is then a statement of the overwhelming probability of irreversible behavior for systems actually occurring in nature. Since the H-theorem is a consequence of statistical mechanics rather than particle dynamics, there is no internal contradiction within the theory.

It is then pertinent to ask in what respects are our statistical methods justified. The collision integral in the Boltzmann equation accounts for collisions by an implied averaging and incorporates the assumption of molecular chaos. As emphasized in the lucid exposition by Born (1949), the resulting irreversibility is a consequence of just such an explicit introduction of uncertainty into the formulation. In our example we were confident that the gas in the first instance would expand once the partition was removed, with only the broadest possible knowledge of the state of the gas. To obtain the "reverse" motion, on the other hand, we had to specify exactly the location and velocity of each molecule. Born shows that it is sufficient to relax our specification for a single molecule to assure irreversibility. Such an uncertainty is of course inherent in the modern description of microscopic motions by means of wave mechanics. It should be evident that this does not mean that fluctuations from the most probable state or motion are not allowed, but to the contrary that the statistical formulation of the principle of increase of entropy is consistent with the statistical nature of wave mechanics.

There remains the question as to whether irreversibility can be satisfactorily understood purely on the basis of classical mechanics. Although the reasons are far too involved to go into here, the answer is yes, provided that we are concerned with a system containing a large number of molecules and adopt the reasonable assumption of equal *a priori* probabilities, in effect neglecting highly improbable "reverse" motions. For a thorough discussion of the classical foundations of irreversibility the reader is referred to Ehrenfest and Ehrenfest (1959), Tolman (1938), and Grad (1961).

Exercise 4.1. Show by direct substitution of equation (4.3) in equation (4.1) that the Maxwellian distribution function f_0, in the second form given by equation (4.3), is a solution of equation (4.1). To obtain the velocities after collision you will have to apply to the two molecules involved in a collision the conditions of conservation of mass, momentum, and energy.

Exercise 4.2. Show that the Maxwellian distribution function is also a solution of the Boltzmann equation for a steady-state isothermal gas with zero mean velocity in the presence of a gravity field. Take the gravity force g per unit mass as acting in the direction of the x_3-axis and show that the Boltzmann equation is satisfied if $dp/dx_3 = \rho g$. (For a general discussion of equilibrium in the presence of external forces see Chapman and Cowling (1952), Sec. 4.14.)

Exercise 4.3. Derive the transformations (4.5c) and (4.5d) by extension of the methods of this section.

Exercise 4.4. Verify that equation (4.20) follows from equations (4.16) and (4.3) and show that this expression (4.20) for the entropy is in agreement with equation (IV 9.5a).

5 THE EQUATIONS OF EQUILIBRIUM FLOW

In the preceding section the Maxwellian distribution function was found to apply to a gas in equilibrium, when all gradients are zero. Aside from the possibility of a translation of the coordinate system, this would require the gas to be at rest. All technically interesting flows, however, involve changes in velocity and gas properties from point to point; such changes provide, in fact, the necessary driving forces for the flow. It seems reasonable to expect, nevertheless, that there will exist certain classes of flows or certain regions in a given flow in which the gradients are sufficiently small that the gas can be considered as if it were in *local equilibrium*. Thus, if the relaxation time for translational nonequilibrium is negligible compared with the characteristic flow time, the Maxwellian distribution function would be expected to apply locally. The situation here is similar to that discussed in Chapter VIII, Sec. 3 with regard to reacting flows in local chemical equilibrium.

An assessment of the validity of the assumption of equilibrium flow can be made with confidence only by considering the limit of the corresponding nonequilibrium situation. In the present case of translational nonequilibrium this will be postponed until Chapter X. Here we shall examine the consequences of such an assumption and compare the resulting form of the conservation equations with the equations of macroscopic gas dynamics.

If the distribution function is locally Maxwellian, explicit expressions can be obtained for the heat-flux vector q_i and the viscous-stress tensor τ_{ij}. In performing this calculation, we must now consider the Maxwellian distribution [equation (4.3)]

$$f_0 = \left(\frac{m}{2\pi kT}\right)^{3/2} e^{-mC^2/2kT} = \left(\frac{m}{2\pi kT}\right)^{3/2} \exp\left[-\frac{m}{2kT}(|c_i - \bar{c}_i|)^2\right]$$

in terms of the local temperature T and, through $C_i = c_i - \bar{c}_i$, the local mean velocity $\bar{c}_i$. Then, for example, when $f = f_0$ we have by definition

$$\begin{aligned}
\tau_{12} &= -\rho\overline{C_1C_2} \\
&= -\rho\left(\frac{m}{2\pi kT}\right)^{3/2}\int_{-\infty}^{\infty}\int_{-\infty}^{\infty}\int_{-\infty}^{\infty} C_1C_2e^{-m(C_1^2+C_2^2+C_3^2)/2kT}\,dc_1\,dc_2\,dc_3 \\
&= -\rho\left(\frac{m}{2\pi kT}\right)^{3/2}\int_{-\infty}^{\infty} C_1e^{-mC_1^2/2kT}\,dC_1 \\
&\qquad\times\int_{-\infty}^{\infty} C_2e^{-mC_2^2/2kT}\,dC_2\int_{-\infty}^{\infty} e^{-mC_3^2/2kT}\,dC_3.
\end{aligned}$$

Here we have used the fact that $dc_1 = dC_1 + d\bar{c}_1 = dC_1$, etc., in an integration over velocity space. Inspection shows that, since C_1 itself is an odd function of C_1 whereas the exponential is an even function, the integration with respect to C_1 must vanish. The same is true for the integration with respect to C_2. Therefore we obtain $\tau_{12} = 0$. By similar evaluations in Exercise 5.1 we find that in general

$$\left.\begin{aligned}\tau_{ij} &= 0\\ q_i &= 0\end{aligned}\right\}\quad \text{when}\quad f = f_0 . \tag{5.1}$$

Applying these results to the conservation equations (2.22), (2.23b), and (2.24b), we obtain the equations of *equilibrium flow* as follows:

$$\frac{\partial\rho}{\partial t} + \frac{\partial}{\partial x_j}(\rho\bar{c}_j) = 0 \tag{5.2a}$$

$$\rho\frac{D\bar{c}_i}{Dt} = -\frac{\partial p}{\partial x_i} + \rho\bar{F}_i \tag{5.2b}$$

$$\rho\frac{D}{Dt}(h + \tfrac{1}{2}\bar{c}^2) = \frac{\partial p}{\partial t} + \rho\overline{F_jc_j} . \tag{5.2c}$$

These equations are seen to be identical (except for the external-force terms) to the equations of macroscopic gas dynamics as used in Chapter VIII, Sec. 2. These equations apply to situations where the gradients in the flow field are small. We can thus interpret such situations as involving negligible departures from the Maxwellian distribution function. By the same token, however, it is apparent that an understanding of flows in translational nonequilibrium, as for example boundary layers and the interior of shock waves, will involve a study of the nonequilibrium distribution function. Before undertaking such a study, we must develop additional tools of analysis.

Exercise 5.1. Calculate τ_{ij} and q_i when $f = f_0$ [see equations (5.1)].

Exercise 5.2. For $f = f_0$, calculate by direct integration (rather than by the manipulation of mean quantities such as $\frac{1}{2}\rho\overline{c^2 c_j}$) the flux of translational energy across a plane that is perpendicular to the j-direction. Express your result in terms of the enthalpy.

6 MOMENTS OF THE BOLTZMANN EQUATION

We have previously discussed moments of the distribution function and moments of the collision integral. Similarly, a moment of the Boltzmann equation is obtained by multiplying the Boltzmann equation (3.3) by any function of molecular velocity and integrating over velocity space. Such equations, also called *moment equations* or *equations of transfer*, play an important part in advanced kinetic theory. In application they can represent a less ambitious formulation than the Boltzmann equation itself. We shall see in Chapter X that in certain approximate methods solutions are obtained to a limited number of moment equations rather than to the Boltzmann equation.

If $Q(c_i)$ is some function of c_i but not of position and time, the equation of transfer of $Q(c_i)$ is

$$\int_{-\infty}^{\infty} Q(c_i)\left[\frac{\partial}{\partial t}(nf) + c_j\frac{\partial}{\partial x_j}(nf) + \frac{\partial}{\partial c_j}(F_j nf)\right] dV_c = \int_{-\infty}^{\infty} Q(c_i)\left[\frac{\partial}{\partial t}(nf)\right]_{\text{coll}} dV_c. \quad (6.1)$$

This equation can be simplified by interchanging the order of integration and differentiation on the left-hand side. If we also define

$$\Delta[Q] \equiv \int_{-\infty}^{\infty} Q(c_i)\left[\frac{\partial}{\partial t}(nf)\right]_{\text{coll}} dV_c,$$

and as before use barred quantities to denote averages over the distribution function, we obtain

$$\boxed{\frac{\partial}{\partial t}(n\bar{Q}) + \frac{\partial}{\partial x_j}(n\overline{c_j Q}) - nF_j\overline{\frac{\partial Q}{\partial c_j}} = \Delta[Q]}\ . \quad (6.2)$$

Here use has been made of the fact that c_i, x_i, and t are all independent variables, so that, for example,

$$Qc_j \frac{\partial}{\partial x_j}(nf) = \frac{\partial}{\partial x_j}(Qc_j nf)$$

and

$$\int_{-\infty}^{\infty} Qc_j \frac{\partial}{\partial x_j}(nf)\, dV_c = \frac{\partial}{\partial x_j}\left[\int_{-\infty}^{\infty} Qc_j nf\, dV_c\right] = \frac{\partial}{\partial x_j}(n\overline{c_j Q}).$$

The third term on the left in equation (6.1) has been simplified by integration by parts with the assumption that $QF_j nf$ approaches zero as the velocity approaches infinity in any direction. Thus we have

$$\int_{-\infty}^{\infty} Q \frac{\partial}{\partial c_1}(F_1 nf)\, dV_c$$

$$= \int_{-\infty}^{\infty}\int_{-\infty}^{\infty}\left[F_1 Qnf\right]_{c_1=-\infty}^{c_1=\infty} dc_2\, dc_3 - \int_{-\infty}^{\infty} F_1\left(\frac{\partial Q}{\partial c_1}\right) nf\, dV_c = -nF_1 \overline{\frac{\partial Q}{\partial c_1}}.$$

The transformations (4.5) of moments of the collision integral give us several alternate expressions for $\Delta[Q]$. The two of greatest interest are

$$\Delta[Q] = \int_{-\infty}^{\infty}\int_{-\infty}^{\infty}\int_{dP_c} \delta_c Q\, n^2 f(c_i) f(\zeta_i) g\, dP_c\, dV_\zeta\, dV_c \tag{6.3a}$$

$$\Delta[Q] = \tfrac{1}{2}\int_{-\infty}^{\infty}\int_{-\infty}^{\infty}\int_{dP_c} (\delta_c Q + \delta_\zeta Q) n^2 f(c_i) f(\zeta_i) g\, dP_c\, dV_\zeta\, dV_c \quad , \tag{6.3b}$$

where as before $\delta_c Q \equiv Q(c_i') - Q(c_i)$ and $\delta_\zeta Q \equiv Q(\zeta_i') - Q(\zeta_i)$.

The equation of transfer in the form (6.2) with $\Delta[Q]$ given by equation (6.3a) or (6.3b) can be considered a continuity equation for $Q(c_i)$, the right-hand side $\Delta[Q]$ being a source term. If equation (6.3b) is used for $\Delta[Q]$, the change in Q is counted for both molecules in a collision. By integration over dV_ζ and dV_c the change in Q in each collision is then counted twice (see Chapter II, Sec. 6); hence the factor $\frac{1}{2}$. With this physical interpretation in mind, the equation of transfer can be derived directly, as were the conservation equations, without recourse to the Boltzmann equation. In fact Maxwell did just this in his original derivation in 1866, prior to the formulation in 1872 of the Boltzmann equation.

In general, the equation of transfer cannot be further simplified without a knowledge of both the distribution function and the details of molecular collisions. If, however, $Q(c_i)$ is taken to be the mass, momentum, or energy per molecule (m, mc_i, or $mc^2/2$), the change in Q for both molecules must be zero in each collision. For this reason m, mc_i, and $mc^2/2$ are called the *collisional invariants.* When Q is one of these invariants we have $\delta_c Q + \delta_\zeta Q = 0$ and, by equation (6.3b), $\Delta[Q] = 0$. The equations of transfer (6.2) for these quantities then reduce to the five conservation equations (2.22), (2.23), and (2.24) (see Exercise 6.1). In verifying this result the reader should note that since we have considered only translational energy in the equation of transfer, the energy, enthalpy, and heat-flux vector in the energy equation (2.24) must be written in a form appropriate to monatomic gases.

Certain additional forms of $\Delta[Q]$ and related integrals are of particular interest to us and will be considered in Sec. 9. To perform the calculations of Sec. 9, however, we must first discuss the details of molecular collisions; such a discussion is the subject of the next section.

Exercise 6.1. Verify that the equation of transfer (6.2) reduces to the conservation equations (2.22), (2.23a), and (2.24a) when $Q = m$, mc_i, and $mc^2/2$, respectively.

7 DYNAMICS OF A BINARY COLLISION

Thus far we have specified a collision only as a rather abrupt change in the trajectories of two molecules as a result of their mutual interaction. We have in addition restricted our attention to binary collisions between electrically neutral molecules in which the force between the molecules depends only on the distance between their centers and in which there is no interchange between translational and internal molecular energy. Collisions satisfying the last restriction are called *elastic collisions*.

Although the elastic-sphere model employed in most of Chapters I and II is qualitatively useful, it is known that a more realistic representation of intermolecular forces is necessary for an accurate description of the motion of gases. A few comments in addition to those already given in Chapter I, Sec. 2 with regard to such forces may clarify our discussion of binary collisions. We shall discuss these forces here largely in terms of the intermolecular potential energy (or simply the intermolecular potential). Such a representation is consistent with modern usage and particularly suitable to the consideration of quantum effects. Since we are concerned with spherically symmetric forces and potentials, the force F between two molecules at any distance of separation r (between centers) acts in the r-direction and is related to the potential V by

$$F = -\frac{dV}{dr}. \tag{7.1}$$

The theoretical and experimental investigation of the nature and magnitude of intermolecular potentials has resulted in a large body of knowledge in the fields of atomic and molecular physics. Because of the difficulties associated with direct evaluation, these potentials are often represented in kinetic theory by simple analytic expressions (the form of which may be based on theoretical considerations) involving one or more unspecified constants. The constants can then be determined, depending on the nature of the molecules involved in the collision, on the basis of empirical information obtained from thermodynamic and transport properties (see, for example, Mason and Rice, 1954). To be definite we shall adopt such models of the intermolecular potential for particular calculations, but the results of our discussion will be expressed largely in terms of certain integrals involving the potential. The evaluation of the potential by experimental and theoretical means for actual gases is beyond our present purposes. It should be reemphasized here that, in accord with observation for neutral gases at sufficiently low densities, we are considering potentials whose effective extent (roughly the molecular "diameter") is much smaller than the molecular spacing or the average distance between collisions. With this restriction on the density, it is possible to discuss a thermally perfect gas on the basis of realistic intermolecular potentials. At higher densities, departures from the elastic-sphere model are also important in the study of imperfect gas effects. Although a fundamental description of the intermolecular potential itself may require application of the methods of wave mechanics, we shall describe the motion of the molecules in classical terms, except when noting certain limitations imposed by wave mechanics. For a qualitative description of the nature of intermolecular potentials the reader is referred to Hirschfelder, Curtiss, and Bird (1964), Sec. 1.3; an extensive account is given in Part III of that book.

In writing the collision integral in the Boltzmann equation, it was understood that for a given intermolecular potential the six components of the molecular velocities c_i' and ζ_i' after a collision can be found as functions of the velocities c_i and ζ_i before collision by means of the four conservation equations of momentum and energy and the two geometric parameters that are needed to specify the relative orientation of the molecules before collision. Since the total momentum of the two molecules is unchanged during a collision, it is convenient in performing this calculation to adopt, as in Chapter II, Sec. 6, a coordinate system moving with the constant velocity w_i of the center of mass:

$$w_i = \frac{m_A c_i + m_B \zeta_i}{m_A + m_B}. \tag{7.2}$$

Here for generality we have allowed for different masses (m_A, m_B) of the colliding molecules. The velocities $\tilde{c}_i$ and $\tilde{\zeta}_i$ in the center-of-mass reference frame are then

$$\tilde{c}_i = c_i - w_i = -\frac{m_B}{m_A + m_B}\, g_i, \tag{7.3a}$$

and

$$\tilde{\zeta}_i = \zeta_i - w_i = \frac{m_A}{m_A + m_B}\, g_i, \tag{7.3b}$$

where the relative velocity g_i is defined by

$$g_i \equiv \zeta_i - c_i. \tag{7.3c}$$

Using equations (7.3) and the reduced mass $m^*_{AB} \equiv m_A m_B/(m_A + m_B)$, we can write the momenta of the two molecules as

$$m_A \tilde{c}_i = -m^*_{AB} g_i, \tag{7.4a}$$

and

$$m_B \tilde{\zeta}_i = m^*_{AB} g_i. \tag{7.4b}$$

The sum of the kinetic energies can be found from equations (7.3) [see Exercise 7.1 and equation (II 6.8)] and is

$$\tfrac{1}{2} m_A \tilde{c}^2 + \tfrac{1}{2} m_B \tilde{\zeta}^2 + \tfrac{1}{2}(m_A + m_B) w^2 = \tfrac{1}{2} m_A c^2 + \tfrac{1}{2} m_B \zeta^2, \tag{7.5}$$

where as before the magnitude of a vector is written without a subscript. Thus the kinetic energy of the two molecules in the center-of-mass coordinate system plus the kinetic energy of the total mass moving with the speed of the center of mass is equal to the total kinetic energy in the laboratory coordinate system.

It can be seen from equations (7.3) that in the center-of-mass reference frame the velocities of the two molecules are parallel and hence their trajectories will lie initially in a single plane, the plane of Fig. 3. The force between the two molecules is assumed to act between their centers. The trajectories will therefore remain in the same plane throughout the collision. The angle ε (which was introduced in Chapter II and used in Sec. 3) is then the angle specifying the orientation of this plane with respect to an arbitrary reference plane parallel to g_i.

Now, in the center-of-mass reference frame the initial momenta of the two molecules are equal and opposite by equations (7.4). Furthermore, the force on each of the two molecules is at every instant equal in magnitude but opposite in direction. As a result the final momenta are also equal and opposite. The molecular velocities therefore remain parallel but deflected at an angle χ, as in Fig. 3. It follows that

$$m_A \tilde{c}' = m_B \tilde{\zeta}', \tag{7.6a}$$

as well as

$$m_A \tilde{c} = m_B \tilde{\zeta}. \tag{7.6b}$$

Referring to equations (7.5) and (7.6), we see that by conservation of total energy the speed of each molecule *in the center-of-mass system* is unchanged by the collision, that is, $\tilde{c}' = \tilde{c}$ and $\tilde{\zeta}' = \tilde{\zeta}$. Since $g = \tilde{g} = \tilde{\zeta} + \tilde{c}$ and $g' = \tilde{g}' = \tilde{\zeta}' + \tilde{c}'$, we then have $g' = g$, as stated without proof in Chapter II and in Secs. 3 and 4. In addition, by conservation of angular momentum, the impact parameter is also unchanged, so that $b' = b$.

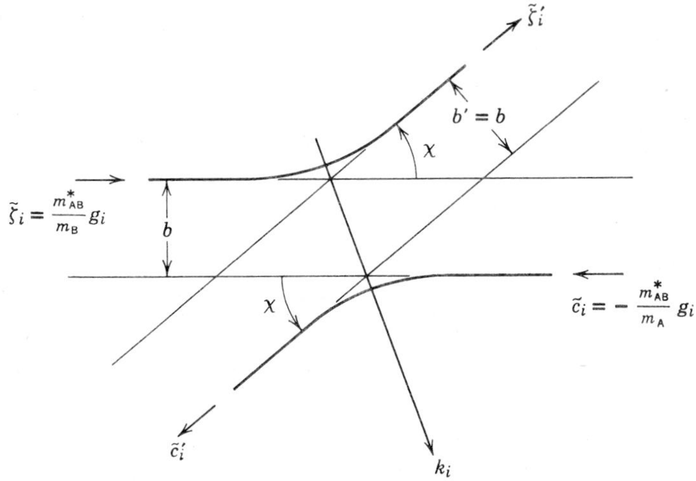

Fig. 3. Center-of-mass representation of the molecular trajectories during a collision.

With these results conservation of total momentum and energy have been satisfied, and the problem reduces to the calculation of the deflection angle χ. Now, χ is an angle defined in the center-of-mass system and therefore for spherically symmetric intermolecular potentials can depend only on the relative speed g and the impact parameter b. With $g' = g$, $b' = b$, and $\chi = \chi(g, b)$, we can see with a little reflection (Exercise 7.3) that an inverse collision with initial velocities c_i' and ζ_i' does in fact result in the final velocities c_i and ζ_i. This fact was used in Sec. 4 and in the derivation of the collision integral in Chapter II.

The velocities after collision can now be written in terms of the initial velocities and the geometry. If the *apse vector* k_i is a unit vector parallel to the line of symmetry of either trajectory (the *apse line*) as shown in Fig. 3, the components of momentum and hence of velocity parallel to

k_i are reversed by the collision, whereas the perpendicular components are unchanged. Therefore we can write

$$c_i' - c_i = \tilde{c}_i' - \tilde{c}_i = -2(\tilde{c}_j k_j)k_i = \frac{2m_B}{m_A + m_B}(g_j k_j)k_i, \tag{7.7a}$$

and

$$\zeta_i' - \zeta_i = \tilde{\zeta}_i' - \tilde{\zeta}_i = -2(\tilde{\zeta}_j k_j)k_i = \frac{-2m_A}{m_A + m_B}(g_j k_j)k_i. \tag{7.7b}$$

This result can be verified by noting that $\tilde{c}_j k_j$ is the dot product of the vectors $\tilde{c}_i$ and k_i, so that $(\tilde{c}_j k_j)k_i$ is the component of $\tilde{c}_i$ parallel to k_i (negative for c_i and positive for ζ_i). The final form of equations (7.7) then follows from equations (7.3). At this point it is also possible to show that $dV_{\zeta'}\,dV_{c'} = dV_\zeta\,dV_c$. The reader is referred to Chapman and Cowling (1952), Sec. 3.52 for such a proof.

For use in our later calculations it will be convenient to rewrite equations (7.7) for the change in velocity of the two molecules in a collision directly in terms of the velocities c_i and ζ_i and the angles χ and ε. Referring to Fig. 3, we see that the vector k_i can be resolved into two vectors, one $k_{i\parallel}$ of magnitude $\cos\frac{1}{2}(\pi - \chi) = \sin(\chi/2)$ and parallel to $g_i = \zeta_i - c_i$ and the other $k_{i\perp}$ of magnitude $\cos(\chi/2)$ and perpendicular to g_i. The dot product of g_i and $k_{i\perp}$ is zero, so that

$$g_j k_j = g_j k_{j\parallel} = g \sin\frac{\chi}{2}.$$

The unit vector in the g_i direction is g_i/g. Therefore we have

$$k_{i\parallel} = \frac{g_i}{g}\sin\frac{\chi}{2},$$

and

$$(g_j k_j)k_i = \left(g \sin\frac{\chi}{2}\right)\left(\frac{g_i}{g}\sin\frac{\chi}{2} + k_{i\perp}\right)$$

Equations (7.7) then become

$$\boxed{\begin{aligned} c_i' - c_i &= \frac{2m_B}{m_A + m_B}\sin\frac{\chi}{2}\left(g_i \sin\frac{\chi}{2} + g k_{i\perp}\right) & (7.8a)\\ \zeta_i' - \zeta_i &= \frac{-2m_A}{m_A + m_B}\sin\frac{\chi}{2}\left(g_i \sin\frac{\chi}{2} + g k_{i\perp}\right) & (7.8b)\end{aligned}}.$$

We shall now take the reference plane from which the angle ε is measured as the plane formed by the directions of the vector g_i and the x_1-axis (see Fig. 4). By purely geometrical reasoning (see Jeans, 1925, Sec. 284), the components of $k_{i\perp}$ can then be written in terms of the variables of interest as

$$k_{1\perp} = \cos\frac{\chi}{2}\sqrt{1 - g_1^2/g^2}\cos\varepsilon \tag{7.9a}$$

$$k_{2\perp} = -\cos\frac{\chi}{2}\left(\frac{g_2g_1\cos\varepsilon + gg_3\sin\varepsilon}{g\sqrt{g^2 - g_1^2}}\right) \tag{7.9b}$$

$$k_{3\perp} = -\cos\frac{\chi}{2}\left(\frac{g_3g_1\cos\varepsilon - gg_2\sin\varepsilon}{g\sqrt{g^2 - g_1^2}}\right). \tag{7.9c}$$

Equations (7.8) and (7.9) give the changes in velocity of the two molecules in terms of c_i, ζ_i, χ, and ε. Since for elastic-sphere molecules the impact angle ψ is related to χ by $\psi = \frac{1}{2}(\pi - \chi)$, this result substantiates the statements made without proof in Chapter II, Sec. 4, and in Sec. 3.

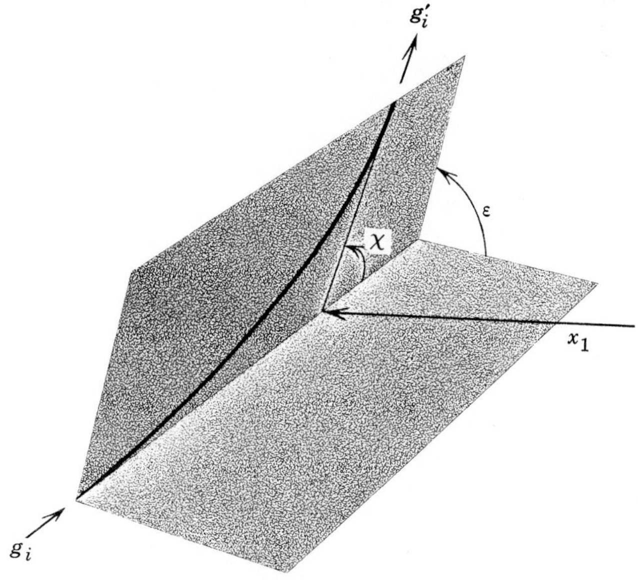

Fig. 4. Specification of the angle ε. The plane of g_i and g_i' is the plane of Fig. 3.

To evaluate the collision integral and integrals related to it, we must relate the deflection angle χ to the intermolecular potential and the relative speed g. This relationship can be made by means of the *collision cross-section* $\sigma(g, \chi)$ for deflection into the differential solid angle $d\Omega$. Let us consider a situation where there are n_c molecules per unit volume with velocity c_i and n_ζ molecules per unit volume with velocity ζ_i. We shall fix our attention on collisions between molecules of these two classes for which the relative velocity g_i' after collision lies in the solid angle $d\Omega$ centered about the direction specified by the angles χ and ε (Fig. 5). The cross-section $\sigma(g, \chi)$ is defined such that the number of collisions per unit volume of the above type per unit time is $n_c n_\zeta g\sigma(g, \chi)\, d\Omega$. Thus $\sigma(g, \chi)$,

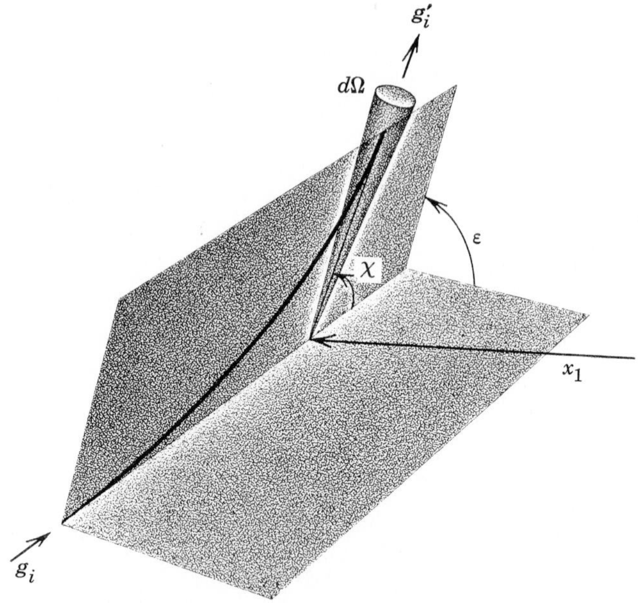

Fig. 5. Specification of the solid angle $d\Omega$.

which has the dimensions of an area, is proportional to the probability that a collision at a relative speed g will result in a deflection χ. As such, σ will depend on the nature of the intermolecular potential. In terms of the distribution function we have $n_c = nf(c_i)\, dV_c$ and $n_\zeta = nf(\zeta_i)\, dV_\zeta$. It then follows that we can write the collision integral in the Boltzmann equation (3.3) as

$$\boxed{\left[\frac{\partial}{\partial t}(nf)\right]_{\text{coll}} = \int_{-\infty}^{\infty}\int_0^{4\pi} n^2[f(c_i')f(\zeta_i') - f(c_i)f(\zeta_i)]g\sigma\, d\Omega\, dV_\zeta}\,. \tag{7.10}$$

Thus the differential cross-section dP_c introduced in Sec. 3 is replaced by $\sigma\, d\Omega$.

The solid angle $d\Omega$ is defined in the center-of-mass reference frame (as is χ); since the magnitude of a solid angle is equal to the area subtended on a unit sphere, $d\Omega = \sin\chi\, d\chi\, d\varepsilon$. In the collision integral (7.10) we then have

$$\int_0^{4\pi} d\Omega = \int_0^{2\pi}\int_0^{\pi} \sin\chi\, d\chi\, d\varepsilon.$$

We also have the following correspondence between $\sigma\, d\Omega$ and the impact-parameter representation of Sec. 3:

$$|\sigma\, d\Omega| = |b\, db\, d\varepsilon|. \tag{7.11a}$$

Equation (7.11a) is written in terms of absolute values to account for the fact that at large values of the impact parameter b the deflection angle χ will be small, whereas at $b = 0$, $\chi = \pi$. Thus, we have

$$\frac{\partial b}{\partial \chi} < 0,$$

and, for the correspondence of the limits of integration,

$$\int_0^{\infty} (\;)\, db \Rightarrow \int_{\pi}^{0} (\;)\, d\chi = -\int_0^{\pi} (\;)\, d\chi.$$

To avoid the awkward double negative implied by these relations we write, as indicated by equation (7.11a),

$$\sigma = \frac{b}{\sin\chi}\left|\frac{\partial b}{\partial \chi}\right| \tag{7.11b}$$

with the understanding that in integrals over χ and b the lower limit is to be taken as zero in both cases.[10]

We shall adopt the form (7.10) of the collision integral in our further discussion of the Boltzmann equation. By using the cross-section σ we take χ and ε as the two geometric parameters that specify a collision, rather than, for example, the impact parameter b and ε. The advantage of using σ is that it lends itself to direct experimental determination or to evaluation for a given intermolecular force by either classical or wave mechanics. An introductory discussion of the collision cross-section which the reader may find helpful is given by Present (1958), Sec. 8.2.

Exercise 7.1. Verify equation (7.5) for the sum of the kinetic energies in the center-of-mass coordinate system and compare this expression with equation (II 6.8).

Exercise 7.2. Sketch the molecular trajectories in a relative coordinate system in which the molecule with mass m_A remains at rest. Relate the velocities and the deflection angle in this system to those in the center-of-mass system.

[10] In some cases, which will not concern us, b is not a single-valued function of χ and the transformations (7.11) must be reconsidered. Nonetheless, the representation $\sigma\, d\Omega$ remains valid.

Exercise 7.3. Show that for a fixed direction of the apse line an inverse collision with initial velocities c_i' and ζ_i' reproduces the velocities c_i and ζ_i with which the molecules entered the corresponding original collision. Note that the apse vector of the inverse collision is the negative of the apse vector of the original collision. Consider the situation first in the center-of-mass coordinate system and then transform your results to the laboratory coordinate system.

8 THE EVALUATION OF COLLISION CROSS-SECTIONS

We next consider the particular forms of the cross-section σ for two simple models of the intermolecular potential, dropping now any distinction between the nature of the two molecules in a collision. If the molecules are represented as elastic spheres of diameters d, the angle of incidence is equal to the angle of reflection and the impact angle ψ (see Fig. 2) is related to χ by $\psi = \frac{1}{2}(\pi - \chi)$. From equation (7.11b) for σ, we then have, with $b = d \sin \psi$ from Fig. 2,

$$\sigma = \frac{d^2}{4} . \qquad \text{(elastic spheres)} \tag{8.1}$$

It can be seen that the cross-section for elastic spheres is independent of both the relative speed g and the deflection angle χ.

A more flexible model, having two free parameters, is one in which the intermolecular potential varies as some inverse power α of the distance r between centers of the molecules, that is,

$$V = \frac{a}{r^\alpha} . \tag{8.2a}$$

The force between the molecules is then

$$F = -\frac{dV}{dr} = \alpha \frac{a}{r^{\alpha+1}} . \tag{8.2b}$$

The constants a and α can be adjusted to conform to experimental data for a given gas over a range of conditions. The properties of many neutral gases can be correlated on the basis of inverse-power repulsive forces with values of α lying between 4 and ∞. (In a certain limit, $\alpha = \infty$ corresponds to the elastic-sphere model.) For this inverse-power model the evaluation of σ is algebraically more involved, although it remains a straightforward problem in two-body dynamics (Chapman and Cowling, 1952, Sec. 10.3). The result cannot be written as a simple explicit function of g and χ, and the deflection angle χ must be related to the relative speed g and the impact parameter b through the integral

$$\chi = \pi - 2\int_0^{b/r_0}\left[1 - \left(\frac{b}{r}\right)^2 - \frac{2}{\alpha}\left(\frac{b}{\beta r}\right)^\alpha\right]^{-1/2} d\left(\frac{b}{r}\right). \tag{8.3a}$$

Here β is a pure number given by

$$\beta \equiv b\left(\frac{mg^2}{2\alpha a}\right)^{1/\alpha}, \tag{8.3b}$$

and r_0, the separation at closest approach, is found by solution of the equation

$$1 - \left(\frac{b}{r_0}\right)^2 - \frac{2}{\alpha}\left(\frac{b}{\beta r_0}\right)^\alpha = 0. \tag{8.3c}$$

The reader should satisfy himself that for a given value of α equations (8.3) give χ as a function of β only [which can also be interpreted as $\chi = \chi(g, b)$]. The differential cross-section itself can be written[11]

$$\sigma\, d\Omega = \left(\frac{2\alpha a}{m}\right)^{2/\alpha} g^{-4/\alpha}\beta\, d\beta\, d\varepsilon. \qquad \text{(inverse-power molecules)} \tag{8.4}$$

It should be noted at this point that we have for inverse-power molecules a different relationship between c_i' and ζ_i' and the variables appearing in $\sigma\, d\Omega$ than when σ is known as a function of g and χ. In the latter case to evaluate, for example, the collision integral (7.10) for a given distribution function we have $c_i' = c_i'(c_i, \zeta_i, \chi, \varepsilon)$ from equations (7.8) and (7.9) and $\sigma = \sigma(g, \chi)$, so that the integrations over χ and ε can be performed directly. For inverse-power molecules the functional dependence of c_i' remains unchanged, but now the intermediate step $\chi = \chi(\beta)$ from equations (8.3) is introduced and the integrations are performed over β and ε as indicated by equation (8.4).

For accurate calculations, more realistic and more complicated models of the intermolecular potential are often required. For example, both the Sutherland model (introduced in Chapter I) of elastic spheres surrounded by an attractive inverse-power potential and the Lennard-Jones model, which superimposes inverse-power attractive and repulsive potentials, allow for long-range attraction and short-range repulsion. The elastic-sphere and inverse-power repulsion models, however, are sufficient to illustrate our discussion.

In addition to the differential cross-section $\sigma\, d\Omega$, several integrated cross-sections are of importance in kinetic theory. The total cross-section

[11] This follows from equation (7.11b) for σ and differentiation of equation (8.3b).

σ_T is given by

$$\boxed{\sigma_T \equiv \int_0^{4\pi} \sigma \, d\Omega = 2\pi \int_0^{\pi} \sigma \sin \chi \, d\chi}. \tag{8.5}$$

For elastic spheres, with $\sigma = d^2/4$ from equation (8.1), we have

$$\sigma_T = \pi \, d^2. \qquad \text{(elastic spheres)} \tag{8.6}$$

This is as would be expected and is in accord with our calculations in Chapters I and II. For inverse-power molecules, $V \sim r^{-\alpha}$, the intermolecular force is not limited to a finite region as with elastic spheres, but is assumed to extend to large separations. Detailed calculations (Kennard, 1938, p. 120) show that for the small deflections corresponding to large separations, σ grows more rapidly than $\sin \chi$ decreases, regardless of the exponent α. Hence *classically* the total cross-section is infinite; the collision frequency and mean free path are then at best ambiguously defined. We shall see shortly that this situation is altered by quantum mechanics.

Since small deflections result in only a small transfer of momentum or energy between the molecules, σ_T will not be as important to us as integrated cross-sections in which significant deflections are more heavily weighted. It follows from Fig. 3 and equations (7.8) that in a collision the momentum change of either molecule in the direction of the initial relative motion is proportional to $1 - \cos \chi = 2 \sin^2 (\chi/2)$. Accordingly, the momentum-transfer cross-section σ_M, which is important in studies of diffusion, is defined by

$$\boxed{\sigma_M \equiv \int_0^{4\pi} (1 - \cos \chi)\sigma \, d\Omega = 2\pi \int_0^{\pi} \sigma(1 - \cos \chi) \sin \chi \, d\chi}. \tag{8.7}$$

In addition, in Chapter X we shall find that the coefficients of viscosity and thermal conductivity are related to the cross-section

$$\boxed{\sigma_\mu \equiv \int_0^{4\pi} \sin^2 \chi \, \sigma \, d\Omega = 2\pi \int_0^{\pi} \sigma \sin^3 \chi \, d\chi}. \tag{8.8}$$

For elastic spheres σ_M and σ_μ are easily evaluated from $\sigma = d^2/4$ and are

$$\sigma_M = \pi d^2, \tag{8.9a}$$

$$\text{(elastic spheres)}$$

$$\sigma_\mu = \tfrac{2}{3}\pi d^2. \tag{8.9b}$$

Using equation (8.4) for $\sigma\, d\Omega$, we obtain for inverse-power molecules

$$\sigma_M = 2\pi\left(\frac{2\alpha a}{m}\right)^{2/\alpha} g^{-4/\alpha} A_1(\alpha), \tag{8.10a}$$

(inverse-power molecules)

$$\sigma_\mu = 2\pi\left(\frac{2\alpha a}{m}\right)^{2/\alpha} g^{-4/\alpha} A_2(\alpha), \tag{8.10b}$$

where

$$A_1(\alpha) \equiv \int_0^\infty (1 - \cos\chi)\,\beta\, d\beta, \tag{8.10c}$$

$$A_2(\alpha) \equiv \int_0^\infty \sin^2\chi\,\beta\, d\beta. \tag{8.10d}$$

$A_1(\alpha)$ and $A_2(\alpha)$ are pure numbers that can be evaluated from equations (8.3); tabulated values are given for various exponents $\nu = \alpha + 1$ by Chapman and Cowling (1952), Sec. 10.3. It is also possible to show (by reference to Kennard, 1938, p. 120) that, unlike σ_T, both σ_M and σ_μ are finite for inverse-power molecules for $\alpha > 1$. For this reason, as well as for the physical arguments mentioned earlier, the momentum-transfer cross-section σ_M is often used in evaluating the collision frequency and the mean free path. Since $\sigma_M = \pi d^2$ for elastic spheres, this interpretation requires no modification of the results of Chapters I and II.

This discussion has been based on a classical interpretation of the molecular trajectories during a collision. In particular, equations (8.1), (8.4), (8.6), (8.9), and (8.10) for the various cross-sections for given models of the intermolecular potential were obtained by the methods of classical mechanics. Although quantum effects under conditions of interest to us require no major modification of kinetic theory, there are quantum restrictions that apply to our cross-section calculations. It is convenient, although somewhat arbitrary, to consider separately the restrictions that arise from the gross effects of diffraction and from the uncertainty in deflection for small values of the deflection angle χ.

Diffraction effects will be small if the molecular de Broglie wavelength $\Lambda = h/mg$ introduced in Chapter IV is small when compared with the molecular dimensions. If we use $\sqrt{\sigma_M}$ as the effective extent of the intermolecular potential and replace g by a characteristic thermal speed $\sqrt{2kT/m}$, this requires that

$$\Lambda \sim \frac{h}{\sqrt{2mkT}} \ll \sqrt{\sigma_M}. \tag{8.11}$$

The restriction (8.11) will ordinarily be satisfied (see Exercise 8.2) except for small m and at low temperatures. Diffraction effects are then unimportant for elastic collisions in neutral gases, except for hydrogen and helium which have small masses and liquify at low temperatures; on the other hand, such effects can be significant in electron-atom collisions. We shall adopt the restriction (8.11) as part of our model of the gas.

For small deflection angles χ, in order for the deflection resulting from a collision to be well defined (see Exercise 8.3 and Wu and Ohmura, 1962, p. 52), the uncertainty principle [equation (IV 3.1)] requires that

$$b(g, \chi)\frac{mg}{2}\chi \gg h, \tag{8.12}$$

where $b(g, \chi)$ is the impact parameter associated with a deflection χ. It can be shown (with reference to Kennard, 1938, p. 120) that, for a given g, the product $b(g, \chi)\chi$ approaches zero as χ approaches zero if the potential falls off more rapidly than r^{-1}. The condition (8.12) is then violated for sufficiently small χ. The total collision cross-section σ_T, which weights small deflections relatively heavily, is significantly affected by the uncertainty in deflection. For inverse-power molecules, the quantum theory (Wu and Ohmura, 1962, p. 9) gives a finite value of σ_T for $\alpha > 2$, whereas classically σ_T is infinite.

Even for elastic spheres in the limit of small de Broglie wavelength, σ_T is found to be twice the classical value of πd^2; for increasing wavelength σ_T is larger. The additional scattering (i.e., deflection) that gives rise to the increase in σ_T is confined to smaller and smaller angles χ as the wavelength decreases. For this reason the cross-sections σ_M and σ_μ, which are weighted with respect to χ^2 for small χ, tend to their classical limits as the de Broglie wavelength decreases.

These qualitative conclusions remain valid for extended-range interactions as well as for elastic spheres. Thus, as long as the restriction (8.11) on the de Broglie wavelength is satisfied, the classical theory can be used in the calculation of σ_M and σ_μ. It also follows that the additional quantum effects of indistinguishability and Pauli exclusion (symmetry) need not be considered.

Since the classical theory is never valid for the total cross-section σ_T and since even the quantum calculation weights small deflections heavily (see Chapman and Cowling, 1952, Sec. 17.31), one approach to the calculation of the collision frequency and mean free path would be to replace σ_T in the definitions of these quantities by the momentum-transfer cross-section σ_M or the cross-section σ_μ. This is motivated not only by necessity (since classically σ_T is infinite except for elastic-sphere molecules) but by the physical argument that in mean-free-path considerations such as those of Chapter I only collisions producing an appreciable change in momentum should be included. Actually, we shall adopt for reasons of convenience a different, but related, convention in Chapter X [equation

(9.5)]. The difficulty associated with infinite values of σ_T is then overcome, except for Coulomb potentials ($V \sim r^{-1}$) in ionized gases for which both the classical and quantum values of σ_T, σ_M, and σ_μ are infinite. The special considerations required for such situations are beyond the scope of this book.

For a more thorough discussion of the influence of quantum effects on collisions, the reader is referred to Mott and Massey (1949), Secs. VII 5 and XII 3, and Massey and Burhop (1952), Sec. VII 3, both of which show graphically the quantum effects on elastic-sphere cross-sections; to Hirschfelder, Curtiss, and Bird (1964), Sec. 10.3, who give the first-order quantum corrections to the transport coefficients; and to Chapman and Cowling (1952), Chapter 17.

With the background acquired in this section, we are now in a position to evaluate the collision integrals that we require for later use.

Exercise 8.1. The Coulomb force ($\alpha = 1$) between electrically charged particles is a relatively simple example of an inverse-power force (although otherwise outside the scope of our discussion). For this case: (a) find integrated expressions for $\chi(\beta)$ and $\chi(g, b)$, and (b) obtain the cross-section in the form $\sigma(g, \chi)$ (known as the *Rutherford cross-section*).

Exercise 8.2. Calculate the ratio $\Lambda/\sqrt{\sigma_M}$ [see equation (8.11)] for helium, argon, hydrogen, and nitrogen at $T = 273$°K and for helium at $T = 20$°K. To estimate σ_M use the mean-free-path theory of Chapter I and the following measured values of viscosity, in dyne sec/cm^2: He, 1875×10^{-7} at 273°K; Ar, 2100×10^{-7} at 273°K; H_2, 840×10^{-7} at 273°K; N_2, 1664×10^{-7} at 273°K; He, 350×10^{-7} at 20°K.

Exercise 8.3. For small deflections, in order that a collision be well defined classically in the center-of-mass coordinate system the following conditions must hold: (i) the uncertainty $\Delta(m\tilde{c}_i')/m\tilde{c}'$ in the deflection angle χ must be small when compared with χ and (ii) the uncertainty Δb in the impact parameter b must be small when compared with b. With the uncertainty principle [equation (IV 3.1)] in the form $\Delta b\, \Delta(m\tilde{c}_i') \simeq h$, show that these conditions result in equation (8.12).

9 THE EVALUATION OF COLLISION INTEGRALS

In this section we are concerned with particular forms of the integrals

$$\Delta[Q] \equiv \int_{-\infty}^{\infty}\int_{-\infty}^{\infty}\int_{0}^{4\pi} \delta_c Q\, n^2 f(c_i) f(\zeta_i) g\sigma\, d\Omega\, dV_\zeta\, dV_c \tag{9.1a}$$

and

$$I[Q] \equiv -\int_{-\infty}^{\infty}\int_{0}^{4\pi}(\delta_c Q + \delta_\zeta Q) f_0(c_i) f_0(\zeta_i) g\sigma \, d\Omega \, dV_\zeta. \qquad (9.1b)$$

As before Q is any function of molecular velocity, $\delta_c Q$ is $Q(c_i') - Q(c_i)$, and $\delta_\zeta Q$ is $Q(\zeta_i') - Q(\zeta_i)$. Although the analyses of this section may seem somewhat tedious, the results will serve us well in Chapter X. We develop these results here both to avoid interrupting the later discussion and to allow the reader to become better acquainted with the nature of collision integrals.

The integral operator $\Delta[Q]$ is one of the alternate forms for the moments of the collision integral [see equations (4.5) and (6.3)]. In the similar operator $I[Q]$ the distribution function is specified as Maxwellian (f_0), and only one integration over velocity space is indicated. Thus $I[Q]$ is a function of c_i, whereas $\Delta[Q]$ is not. Since $\Delta[Q]$ can be written with $\frac{1}{2}(\delta_c Q + \delta_\zeta Q)$ in place of $\delta_c Q$ with the transformation (4.5c) rather than (4.5d), both $\Delta[Q]$ and $I[Q]$ are necessarily zero when $\delta_c Q + \delta_\zeta Q = 0$. This will occur when Q is one of the collisional invariants m, mc_i, or $mc^2/2$. It also follows from the definitions (9.1) that if a and b are independent of molecular velocity we have

$$\Delta[aQ_1 + bQ_2] = a\,\Delta[Q_1] + b\,\Delta[Q_2],$$
$$I[aQ_1 + bQ_2] = aI[Q_1] + bI[Q_2]. \qquad (9.2)$$

The particular values of Q of interest to us are $c_i c_j$ and $c_i c^2$. It is easy to show from equations (9.2) and the properties of the collisional invariants (Exercise 9.1) that

$$\Delta[c_i c_j] = \Delta[C_i C_j], \qquad (9.3a)$$
$$I[c_i c_j] = I[C_i C_j], \qquad (9.3b)$$

and

$$\Delta[c_i c^2] = \Delta[C_i C^2] + 2\bar{c}_j\,\Delta[C_i C_j], \qquad (9.3c)$$
$$I[c_i c^2] = I[C_i C^2] + 2\bar{c}_j I[C_i C_j]. \qquad (9.3d)$$

Here it is important to note that the summation convention applies to the last term in equations (9.3c) and (9.3d). This is in contrast to the exclusion from the summation convention of the argument c_i in $f(c_i)$.

We shall first evaluate the integral $\Delta[Q]$ for the special case of molecules that are assumed to repel each other with a force inversely proportional to the fifth power of the distance between their centers; such molecules are called *Maxwellian molecules*. Although this model is not a particularly accurate representation of the intermolecular forces in actual gases, it affords a remarkable simplification in the calculation of $\Delta[Q]$. As we shall see, in this case $\Delta[Q]$ can be expressed in terms of moments of the distribution function, independently of any explicit knowledge of the distribution function itself. Since for Maxwellian molecules the intermolecular force varies as r^{-5}, the intermolecular potential is proportional to r^{-4} [see equations (8.2)]. For inverse-power molecules with the exponent $\alpha = 4$, the product of the relative speed g and the differential cross-section (8.4) is

$$g\sigma\, d\Omega = \left(\frac{8a}{m}\right)^{1/2} \beta\, d\beta\, d\varepsilon. \qquad \text{(Maxwellian molecules)} \qquad (9.4)$$

The essence of the simplification is thus the elimination of the relative speed in $g\sigma\, d\Omega$.

For an example of the calculation of $\Delta[Q]$ for Maxwellian molecules we take $Q = c_1^2$. Since by equation (9.3a), $\Delta[c_1^2] = \Delta[C_1^2]$, we can as well work with the somewhat simpler form $Q = C_1^2$. Equations (7.8a) and (7.9a) give the change in C_1 during a collision as

$$\begin{aligned} c_1' - c_1 = C_1' - C_1 &= \sin\frac{\chi}{2}\left(g_1 \sin\frac{\chi}{2} + \cos\frac{\chi}{2}\sqrt{g^2 - g_1^2}\cos\varepsilon\right) \\ &= g_1 \sin^2\frac{\chi}{2} + \sqrt{g_2^2 + g_3^2}\sin\frac{\chi}{2}\cos\frac{\chi}{2}\cos\varepsilon. \end{aligned}$$

It follows that in this case

$$\begin{aligned} \delta_c Q = C_1'^2 - C_1^2 &= 2C_1\left(g_1 \sin^2\frac{\chi}{2} + \sqrt{g_2^2 + g_3^2}\sin\frac{\chi}{2}\cos\frac{\chi}{2}\cos\varepsilon\right) + g_1^2 \sin^4\frac{\chi}{2} \\ &\quad + 2g_1\sqrt{g_2^2 + g_3^2}\sin^3\frac{\chi}{2}\cos\frac{\chi}{2}\cos\varepsilon + (g_2^2 + g_3^2)\sin^2\frac{\chi}{2}\cos^2\frac{\chi}{2}\cos^2\varepsilon. \end{aligned}$$

This can be simplified by anticipating from equations (9.1a) and (9.4) the integrations in $\Delta[Q]$, and noting that

$$\int_0^{2\pi} \cos\varepsilon\, d\varepsilon = 0, \quad \int_0^{2\pi} \cos^2\varepsilon\, d\varepsilon = \pi, \quad \text{and} \quad \sin^4\frac{\chi}{2} = \sin^2\frac{\chi}{2} - \frac{1}{4}\sin^2\chi.$$

The integration over ε then gives, for $Q = C_1^2$,

$$\int_0^{2\pi} \delta_c Q\, d\varepsilon = \pi(2g_1^2 + 4C_1 g_1)\sin^2\frac{\chi}{2} - \frac{\pi}{4}(3g_1^2 - g^2)\sin^2\chi. \qquad (9.5)$$

Introducing $Z_i \equiv \zeta_i - \bar{c}_i$, the thermal velocity corresponding to ζ_i, and recalling that the form (9.4) of $g\sigma\, d\Omega$ for Maxwellian molecules does not depend explicitly on velocity, we can write the integrations with respect to velocity as

$$\int_{-\infty}^{\infty}\int_{-\infty}^{\infty} C_1 g_1 f(c_i) f(\zeta_i)\, dV_\zeta\, dV_c = \int_{-\infty}^{\infty}\int_{-\infty}^{\infty} (C_1 Z_1 - C_1^2) f(c_i) f(\zeta_i)\, dV_\zeta\, dV_c$$

$$= \bar{C}_1 \bar{Z}_1 - \overline{C_1^2} = -\overline{C_1^2},$$

$$\int_{-\infty}^{\infty}\int_{-\infty}^{\infty} g_1^2 f(c_i) f(\zeta_i)\, dV_\zeta\, dV_c = \overline{Z_1^2} - 2\bar{Z}_1\bar{C}_1 + \overline{C_1^2} = 2\overline{C_1^2}, \text{ etc.}$$

Here we have used the fact that, since $f(\zeta_i)$ is the same *function* as $f(c_i)$, we have by definition $\bar{Z}_i = \bar{C}_i = 0$ and $\overline{Z_1^2} = \overline{C_1^2}$. With some algebraic manipulation, equation (9.1a) then becomes

$$\Delta[C_1^2] = \pi\sqrt{\frac{2a}{m}}\, n^2(\overline{C^2} - 3\overline{C_1^2}) \int_0^\infty \sin^2\chi\, \beta\, d\beta.$$

With the definitions (8.10d) of $A_2(\alpha)$ (a pure number) and (2.17) of τ_{11}, we obtain the final result

$$\Delta[c_1^2] = \Delta[C_1^2] = 3\pi A_2(4)\sqrt{\frac{2a}{m}}\frac{n}{m}\tau_{11}. \quad \text{(Maxwellian molecules)} \tag{9.6}$$

By similar integrations (Exercise 9.2) it can be shown that[12]

$$\Delta[c_i c_j] = 3\pi A_2(4)\sqrt{\frac{2a}{m}}\frac{n}{m}\tau_{ij} \tag{9.7a}$$

$$\Delta[c_i c^2] = 6\pi A_2(4)\sqrt{\frac{2a}{m}}\frac{n}{m}(\tau_{ij}\bar{c}_j - \tfrac{2}{3}q_i). \quad \text{(Maxwellian molecules)} \tag{9.7b}$$

Equation (9.6) then becomes a special case of equation (9.7a).

These values of $\Delta[Q]$ can now be used with the equations of transfer (6.2) for moments higher than those of the collisional invariants. A significant simplification is thus affected for Maxwellian molecules by writing the equations of transfer independently of an explicit knowledge of the distribution function. This simplification is incorporated in several approximate theories of rarefied-gas dynamics (e.g., Lees, 1959, and Lees and Liu, 1962) and will be used in our discussion in Chapter X of the structure of a strong shock wave. Examination of the equations of transfer will show, however, that the number of moments appearing therein always exceeds the number of equations, so that the formulation for Maxwellian molecules is not complete within itself.

[12] See Jeans (1925), p. 240. Lees (1959), p. 30, corrected numerical errors in Jeans' expression corresponding to our equation (9.7b).

We next consider, for an arbitrary intermolecular potential, the integral $I[Q]$ defined by equation (9.1b). Again we shall take $Q = C_1^2$. Here the quantity $\int_0^{2\pi} \delta_c Q \, d\varepsilon$ has been previously evaluated [equation (9.5)]; similarly, it is easy to show that

$$\int_0^{2\pi} \delta_\zeta Q \, d\varepsilon = \int_0^{2\pi} (Z_1'^2 - Z_1^2) \, d\varepsilon$$
$$= \pi(2g_1^2 - 4Z_1g_1) \sin^2 \frac{\chi}{2} - \frac{\pi}{4}(3g_1^2 - g^2) \sin^2 \chi.$$

Therefore, since $Z_1 - C_1 = g_1$, we have

$$\int_0^{2\pi} (\delta_c Q + \delta_\zeta Q) \, d\varepsilon = -\frac{\pi}{2}(3g_1^2 - g^2) \sin^2 \chi.$$

With $d\Omega = \sin \chi \, d\chi \, d\varepsilon$, $I[C_1^2]$ then becomes

$$I[C_1^2] = \frac{\pi}{2} \int_{-\infty}^{\infty} (3g_1^2 - g^2) g f_0(c_i) f_0(\zeta_i) \left[\int_0^{\pi} \sigma \sin^3 \chi \, d\chi \right] dV_\zeta. \tag{9.8}$$

For a given $\sigma(g, \chi)$ the integral over χ in brackets is a function of g. Equation (9.8) and hence $I[C_1^2]$ can therefore be evaluated by performing the indicated integration over dV_ζ. By a similar analysis (Exercise 9.3) it is found that

$$I[C_iC_j] = \frac{\pi}{2} \int_{-\infty}^{\infty} (3g_ig_j - g^2\delta_{ij}) g f_0(c_i) f_0(\zeta_i) \left[\int_0^{\pi} \sigma \sin^3 \chi \, d\chi \right] dV_\zeta. \tag{9.9}$$

Although $I[Q]$ itself appears in the theoretical development in Chapter X, to find the coefficients of thermal conductivity and viscosity we shall require actual values of the complete integrals

$$a_{11} \equiv \left(\frac{m}{2kT}\right)^3 \int_{-\infty}^{\infty} \left(\frac{5kT}{m} - C^2\right) C_j I\left[\left(\frac{5kT}{m} - C^2\right) C_j\right] dV_c, \tag{9.10}$$

and

$$b_{11} \equiv \left(\frac{m}{2kT}\right)^2 \int_{-\infty}^{\infty} \left(C_jC_k - \frac{1}{3} C^2\delta_{jk}\right) I\left[C_jC_k - \frac{1}{3} C^2\delta_{jk}\right] dV_c. \tag{9.11}$$

In spite of the somewhat imposing appearance of these integrals, their evaluation is conceptually straightforward. We shall perform a part of the lengthy calculation of a_{11} and b_{11} to acquaint the reader further with the structure of collision integrals and the methods used in their evaluation. Because of the rather specialized nature of these calculations, however, some readers may wish to omit from their consideration the remaining part of this section.

Applying the summation convention and using the fact that $I[Q]$ is a linear operator [equation (9.2)], we can write a_{11} and b_{11} as sums of integrals involving $I[C_iC_j]$ and $I[C_iC^2]$. One such integral is

$$\left(\frac{m}{2kT}\right)^2 \int_{-\infty}^{\infty} C_1^2 I[C_1^2]\, dV_c. \tag{9.12}$$

We shall consider in detail the sum of integrals that make up b_{11}. These integrals can be reduced to two terms by use of $I[C^2] = 0$ [from equation (9.3a) and the fact that $mc^2/2$ is a collisional invariant] and the symmetry relations

$$\int_{-\infty}^{\infty} C_1^2 I[C_1^2]\, dV_c = \int_{-\infty}^{\infty} C_2^2 I[C_2^2]\, dV_c = \int_{-\infty}^{\infty} C_3^2 I[C_3^2]\, dV_c,$$

$$\int_{-\infty}^{\infty} C_1C_2 I[C_1C_2]\, dV_c = \int_{-\infty}^{\infty} C_1C_3 I[C_1C_3]\, dV_c = \int_{-\infty}^{\infty} C_2C_3 I[C_2C_3]\, dV_c,$$

and

$$\int_{-\infty}^{\infty} C^2 I[C_1^2]\, dV_c = \int_{-\infty}^{\infty} C^2 I[C_2^2]\, dV_c$$

$$= \int_{-\infty}^{\infty} C^2 I[C_3^2]\, dV_c = \tfrac{1}{3}\int_{-\infty}^{\infty} C^2 I[C^2]\, dV_c = 0.$$

Therefore the integral b_{11} can be written

$$b_{11} = \left(\frac{m}{2kT}\right)^2 \left\{3\int_{-\infty}^{\infty} C_1^2 I[C_1^2]\, dV_c + 6\int_{-\infty}^{\infty} C_1C_2 I[C_1C_2]\, dV_c\right\}. \tag{9.13}$$

Using equation (9.8) for $I[C_1^2]$ and the definition (8.8) of σ_μ, we can write the integral (9.12) appearing in expression (9.13) for b_{11} as

$$\left(\frac{m}{2kT}\right)^2 \int_{-\infty}^{\infty}\int_{-\infty}^{\infty} \tfrac{1}{4} C_1^2(3g_1^2 - g^2) g\sigma_\mu(g) f_0(c_i) f_0(\zeta_i)\, dV_\zeta\, dV_c. \tag{9.14}$$

Here, as is often the case, it is convenient to rewrite $f_0(c_i)f_0(\zeta_i)$ in terms of the relative and center-of-mass velocities g_i and $W_i \equiv w_i - c_i$. Equating $\frac{1}{2}mC^2 + \frac{1}{2}mZ^2 = \frac{1}{2}(2m)W^2 + \frac{1}{2}\left(\frac{m}{2}\right)g^2$ as in equation (II 6.8), we have

$$f_0(c_i)f_0(\zeta_i) = \left(\frac{m}{2\pi kT}\right)^3 e^{-mW^2/kT}\, e^{-mg^2/4kT}.$$

From equation (7.3a) we can also write

$$C_1 = W_1 - \frac{g_1}{2}.$$

In addition, as in Chapter II, Sec. 6, we have $dV_\zeta\, dV_c = dV_W\, dV_g$. With these transformations the integral (9.14) becomes

$$\frac{1}{4\pi^3}\left(\frac{m}{2kT}\right)^5 \int_{-\infty}^{\infty}\int_{-\infty}^{\infty}\left(W_1{}^2 - g_1 W_1 + \frac{g_1{}^2}{4}\right) \times (3g_1^2 - g^2) g\sigma_\mu(g) e^{-mW^2/kT} e^{-mg^2/4kT}\, dV_W\, dV_g. \quad (9.15)$$

It is frequently useful to perform the integrations over velocity using spherical coordinates in the W_i and g_i velocity spaces. As in Chapter II, Sec. 6, we now let

$$\begin{aligned} W_1 &= W\cos\phi_W, & g_1 &= g\cos\phi_g,\\ W_2 &= W\sin\phi_W\cos\theta_W, & g_2 &= g\sin\phi_g\cos\theta_g,\\ W_3 &= W\sin\phi_W\sin\theta_W, & g_3 &= g\sin\phi_g\sin\theta_g, \end{aligned}$$

and

$$dV_W = W^2\sin\phi_W\, d\phi_W\, d\theta_W\, dW \qquad dV_g = g^2\sin\phi_g\, d\phi_g\, d\theta_g\, dg.$$

By integration with respect to θ_w and θ_g from 0 to 2π and ϕ_w and ϕ_g from 0 to π, all terms in (9.15) vanish except for the one involving $g_1^2(3g_1^2 - g^2)$. Since

$$\int_0^\infty\int_0^{2\pi}\int_0^\pi e^{-mW^2/kT}\, W^2\sin\phi_W\, d\phi_W\, d\theta_W\, dW = \left(\frac{\pi kT}{m}\right)^{3/2}$$

and

$$\int_0^{2\pi}\int_0^\pi g^2\cos^2\phi_g(3g^2\cos^2\phi_g - g^2)\sin\phi_g\, d\phi_g\, d\theta_g = \tfrac{16}{15}\pi g^4,$$

we finally obtain for the integral (9.15), which was originally introduced as expression (9.12),

$$\left(\frac{m}{2kT}\right)^2\int_{-\infty}^{\infty} C_1^2 I[C_1^2]\, dV_c = \frac{4}{15\sqrt{\pi}}\left(\frac{m}{4kT}\right)^{7/2}\int_0^\infty g^7\sigma_\mu(g)e^{-mg^2/4kT}\, dg. \quad (9.16)$$

Similarly it is found (Exercise 9.4), with equation (9.9) for $I[C_1C_2]$, that

$$\left(\frac{m}{2kT}\right)^2\int_{-\infty}^{\infty} C_1C_2 I[C_1C_2]\, dV_c = \frac{3}{15\sqrt{\pi}}\left(\frac{m}{4kT}\right)^{7/2}\int_0^\infty g^7\sigma_\mu(g)e^{-mg^2/4kT}\, dg. \quad (9.17)$$

Incorporating equations (9.16) and (9.17) into the expression (9.13) for b_{11}, we have

$$\boxed{b_{11} = \frac{2}{\sqrt{\pi}}\left(\frac{m}{4kT}\right)^{7/2}\int_0^\infty g^7\sigma_\mu(g)e^{-mg^2/4kT}\, dg}. \quad (9.18)$$

This result, obtained by a different method, is given by Chapman and Cowling (1952), Sec. 9.7, where it is also shown that

$$\boxed{a_{11} = b_{11}} \;. \tag{9.19}$$

We now have values of the collision integrals that will be required in Chapter X and can return to a more physically oriented discussion.

Exercise 9.1. Derive equations (9.3).

Exercise 9.2. Derive equation (9.7a).

Exercise 9.3. Perform an analysis similar to that of this section to obtain the quantity $I[C_1C_2]$, as given by equation (9.9) (see Exercise 9.2).

Exercise 9.4. Obtain equation (9.17) from equation (9.9). [Exercises 9.3 and 9.4 complete the derivation of expression (9.18) for the quantity b_{11}.]

10 GAS MIXTURES

We shall now give a brief development of some of the important points that arise when the discussion of this chapter is extended to gas mixtures. For a more detailed account of the kinetic theory of mixtures the reader is referred to Hirschfelder, Curtiss, and Bird (1964).

It is first necessary to generalize certain of our definitions. If we denote the differing chemical species in a gas mixture by the subscript s, then n_s, m_s, f_s, etc., will in general be different for each species. For simplicity we assume that the molecules are monatomic and possess only translational energy. For the mixture as a whole we have

$$n \equiv \sum_s n_s, \tag{10.1}$$

and

$$\rho \equiv \sum_s \rho_s. \tag{10.2}$$

The mean velocity $(\bar{c}_i)_s$ of the species s is[13]

$$(\bar{c}_i)_s \equiv \int_{-\infty}^{\infty} c_i f_s \, dV_c. \tag{10.3}$$

[13] In the discussion of mixtures a subscript denoting the species is often affixed to c_i as well as f in an expression such as (10.3). This is not necessary here and would be awkward with our subscript notation for vectors and tensors.

Similarly for any function $Q(c_i)$ of molecular velocity, the mean values $(\bar{Q})_s$ and $\bar{Q}$ for the species s and the mixture are given respectively by

$$(\bar{Q})_s \equiv \int_{-\infty}^{\infty} Q f_s \, dV_c, \tag{10.4}$$

and

$$n\bar{Q} \equiv \sum_s \int_{-\infty}^{\infty} n_s Q f_s \, dV_c = \sum_s n_s(\bar{Q})_s. \tag{10.5}$$

The *mass velocity* c_{0_i} *of the mixture* is defined by

$$\rho c_{0_i} \equiv \sum_s \rho_s(\bar{c}_i)_s, \tag{10.6}$$

so that ρc_{0_i} is the total mass flux. It can be seen from equations (10.5) and (10.6) that the *mass* velocity c_{0_i} is equal to the *mean* velocity $\bar{c}_i$ for a single-species gas but not for a mixture. The thermal or peculiar velocity is defined in terms of c_{0_i} rather than $(\bar{c}_i)_s$ and is

$$C_i \equiv c_i - c_{0_i}. \tag{10.7}$$

By extension of our definitions in Sec. 2 and use of equations (10.5) and (10.7), the temperature, pressure, viscous stress, and heat flux for the mixture are

$$\tfrac{3}{2}kT \equiv \tfrac{1}{2}\overline{mC^2} = \frac{1}{n}\sum_s \tfrac{1}{2}n_s m_s(\overline{C^2})_s, \tag{10.8}$$

$$p \equiv \sum_s p_s \equiv \tfrac{1}{3}\sum_s n_s m_s(\overline{C^2})_s = \tfrac{1}{3}\overline{nmC^2} = nkT, \tag{10.9}$$

$$\begin{aligned}\tau_{ij} &\equiv \sum_s (\tau_{ij})_s \equiv -\sum_s [n_s m_s(\overline{C_i C_j})_s - p_s\,\delta_{ij}] \\ &= -(\overline{nmC_iC_j} - p\,\delta_{ij}),\end{aligned} \tag{10.10}$$

and

$$q_i \equiv \sum_s (q_i)_s \equiv \sum_s \tfrac{1}{2}n_s m_s(\overline{C_i C^2})_s = \tfrac{1}{2}\overline{nmC_iC^2}. \tag{10.11}$$

These quantities will appear in the conservation equations for the mixture.

The motion of each species is described by a separate Boltzmann equation

$$\begin{aligned}&\frac{\partial}{\partial t}(n_s f_s) + c_j\frac{\partial}{\partial x_j}(n_s f_s) + \frac{\partial}{\partial c_j}(F_{s_j} n_s f_s) \\ &\quad = \sum_r \int_{-\infty}^{\infty}\int_0^{4\pi} n_s n_r[f_s(c_i')f_r(\zeta_i') - f_s(c_i)f_r(\zeta_i)]g\sigma_{sr}\,d\Omega\,dV_\zeta.\end{aligned} \tag{10.12}$$

Here r is a dummy index that, like s, denotes any species in the mixture. The summation over r then includes s-s collisions as well as collisions between species s and all other species. The differential cross-section for s-r collisions is written $\sigma_{sr}\,d\Omega$ and depends on the nature of the two molecules involved. It should be noted that the summation convention does not apply to the subscripts s and r. Equation (10.12) can be derived by including in the discussion in Sec. 3 and in Chapter II, Sec. 4, all possible collisions for the species s; hence the summation of collision integrals.

When all derivatives $\partial/\partial t$ and $\partial/\partial x_j$ are zero and external forces are absent, it can be shown (Hirschfelder, Curtiss, and Bird, 1964, Sec. 7.3a) as in Sec. 4 that the distribution function of each of the various species is separately Maxwellian, that is,

$$f_{s_0} = \left(\frac{m_s}{2\pi kT}\right)^{3/2} e^{-m_s C^2/2kT}. \tag{10.13}$$

From this result it follows that at equilibrium the mean kinetic energy of thermal motion is the same for all species. The fact that the mean thermal kinetic energies remain equal for small departures from equilibrium (see Chapman and Cowling, 1952, Sec. 8.31) is in agreement with the use in macroscopic gas dynamics of equal temperatures for the different species in a mixture.

The equation of transfer for the species s is obtained as in Sec. 6 by multiplying the Boltzmann equation (10.12) by any function $Q(c_i)$ of velocity and integrating over velocity space. The resulting equation is

$$\frac{\partial}{\partial t}[n_s(\bar{Q})_s] + \frac{\partial}{\partial x_j}[n_s(\overline{c_j Q})_s] - n_s\left(\overline{F_{s_j}\frac{\partial Q}{\partial c_j}}\right)_s = \sum_r \Delta_{sr}[Q], \tag{10.14a}$$

where

$$\Delta_{sr}[Q] \equiv \int_{-\infty}^{\infty}\int_{-\infty}^{\infty}\int_0^{4\pi} Q(c_i) n_s n_r [f_s(c_i')f_r(\zeta_i') - f_s(c_i)f_r(\zeta_i)] g\sigma_{sr}\,d\Omega\,dV_\zeta\,dV_c. \tag{10.14b}$$

We have written the left-hand side of equation (10.14a) directly in terms of mean quantities by recognizing that for these terms the derivation leading to the equation of transfer (6.2) for a simple gas remains valid. The collision term $\Delta_{sr}[Q]$ can be transformed by arguments similar to those of Sec. 4 (see Chapman and Cowling, 1952, Sec. 3.53) to give

$$\Delta_{sr}[Q] = \int_{-\infty}^{\infty}\int_{-\infty}^{\infty}\int_0^{4\pi} \delta_c Q\; n_s n_r f_s(c_i) f_r(\zeta_i) g\sigma_{sr}\,d\Omega\,dV_\zeta\,dV_c. \tag{10.15}$$

This expression is analogous to the form (6.3a) for a simple gas. It is important to note, however, that the form (6.3b) involving $\delta_c Q + \delta_\zeta Q$ cannot be applied to $\Delta_{sr}[Q]$ for a gas mixture.

As before, the conservation equations result from the identification of Q with m_s, $m_s c_i$, and $\frac{1}{2}m_s c^2$. In these cases, since the integral $\Delta_{sr}[Q]$ in equation (10.15) is the change in Q in a single collision integrated over all

s-r collisions, then $\sum_r \Delta_{sr}[Q]$ in equation (10.14a) can be directly interpreted as the collisional rate of change per unit volume of the mass, momentum, and energy associated with the species s. The conservation equations for the species s are then

$$\frac{\partial \rho_s}{\partial t} + \frac{\partial}{\partial x_j}[\rho_s(\bar{c}_j)_s] = \sum_r \Delta_{sr}[m_s] = 0, \tag{10.16a}$$

$$\frac{\partial}{\partial t}[\rho_s(\bar{c}_i)_s] + \frac{\partial}{\partial x_j}[\rho_s(\overline{c_i c_j})_s] - \rho_s(\bar{F}_{s_i})_s = \sum_r \Delta_{sr}[m_s c_i], \tag{10.16b}$$

and

$$\frac{1}{2}\frac{\partial}{\partial t}[\rho_s(\overline{c^2})_s] + \frac{1}{2}\frac{\partial}{\partial x_j}[\rho_s(\overline{c_j c^2})_s] - \rho_s(\overline{F_{s_j} c_j})_s = \sum_r \Delta_{sr}[\tfrac{1}{2}m_s c^2]. \tag{10.16c}$$

The species s cannot exchange mass with any species of the gas (we have assumed here a chemically inert gas), nor can it, as a whole, exchange momentum and energy with itself. Therefore all the collision terms in the continuity equation (10.16a) and the collision terms for $r = s$ in the momentum and energy equations (10.16b) and (10.16c) must be zero. On the other hand, the species s can exchange momentum and energy with the other species in the gas as a result of collisions. This is reflected in non-zero collision terms for $r \neq s$ in the momentum and energy equations. Because of the presence of these collision terms and because of the definitions (10.10) and (10.11) for $(\tau_{ij})_s$ and $(q_i)_s$ in terms of $C_i \equiv c_i - c_{0_i}$ [rather than $c_i - (\bar{c}_i)_s$], it is not possible in general to write equations (10.16b) and (10.16c) for a single species in a mixture in the form of the corresponding conservation equations of Sec. 2 for a simple gas.

To obtain the conservation equations of macroscopic gas dynamics for a mixture we must sum equations (10.16) over all species. Physically, since collisions can produce no change in the momentum and energy of the gas as a whole, $\sum_s \sum_r \Delta_{sr}[m_s c_i]$ and $\sum_s \sum_r \Delta_{sr}[\frac{1}{2}m_s c^2]$ must be zero. Mathematically this can be shown by a transformation proved by Hirschfelder, Curtiss, and Bird (1964), p. 461. The conservation equations of a monatomic gas mixture are then, after some manipulation,

$$\boxed{\begin{aligned}
&\frac{\partial \rho}{\partial t} + \frac{\partial}{\partial x_j}(\rho c_{0_j}) = 0 && (10.17\text{a})\\
&\rho\frac{Dc_{0_i}}{Dt} = -\frac{\partial p}{\partial x_i} + \frac{\partial \tau_{ij}}{\partial x_j} + \sum_s \rho_s(\bar{F}_{s_i})_s && (10.17\text{b})\\
&\rho\frac{D}{Dt}\left(\frac{5}{2}\frac{nkT}{\rho} + \frac{c_0^2}{2}\right) = \frac{\partial p}{\partial t} + \frac{\partial}{\partial x_j}(\tau_{kj}c_{0_k} - q_j) + \sum_s \rho_s(\overline{F_{s_j} c_j})_s
\end{aligned}}, \quad (10.17\text{c})$$

where now

$$\frac{D}{Dt} \equiv \frac{\partial}{\partial t} + c_{0_j} \frac{\partial}{\partial x_j}.$$

The great similarity of the foregoing equations to the conservation equations (2.22), (2.23b), and (2.24b) for a simple gas is to some extent misleading. In a gas mixture in addition to momentum and energy transfer we must consider the transfer of mass (diffusion) of the separate species relative to the gas as a whole; there exists in fact a coupling between mass and energy transport. Such effects can be described in terms of the *diffusion velocities* $(\bar{C}_i)_s$. It can be shown, for example, that the heat flux q_i depends on the diffusion velocities as well as the temperature gradient. Thus for a gas mixture, as compared with a simple gas, the additional quantities $(\bar{C}_i)_s$ are required to completely describe a given physical situation. In agreement with our approximate treatment of self-diffusion in Chapter I, Sec. 5, the diffusion velocities depend, in part, on the gradients of the species densities ρ_s. (For small departures from equilibrium an explicit formula for the diffusion velocities is given in Chapter X.) Thus to obtain a description of the mixture in terms of the conservation equations, we must consider the species continuity equations (10.16a) as well as the overall momentum and energy equations (10.17b) and (10.17c). Since by equation (10.7)

$$(\bar{C}_i)_s = (\bar{c}_i)_s - c_{0_i}, \tag{10.18}$$

the continuity equation (10.16a) for species s can be written in terms of the diffusion velocity as

$$\boxed{\frac{\partial \rho_s}{\partial t} + \frac{\partial}{\partial x_j} [\rho_s c_{0_j} + \rho_s (\bar{C}_j)_s] = 0}\,. \tag{10.19}$$

Here the quantity $\rho_s(\bar{C}_i)_s$ is the mass flux of species s in a coordinate system moving with the mass velocity c_{0_i}. From the definition (10.6) of c_{0_i}, we have $\sum_s \rho_s(\bar{C}_i)_s = 0$. We then recover the continuity equation (10.17a) for the mixture as a whole by summing equation (10.19) over s.

It is instructive to examine the relationship between the conservation equations of this section and the corresponding equations of Chapter VIII, Sec. 2. For a chemically reacting mixture, such as that considered in Chapter VIII, a source term must be added to the right-hand side of the continuity equation (10.19) for species s. The resulting equation is then similar to equation (VIII 2.11) (the diffusion velocities are taken as zero in that chapter). The overall momentum and energy equations (10.17b) and (10.17c) are unchanged for a reacting flow, except that, as discussed in Sec. 2, for gases with internal molecular structure suitable forms of the enthalpy [written as $5nkT/2\rho$ for monatomic gases in equation (10.17c)] and the heat flux must be used in the energy equation. With these modifications the conservation equations of this section describe the flow of a chemically reacting mixture of gases. If we now set $(\bar{C}_i)_s$, τ_{ij}, q_i, and F_{s_i} equal to zero, we recover the conservation equations (VIII 2.11), (VIII 2.3), and (VIII 2.4b). (The form of equations (10.17b) and (10.17c) was in fact chosen with this purpose in mind.) We also note from equation (10.18) that in the absence of diffusion, that is, with $(\bar{C}_i)_s = 0$, the distinction between c_{0_i} and $(\bar{c}_i)_s$ disappears, so that u_i in Chapter VIII is both the mass velocity and the mean velocity of the individual components.

Returning now to the conservation equations (10.19), (10.17b), and (10.17c) for a chemically inert monatomic gas, we see that these equations become a determinate set if τ_{ij}, q_i, and $(\bar{C}_i)_s$ can be related to c_{0_i}, T, and ρ_s. For small departures from equilibrium this relationship can be accurately provided by kinetic theory. In the next chapter we shall discuss this analysis for a single-species gas, quoting there the results for a gas mixture.

Exercise 10.1. By comparison of equation (10.9) with the expression $p = \rho RT$, find the perfect-gas constant R for a mixture. Write the energy equation (10.17c) in terms of enthalpy and find in terms of R the specific heat $c_{p_{tr}}$ for a mixture.

Exercise 10.2. Show that the conservation equations (10.17) for a mixture reduce for a single species to the corresponding equations (2.22), (2.23b), and (2.24b).

Exercise 10.3. If the external forces F_{s_i} are independent of the molecular velocities, express the external-force term in the energy equation (10.17c) in terms of the mass velocity c_{0_i} and the diffusion velocities $(\bar{C}_i)_s$.

References

Born, M., 1949, *Natural Philosophy of Cause and Chance*, Oxford.

Chapman, S., and T. G. Cowling, 1952, *The Mathematical Theory of Non-Uniform Gases*, 2nd ed., Cambridge.

Ehrenfest, P., and T. Ehrenfest, 1959, *The Conceptual Foundations of the Statistical Approach in Mechanics*, Cornell.

Grad, H., 1958, Principles of the Kinetic Theory of Gases; in *Handbuch der Physik*, vol. 12, p. 205, Springer.

Grad, H., 1961, The Many Faces of Entropy. *Comm. Pure. Appl. Math.*, vol. 14, no. 3, p. 323.

Hirschfelder, J. O., C. F. Curtiss, and R. B. Bird, 1964, *Molecular Theory of Gases and Liquids*, corrected printing, Wiley.

Jeans, J. H., 1925, *The Dynamical Theory of Gases*, 4th ed., Cambridge. (Also Dover 1954.)

Jeffreys, H., 1957, *Cartesian Tensors*, Cambridge.
Kennard, E. H., 1938, *Kinetic Theory of Gases*, McGraw-Hill.
Landau, L. D., and E. M. Lifshitz, 1958, *Statistical Physics*, Addison-Wesley.
Lees, L., 1959, A Kinetic Theory Description of Rarefied Gas Flows. *GALCIT Hypersonic Research Project*, Memo. No. 51.
Lees, L., and C. Liu, 1962, Kinetic-Theory Description of Conductive Heat Transfer from a Fine Wire. *Phys. Fluids*, vol. 5, no. 10, p. 1137.
Mason, E. A., and W. E. Rice, 1954, The Intermolecular Potentials for Some Simple Nonpolar Molecules. *J. Chem. Phys.*, vol. 22, no. 5, p. 843.
Massey, H. S. W., and E. H. S. Burhop, 1952, *Electronic and Ionic Impact Phenomena*, Oxford.
Mott, N. F., and H. S. W. Massey, 1949, *The Theory of Atomic Collisions*, 2nd ed., Oxford.
Present, R. D., 1958, *Kinetic Theory of Gases*, McGraw-Hill.
Tolman, R. C., 1938, *The Principles of Statistical Mechanics*, Oxford.
Wu, T.-Y., and T. Ohmura, 1962, *Quantum Theory of Scattering*, Prentice-Hall.

Chapter X

Flow with Translational Nonequilibrium

1 INTRODUCTION

The theoretical study of flow with translational nonequilibrium is largely concerned with attempts to solve the Boltzmann equation. Because of the difficulties associated with this equation, exact solutions are rare, and approximate methods are required even for apparently simple physical situations. We shall examine several approaches to the Boltzmann equation, each with its own usefulness and limitations. The state of affairs is somewhat analogous to the study of the Navier-Stokes equations of macroscopic gas dynamics, for which no general method of solution is available and various solutions and simplifications have been developed for particular classes of flows. This analogy breaks down, however, when it is realized that the Boltzmann equation is more complex and yields more information than the Navier-Stokes equations. Hence much of our discussion will be directed toward the reduction of the kinetic formulation to a more manageable level, with a particular emphasis on the molecular interpretation of macroscopic gas dynamics. This chapter is thus somewhat misleadingly titled, since we shall be left with limited space to consider individual flow problems and shall be forced to exclude many interesting applications. For a comparison of the various methods, however, we shall discuss in some detail the structure of shock waves and linearized Couette flow. As in Chapter IX our attention will be directed almost exclusively to simple monatomic gases, although we shall give an indication of the applicability and extension of the theory to gases with internal degrees of freedom and to gas mixtures.

Although it would be impractical to summarize all approaches to the Boltzmann equation, it is instructive to consider examples of three representative but not necessarily exclusive classes of solution. Of these, the most direct is the analysis of situations for which the departure from a known solution is small. We shall consider in particular the Chapman-Enskog expansion, which for small departures from local translational equilibrium leads to the Navier-Stokes equations. A less exact method, which can be useful for particular problems over a wider range of physical parameters, is that in which the functional form of the distribution function is assumed *a priori*. This approach, which is analogous to boundary-layer integral methods, will be illustrated by the Mott-Smith investigation of shock-wave structure and by Lees' analysis of the Couette flow between parallel plates with an arbitrary ratio of spacing to mean free path. In the third approach, the Boltzmann equation itself is avoided and a simpler formulation is constructed. In this connection we shall discuss both the use of the moments of the Boltzmann equation and the Bhatnagar-Gross-Krook collision model in which the collision integral is replaced by a simple relaxation approximation.

For reasons of simplicity and clarity we shall discuss first the kinetic equation resulting from the Bhatnagar-Gross-Krook collision model and apply to it the Chapman-Enskog expansion. Having established the pertinent concepts we shall then consider the Chapman-Enskog method in its original and accurate form as applied to the Boltzmann equation itself. Finally we shall discuss the moment methods of Mott-Smith and Lees, and a comparison will be made with the results of the other approaches.

2 THE BHATNAGAR-GROSS-KROOK COLLISION MODEL

The difficulties encountered in the solution of the Boltzmann equation are largely associated with the nonlinear integral nature of the collision term [equation (IX 7.10)]. Bhatnagar, Gross, and Krook (1954) have proposed that this term be replaced by an approximation that simplifies the mathematics but retains many of the qualitative features of the true collision integral. Although several such representations have been suggested, we shall consider only the frequently used approximation in which the Boltzmann collision integral is replaced by

$$\boxed{\left[\frac{\partial}{\partial t}(nf)\right]_{\text{coll}} = n\nu(f_0 - f)} \, . \tag{2.1}$$

Here $f_0 = f_0(\bar{c}_i, T)$ is the Maxwellian distribution expressed in terms of the local mean velocity and the local temperature and ν is a collision frequency *that is proportional to density* and that may depend on temperature but not on molecular velocity. This approximation to the collision integral is designated by a variety of terms that include the B-G-K model, the relaxation model, and the statistical model. The kinetic equation that is obtained by substituting equation (2.1) into the Boltzmann equation is known variously as the Krook equation, the model equation, and the "Krooked" Boltzmann equation. The term relaxation model stems from the fact that if the left-hand side of the Boltzmann equation is considered as a time derivative—either with the convective and external-force terms zero or as a substantial derivative—the equation describes the relaxation of an initially nonequilibrium distribution toward the equilibrium distribution f_0. In this respect the B-G-K model (2.1) bears a close resemblance to the vibrational relaxation equation [equation (VII 2.11)]. In the B-G-K model the local relaxation time is $1/\nu$.

If the collision integral is replaced by the B-G-K model (2.1), the Boltzmann equation [equation (IX 3.2a)] becomes

$$\frac{\partial}{\partial t}(nf) + c_j\frac{\partial}{\partial x_j}(nf) + \frac{\partial}{\partial c_j}(F_j nf) = n\nu(f_0 - f). \tag{2.2}$$

Since $\bar{c}_i$ and T, which appear in f_0, are defined in terms of moments of the distribution function [see equations (2.4) below], the Krook equation (2.2) remains a nonlinear integro-differential equation. Although this equation is considerably simpler than the Boltzmann equation, its solution in most cases is still a matter of considerable difficulty.

Examination of equation (2.2) shows that it gives the correct solution, $f = f_0$, at equilibrium. We shall see in the next section that it leads to a kinetic-theory formulation of the Navier-Stokes equations when used with the Chapman-Enskog expansion. In the other limiting situation, that of very low densities (free-molecular flow), collisions are unimportant in the region of interest, and the structure of the collision term is irrelevant. It can also be shown that the B-G-K model reproduces the conservation equations. If as in Chapter IX, Sec. 6, we take the moments of equation (2.2) that correspond to the collisional invariants m, mc_i, and $mc^2/2$, the right-hand sides $\Delta[Q]$ of the equations of transfer thus obtained are zero, as follows:

$$\begin{aligned}
\int_{-\infty}^{\infty} mn\nu(f_0 - f)\,dV_c &= mn\nu\left[\int_{-\infty}^{\infty} f_0\,dV_c - \int_{-\infty}^{\infty} f\,dV_c\right] = 0,\\
\int_{-\infty}^{\infty} mc_i n\nu(f_0 - f)\,dV_c &= mn\nu[\bar{c}_i - \bar{c}_i] = 0, \qquad (2.3)\\
\int_{-\infty}^{\infty} \frac{1}{2}mc^2 n\nu(f_0 - f)\,dV_c &= \frac{1}{2}mn\nu\left[\left(\bar{c}^2 + 3\frac{k}{m}T\right) - \left(\bar{c}^2 + 3\frac{k}{m}T\right)\right] = 0.
\end{aligned}$$

Satisfaction of these conditions rests on the relation $\overline{c^2} = \bar{c}^2 + \overline{C^2}$ and the definitions

$$\int_{-\infty}^{\infty} f\,dV_c \equiv 1,$$

$$\int_{-\infty}^{\infty} c_i f\,dV_c \equiv \bar{c}_i, \tag{2.4}$$

and

$$\int_{-\infty}^{\infty} C^2 f\,dV_c \equiv 3\,\frac{k}{m}\,T,$$

which are an implicit part of the model. The derivative terms in the equation of transfer that is associated with the B-G-K model are unchanged from those found in Chapter IX, Sec. 6. Thus, on the basis of the collisional invariants we obtain, as before, the conservation equations (IX 2.22), (IX 2.23), and (IX 2.24).

The precise relationship between the Boltzmann collision integral and the B-G-K collision model is not so easily established. The collision integral is the sum of two terms, one that involves $f(c_i)f(\zeta_i)$ and results from depleting collisions and the other that involves $f(c_i')f(\zeta_i')$ and results from inverse, replenishing collisions. The B-G-K model replaces the first by $-n\nu f$ and the second by $n\nu f_0$.

The first replacement does in fact give the correct form, since this term in the collision integral is directly proportional to nf and to a collision frequency—the number of collisions per unit time between molecules of class c_i and all other molecules. In general, however, this collision frequency depends on the molecular velocity c_i. In the special case of Maxwellian molecules we have seen that the relative speed can be eliminated from the product $g\sigma\,d\Omega$. It then follows (see Exercise 2.1) that the collision frequency in this case is independent of c_i. For other intermolecular potentials, if we restrict our attention to the form of the collision model for which $\nu = \nu(n, T)$, the collision frequency must be taken as an average value that depends only on the local state of the gas. At this point one possibility is to adopt a definite expression for the average collision frequency ν (see Exercise 2.2). For greater flexibility it is more useful to defer this choice so that the exact form of ν can be selected to meet the requirements of the particular problem under consideration. One method used to prescribe ν is discussed in the next section.

The inverse-collision term is the greatest source of difficulty both here in interpretation and in any attempt to solve the Boltzmann equation itself. Physically the assumption that the number of inverse collisions is equal to $n\nu f_0$ can be considered equivalent to the approximation that molecules after a collision are instantaneously accommodated to a Maxwellian distribution at the local values of $\bar{c}_i$ and T. Liepmann, Narasimha, and Chahine (1962) have suggested that this use of the isotropic distribution f_0 can be supported by the fact that elastic spheres scatter isotropically in the center-of-mass reference frame [i.e., by equation (IX 8.1), $\sigma \neq \sigma(\chi)$].

A more detailed and mathematical comparison of the B-G-K collision model and the linearized form of the Boltzmann collision integral has been made by Gross and Jackson (1959) who suggest extensions of the model. In such an assessment of the utility of the B-G-K model we must bear in mind the alternatives available and the fact that for a wide range of physical parameters no exact solutions of the Boltzmann equation have thus far been obtained.

Exercise 2.1. (a) Find an expression for the collision frequency $\nu(c_i)$ that would make the term $-nf\nu$ in the B-G-K model an exact representation of the corresponding term in the Boltzmann collision integral.

(b) Show that for Maxwellian molecules the collision frequency of part (a) is independent of the molecular velocity c_i.

Exercise 2.2. A direct specification of $\nu(n, T)$ can be obtained from the average value of $\nu(c_i)$ [i.e., $\nu(n, T) = \overline{\nu(c_i)}$, where $\nu(c_i)$ is given by Exercise 2.1]. With the *ad hoc* substitution of the momentum-transfer cross-section σ_M for the total cross-section σ_T (as discussed in Chapter IX, Sec. 8) and with $f = f_0$, find $\nu(n, T)$ in this way for elastic-sphere and inverse-power molecules.

3 THE CHAPMAN-ENSKOG SOLUTION OF THE KROOK EQUATION

The equilibrium solution of both the Boltzmann equation and the Krook equation is $f = f_0$. The Chapman-Enskog solution, which gives the Navier-Stokes equations of macroscopic gas dynamics, results from an expansion of the distribution function for small departures from f_0. To develop the expansion for the Krook equation we shall first write equation (2.2) in a nondimensional form. We obtain the nondimensional variables by dividing c_i by a local reference speed c_r, x_i by a characteristic length L required for a significant change in nf, time by L/c_r, F_i by c_r^2/L, ν by a reference collision frequency ν_r, n by a reference density n_r, and f by c_r^{-3}. In this chapter we shall consider only external forces F_i that are independent of molecular velocity. If we denote the nondimensional variables by a caret placed over the symbol (e.g., $\hat{n}$), the nondimensional Krook equation then becomes

$$\xi\left[\frac{\partial}{\partial \hat{t}}(\hat{n}\hat{f}) + \hat{c}_j\frac{\partial}{\partial \hat{x}_j}(\hat{n}\hat{f}) + \hat{F}_j\frac{\partial}{\partial \hat{c}_j}(\hat{n}\hat{f})\right] = \hat{n}\hat{\nu}(\hat{f}_0 - \hat{f}), \tag{3.1}$$

where

$$\xi \equiv \frac{c_r}{L\nu_r}.$$

We assume that the reference quantities c_r, L, ν_r, and n_r can be chosen in such a way that each term in equation (3.1), with the exception of the parameter ξ, is at most of order unity.[1] An appropriate choice for c_r is $\sqrt{2kT/m}$ for low-speed flows and $\bar{c}$ at high Mach numbers.

The parameter ξ is inversely proportional to density through ν_r and is a measure of the degree of departure from local translational equilibrium.[2] The collision frequency ν_r that appears in ξ can be related to a reference mean free path λ_r by the expression $\lambda_r \nu_r = \sqrt{8kT/\pi m}$ [see equation (II 6.16)]. We can then give two equivalent physical interpretations of ξ by writing

$$\xi = \left(\frac{c_r}{\bar{c}}\right)\left(\frac{1/\nu_r}{L/\bar{c}}\right) = \left(\frac{c_r}{\sqrt{\dfrac{8kT}{\pi m}}}\right)\left(\frac{\lambda_r}{L}\right). \tag{3.2}$$

In the first expression, ξ is proportional to the ratio of a characteristic time between collisions $1/\nu_r$ (the relaxation time) to a characteristic flow time $L/\bar{c}$. In the alternate expression, ξ is proportional to the ratio of the mean free path λ_r to a characteristic length L. The ratio λ/L appears in the study of rarefied-gas dynamics and is called the Knudsen number.

Both of the foregoing interpretations indicate that for most situations encountered at ordinary densities in gas dynamics the parameter ξ will be very small (see Chapter I, Sec. 6, and Exercise 3.3). In such cases it can be said that the flow is collision-dominated. We shall consider in Secs. 9 and 10 limitations of the Chapman-Enskog theory in situations in which ξ is not small.

From equation (3.1) it can be seen that as the parameter ξ approaches zero f approaches f_0. In this situation the conservation equations reduce to the equations of equilibrium flow discussed in Chapter IX, Sec. 5. From our preceding discussion of the significance of ξ, the condition for the validity of these equations is the approach to zero of the ratio of the relaxation time to the flow time. Thus in this respect an analogy exists between the present situation and that encountered in the study of flows

[1] The time-derivative and external-force terms on the left-hand side of the equation (2.2) have been taken as of no greater order than the convective term. For problems where this is not so the choice of reference quantities must be modified. The results obtained by the expansion, however, would be unchanged.

[2] Here we consider small departures from translational equilibrium in the special sense implied by $\xi \ll 1$. In general, however, the departure from equilibrium can be small when the *magnitude* of the disturbance from equilibrium is small, independently of the value of ξ. We shall discuss such a situation, that of linearized Couette flow, in Sec. 10.

with chemical and vibrational nonequilibrium (see Chapter VIII, Sec. 3). It is also apparent from equation (3.1) that for nonnegligible values of ξ, gradients in the flow must be considered in the calculation of the distribution function.

The Chapman-Enskog expansion is an expansion of the distribution function for small values of ξ. For practical purposes the expansion is truncated after only two or, on occasion, three terms. We shall write f as a power series for small values of the parameter ξ in the forms

$$f = f_0(1 + \Phi_1 + \Phi_2 + \cdots) \tag{3.3a}$$

and

$$\hat{f} = \hat{f}_0(1 + \xi\phi_1 + \xi^2\phi_2 + \cdots) \tag{3.3b}$$

The alternate expressions (3.3a) and (3.3b) of this expansion will be used when we are considering dimensional and nondimensional variables, respectively (note that $\Phi_n = \xi^n\phi_n$). To obtain the functions ϕ_1, ϕ_2, etc., we substitute the expansion (3.3b) into the Krook equation (3.1) and equate like powers of ξ. Since ξ appears on the left-hand side of equation (3.1), the gradient terms on the left will affect the nth approximation ϕ_n only through the previously calculated ϕ_{n-1}.

If we retain only terms of first order in ξ, substitution of the expansion (3.3b) into the Krook equation (3.1) gives

$$\xi\left[\frac{\partial}{\partial \hat{t}}(\hat{n}\hat{f}_0) + \hat{c}_j\frac{\partial}{\partial \hat{x}_j}(\hat{n}\hat{f}_0) + \hat{F}_j\frac{\partial}{\partial \hat{c}_j}(\hat{n}\hat{f}_0)\right] = -\hat{n}\hat{v}\hat{f}_0\xi\phi_1. \tag{3.4}$$

With equation (IX 4.3) for f_0, the indicated differentiations on the left-hand side of equation (3.4) can be performed directly. If we consider f_0 a function of x_i and t through the dependence of $\bar{c}_i$ and T on x_i and t, we can write

$$\frac{\partial}{\partial \hat{t}}(\hat{n}\hat{f}_0) = \frac{\partial}{\partial \hat{n}}(\hat{n}\hat{f}_0)\frac{\partial \hat{n}}{\partial \hat{t}} + \frac{\partial}{\partial \hat{\bar{c}}_j}(\hat{n}\hat{f}_0)\frac{\partial \hat{\bar{c}}_j}{\partial \hat{t}} + \frac{\partial}{\partial \hat{T}}(\hat{n}\hat{f}_0)\frac{\partial \hat{T}}{\partial \hat{t}}, \tag{3.5a}$$

$$\frac{\partial}{\partial \hat{x}_i}(\hat{n}\hat{f}_0) = \frac{\partial}{\partial \hat{n}}(\hat{n}\hat{f}_0)\frac{\partial \hat{n}}{\partial \hat{x}_i} + \frac{\partial}{\partial \hat{\bar{c}}_j}(\hat{n}\hat{f}_0)\frac{\partial \hat{\bar{c}}_j}{\partial \hat{x}_i} + \frac{\partial}{\partial \hat{T}}(\hat{n}\hat{f}_0)\frac{\partial \hat{T}}{\partial \hat{x}_i}. \tag{3.5b}$$

Here $\hat{\bar{c}}_i$ and $\hat{T}$ are normalized by c_r and mc_r^2/k, respectively. Explicit differentiation of $\hat{n}\hat{f}_0$ with respect to $\hat{c}_i$, $\hat{n}$, $\hat{\bar{c}}_i$, and $\hat{T}$ then gives

$$\frac{\partial}{\partial \hat{c}_i}(\hat{n}\hat{f}_0) = -\hat{n}\hat{f}_0\frac{\hat{C}_i}{\hat{T}}, \tag{3.6a}$$

$$\frac{\partial}{\partial \hat{n}}(\hat{n}\hat{f}_0) = \hat{f}_0, \tag{3.6b}$$

$$\frac{\partial}{\partial \hat{\bar{c}}_i}(\hat{n}\hat{f}_0) = \hat{n}\hat{f}_0\frac{\hat{C}_i}{\hat{T}}, \tag{3.6c}$$

and

$$\frac{\partial}{\partial \hat{T}}(\hat{n}\hat{f}_0) = \hat{n}\hat{f}_0\left[\frac{\hat{C}^2}{2\hat{T}^2} - \frac{3}{2\hat{T}}\right]. \tag{3.6d}$$

If we were now to substitute equations (3.5) and (3.6) into equation (3.4) we would obtain an expression for the unknown function ϕ_1 in terms of the derivatives with respect to x_i and t of n, $\bar{c}_i$, and T. These derivatives, however, are not independent but are related by the conservation equations. An essential feature of the Chapman-Enskog expansion is the elimination of the time derivatives by means of the conservation equations. In our present notation the conservation equations (IX 2.22), (IX 2.23b), and (IX 2.24) can be written

$$\frac{\partial \hat{n}}{\partial \hat{t}} = -\frac{\partial}{\partial \hat{x}_j}(\hat{n}\hat{\bar{c}}_j), \tag{3.7a}$$

$$\hat{n}\frac{\partial \hat{\bar{c}}_i}{\partial \hat{t}} = -\hat{n}\hat{\bar{c}}_j\frac{\partial \hat{\bar{c}}_i}{\partial \hat{x}_j} - \frac{\partial \hat{p}}{\partial \hat{x}_i} + \frac{\partial \hat{\tau}_{ij}}{\partial \hat{x}_j} + \hat{n}\hat{F}_i, \tag{3.7b}$$

and

$$\hat{n}\frac{3}{2}\frac{\partial \hat{T}}{\partial \hat{t}} = -\hat{n}\hat{\bar{c}}_j\frac{3}{2}\frac{\partial \hat{T}}{\partial \hat{x}_j} - \hat{p}\frac{\partial \hat{\bar{c}}_j}{\partial \hat{x}_j} + \hat{\tau}_{kj}\frac{\partial \hat{\bar{c}}_k}{\partial \hat{x}_j} - \frac{\partial \hat{q}_j}{\partial \hat{x}_j}, \tag{3.7c}$$

where $\hat{p}$, $\hat{\tau}_{ij}$, and $\hat{q}_i$ are normalized by $mn_rc_r^2$, $mn_rc_r^2$, and $mn_rc_r^3$, respectively. The energy equation (3.7c) has been written in a form convenient for present purposes by combination with the continuity and momentum equations.

The viscous stress τ_{ij} and heat flux q_i are zero when $f = f_0$. With the distribution function $\hat{f} = \hat{f}_0(1 + \xi\phi_1)$, these quantities are given by

$$\hat{\tau}_{ij} = -\left[\int_{-\infty}^{\infty} \hat{n}\hat{C}_i\hat{C}_j(1 + \xi\phi_1)\hat{f}_0\,d\hat{V}_c - \hat{p}\,\delta_{ij}\right]$$

$$= -\xi\hat{n}\int_{-\infty}^{\infty} \hat{C}_i\hat{C}_j\phi_1\hat{f}_0\,dV_{\hat{c}}, \tag{3.8a}$$

and

$$\hat{q}_i = \xi\hat{n}\int_{-\infty}^{\infty} \tfrac{1}{2}\hat{C}_i\hat{C}^2\phi_1\hat{f}_0\,dV_{\hat{c}}. \tag{3.8b}$$

Thus $\hat{\tau}_{ij}$ and $\hat{q}_i$ are themselves of order ξ in our expansion in nondimensional variables. The terms involving $\hat{\tau}_{ij}$ and $\hat{q}_j$ that enter the left-hand side of equation (3.4) through equations (3.7) are then of second order in ξ and hence do not contribute to the present solution.

With the foregoing result the solution for ϕ_1 is obtained by substitution of equations (3.5), (3.6), and (3.7) [with τ_{ij} and q_i omitted from (3.7)] into equation (3.4). If we now return to our original dimensional variab have, after some manipulation,

$$\Phi_1 = \xi\phi_1 = -\frac{1}{n\nu f_0}\left[\frac{\partial}{\partial t}(nf_0) + c_j\frac{\partial}{\partial x_j}(nf_0) + F_j\frac{\partial}{\partial c_j}(nf_0)\right] \tag{3.9a}$$

$$= -\frac{1}{\nu}\left[C_j\left(\frac{mC^2}{2kT} - \frac{5}{2}\right)\frac{\partial}{\partial x_j}(\ln T) + \frac{m}{kT}(C_jC_k - \tfrac{1}{3}C^2\delta_{jk})\frac{\partial \bar{c}_j}{\partial x_k}\right]. \tag{3.9b}$$

A tensor such as $C_iC_j - \frac{1}{3}C^2\,\delta_{ij}$ for which the sum of the diagonal $(i = j)$ components is zero is said to be *traceless* or *nondivergent* and is given a special symbol as follows:

$$C_i\overset{\circ}{}C_j \equiv C_iC_j - \tfrac{1}{3}C^2\,\delta_{ij}. \tag{3.10}$$

With this notation and equation (3.9b) for ϕ_1, the distribution function (3.3) is

$$\boxed{f = f_0\left\{1 - \frac{1}{\nu}\left[C_j\left(\frac{mC^2}{2kT} - \frac{5}{2}\right)\frac{\partial}{\partial x_j}(\ln T) + \frac{m}{kT}\,C_j\overset{\circ}{}C_k\frac{\partial \bar{c}_j}{\partial x_k}\right]\right\}}. \tag{3.11}$$

To give a brief summary of this calculation, we note that equation (3.11) is obtained formally if on the left-hand side of the Krook equation (2.2) we replace f by f_0 and eliminate time derivatives by means of the conservation equations in the form consistent with $f = f_0$.

In spite of the relatively simple nature of the B-G-K collision model, the distribution function (3.11) is similar to that obtained by the first approximation in the Chapman-Enskog solution of the Boltzmann equation. Substitution in equations (2.4) shows that equation (3.11) satisfies the normalization condition and the definitions of mean velocity and temperature. The first-order correction (3.9b) to the Maxwellian distribution function is proportional to the gradients of T and $\bar{c}_i$. Such a solution of the Boltzmann equation for which the state of the gas depends only on the parameters n, T, and $\bar{c}_i$ is called a normal solution. The relationship between normal and general solutions is discussed by Chapman and Cowling (1952), Sec. 7.2 and by Grad (1958), Chapter II.

With the nonequilibrium distribution function (3.11) we can now find the viscous-stress tensor τ_{ij} and the heat-flux vector q_i. By substitution of equation (3.9b) for $\xi\phi_1$ into equations (3.8) we obtain

$$\tau_{ij} = \frac{nkT}{\nu}\left[\frac{\partial \bar{c}_i}{\partial x_j} + \frac{\partial \bar{c}_j}{\partial x_i} - \frac{2}{3}\frac{\partial \bar{c}_k}{\partial x_k}\delta_{ij}\right], \tag{3.12a}$$

and

$$q_i = -\frac{5}{2}\left(\frac{k}{m}\right)\frac{nkT}{\nu}\frac{\partial T}{\partial x_i}. \tag{3.12b}$$

The foregoing expressions become identical to their continuum equivalents if for the coefficients of viscosity μ and thermal conductivity K we write

$$\boxed{\mu = \frac{nkT}{\nu}} \tag{3.13a}$$

$$\boxed{K = \frac{5}{2}\left(\frac{k}{m}\right)\frac{nkT}{\nu}}. \tag{3.13b}$$

These results are similar in form to those obtained by mean-free-path arguments [see Chapter I, Sec. 5, and equation (II 6.16)]. It can be seen that the dependence on temperature given by equations (3.13) is the same for both μ and K and that the Prandtl number $\text{Pr} = c_p\mu/K$ is unity. The observed Prandtl number for monatomic gases varies little with temperature, as predicted here, but is more nearly equal to $\frac{2}{3}$.

The Navier-Stokes equations now can be obtained from the conservation equations (IX 2.22), (IX 2.23), (IX 2.24) with τ_{ij} and q_i given in terms of μ and K by equations (3.12) and (3.13). Since the rigorous kinetic-theory derivation of the Navier-Stokes equations must be based on the full Boltzmann equation, we shall defer writing these equations until a later section.

It is appropriate to indicate at this point a method useful for specifying the collision frequency ν in the B-G-K model. The coefficient of viscosity is known for most gases as a function of temperature and pressure (either from experiment or by calculations based on the application of the Chapman-Enskog expansion to the Boltzmann equation). By equation (3.13a) we can relate ν to the known coefficient of viscosity. We then assume that this relationship is reasonable whether or not equation (3.12a) for τ_{ij} is applicable. The value of ν thus found can be used with the B-G-K model to investigate physical situations that are not readily analyzed with the Boltzmann equation itself. Thus the results of this section are used to provide the connection between the parameter ν in the B-G-K model and the intermolecular potential for a particular gas. This procedure should not be taken as definitive, however, particularly since the Prandtl number predicted by equations (3.13) is unity rather than the correct value of $\frac{2}{3}$ and since ν could be taken as specified by K rather than by μ.

The B-G-K model has been applied to a number of problems of translational nonequilibrium. In later sections we shall quote the results from the application of this model to the study of both the structure of shock waves and linearized Couette flow. Our purpose here has been to introduce this useful method and at the same time to discuss the Chapman-Enskog expansion within a relatively simple context in which the general approach is not obscured by extensive mathematics. In the next section we shall apply the Chapman-Enskog expansion to the Boltzmann equation.

Exercise 3.1. Verify: (a) equations (3.6) and (b) equations (3.12).

Exercise 3.2. Use the results of Exercise 2.2 to predict with the B-G-K model the dependence of the coefficients of viscosity and thermal conductivity on temperature for inverse-power and elastic-sphere molecules. Compare your results with equations (5.9) below (elastic-sphere molecules correspond to $\alpha \to \infty$).

Exercise 3.3. A circular cylinder 1 cm in diameter moves at 10^3 cm/sec through argon at standard temperature and pressure. For positions removed from the stagnation point by an angle of, say, 45°, estimate the value of the parameter ξ (a) at the outer edge of the boundary layer and (b) in the boundary layer near the wall. Assume that the boundary layer is laminar and neglect density changes in the flow. (See Chapter I, Exercise 5.1.)

Exercise 3.4. Outline the procedure by which you would obtain the second approximation $f = f_0(1 + \Phi_1 + \Phi_2)$ to the distribution function. Indicate in what form τ_{ij} and q_i enter into the elimination of time derivatives in the equation for Φ_2. Find the order of the derivatives (e.g., second derivative) of $\bar{c}_i$ and T that appear in Φ_2, τ_{ij}, q_i and the conservation equations. (The conservation equations that correspond to this approximation are called the *Burnett equations*.)

4 THE CHAPMAN-ENSKOG SOLUTION OF THE BOLTZMANN EQUATION

In its accurate and original form the Chapman-Enskog expansion is used to obtain a solution of the Boltzmann equation. Such a solution was found independently by Chapman and by Enskog, whose methods differed widely but gave identical results. Our discussion will follow Enskog's approach, which is based on the mathematical theory of Hilbert and is the less intuitive and more systematic. An interesting historical summary of the development of the theory is given by Chapman and Cowling (1952), p. 380.

Proceeding as in Sec. 3, we can write the Boltzmann equation in non-dimensional form if we divide the differential collision frequency $ng\sigma\, d\Omega$

by ν_r and normalize the other variables as before. We then obtain[3]

$$\xi\left[\frac{\partial}{\partial \hat{t}}(\hat{n}\hat{f}) + \hat{c}_j\frac{\partial}{\partial \hat{x}_j}(\hat{n}\hat{f}) + \hat{F}_j\frac{\partial}{\partial \hat{c}_j}(\hat{n}\hat{f})\right] = \left[\frac{\partial}{\partial \hat{t}}(\hat{n}\hat{f})\right]_{\text{coll}}, \tag{4.1}$$

with

$$\xi \equiv \frac{c_r}{L\nu_r}.$$

The parameter ξ, which is as before a measure of the degree of departure from local translational equilibrium, has the same physical interpretation that it did in Sec. 3. For small values of ξ we write f as the power series [cf. equations (3.3)]

$$f = f_0(1 + \Phi_1 + \Phi_2 + \cdots), \tag{4.2a}$$

and

$$\hat{f} = \hat{f}_0(1 + \xi\phi_1 + \xi^2\phi_2 + \cdots). \tag{4.2b}$$

Again we note that $\Phi_n = \xi^n\phi_n$. When we substitute the expansion (4.2b) into the Boltzmann equation (4.1) and equate like powers of ξ, we obtain integral equations for the succeeding approximations to the distribution function. The zero-order solution, which corresponds to $\xi \to 0$, is $f = f_0$. [This was proved in Chapter IX, Sec. 4, and justifies the form of the expansion (4.2b).] As we have seen the conservation equations in this case reduce to the equations of equilibrium flow. Retention of first-order terms in ξ leads to the Navier-Stokes equations of macroscopic gas dynamics. As with the B-G-K model, to first order in ξ only those terms involving f_0 and not ϕ_1 contribute to the left-hand side of the Boltzmann equation (4.1). In the collision integral [equation (IX 7.10)] on the right-hand side of equation (4.1) we have from Chapter IX, Sec. 4,

$$\hat{f}_0(\hat{c}_i')\hat{f}_0(\hat{\zeta}_i') - \hat{f}_0(\hat{c}_i)\hat{f}_0(\hat{\zeta}_i) = 0. \tag{4.3}$$

With $\hat{f} = \hat{f}_0(1 + \xi\phi_1)$ we then have, to first order in ξ,

$$\begin{aligned}\hat{f}(\hat{c}_i')\hat{f}(\hat{\zeta}_i') - \hat{f}(\hat{c}_i)\hat{f}(\hat{\zeta}_i) &= \xi\{\hat{f}_0(\hat{c}_i')[\hat{f}_0(\hat{\zeta}_i')\phi_1(\hat{\zeta}_i')] + [\hat{f}_0(\hat{c}_i')\phi_1(\hat{c}_i')]\hat{f}_0(\hat{\zeta}_i')\}\\ &\quad -\xi\{\hat{f}_0(\hat{c}_i)[\hat{f}_0(\hat{\zeta}_i)\phi_1(\hat{\zeta}_i)] + [\hat{f}_0(\hat{c}_i)\phi_1(\hat{c}_i)]\hat{f}_0(\hat{\zeta}_i)\}\\ &= \xi[\phi_1(\hat{c}_i') - \phi_1(\hat{c}_i) + \phi_1(\hat{\zeta}_i') - \phi_1(\hat{\zeta}_i)]\hat{f}_0(\hat{c}_i)\hat{f}_0(\hat{\zeta}_i),\end{aligned} \tag{4.4}$$

[3] As stated in Sec. 3, in this chapter we consider only external forces F_i that are independent of molecular velocity.

where equation (4.3) has been used to obtain the last expression. We can now write the Boltzmann equation (4.1) as

$$\xi\left[\frac{\partial}{\partial \hat{t}}(\hat{n}\hat{f}_0) + \hat{c}_j\frac{\partial}{\partial \hat{x}_j}(\hat{n}\hat{f}_0) + \hat{F}_j\frac{\partial}{\partial \hat{c}_j}(\hat{n}\hat{f}_0)\right]$$
$$= \xi\int_{-\infty}^{\infty}\int_0^{4\pi} \hat{n}^2[\phi_1(\hat{c}_i') - \phi_1(\hat{c}_i) + \phi_1(\hat{\zeta}_i')$$
$$- \phi_1(\hat{\zeta}_i)]\hat{f}_0(\hat{c}_i)\hat{f}_0(\hat{\zeta}_i)\hat{g}\hat{\sigma}\, d\Omega\, dV_{\hat{\zeta}}, \quad (4.5)$$

plus terms of order ξ^2.

The left-hand side of equation (4.5) can be evaluated by explicit differentiation of $\hat{n}\hat{f}_0$ and the time derivatives eliminated by use of the conservation equations in the form consistent with $\hat{f} = \hat{f}_0$.[4] The result of these calculations is obtained directly from equations (3.9), which were derived in connection with the Chapman-Enskog solution of the Krook equation. If we return now to our original dimensional variables and use the notation of equation (3.10), we have

$$\frac{\partial}{\partial t}(nf_0) + c_j\frac{\partial}{\partial x_j}(nf_0) + F_j\frac{\partial}{\partial c_j}(nf_0)$$
$$= nf_0\left[C_j\left(\frac{mC^2}{2kT} - \frac{5}{2}\right)\frac{\partial}{\partial x_j}(\ln T) + \frac{m}{kT}\overset{\circ}{C_jC_k}\frac{\partial \bar{c}_j}{\partial x_k}\right]. \quad (4.6)$$

The right-hand side of equation (4.5) is, to within a factor, the integral $I[Q]$ that was defined in Chapter IX, Sec. 9, with $Q = \Phi_1$. It is convenient to express this integral in terms of the variable C_i that appears on the right-hand side of equation (4.6). With $g = |\zeta_i - c_i| = |Z_i - C_i|$ and $dV_\zeta = dV_Z$, we have

$$I[\Phi_1] = -\int_{-\infty}^{\infty}\int_0^{4\pi}(\delta_C\Phi_1 + \delta_Z\Phi_1)f_0(C_i)f_0(Z_i)g\sigma\, d\Omega\, dV_Z, \quad (4.7a)$$

where Φ_1 is now considered to be a function of the thermal velocity C_i and

$$\delta_C\Phi_1 = \Phi_1(C_i') - \Phi_1(C_i),$$
$$\delta_Z\Phi_1 = \Phi_1(Z_i') - \Phi_1(Z_i). \quad (4.7b)$$

We note that equations (4.7) are simply a change in notation and that $\delta_C\Phi_1 = \delta_c\Phi_1$, just as the two forms for f_0 given by equation (IX 4.3) are equivalent. With equations (4.6) and (4.7) and $\Phi_1 = \xi\phi_1$, the linearized

[4] The relationship of these calculations to the mathematical conditions of solubility of equation (4.5) is discussed by Chapman and Cowling (1952), Sec. 7.13 and pp. 114 and 120, and by Uhlenbeck and Ford (1963), p. 105.

Boltzmann equation (4.5) becomes

$$nf_0\left[C_j\left(\frac{mC^2}{2kT}-\frac{5}{2}\right)\frac{\partial}{\partial x_j}(\ln T)+\frac{m}{kT}\overset{\circ}{C_jC_k}\frac{\partial \bar{c}_j}{\partial x_k}\right]=-n^2I[\Phi_1]\,. \tag{4.8}$$

Thus for small departures from local equilibrium, the mathematical description of a gas is transformed from the nonlinear integro-differential Boltzmann equation for f to the linear integral equation (4.8) for Φ_1.

From equation (4.8) we can establish the form of Φ_1 without actually obtaining a complete solution. Equations (4.7) and (4.8) have been written so that the molecular velocity appears through the variable C_i rather than the variable c_i. (The velocity variables C_i and $\bar{c}_i$ taken together are completely equivalent to c_i and $\bar{c}_i$.) From equations (IX 7.8) and (IX 7.9) with $C_i' - C_i = c_i' - c_i$ and $Z_i' - Z_i = \zeta_i' - \zeta_i$, the velocities C_i' and Z_i' in equations (4.7) are functions only of C_i, Z_i, χ, and ε. Therefore $\bar{c}_i$ does not appear explicitly in equation (4.8) except in $\partial\bar{c}_i/\partial x_j$. The temperature T enters both directly and in $\partial \ln T/\partial x_i$. It then follows from equation (4.8) that Φ_1 depends on C_i and on space and time only through n, T, and the derivatives of $\bar{c}_i$ and T. Since $I[\Phi_1]$ is linear in Φ_1, equation (4.8) can be written in terms of the variable $n\Phi_1$, eliminating n, so that Φ_1 must be inversely proportional to n. In addition, the left-hand side of equation (4.8) is linear in the components of $\partial \ln T/\partial x_i$ and $\partial\bar{c}_i/\partial x_j$. Therefore the product $n\Phi_1$ must be a function of only C_i, T, and a linear combination of the components of $\partial \ln T/\partial x_i$ and $\partial\bar{c}_i/\partial x_j$. A partial solution of equation (4.8) can then be written in the form

$$\Phi_1=-\frac{1}{n}\left[\sqrt{2kT/m}\,A_j\frac{\partial}{\partial x_j}(\ln T)+B_{jk}\frac{\partial \bar{c}_j}{\partial x_k}+\psi\right], \tag{4.9}$$

where A_i, B_{ij}, and ψ are unknown functions of C_i and T. The minus sign and the factor $(2kT/m)^{1/2}$ have been introduced in equation (4.9) for later convenience. The terms in Φ_1 involving A_i and B_{ij} on the one hand and ψ on the other correspond respectively to the particular and complementary solutions of a differential equation. The reader who has not been convinced by the discussion that led to this result may wish to consider equation (4.9) an assumed form that can be confirmed by substitution into equation (4.8).

When equation (4.9) for Φ_1 is substituted into the linearized Boltzmann equation in the form (4.8), we find that A_i, B_{ij}, and ψ must satisfy the equations

$$I[A_i]=f_0\mathscr{C}_i(\mathscr{C}^2-\tfrac{5}{2}), \tag{4.10a}$$

$$I[B_{ij}]=2f_0\overset{\circ}{\mathscr{C}_i\mathscr{C}_j}, \tag{4.10b}$$

and

$$I[\psi] = 0, \tag{4.10c}$$

where

$$\mathscr{C}_i \equiv \frac{C_i}{\sqrt{2kT/m}}. \tag{4.11}$$

In addition to equations (4.10), which are equivalent to equation (4.8), the distribution function $f = f_0(1 + \Phi_1)$ must satisfy the conditions

$$\int_{-\infty}^{\infty} f\, dV_c \equiv 1, \qquad \int_{-\infty}^{\infty} c_i f\, dV_c \equiv \bar{c}_i, \qquad \text{and} \qquad \frac{m\overline{C^2}}{2} \equiv \tfrac{3}{2}kT. \tag{4.12a}$$

Since equations (4.12a) are satisfied by f_0 alone, we must have

$$\int_{-\infty}^{\infty} f_0\Phi_1\, dV_c = 0, \qquad \int_{-\infty}^{\infty} c_i f_0 \Phi_1\, dV_c = 0,$$

and

$$\int_{-\infty}^{\infty} C^2 f_0 \Phi_1\, dV_c = 0. \tag{4.12b}$$

At this point the equations that govern the Chapman-Enskog solution have been established. We shall next outline the reasoning used to find the structure of A_i and B_{ij}, and the solution for ψ. For a more detailed analysis, the reader is referred to Chapman and Cowling (1952), Sec. 7.31. We first note that A_i is a vector function of $\mathscr{C}_i$ and T. The only vectors that can be formed from these quantities are products of $\mathscr{C}_i$ itself and scalar functions $A(\mathscr{C}, T)$. Hence we can write

$$A_i = A(\mathscr{C}, T)\mathscr{C}_i. \tag{4.13}$$

Similarly, the component equations (4.10b) show that $B_{ij} = B_{ji}$ and that $B_{ii} = 0$. That is, B_{ij} is a symmetrical traceless tensor. The only second-order symmetrical traceless tensors that can be formed from $\mathscr{C}_i$ and T are products of $\overset{\circ}{\mathscr{C}_i\mathscr{C}_j}$ and scalar functions $B(\mathscr{C}, T)$. We thus have

$$B_{ij} = B(\mathscr{C}, T)\overset{\circ}{\mathscr{C}_i\mathscr{C}_j}. \tag{4.14}$$

Chapman (1915) obtained essentially these results [equations (4.13) and (4.14)] by the less abstract requirement that the form of Φ_1 be invariant with respect to a rotation of the coordinate axes.

We consider now the function ψ. From equations (4.7) we see that $\delta_C\psi + \delta_Z\psi = \delta_c\psi + \delta_\zeta\psi = 0$ when ψ is one of the collisional invariants. Hence a linear combination of the collisional invariants with arbitrary coefficients is a solution of $I[\psi] = 0$ [equation (4.10c)]. By means of an analysis similar to that of Chapter IX, Sec. 4, it is possible to show that this is the only solution (see Chapman and Cowling, 1952, Sec. 4.41). When the conditions (4.12b) are imposed on Φ_1 with this form for ψ, the coefficients of all terms in ψ except one must be zero. The term that remains can be absorbed, however, in the as yet unspecified term in Φ_1 involving A. Thus, we can set $\psi = 0$. In addition, to satisfy the conditions (4.12b) we must have (Exercise 4.2)

$$\int_{-\infty}^{\infty} A(\mathscr{C}, T)\mathscr{C}^2 f_0 \, dV_C = 0. \tag{4.15}$$

These results now can be combined and the partial solution summarized as follows. For small values of the parameter ξ, the distribution function can be written

$$\boxed{f = f_0(1 + \Phi_1)}\,, \tag{4.16}$$

where $\Phi_1 \ll 1$. The linearized Boltzmann equation (4.8) and the conditions (4.12) that result from the definitions of $\bar{c}_i$ and T and the normalization of f require that Φ_1 be of the form

$$\boxed{\Phi_1 = -\frac{1}{n}\left[\sqrt{2kT/m}\, A_j \frac{\partial}{\partial x_j}(\ln T) + B_{jk}\frac{\partial \bar{c}_j}{\partial x_k}\right]}\,, \tag{4.17a}$$

with

$$A_i = A(\mathscr{C}, T)\mathscr{C}_i \qquad \text{and} \qquad B_{ij} = B(\mathscr{C}, T)\overset{\circ}{\mathscr{C}_i \mathscr{C}_j}. \tag{4.17b}$$

The two remaining unknown functions $A(\mathscr{C}, T)$ and $B(\mathscr{C}, T)$ must satisfy the equations

$$I[A\mathscr{C}_i] = f_0\mathscr{C}_i(\mathscr{C}^2 - \tfrac{5}{2}) \tag{4.18a}$$

$$I[B\overset{\circ}{\mathscr{C}_i\mathscr{C}_j}] = 2f_0\overset{\circ}{\mathscr{C}_i\mathscr{C}_j}\,, \tag{4.18b}$$

subject to the condition (4.15).

Exercise 4.1. Derive equations (4.10) for A_i, B_{ij}, and ψ from equations (4.8) and (4.9).

Exercise 4.2. With Φ_1 given by equations (4.17), show that the distribution function (4.16) satisfies the conditions (4.12) if equation (4.15) holds.

5 THE NAVIER-STOKES EQUATIONS

The partial solution (4.17) now can be used to find expressions for the viscous-stress tensor τ_{ij} and the heat-flux vector q_i. When equations (4.16) and (4.17) for the distribution function are substituted into the definitions (IX 2.17) and (IX 2.19) for τ_{ij} and q_i, we obtain after some manipulation (see Exercise 5.1)

$$\tau_{ij} = \frac{kT}{5}\left\{\int_{-\infty}^{\infty} \mathscr{C}_l \overset{\circ}{\mathscr{C}}_m B_{lm} f_0 \, dV_C\right\}\left(\frac{\partial \bar{c}_i}{\partial x_j} + \frac{\partial \bar{c}_j}{\partial x_i} - \frac{2}{3}\frac{\partial \bar{c}_k}{\partial x_k}\delta_{ij}\right) \tag{5.1a}$$

$$= \frac{2kT}{15}\left\{\int_{-\infty}^{\infty} B(\mathscr{C}, T)\mathscr{C}^4 f_0 \, dV_C\right\}\left(\frac{\partial \bar{c}_i}{\partial x_j} + \frac{\partial \bar{c}_j}{\partial x_i} - \frac{2}{3}\frac{\partial \bar{c}_k}{\partial x_k}\delta_{ij}\right), \tag{5.1b}$$

and

$$q_i = -\frac{2}{3}\frac{k^2T}{m}\left\{\int_{-\infty}^{\infty} \mathscr{C}_j \mathscr{C}^2 A_j f_0 \, dV_C\right\}\frac{\partial T}{\partial x_i} \tag{5.2a}$$

$$= -\frac{2}{3}\frac{k^2T}{m}\left\{\int_{-\infty}^{\infty} A(\mathscr{C}, T)\mathscr{C}^4 f_0 \, dV_C\right\}\frac{\partial T}{\partial x_i}. \tag{5.2b}$$

These expressions are similar in form to those obtained in Sec. 3 on the basis of the B-G-K model. We now can find, however, accurate formulas for the coefficients of viscosity μ and thermal conductivity K. To this end we note that equations (5.1) and (5.2) will agree with the corresponding expressions

$$\tau_{ij} = \mu\left(\frac{\partial \bar{c}_i}{\partial x_j} + \frac{\partial \bar{c}_j}{\partial x_i} - \frac{2}{3}\frac{\partial \bar{c}_k}{\partial x_k}\delta_{ij}\right) + \mu_B \frac{\partial \bar{c}_k}{\partial x_k}\delta_{ij} \tag{5.3a}$$

and

$$q_i = -K\frac{\partial T}{\partial x_i} \tag{5.3b}$$

that are derived by the methods of continuum gas dynamics (see, for example, Serrin, 1959, Chapter G) if we write

$$\boxed{\begin{aligned} \mu &= \frac{2kT}{15}\int_{-\infty}^{\infty} B(\mathscr{C}, T)\mathscr{C}^4 f_0 \, dV_C \qquad (5.4a) \\ K &= \frac{2k^2T}{3m}\int_{-\infty}^{\infty} A(\mathscr{C}, T)\mathscr{C}^4 f_0 \, dV_C \qquad (5.4b) \end{aligned}},$$

and

$$\mu_B = 0. \tag{5.5}$$

Thus μ and K are given in terms of the functions $B(\mathscr{C}, T)$ and $A(\mathscr{C}, T)$ that are the solutions of equations (4.18). The quantity μ_B that enters the continuum formulation (5.3) is the *bulk-viscosity coefficient.* From equation (5.5) it can be seen that for perfect monatomic gases kinetic theory predicts that $\mu_B = 0$. We shall examine the relationship between rotational nonequilibrium and bulk viscosity in Sec. 8.[5]

[5] When μ_B is not zero care must be exercised in the interpretation of the continuum result (5.3a), particularly with regard to our definition of pressure (see Sec. 8).

When equations (5.1), (5.2), and (5.4) are used for τ_{ij} and q_i in the conservation equations (IX 2.22), (IX 2.23b), (IX 2.24b), we obtain the Navier-Stokes equations (with $\mu_B = 0$) as follows:

$$\frac{\partial \rho}{\partial t} + \frac{\partial}{\partial x_j}(\rho \bar{c}_j) = 0, \tag{5.6a}$$

$$\rho \frac{D\bar{c}_i}{Dt} = -\frac{\partial p}{\partial x_i} + \frac{\partial}{\partial x_j}\left[\mu\left(\frac{\partial \bar{c}_i}{\partial x_j} + \frac{\partial \bar{c}_j}{\partial x_i}\right)\right] - \frac{2}{3}\frac{\partial}{\partial x_i}\left[\mu \frac{\partial \bar{c}_j}{\partial x_j}\right] + \rho F_i, \tag{5.6b}$$

and

$$\rho \frac{D}{Dt}(h + \tfrac{1}{2}\bar{c}^2) = \frac{\partial p}{\partial t} + \frac{\partial}{\partial x_j}\left[\bar{c}_k \mu\left(\frac{\partial \bar{c}_j}{\partial x_k} + \frac{\partial \hat{c}_k}{\partial x_j}\right) - \bar{c}_j \tfrac{2}{3}\mu \frac{\partial \bar{c}_k}{\partial x_k}\right]$$
$$+ \frac{\partial}{\partial x_j}\left(K \frac{\partial T}{\partial x_j}\right) + \rho F_j \bar{c}_j. \tag{5.6c}$$

Here we have imposed the condition $F_i \neq F_i(c_i)$. Thus, in the present case of monatomic, nonreacting gases, kinetic theory is able to establish from molecular considerations the equations of macroscopic gas dynamics. As we shall see in the next section, it also provides the means whereby the values of the transport properties can be calculated for given intermolecular potentials.

In general, to evaluate the transport properties μ and K from equations (5.4) we must solve equations (4.18) for $A(\mathscr{C}, T)$ and $B(\mathscr{C}, T)$. For inverse-power and elastic-sphere molecules, however, we can obtain μ and K to within numerical factors from the partial solution. To do this we must find the temperature dependence of $A(\mathscr{C}, T)$ and $B(\mathscr{C}, T)$.

We shall consider directly only inverse-power molecules, for which the intermolecular potential varies as $r^{-\alpha}$, since the elastic-sphere result is obtained when $\alpha \to \infty$. With equations (IX 4.3) and (IX 8.4), we can write the quantity $f_0(Z_i) g\sigma \, d\Omega \, dV_Z$ in equation (4.7a) for $I[\Phi_1]$ as

$$\left(\frac{m}{2\pi kT}\right)^{3/2} e^{-mZ^2/2kT} g \left(\frac{2\alpha a}{m}\right)^{2/\alpha} g^{-4/\alpha} \beta \, d\beta \, d\varepsilon \, dV_Z$$
$$= \left(\frac{2kT}{m}\right)^{(\alpha-4)/2\alpha} \left[\frac{1}{\pi^{3/2}} e^{-\hat{Z}^2} \hat{g} \left(\frac{2\alpha a}{m}\right)^{2/\alpha} \hat{g}^{-4/\alpha} \beta \, d\beta \, d\varepsilon \, dV_{\hat{Z}}\right], \tag{5.7}$$

where, as before, a velocity written with a caret (e.g., $\hat{Z}$) is normalized by $\sqrt{2kT/m}$, as is $\mathscr{C}_i$. We then note that, through equations (IX 7.8), (IX 7.9), and (IX 8.3), the relationship implicit in equations (4.7) between C_i' and Z_i', and C_i, Z_i, β, and ε can be expressed solely in terms of the nondimensional form of the variables and does not involve T. Therefore, since I is a linear operator, equations (4.18) can be written with the use of equation (5.7) as

$$I\left[\left(\frac{2kT}{m}\right)^{(\alpha-4)/2\alpha} A(\mathscr{C}, T)\mathscr{C}_i\right] = \hat{f}_0\mathscr{C}_i(\mathscr{C}^2 - \tfrac{5}{2}), \tag{5.8a}$$

$$I\left[\left(\frac{2kT}{m}\right)^{(\alpha-4)/2\alpha} B(\mathscr{C}, T)\overset{\circ}{\mathscr{C}_i\mathscr{C}_j}\right] = 2\hat{f}_0\overset{\circ}{\mathscr{C}_i\mathscr{C}_j}, \tag{5.8b}$$

where f_0 now appears as $\hat{f}_0$ in I. In this form, T does not enter equations (5.8) except as a multiplier of A and B. It then follows that A and B are each inversely proportional to $T^{(\alpha-4)/2\alpha}$. Now, from equations (5.4) we see that μ and K are independent of density and that, if we integrate over the variable $\mathscr{C}_i$ rather than C_i, $\mathscr{C}^4 f_0\, dV_C$ depends only on $\mathscr{C}$ and not explicitly on T. With the foregoing temperature dependence of A and B, we then have the result that both μ and K depend on temperature to the power $1 - (\alpha - 4)/2\alpha = (\alpha + 4)/2\alpha$. That is, we can write

$$\mu = \mu_r\left(\frac{T}{T_r}\right)^{(\alpha+4)/2\alpha} \tag{5.9a}$$

and

$$K = K_r\left(\frac{T}{T_r}\right)^{(\alpha+4)/2\alpha}, \tag{5.9b}$$

where μ_r, K_r, and T_r are reference quantities. For example, for Maxwellian molecules $\alpha = 4$ and μ and K are proportional to temperature. The elastic-sphere result is obtained when $\alpha \to \infty$; μ and K then vary as the square root of the temperature (see Exercise 5.3).

In this section we have seen some of the consequences of the Chapman-Enskog theory and have found to within numerical constants the transport properties for certain representations of the intermolecular potential. To proceed further and consider general intermolecular potentials we must go beyond the partial solution.

Exercise 5.1. Derive equations (5.1) and (5.2) for τ_{ij} and q_i. The following (inverse) approach is suggested: (a) Show that (5.2b) follows from (5.2a). (b) Show that $(\overset{\circ}{\mathscr{C}_l\mathscr{C}_m})(\overset{\circ}{\mathscr{C}_l\mathscr{C}_m}) = \frac{2}{3}\mathscr{C}^4$ and hence that (5.1b) follows from (5.1a). (c) Derive equation (5.2a). (d) Derive equation (5.1a), first for $i = j$, then for $i \neq j$.

In parts (c) and (d) it is helpful to use the fact that certain integrals, for which the integrands are odd functions of the components of velocity, can be set equal to zero by inspection.

Exercise 5.2. The energy equation (5.6c) corresponds to equation (VIII 2.4b). Find the energy equation that incorporates the results of this section and corresponds to equation (VIII 2.4a) (see Chapter IX, Exercise 2.3).

Exercise 5.3. By arguments similar to those used for inverse-power molecules, obtain *directly* the temperature dependence of the coefficients of viscosity and thermal conductivity for elastic-sphere molecules.

6 EXPANSION IN SONINE POLYNOMIALS

The most difficult part of the Chapman-Enskog theory is the solution of equations (4.18) for $A(\mathscr{C}, T)$ and $B(\mathscr{C}, T)$. With equations (5.4), such a solution gives the values of the coefficients of thermal conductivity K and viscosity μ. A closed-form solution of equations (4.18) exists only for Maxwellian molecules. Fortunately a series solution has been developed for other intermolecular potentials. This method gives K and μ accurately in many cases with a few terms in the series, even though only approximate solutions are found for $A(\mathscr{C}, T)$ and $B(\mathscr{C}, T)$ (and hence for the distribution function). The method was introduced into the theory by Burnett (1935) and consists of the expansion of $A(\mathscr{C}, T)$ and $B(\mathscr{C}, T)$ in series of Sonine polynomials [defined by equation (6.1) below]. The discussion of the Sonine-polynomial expansions becomes involved and is rather specialized and mathematical. For this reason, some readers who are more interested in the application of kinetic theory to flow problems, such as those considered in Secs. 9 and 10, may wish to omit this section.

Although other series solutions are possible, as for example the power series used originally by Chapman, the use of the Sonine-polynomial series brings about a great simplification in the accurate evaluation of transport properties. The reader for whom this method is less than obvious may be consoled by the fact that over twenty years elapsed between the publication of the original papers of Chapman and of Enskog and the work of Burnett. In fact, certain properties of the Sonine polynomials that clarify the interpretation of the method have been found only recently (see Uhlenbeck and Ford, 1963, and Wang Chang and Uhlenbeck, 1952).

The Sonine polynomials $S_m^{(n)}$ depend on the two indices m and n and are defined by

$$S_m^{(n)}[x] = \sum_{p=0,1,2,\ldots}^{n} \frac{(m+n)!}{(m+p)!\,p!\,(n-p)!}(-x)^p. \tag{6.1}$$

In particular we have

$$S_m^{(0)}[x] = 1, \qquad S_m^{(1)}[x] = m + 1 - x,$$

and

$$S_m^{(2)}[x] = \frac{(m+1)(m+2)}{2} + (m+2)(-x) + \frac{x^2}{2}. \tag{6.2}$$

The polynomials have the useful orthogonality property

$$\int_0^\infty x^m e^{-x} S_m^{(n)}[x] S_m^{(q)}[x]\, dx = 0 \qquad \text{when} \quad n \neq q,$$
$$= \frac{(m+n)!}{n!} \qquad \text{when} \quad n = q. \quad (6.3a)$$

In equation (6.3a) the factorial of a number other than an integer must be interpreted as a Γ-function.[6] We shall need, however, only the rule that, if q is an integer,

$$\left(\frac{q}{2}\right)! = \left(\frac{q}{2}\right)\left(\frac{q}{2} - 1\right) \cdots (\tfrac{5}{2})(\tfrac{3}{2})(\tfrac{1}{2})\sqrt{\pi}. \tag{6.3b}$$

Our purpose is to obtain the solutions of equations (4.18) subject to the condition (4.15). In terms of Sonine polynomials these equations can be written [see equations (6.2)]

$$I[A\mathscr{C}_i] = -f_0 \mathscr{C}_i S_{3/2}^{(1)}[\mathscr{C}^2], \tag{6.4a}$$

$$I[B\overset{\circ}{\mathscr{C}_i\mathscr{C}_j}] = 2f_0 \overset{\circ}{\mathscr{C}_i\mathscr{C}_j}\, S_{5/2}^{(0)}[\mathscr{C}^2], \tag{6.4b}$$

and

$$\int_0^\infty A(\mathscr{C}, T) S_{3/2}^{(0)}[\mathscr{C}^2](\mathscr{C}^2)^{3/2} e^{-\mathscr{C}^2}\, d(\mathscr{C}^2) = 0. \tag{6.4c}$$

Although the index $m = \frac{3}{2}$ in equation (6.4a) follows from equations (4.18a) and (6.2), the motivation for $m = \frac{5}{2}$ in equation (6.4b) will be apparent only later, in connection with the use of the orthogonality property (6.3a).

For Maxwellian molecules, there exists an exact, closed-form solution of equations (6.4). Here it can be shown (Uhlenbeck and Ford, 1963) that the Sonine polynomials are eigenfunctions of the integral operator $(1/f_0)I$. For other intermolecular potentials, the functions A and B can be developed in terms of these eigenfunctions. Burnett proved, for certain molecular models, the convergence of the following expansions:

$$A(\mathscr{C}, T) = \sum_{r=0}^{\infty} a_r S_{3/2}^{(r)}[\mathscr{C}^2] \tag{6.5a}$$

$$B(\mathscr{C}, T) = \sum_{r=1}^{\infty} b_r S_{5/2}^{(r-1)}[\mathscr{C}^2]\,. \tag{6.5b}$$

[6] This question does not arise in the earlier equation (6.1) since there the ratio $(m+n)!/(m+p)!$ is in any case equal to $(m+n)(m+n-1)\cdots(m+p+1)$.

The coefficients a_r and b_r may depend on T but not on $\mathscr{C}$ and are to be determined subject to equations (6.4). With equations (6.1) and (6.2), it can be seen that the series (6.5) are essentially power series. The convenience of the present form stems from the orthogonality property (6.3a), which greatly simplifies the calculations. When the series (6.5) are substituted into equations (6.4a) and (6.4b) for A and B, we have

$$\sum_{r=0}^{\infty} a_r I[S_{3/2}^{(r)}[\mathscr{C}^2]\mathscr{C}_i] = -f_0\mathscr{C}_i S_{3/2}^{(1)}[\mathscr{C}^2] \tag{6.6a}$$

and

$$\sum_{r=1}^{\infty} b_r I[S_{5/2}^{(r-1)}[\mathscr{C}^2]\overset{\circ}{\mathscr{C}_i\mathscr{C}_j}] = 2f_0\overset{\circ}{\mathscr{C}_i\mathscr{C}_j}S_{5/2}^{(0)}[\mathscr{C}^2]. \tag{6.6b}$$

Before we discuss the solution for a_r and b_r we shall note two immediate consequences of the orthogonality property (Exercise 6.1). The first is that to satisfy the condition (6.4c) we must have

$$a_0 = 0. \tag{6.7}$$

[The choice of indices in equations (6.5) is then such that the first nonzero term in each corresponds to $r = 1$.] The second result follows from equations (5.4) for K and μ. In the present notation, and with the use of equation (6.4c), these equations are

$$K = -\frac{2}{3}\frac{k^2T}{m}\frac{2}{\sqrt{\pi}}\int_0^{\infty} A(\mathscr{C}, T)S_{3/2}^{(1)}[\mathscr{C}^2](\mathscr{C}^2)^{3/2}e^{-\mathscr{C}^2}\, d(\mathscr{C}^2) \tag{6.8a}$$

and

$$\mu = \frac{2kT}{15}\frac{2}{\sqrt{\pi}}\int_0^{\infty} B(\mathscr{C}, T)S_{5/2}^{(0)}[\mathscr{C}^2](\mathscr{C}^2)^{5/2}e^{-\mathscr{C}^2}\, d(\mathscr{C}^2). \tag{6.8b}$$

Then with the orthogonality property (6.3a) and equations (6.5) for A and B, we have

$$\boxed{K = -\frac{5}{2}\frac{k}{m}kT\,a_1} \tag{6.9a}$$

$$\boxed{\mu = \tfrac{1}{2}kTb_1} . \tag{6.9b}$$

(As we shall see, the coefficient a_1 is negative.) Thus to find the transport properties we must know only the first (nonzero) terms in the series (6.5). The simplicity of this result particularly supports the use of the Sonine-polynomial expansion.

We consider next the formal solutions of equations (6.6) for the coefficients a_r and b_r. These solutions can be found most directly by the use of the following additional expansions, in terms of Sonine polynomials, of the integrals I that appear in equations (6.6):

$$I[S_{3/2}^{(r)}[\mathscr{C}^2]\mathscr{C}_i] = \sum_{s=0}^{\infty} \frac{\sqrt{\pi}}{2} \frac{s!}{(\frac{3}{2}+s)!} a_{rs} S_{3/2}^{(s)}[\mathscr{C}^2]\mathscr{C}_i f_0, \tag{6.10a}$$

$$I[S_{5/2}^{(r-1)}[\mathscr{C}^2]\overset{\circ}{\mathscr{C}_i\mathscr{C}_j}] = \sum_{s=1}^{\infty} \frac{3\sqrt{\pi}}{4} \frac{(s-1)!}{(\frac{3}{2}+s)!} b_{rs} S_{5/2}^{(s-1)}[\mathscr{C}^2]\overset{\circ}{\mathscr{C}_i\mathscr{C}_j} f_0. \tag{6.10b}$$

The factorial expressions on the right in these expansions have been included so as to simplify certain of the equations that appear below and so that the notation conforms to that of Chapman and Cowling (1952).

The coefficients a_{rs} and b_{rs} in equations (6.10) can be found by the multiplication (with the summation convention) of these equations by $S_{3/2}^{(t)}[\mathscr{C}^2]\mathscr{C}_i$ and $S_{5/2}^{(t)}[\mathscr{C}^2]\overset{\circ}{\mathscr{C}_i\mathscr{C}_j}$, respectively, and the integration of the resulting equations with respect to velocity. With the orthogonality property (6.3a), we then have for a_{rs}

$$\int_{-\infty}^{\infty} S_{3/2}^{(t)}[\mathscr{C}^2]\mathscr{C}_j I[S_{3/2}^{(r)}[\mathscr{C}^2]\mathscr{C}_j]\, dV_C$$
$$= \sum_{s=0}^{\infty} \frac{\sqrt{\pi}}{2} \frac{s!}{(\frac{3}{2}+s)!} a_{rs} \int_{-\infty}^{\infty} S_{3/2}^{(t)}[\mathscr{C}^2] S_{3/2}^{(s)}[\mathscr{C}^2]\mathscr{C}_j\mathscr{C}_j f_0\, dV_C$$
$$= \sum_{s=0}^{\infty} \frac{\sqrt{\pi}}{2} \frac{s!}{(\frac{3}{2}+s)!} a_{rs} \int_0^{\infty} S_{3/2}^{(t)}[\mathscr{C}^2] S_{3/2}^{(s)}[\mathscr{C}^2] \frac{2}{\sqrt{\pi}} (\mathscr{C}^2)^{3/2} e^{-\mathscr{C}^2}\, d(\mathscr{C}^2)$$
$$= a_{rt},$$

where by equation (6.3a) all terms on the right are zero except the one for which $s = t$. With a trivial change in notation we can then write

$$a_{rs} = \int_{-\infty}^{\infty} S_{3/2}^{(s)}[\mathscr{C}^2]\mathscr{C}_j I[S_{3/2}^{(r)}[\mathscr{C}^2]\mathscr{C}_j]\, dV_C. \tag{6.11a}$$

Similarly, with $\overset{\circ}{\mathscr{C}_j\mathscr{C}_k}\overset{\circ}{\mathscr{C}_j\mathscr{C}_k} = \frac{2}{3}\mathscr{C}^4$ from Exercise 5.1, we have

$$b_{rs} = \int_{-\infty}^{\infty} S_{5/2}^{(s-1)}[\mathscr{C}^2]\overset{\circ}{\mathscr{C}_j\mathscr{C}_k} I[S_{5/2}^{(r-1)}[\mathscr{C}^2]\overset{\circ}{\mathscr{C}_j\mathscr{C}_k}]\, dV_C. \tag{6.11b}$$

The integrals (6.11) involve only known functions and hence can be directly evaluated. We note that a_{rs} and b_{rs} depend on the intermolecular potential. Since by the first of equations (6.2) $S_{3/2}^{(0)} = 1$, it is easily seen, with the use of the properties of the collisional invariants, that

$$a_{r0} = 0. \tag{6.11c}$$

It can also be shown [see equation (6.20a)] that

$$a_{rs} = a_{sr}$$

and (6.11b)

$$b_{rs} = b_{sr}.$$

We now substitute equations (6.10) into equations (6.6) and, with $a_0 = 0$ and $a_{r0} = 0$ from equations (6.7) and (6.11c), obtain

$$\sum_{r=1}^{\infty} a_r \sum_{s=1}^{\infty} \frac{\sqrt{\pi}}{2} \frac{s!}{(\frac{3}{2}+s)!} a_{rs} S_{3/2}^{(s)}[\mathscr{C}^2]\mathscr{C}_i f_0 = -f_0 \mathscr{C}_i S_{3/2}^{(1)}[\mathscr{C}^2]$$

and hence

$$\sum_{s=1}^{\infty} \frac{\sqrt{\pi}}{2} \frac{s!}{(\frac{3}{2}+s)!} S_{3/2}^{(s)}[\mathscr{C}^2] \sum_{r=1}^{\infty} a_r a_{rs} = -S_{3/2}^{(1)}[\mathscr{C}^2]. \tag{6.12a}$$

Similarly we have

$$\sum_{s=1}^{\infty} \frac{3\sqrt{\pi}}{4} \frac{(s-1)!}{(\frac{3}{2}+s)!} S_{5/2}^{(s-1)}[\mathscr{C}^2] \sum_{r=1}^{\infty} b_r b_{rs} = 2S_{5/2}^{(0)}[\mathscr{C}^2]. \tag{6.12b}$$

The coefficients a_r and b_r can be found from equations (6.12) by a further use of the orthogonality property (6.3a), in a manner similar to that used earlier to find a_{rs} and b_{rs}. The result is that the coefficients in equations (6.12) of the Sonine polynomials of all orders must be zero. For example, we must have

$$\left[\frac{\sqrt{\pi}}{2} \frac{1!}{(\frac{3}{2}+1)!} \sum_{r=1}^{\infty} a_r a_{r1} + 1\right] S_{3/2}^{(1)}[\mathscr{C}^2] = 0,$$

and thus, with the use of equation (6.3b), we find that

$$\sum_{r=1}^{\infty} a_r a_{r1} = -\frac{2}{\sqrt{\pi}} (\tfrac{3}{2}+1)! = -\tfrac{15}{4}.$$

With these conditions on equations (6.12) we obtain the exact solutions for a_r and b_r as follows:

$$\sum_{r=1}^{\infty} a_r a_{rs} = \alpha_s \tag{6.13a}$$

and

$$\sum_{r=1}^{\infty} b_r b_{rs} = \beta_s, \tag{6.13b}$$

where

$$\left.\begin{aligned} \alpha_s &= -\tfrac{15}{4} \\ \beta_s &= 5 \end{aligned}\right\} \quad (s = 1), \qquad \alpha_s = \beta_s = 0 \quad (s > 1). \tag{6.13c}$$

Equations (6.13) are infinite sets ($s = 1, 2, 3, \ldots$) of linear equations for the unknown quantities a_r and b_r. In principle these equations can be solved by Cramer's rule to obtain a_1 and b_1 and, by equations (6.9), K and μ. In practice, of course, approximate solutions are obtained by the use of a finite number of terms in the Sonine-polynomial expansions (6.5). In many cases K and μ are given accurately with only a few terms.

Several equivalent methods are used to find approximate solutions to equations (6.4) for A and B. These methods give at each stage of approximation identical results for a_1 and b_1 and hence K and μ. If the series (6.5) for A and B are truncated after m terms and if, correspondingly, m (nonzero) terms are taken in the series (6.10) for the integrals I, the approximate solutions are given by

$$A^{(m)} = \sum_{r=1}^{m} a_r^{(m)} S_{3/2}^{(r)}[\mathscr{C}^2], \tag{6.14a}$$

$$B^{(m)} = \sum_{r=1}^{m} b_r^{(m)} S_{5/2}^{(r-1)}[\mathscr{C}^2], \tag{6.14b}$$

and

$$\boxed{\begin{aligned} \sum_{r=1}^{m} a_r^{(m)} a_{rs} &= \alpha_s \quad &(6.15a)\\ \sum_{r=1}^{m} b_r^{(m)} b_{rs} &= \beta_s \quad &(6.15b)\end{aligned}}\;.$$

Here the superscript m refers to the mth stage of approximation. The solutions to equations (6.15) (obtained, for example, by Cramer's rule), with equations (6.11) and (6.13c) for a_{rs}, b_{rs}, α_r, and β_r and with equations (6.9) in the form

$$K^{(m)} = -\frac{5}{2}\frac{k}{m}\,kT\,a_1^{(m)} \tag{6.16a}$$

and

$$\mu^{(m)} = \tfrac{1}{2}kTb_1^{(m)}, \tag{6.16b}$$

are used in the practical calculation of the transport properties K and μ. The results thus obtained are identical for every m to the values given by equations (7.51,13) and (7.52,7) of Chapman and Cowling (1952). We also note that in the limit $m \to \infty$ the exact solutions (6.13) are recovered.

In the next section we shall consider in greater detail the evaluation of the transport properties. It is instructive first, however, to summarize an alternate approach to the approximate solutions (6.15), the *variational method* used by Hirschfelder, Curtiss, and Bird (1954).

This method serves to clarify the interpretation of the approximate solutions. As before we take $A^{(m)}$ and $B^{(m)}$ as each given by m terms in the finite series (6.14) in Sonine polynomials. The approximations $K^{(m)}$ and $\mu^{(m)}$ to the transport properties are given by equations (5.4) with A and B replaced by $A^{(m)}$ and $B^{(m)}$. With the identity $\overset{\circ}{\mathscr{C}_j\mathscr{C}_k}\,\overset{\circ}{\mathscr{C}_j\mathscr{C}_k} = \frac{2}{3}\mathscr{C}^4$ from Exercise 5.1 and the condition (4.15) in the form

$$\int_{-\infty}^{\infty} A^{(m)}\mathscr{C}^2 f_0\, dV_C = 0,$$

we can then write

$$\begin{aligned} K^{(m)} &= \frac{2}{3}\frac{k^2T}{m}\int_{-\infty}^{\infty} A^{(m)}\mathscr{C}^4 f_0\, dV_C \\ &= \frac{2}{3}\frac{k^2T}{m}\int_{-\infty}^{\infty} A^{(m)}\mathscr{C}_j f_0 \mathscr{C}_j(\mathscr{C}^2 - \tfrac{5}{2})\, dV_C \end{aligned} \tag{6.17a}$$

and

$$\begin{aligned} \mu^{(m)} &= \frac{2kT}{15}\int_{-\infty}^{\infty} B^{(m)}\mathscr{C}^4 f_0\, dV_C \\ &= \frac{kT}{10}\int_{-\infty}^{\infty} B^{(m)}\overset{\circ}{\mathscr{C}_j\mathscr{C}_k}\, 2f_0\overset{\circ}{\mathscr{C}_j\mathscr{C}_k}\, dV_C. \end{aligned} \tag{6.17b}$$

If we now use the *exact* integral equations (4.18) to express $2f_0\overset{\circ}{\mathscr{C}_j\mathscr{C}_k}$ and $f_0\mathscr{C}_j(\mathscr{C}^2 - \frac{5}{2})$ in equations (6.17) in terms of A and B we obtain

$$\begin{aligned} \frac{3mK^{(m)}}{2k^2T} &= \int_{-\infty}^{\infty} A^{(m)}\mathscr{C}_j f_0\mathscr{C}_j(\mathscr{C}^2 - \tfrac{5}{2})\, dV_C \\ &= \int_{-\infty}^{\infty} A^{(m)}\mathscr{C}_j I[A\mathscr{C}_j]\, dV_C \end{aligned} \tag{6.18a}$$

and

$$\begin{aligned} \frac{10\mu^{(m)}}{kT} &= \int_{-\infty}^{\infty} B^{(m)}\overset{\circ}{\mathscr{C}_j\mathscr{C}_k}\, 2f_0\overset{\circ}{\mathscr{C}_j\mathscr{C}_k}\, dV_C \\ &= \int_{-\infty}^{\infty} B^{(m)}\overset{\circ}{\mathscr{C}_j\mathscr{C}_k}\, I[B\overset{\circ}{\mathscr{C}_j\mathscr{C}_k}]\, dV_C. \end{aligned} \tag{6.18b}$$

Here exact quantities such as A and B are distinguished from the approximations such as $A^{(m)}$ and $B^{(m)}$ by the omission of the superscript m.

We now require that $A^{(m)}$ and $B^{(m)}$ satisfy the second equalities in equations (6.18), namely,

$$\int_{-\infty}^{\infty} A^{(m)}\mathscr{C}_j f_0\mathscr{C}_j(\mathscr{C}^2 - \tfrac{5}{2})\, dV_C = \int_{-\infty}^{\infty} A^{(m)}\mathscr{C}_j I[A^{(m)}\mathscr{C}_j]\, dV_C \tag{6.19a}$$

$$\int_{-\infty}^{\infty} B^{(m)}\overset{\circ}{\mathscr{C}_j\mathscr{C}_k}\, 2f_0\overset{\circ}{\mathscr{C}_j\mathscr{C}_k}\, dV_C = \int_{-\infty}^{\infty} B^{(m)}\overset{\circ}{\mathscr{C}_j\mathscr{C}_k}\, I[B^{(m)}\overset{\circ}{\mathscr{C}_j\mathscr{C}_k}]\, dV_C. \tag{6.19b}$$

These conditions are not sufficient to determine $A^{(m)}$ and $B^{(m)}$. The variational method is used to find $A^{(m)}$ and $B^{(m)}$, subject to equations (6.19), such that the differences between $K^{(m)}$ and K, and $\mu^{(m)}$ and μ are minimized.

To derive the variational method we shall use the general relations

$$\int_{-\infty}^{\infty} FI[G]\,dV_C = \int_{-\infty}^{\infty} GI[F]\,dV_C \tag{6.20a}$$

and

$$\int_{-\infty}^{\infty} FI[F]\,dV_C \geq 0. \tag{6.20b}$$

These relations are derived by Chapman and Cowling (1952), Secs. 4.4 and 4.41, by arguments similar to those used in Chapter IX, Sec. 4 in our discussion of the approach to equilibrium. Here F and G are any functions of molecular velocity and can be vectors or tensors [the summation convention is then implied in (6.20)]. If we set F in (6.20b) equal first to $(A^{(m)}\mathscr{C}_i - A\mathscr{C}_i)$ and then to $(B^{(m)}\overset{\circ}{\mathscr{C}_i\mathscr{C}_j} - B\overset{\circ}{\mathscr{C}_i\mathscr{C}_j})$, we have, with the use of (6.20a),

$$\begin{aligned}\int_{-\infty}^{\infty} (A^{(m)}\mathscr{C}_j - A\mathscr{C}_j)I[A^{(m)}\mathscr{C}_j - A\mathscr{C}_j]\,dV_C \\ = \int_{-\infty}^{\infty} A^{(m)}\mathscr{C}_j I[A^{(m)}\mathscr{C}_j]\,dV_C - 2\int_{-\infty}^{\infty} A^{(m)}\mathscr{C}_j I[A\mathscr{C}_j]\,dV_C \\ + \int_{-\infty}^{\infty} A\mathscr{C}_j I[A\mathscr{C}_j]\,dV_C \geq 0\end{aligned} \tag{6.21a}$$

and

$$\begin{aligned}\int_{-\infty}^{\infty} (B^{(m)}\overset{\circ}{\mathscr{C}_j\mathscr{C}_k} - B\overset{\circ}{\mathscr{C}_j\mathscr{C}_k})I[B^{(m)}\overset{\circ}{\mathscr{C}_j\mathscr{C}_k} - B\overset{\circ}{\mathscr{C}_j\mathscr{C}_k}]\,dV_C \\ = \int_{-\infty}^{\infty} B^{(m)}\overset{\circ}{\mathscr{C}_j\mathscr{C}_k} I[B^{(m)}\overset{\circ}{\mathscr{C}_j\mathscr{C}_k}]\,dV_C - 2\int_{-\infty}^{\infty} B^{(m)}\overset{\circ}{\mathscr{C}_j\mathscr{C}_k} I[B\overset{\circ}{\mathscr{C}_j\mathscr{C}_k}]\,dV_C \\ + \int_{-\infty}^{\infty} B\overset{\circ}{\mathscr{C}_j\mathscr{C}_k} I[B\overset{\circ}{\mathscr{C}_j\mathscr{C}_k}]\,dV_C \geq 0.\end{aligned} \tag{6.21b}$$

If we now use the fact that the right-hand sides of equations (6.18a) and (6.19a) and (6.18b) and (6.19b) are equal, the inequalities (6.21) become

$$\int_{-\infty}^{\infty} A^{(m)}\mathscr{C}_j I[A^{(m)}\mathscr{C}_j]\,dV_C \leq \int_{-\infty}^{\infty} A\mathscr{C}_j I[A\mathscr{C}_j]\,dV_C \tag{6.22a}$$

and

$$\int_{-\infty}^{\infty} B^{(m)}\overset{\circ}{\mathscr{C}_j\mathscr{C}_k} I[B^{(m)}\overset{\circ}{\mathscr{C}_j\mathscr{C}_k}]\,dV_C \leq \int_{-\infty}^{\infty} B\overset{\circ}{\mathscr{C}_j\mathscr{C}_k} I[B\overset{\circ}{\mathscr{C}_j\mathscr{C}_k}]\,dV_C. \tag{6.22b}$$

From this result, with equations (6.18) and (6.19) for $K^{(m)}$ and $\mu^{(m)}$ and the corresponding equations with the superscript m omitted for K and μ, we obtain

$$\boxed{\begin{aligned} K^{(m)} &\leq K \\ \mu^{(m)} &\leq \mu \end{aligned}} \;. \qquad \begin{aligned} &(6.23a) \\ &(6.23b) \end{aligned}$$

Thus the variational method can be stated as follows: To find the best approximations to the transport properties with $A^{(m)}$ and $B^{(m)}$ given by a finite series of Sonine polynomials we must maximize $K^{(m)}$ and $\mu^{(m)}$, subject to the conditions (6.19).

In terms of our previous notation, with equations (6.14), (6.11), and (6.13c) and the straightforward use of the orthogonality property (6.3a), the conditions (6.19) become

$$\alpha_1 a_1^{(m)} = \sum_{r=1}^{m} \sum_{s=1}^{m} a_r^{(m)} a_s^{(m)} a_{rs} \qquad (6.24a)$$

and

$$\beta_1 b_1^{(m)} = \sum_{r=1}^{m} \sum_{s=1}^{m} b_r^{(m)} b_s^{(m)} b_{rs}. \qquad (6.24b)$$

By equations (6.16), $K^{(m)}$ and $\mu^{(m)}$ are proportional to the positive quantities $-a_1^{(m)}$ and $b_1^{(m)}$. We then must maximize $-a_1^{(m)}$ and $b_1^{(m)}$ subject to the conditions (6.24). It can easily be shown, for example by the use of Lagrange multipliers, that these requirements give precisely our previous form (6.15) of the approximate solutions (Exercise 6.3). We then see that the approximate solutions (6.15) give the most accurate values of the transport properties for a given number of terms in the Sonine-polynomial expansions. The approximate values $K^{(m)}$ and $\mu^{(m)}$ that are obtained in this way can not exceed the exact values and can not decrease as m increases.

At this point we note that the Chapman-Enskog solution has been obtained by means of two successive expansions, one the expansion of the distribution function for small departures from equilibrium and the other the expansion of the quantities $A(\mathscr{C}, T)$ and $B(\mathscr{C}, T)$ in terms of Sonine polynomials. The distinction between the two must be kept clearly in mind.

Exercise 6.1. (a) Prove that to satisfy the condition (6.4c) we must have $a_0 = 0$.

(b) Verify equations (6.9).

Exercise 6.2. (a) Derive equation (6.11b) for b_{rs}.

(b) Show that $a_{r0} = 0$.

Exercise 6.3. For $m = 2$, show that the variational method leads to the approximate solutions (6.15). (The use of Lagrange multipliers is not necessary.)

Exercise 6.4. Show that the approximate solutions (6.15) can be obtained from the moment equations that are formed by the multiplication of equations (6.4a) and (6.4b), with equations (6.14) for A and B, by

$$S_{3/2}^{(s)}[\mathscr{C}^2]\mathscr{C}_i \quad \text{and} \quad S_{5/2}^{(s-1)}[\mathscr{C}^2]\overset{\circ}{\mathscr{C}_i\mathscr{C}_j},$$

respectively, and the integration of the resulting equations with respect to molecular velocity.

7 TRANSPORT PROPERTIES

It is hoped that our preceding discussion of the Chapman-Enskog expansion has provided an understanding of the theory on which transport-property calculations are based. In this section we shall indicate some of the consequences of the theory. The reader who is interested in a detailed discussion of the evaluation of transport properties is referred to Chapman and Cowling (1952) and, in particular, to Hirschfelder, Curtiss, and Bird (1964). In these references general techniques are developed for the calculation of the coefficients of viscosity, thermal conductivity, and diffusion for simple gases and gas mixtures. The calculations involve determinants of certain collision integrals. These integrals have been reduced to standard forms that have been evaluated for a number of intermolecular potentials. The representation of the intermolecular potential that is appropriate to a particular gas can be found by both empirical and theoretical methods. A discussion of such considerations for a variety of molecular interactions is given by Hirschfelder, Curtiss, and Bird (1964). Simply speaking, it is possible to find analytical representations of the intermolecular potential for which agreement is obtained between the calculated and measured values of the thermodynamic and transport properties of a given gas or mixture.[7] Such potentials can then be associated with a particular molecular interaction (for example, argon-argon collisions).

Because of the rapid convergence of the Sonine-polynomial expansions (6.14), remarkably accurate values of the transport properties μ and K

[7] For a dense gas the equations of state are influenced by the intermolecular potential. Such thermodynamic data can be used in the prediction of the transport properties of dilute (perfect) gases. The intermolecular potential can be inferred also, of course, from measured transport properties.

are obtained in many cases by the retention of only the first terms in these expansions. To illustrate the calculations we shall evaluate $\mu^{(1)}$ and $K^{(1)}$ for elastic-sphere and inverse-power molecules. It must be emphasized that more realistic models, such as the Lennard-Jones and Buckingham potentials, are often required for practical calculations that are intended to apply over a wide range of temperature.

From equations (6.16), (6.15), and (6.13c), $\mu^{(1)}$ and $K^{(1)}$ for a simple gas are given by

$$\mu^{(1)} = \frac{1}{2} kTb_1^{(1)} = \frac{5}{2} kT\left(\frac{1}{b_{11}}\right) \tag{7.1a}$$

and

$$K^{(1)} = -\frac{5}{2}\frac{k}{m} kTa_1^{(m)} = \frac{75}{8}\frac{k}{m} kT\left(\frac{1}{a_{11}}\right). \tag{7.1b}$$

The quantities a_{11} and b_{11} were evaluated in Chapter IX, Sec. 9.[8] There [equations (IX 9.18) and (IX 9.19)] we found that

$$a_{11} = b_{11} = \frac{2}{\sqrt{\pi}}\left(\frac{m}{4kT}\right)^{7/2} \int_0^\infty g^7 \sigma_\mu(g) e^{-mg^2/4kT}\, dg.$$

With this result we have

$$\boxed{\begin{aligned} \mu^{(1)} &= \frac{\frac{5}{8}\sqrt{\pi mkT}}{\left(\frac{m}{4kT}\right)^4 \int_0^\infty g^7 \sigma_\mu(g) e^{-mg^2/4kT}\, dg} \\ K^{(1)} &= \frac{15}{4}\frac{k}{m}\mu^{(1)} \end{aligned}} . \tag{7.2}$$

From equations (7.2) it can be seen that the Prandtl number is

$$\frac{c_p \mu^{(1)}}{K^{(1)}} = \frac{5}{2}\frac{k}{m}\frac{\mu^{(1)}}{K^{(1)}} = \frac{2}{3}. \tag{7.3}$$

With a given representation of the intermolecular potential and hence a knowledge of the cross-section $\sigma_\mu(g)$, the transport properties $\mu^{(1)}$ and $K^{(1)}$ can now be calculated. For elastic spheres of diameter d and inverse-power molecules, $\sigma_\mu(g)$ is given by equations (IX 8.9b) and (IX 8.10b). With these values, equations (7.2) yield

$$\mu^{(1)} = \frac{5}{16d^2}\sqrt{\frac{mkT}{\pi}} \qquad \text{(elastic spheres)} \tag{7.4a}$$

[8] By equations (6.11) and (6.20b), a_{11} and b_{11} must be positive. It then follows from the variational method of Sec. 6 that to any order m of approximation $\mu^{(m)}$ and $K^{(m)}$ must also be positive.

and

$$\mu^{(1)} = \frac{5\sqrt{mkT/\pi}\,(2kT/\alpha a)^{2/\alpha}}{8A_2(\alpha)\Gamma(4 - 2/\alpha)}, \qquad \begin{pmatrix}\text{inverse-power}\\ \text{molecules}\end{pmatrix} \tag{7.4b}$$

where we have written only $\mu^{(1)}$ since $K^{(1)}$ can be obtained from equation (7.3). In equation (7.4b) for inverse-power molecules Γ is the gamma function and, as explained in Chapter IX, Sec. 8, $A_2(\alpha)$ is a pure number tabulated in the form $A_2(\nu = \alpha + 1)$ by Chapman and Cowling (1952), Sec. 10.3. The elastic-sphere result (7.4a) differs by only a numerical factor close to unity from the mean-free-path formula [see Chapter I, Sec. 5, and equations (II 5.12) and (II 6.16)]

$$\mu = \tfrac{1}{2}\rho\bar{C}\lambda = \frac{1}{\pi d^2}\sqrt{mkT/\pi}. \tag{7.4c}$$

It can be seen that the present result (7.4b) for inverse-power molecules is in agreement with the temperature dependence (5.9) of μ and K found by means of the partial solution of Sec. 4.

In the first approximation ($m = 1$) the distribution function (4.16) is

$$f = f_0\left[1 - \frac{2\sqrt{2}}{5}\sqrt{m/kT}\,\mathscr{C}_j(\mathscr{C}^2 - \tfrac{5}{2})\frac{K^{(1)}}{nkT}\frac{\partial T}{\partial x_j} - 2\overset{\circ}{\mathscr{C}_j\mathscr{C}_k}\frac{\mu^{(1)}}{nkT}\frac{\partial \bar{c}_j}{\partial x_k}\right]. \tag{7.5}$$

It is interesting to compare this result with the distribution function (3.11) that was obtained with the B-G-K model. Although the values of $\mu^{(m)}$ and $K^{(m)}$ converge rapidly in many cases for a few terms in the Sonine-polynomial expansion, the convergence is not so rapid for the distribution function itself. For this reason equation (7.5) should be taken as only a rough approximation. We note from this result, however, that the condition $\Phi_1 \ll 1$ [see equation (4.16)] will be violated for sufficiently large values of the molecular velocity. Since the relative number of molecules with such velocities is exponentially small, this need not concern us here.

For Maxwellian molecules ($\alpha = 4$), the values of μ and K given by equations (7.4b) and (7.3) are exact. With other intermolecular potentials the inclusion of additional terms in the Sonine-polynomial expansions (6.14) results in more accurate approximations. To any order m, μ and K are obtained by the use in equations (6.16) of the appropriate solutions to equations (6.15). These calculations involve the evaluation of additional collision integrals a_{rs} and b_{rs} [see equations (6.11)]. Although it is not always the case (for example with diffusion coefficients in a gas mixture), for elastic-sphere and inverse-power molecules the exact values of μ and K in a simple gas differ from $\mu^{(1)}$ and $K^{(1)}$ by only a few percent. Accordingly, the exact value of the Prandtl number is very nearly $\frac{2}{3}$.

When the Chapman-Enskog theory is applied to a *mixture* of monatomic gases (see Hirschfelder, Curtiss, and Bird, 1964, Chapter VII) it is found that the viscous stress τ_{ij}, heat flux q_j, and diffusion velocity $(\bar{C}_j)_s$ are given by

$$\tau_{ij} = \mu\left(\frac{\partial c_{0_i}}{\partial x_j} + \frac{\partial c_{0_j}}{\partial x_i}\right) - \frac{2}{3}\mu\frac{\partial c_{0_k}}{\partial x_k}\delta_{ij}, \tag{7.6a}$$

$$q_i = -K'\frac{\partial T}{\partial x_i} + \frac{5}{2}kT\sum_s n_s(\bar{C}_i)_s - nkT\sum_s \frac{1}{\rho_s} D_s^T d_{s_i}, \tag{7.6b}$$

and

$$(\bar{C}_i)_s = \frac{n^2}{n_s\rho}\sum_r m_r D_{sr} d_{r_i} - \frac{1}{\rho} D_s^T \frac{\partial \ln T}{\partial x_i}, \tag{7.6c}$$

where the notation is that of Chapter IX, Sec. 10, and

$$d_{s_i} \equiv \frac{\partial}{\partial x_i}\left(\frac{n_s}{n}\right) + \left(\frac{n_s}{n} - \frac{\rho_s}{\rho}\right)\frac{\partial \ln p}{\partial x_i} - \frac{\rho_s}{p}\left[F_{s_i} - \sum_r \frac{\rho_r}{\rho} F_{r_i}\right].$$

The four transport properties—the coefficients of viscosity μ, thermal conductivity K', diffusion D_{sr}, and thermal diffusion D_s^T—can be calculated by methods similar to (but more involved than) those that we have discussed in connection with a simple gas.[9] The thermal-diffusion coefficient D_s^T is associated with the diffusion flux that results from a temperature gradient [see equation (7.6c)]. The value of D_s^T is zero for Maxwellian molecules. In the calculation of D_s^T for other intermolecular potentials, terms of at least second order must be retained in the Sonine-polynomial expansions that are used for mixtures. The thermal-diffusion effect was not known prior to the development of the Chapman-Enskog theory. Its existence was later confirmed by experiment. The purely theoretical prediction of this effect stands as one of the significant accomplishments of the theory.

Thus far we have discussed only monatomic gases, for which the theory is accurate and self-consistent. For gases with internal energies our treatment of the collision integral breaks down, since we have allowed no mechanism for the interchange of the translational and internal energies of the molecules. Hirschfelder, Curtiss, and Bird (1964), Sec. 7.6, discuss

[9] Different definitions of the transport properties of a multicomponent mixture are used by various authors. Here we have followed Hirschfelder, Curtiss, and Bird (1964), and have written μ and K' for their η and λ'. Our purpose is to indicate only the general structure of the results of the Chapman-Enskog theory for gas mixtures. The reader interested in the interpretation and evaluation of the transport properties of mixtures and the relationship between K' and the more usual coefficient of thermal conductivity should consult the foregoing reference.

modifications to the theory that have been developed to overcome this difficulty. We do not wish to go into the details of this complicated subject here. With certain corrections and limitations, however, the theory for monatomic gases can be used in the calculation of the transport properties of other gases. As might be expected the coefficients of viscosity and diffusion, which are associated with the transfer of momentum and mass by the translational motion of the molecules, are predicted successfully for many diatomic and polyatomic gases by the monatomic theory. The state of affairs with regard to thermal conductivity is not as satisfactory. It is found, however, that in many cases reasonable estimates of the coefficient of thermal conductivity are given by Eucken's relation [equation (I 5.12)] or modifications thereof.[10] Several more sophisticated modifications of the theory have been suggested (Wang Chang and Uhlenbeck, 1951, and Mason and Monchick, 1962), but these methods are beyond the scope of our discussion. The transfer of energy between translational and internal molecular motions is responsible also, at least in part, for the effect of bulk viscosity. This effect is discussed in the next section.

Exercise 7.1. (a) Find $\mu^{(2)}$ and $K^{(2)}$ in terms of the collision integrals a_{rs} and b_{rs}.

(b) For elastic spheres $a_{12} = b_{12} = -\frac{1}{4}a_{11}$, $a_{22} = \frac{45}{16}a_{11}$, and $b_{22} = \frac{205}{48}a_{11}$. In this case find $\mu^{(2)}/\mu^{(1)}$, $K^{(2)}/K^{(1)}$, and $c_p\mu^{(2)}/K^{(2)}$.

8 BULK VISCOSITY

In Sec. 5 we indicated that two coefficients of viscosity, μ and μ_B, result from the continuum derivation of the Navier-Stokes equations. Since the bulk-viscosity coefficient μ_B in the continuum expression (5.3a) for the viscous stress is multiplied by the divergence of the velocity $\partial \bar{c}_j/\partial x_j$, then, by the continuity equation, bulk viscosity can be important only when compressibility effects are important. The Chapman-Enskog theory gives a rigorous interpretation and method of evaluation for μ by means of kinetic theory and shows that for a perfect monatomic gas μ_B is zero. It is sometimes argued that $\mu_B = 0$ in general. Stokes set forth certain reasons why this might be so, but he apparently had reservations regarding this conclusion.[11] It turns out that both experiment and kinetic theory

[10] See, for example, Hirschfelder, Curtiss, Bird, and Spotz (1955), Sec. 11 and Lighthill (1956), Sec. 2.4.

[11] The *Stokes relation* $3\lambda + 2\mu = 0$, where $\lambda \equiv \mu_B - \frac{2}{3}\mu$, is equivalent to $\mu_B = 0$. We shall here avoid the use of the so-called "second viscosity coefficient" λ.

show that μ_B is not always zero. The effect of bulk viscosity can, in fact, have a significant influence on sound propagation and shock-wave structure.

There exists a certain amount of confusion with regard to bulk viscosity. One of the reasons for this is that, as we shall see, in the interpretation of bulk viscosity care must be exercised in the matter of definitions, particularly that of pressure. Another reason is that a bulk-viscosity effect enters the discussion of two *separate* phenomena. One is the modification of the stress tensor in a dense gas to take account of the finite range of the intermolecular forces (see Chapman and Cowling, 1952, Chapter 16). This correction is proportional to the fraction of the volume of the gas that is actually occupied by the molecules. If as in Chapter IX, Sec. 8 we take $\sqrt{\sigma_M}$ to be the effective extent of the intermolecular force, this fraction is approximately $n(\sigma_M)^{3/2}$. We shall continue to consider only the perfect-gas limit in which $n(\sigma_M)^{3/2} \ll 1$ and hence shall not discuss bulk viscosity in dense gases.

The second phenomenon that gives rise to a bulk-viscosity effect is the departure from local equilibrium of the distribution of internal molecular energy in a diatomic or polyatomic gas. In Chapter VIII we considered vibrational nonequilibrium, where the relaxation times are great when compared to the time between collisions. In rotational nonequilibrium the relaxation times at moderate temperatures are of the same order, but somewhat greater, than the times required for the adjustment of the translational motions of the molecules. (The latter requires only a few collisions.) For this reason the effects of small departures from rotational equilibrium can be incorporated in the dissipative terms in the Navier-Stokes equations if the bulk-viscosity term is retained in the stress tensor, this being the only term in the model that can be adjusted after μ and K are known. We shall consider in this section the role of bulk viscosity in such a representation of rotational nonequilibrium. We note, however, that there exists a difference of opinion with regard to the accuracy and utility of this representation (compare Lighthill, 1956, and Talbot and Scala, 1961). Our intention here is to give a physical interpretation of bulk viscosity rather than a derivation. Reference to more rigorous analyses will be made later in the section.[12]

The physical mechanism whereby rotational nonequilibrium gives rise to bulk viscosity can be described as follows. When a gas is compressed,

[12] It has been suggested that in some situations the effects of vibrational and chemical nonequilibrium can also be represented by means of bulk viscosity (see De Groot and Mazur, 1962, p. 329, and Lighthill, 1956). In this context it is important to note that if the methods of Chapter VIII are used to account explicitly for the effects of vibrational or chemical nonequilibrium, these effects should not be included in bulk viscosity.

as, for example, by a sound wave, the thermal energy per unit mass increases. This increase in energy appears first as translational energy and, through collisions, the rotational energy tends to rise only after a certain relaxation time. Under these conditions the translational energy at first comprises a larger fraction of the total thermal energy than it would in complete equilibrium at the same density and total energy. Since the kinetic pressure depends only on the translational molecular motions, it too is larger than if the rotational energy were equilibrated. Conversely, when the gas expands and cools the translational energy and kinetic pressure tend at first to be smaller than in equilibrium, while the rotational energy is greater.

To put this in more quantitative terms we define a thermodynamic or equilibrium pressure p^* as the pressure that would exist in equilibrium at a given density ρ and total thermal energy e per unit mass. Thus p^* is a known function of ρ and e, which themselves have unambiguous physical definitions. For example, for conditions such that the gas is calorically perfect at equilibrium we have

$$p^* = (\gamma - 1)\rho e. \tag{8.1}$$

(From Chapter IV we find that equation (8.1) with $\gamma = \frac{7}{5}$ applies to air at moderate temperatures, where the rotational energy is fully excited and the vibrational energy is small.)

Bulk viscosity accounts for the difference between the equilibrium pressure p^* and the kinetic pressure $\frac{1}{3}\rho\overline{C^2}$ [see equation (IX 2.15)]. It follows from our foregoing discussion that this difference is negative when the gas is being compressed and positive when it is being expanded. For small departures from rotational equilibrium, the bulk-viscosity coefficient μ_B linearly relates $p^* - \frac{1}{3}\rho\overline{C^2}$ to the rate of relative change of density $\rho^{-1}D\rho/Dt$ following the fluid, that is,[13]

$$p^* - \tfrac{1}{3}\rho\overline{C^2} = -\frac{\mu_B}{\rho}\frac{D\rho}{Dt}. \tag{8.2a}$$

With the continuity equation (5.6a) in the form

$$\frac{D\rho}{Dt} = -\rho\frac{\partial \bar{c}_j}{\partial x_j},$$

[13] If our restriction of small departures from rotational equilibrium is not satisfied, as would be the case if the relaxation time or relative density change were large, rotational nonequilibrium must be treated by methods similar to those of Chapter VIII (see Lighthill, 1956, Talbot and Scala, 1961, and Wang Chang and Uhlenbeck, 1951).

this relation (8.2a) becomes

$$\boxed{p^* - \tfrac{1}{3}\rho\overline{C^2} = \mu_B \frac{\partial \bar{c}_j}{\partial x_j}} \,. \tag{8.2b}$$

We are now in a position to combine the effect of bulk viscosity with the viscous stress that we have found previously. For clarity we shall express this result in terms of the *pressure tensor* p_{ij} defined by

$$p_{ij} \equiv \rho\overline{C_i C_j}. \tag{8.3}$$

By definition, the kinetic pressure $\frac{1}{3}\rho\overline{C^2}$ is equal to $\frac{1}{3}p_{jj}$, the mean of the normal components of p_{ij}. From Sec. 5 the pressure tensor for a monatomic gas is

$$p_{ij} = \tfrac{1}{3}\rho\overline{C^2}\,\delta_{ij} - \mu\left(\frac{\partial \bar{c}_i}{\partial x_j} + \frac{\partial \bar{c}_j}{\partial x_i} - \frac{2}{3}\frac{\partial \bar{c}_k}{\partial x_k}\,\delta_{ij}\right). \tag{8.4}$$

We now assume equation (8.4) remains valid for a diatomic gas in rotational nonequilibrium and express $\frac{1}{3}\rho\overline{C^2}$ in terms of p^* by means of equation (8.2b). The pressure tensor can then be written

$$\boxed{p_{ij} = \left(p^* - \mu_B \frac{\partial \bar{c}_k}{\partial x_k}\right)\delta_{ij} - \mu\left(\frac{\partial \bar{c}_i}{\partial x_j} + \frac{\partial \bar{c}_j}{\partial x_i} - \frac{2}{3}\frac{\partial \bar{c}_k}{\partial x_k}\,\delta_{ij}\right)} \,. \tag{8.5}$$

This result has been obtained rigorously by Wang Chang and Uhlenbeck (1951) by means of a generalization of the Chapman-Enskog expansion. (For the special model of a rough spherical molecule see also Chapman and Cowling, 1952, Note B.)

To find the significance of the bulk-viscosity expression (8.5) for the pressure tensor we first note that the momentum and energy conservation equations can be written in the forms (see Chapter IX, Exercise 2.3)

$$\rho \frac{D\bar{c}_i}{Dt} = -\frac{\partial p_{ij}}{\partial x_j} + \rho F_i \tag{8.6}$$

and

$$\rho \frac{De}{Dt} = -p_{ij}\frac{\partial \bar{c}_i}{\partial x_j} - \frac{\partial q_j}{\partial x_j}\,. \tag{8.7}$$

Now, by equation (8.2b), equations (8.4) and (8.5) for p_{ij} are identical in terms of $\frac{1}{3}\rho\overline{C^2}$. At first it might appear that nothing has been gained in equation (8.5) by the introduction of bulk viscosity. For a monatomic gas, however, we have

$$p^* = \tfrac{2}{3}\rho e = \tfrac{1}{3}\rho\overline{C^2} = nkT.$$

Thus for a monatomic gas, $\frac{1}{3}\rho\overline{C^2}$ in equation (8.4) for p_{ij} is a function of the other variables (ρ and e in this case) that appear in the conservation equations. When rotational nonequilibrium is considered, equation (8.2b) is required to provide such a relation, and results in equation (8.5). In both cases then, if we assume that q_i is known in terms of the other variables and use equation (8.1) for p^*, the number of independent variables (ρ, $\bar{c}_i$, and e) is equal to the number of conservation equations [equations (8.6), (8.7), and the continuity equation]. We also note that nothing in this section contradicts the formulation of the conservation equations in Chapter IX, Sec. 2 (see Exercise 8.1). Our present discussion, however, bears a closer relationship to the continuum description.

We have avoided, up to now, the use of temperature within the context of rotational nonequilibrium. With regard to the heat flux q_i, however, it is necessary to observe that in general the kinetic temperature $T \equiv m\overline{C^2}/3k$ is not equal to the equilibrium temperature $T^* = T^*(\rho, e)$. It can be seen from the relations

$$p \equiv \tfrac{1}{3}\rho\overline{C^2} = nkT,$$
$$p^* \equiv p^*(\rho, e) = nkT^*,$$

and equation (8.2b) that $T^* - T$ is proportional to $\partial\bar{c}_j/\partial x_j$. In the heat flux

$$q_i = -K\frac{\partial T}{\partial x_i}, \tag{8.8}$$

this temperature difference is of second order and can be neglected as long as the departure from rotational equilibrium is small.

The practical importance of bulk viscosity depends on the relative magnitudes of μ and μ_B. From our qualitative discussion at the beginning of this section it would be reasonable to expect that μ_B is proportional to the product of the rotational thermal energy per unit volume $\rho e c_{v_{\text{rot}}}/c_v$ and the mean relaxation time $\bar{\tau}$ for rotational energy. Wang Chang and Uhlenbeck (1951) and Lighthill (1956) find by detailed analyses that

$$\mu_B = (\gamma - 1)^2\rho e\frac{c_{v_{\text{rot}}}}{c_v}\bar{\tau}. \tag{8.9}$$

From this result it can be seen that, in agreement with arguments based on the second law of thermodynamics, μ_B must be positive. Although Wang Chang and Uhlenbeck (1951) give for $\bar{\tau}$ an explicit formula that is similar to the viscosity and thermal-conductivity collision integral (7.2), the cross-sections required for the evaluation of their expression are not known. Hence $\bar{\tau}$ and μ_B must be found by experiment.

Lighthill (1956) describes the effects of the dissipative mechanisms of viscosity and thermal conductivity on the propagation of sound. Here translational and rotational nonequilibrium result in absorption and dispersion analogous to that discussed in Chapter VIII, Sec. 6 with reference to chemical and vibrational nonequilibrium. Measurements of the absorption of sound waves at frequencies below the relaxation frequency of rotational energy can be used to find μ_B. In nitrogen, for example, it is found from such measurements that at moderate temperatures μ_B/μ is about 0.8, so that the effects of bulk viscosity cannot be neglected.

We now return, for the remainder of this chapter, to the consideration of simple monatomic gases.

Exercise 8.1. Show that it is possible to account for the effects of bulk viscosity within the framework of the conservation equations (IX 2.22), (IX 2.23), and (IX 2.24). Use the definitions of that section for p, T, and τ_{ij} and equations (8.5) and (8.2b) to express τ_{ij} in terms of the other variables. Note that the essence of the problem lies in the relationship between p (or T) and e. It is not necessary to perform any involved manipulations of the equations. You should show, however, that with equation (8.8) for q_i the number of independent variables is equal to the number of equations. Relate your results to equation (5.3).

Exercise 8.2. If in N_2 at moderate temperatures $\mu_B/\mu = 0.8$, find the average number of collisions that take place in the rotational relaxation time $\bar{\tau}$. Take the average time between collisions as $\lambda\sqrt{\pi m/8kT}$, where λ is the mean free path.

9 THE STRUCTURE OF SHOCK WAVES

In this and the following section we consider two physical situations that serve to illustrate the application of kinetic theory to flows in translational nonequilibrium. The internal structure of shock waves has been studied by various investigators since the work of Rankine in 1870. This problem has the distinct advantage that it does not involve the complicating effect of molecular interactions with solid surfaces. Although the structure of shock waves would be given by a solution to the Boltzmann equation, an exact solution valid for a general range of Mach number has not yet been obtained. We shall here discuss three different approaches to the problem.

Regardless of the method that we use, the conservation equations of mass, momentum, and energy must be satisfied. We consider a steady, plane (that is, one-dimensional) shock wave and assume that external forces are absent and that conditions both far upstream and far downstream are uniform. If we adopt a coordinate system in which the shock wave is stationary, the conservation equations (IX 2.22), (IX 2.23a), and (IX 2.24a) can immediately be integrated once to yield the following:

$$\rho u = \rho_\alpha u_\alpha, \tag{9.1a}$$

$$p + \rho u^2 - \tau_{xx} = p_\alpha + \rho_\alpha u_\alpha^2, \tag{9.1b}$$

and

$$\rho u(h + \tfrac{1}{2}u^2) + q_x - u\tau_{xx} = \rho_\alpha u_\alpha(h_\alpha + \tfrac{1}{2}u_\alpha^2). \tag{9.1c}$$

These equations apply locally within the shock wave. Here we have used the subscript α to denote upstream, supersonic conditions and, as in Chapter VIII, have written u and x for the velocity and coordinate in the direction of flow. The overall conservation relations can be obtained directly from equations (9.1) and give the conditions on the downstream, subsonic side of the wave (subscript β) as follows:

$$\rho_\beta u_\beta = \rho_\alpha u_\alpha, \tag{9.2a}$$

$$p_\beta + \rho_\beta u_\beta^2 = p_\alpha + \rho_\alpha u_\alpha^2, \tag{9.2b}$$

and

$$\rho_\beta u_\beta(h_\beta + \tfrac{1}{2}u_\beta^2) = \rho_\alpha u_\alpha(h_\alpha + \tfrac{1}{2}u_\alpha^2),$$

or, with equation (9.2a),

$$h_\beta + \tfrac{1}{2}u_\beta^2 = h_\alpha + \tfrac{1}{2}u_\alpha^2. \tag{9.2c}$$

Equations (9.2) are equivalent to the well-known Rankine-Hugoniot equations and give the jump conditions that are used when the shock wave is represented by a discontinuity (cf. Chapter VI, Sec. 2). As a matter of fact, of course, shock waves are not discontinuities. Density, velocity, temperature, and the other quantities of interest vary continuously with position through the wave (see, for example, Figs. 2 and 3 later in this section). The problem of this section is to use equations (9.1) to find this variation.

Since the Chapman-Enskog theory leads to the Navier-Stokes equations, the application of the results of this theory to many physical situations is familiar through the continuum interpretation. Although the thickness of shock waves is characteristically measured in terms of mean free paths, the thickness becomes large and gradients become small as the Mach number M_α approaches unity. For this reason the Navier-Stokes equations give an accurate description of the structure of weak shock waves. The Navier-Stokes formulation is given by the conservation equations (9.1) with

$$\tau_{xx} = \tfrac{4}{3}\mu \frac{du}{dx} \tag{9.3a}$$

and

$$q_x = -K\frac{dT}{dx} \tag{9.3b}$$

from equations (5.1), (5.2), and (5.4). Equations (9.1) with equations (9.3) have been numerically integrated over a range of Mach number by Gilbarg and Paolucci (1953) for several representations of the temperature variation of the transport properties μ and K. An analytic solution can be obtained for weak shock waves ($M_\alpha^2 - 1 \ll 1$), in which case the equations of motion (9.1) and (9.3) can be linearized. We omit the algebraic details of this solution and quote from Lighthill (1956), Sec. 5.3, the result that

$$\frac{du}{dx} = -\frac{\frac{1}{2}(\gamma + 1)(u_\alpha - u)(u - u_\beta)}{\dfrac{4}{3}\dfrac{\mu_\alpha}{\rho_\alpha} + \dfrac{(\gamma - 1)}{c_p}\dfrac{K_\alpha}{\rho_\alpha}}. \tag{9.4}$$

It is useful to introduce here an effective mean free path λ by means of the relation $\mu = \frac{1}{2}\rho\bar{C}\lambda$ from equation (7.4c). It can be shown that with the Navier-Stokes (Chapman-Enskog) distribution function [equations (4.16) and (4.17)] $\bar{C}$ is unchanged from its equilibrium value $\sqrt{8kT/\pi m}$ given by equation (II 5.12). We then have

$$\lambda = \frac{\mu}{\rho}\sqrt{\frac{\pi m}{2kT}}. \tag{9.5}$$

We shall find it convenient to use λ from equation (9.5) as a microscopic length scale, regardless of the degree of nonequilibrium. That is, we shall adopt the convention in this and the next section that λ is defined by equation (9.5) and is related to the intermolecular potential by means of the Chapman-Enskog viscosity (5.4a), whether or not the Navier-Stokes relation (5.1b) between viscous stress and velocity gradients remains applicable.

If we then write μ_α in terms of the mean free path λ_α and use the Prandtl number $\Pr = \mu c_p/K$, equation (9.4) becomes

$$\frac{du}{dx} = -G\frac{(u_\alpha - u)(u - u_\beta)}{\bar{C}_\alpha\lambda_\alpha}, \tag{9.6}$$

where

$$G \equiv \frac{\gamma + 1}{\dfrac{4}{3} + \dfrac{\gamma - 1}{\Pr}}.$$

The constant G is $\frac{8}{7}$ for monatomic gases. The solution to equation (9.6) is easily found as

$$\boxed{\frac{u - u_\beta}{u_\alpha - u_\beta} = \left\{1 + \exp\left[G\left(\frac{u_\alpha - u_\beta}{\bar{C}_\alpha}\right)\frac{x}{\lambda_\alpha}\right]\right\}^{-1}} \quad . \tag{9.7}$$

It can be seen that as $x \to -\infty$, $u \to u_\alpha$ and as $x \to +\infty$, $u \to u_\beta$. The density and temperature profiles within the shock wave can be obtained from the velocity profile (9.7) by means of the conservation equations.

Several measures are used to characterize the thickness of a shock wave. For example, the velocity-gradient thickness $\bar{x}_u$ is defined by

$$\bar{x}_u \equiv \frac{u_\alpha - u_\beta}{|du/dx|_{\max}} . \tag{9.8}$$

Although the thickness $\bar{x}_u$ can be misleading since it depends only on local information, it is often used in comparisons with experiment. From equation (9.6) we find that for weak shock waves

$$\frac{\bar{x}_u}{\lambda_\alpha} = \frac{4}{G}\frac{\bar{C}_\alpha}{(u_\alpha - u_\beta)} . \tag{9.9}$$

Thus when $M_\alpha \to 1$ the thickness $\bar{x}_u$ becomes large as $u_\alpha \to u_\beta$. If we use $\bar{x}_u$ for L in the expansion parameter ξ [equation (3.2)] in the Chapman-Enskog expansion, we have in this situation

$$\xi = \frac{u_\alpha}{\bar{C}_\alpha}\frac{\lambda_\alpha}{\bar{x}_u} = \frac{G}{4}\left(\frac{u_\alpha}{\bar{C}_\alpha}\right)^2\left(1 - \frac{u_\beta}{u_\alpha}\right).$$

From this result we see that for weak shock waves $\xi \ll 1$ and hence that the applicability of the Navier-Stokes equations is assured. In this limit, equation (9.7) and (9.9) are then exact and can be taken as standards of comparison for other theories.

Actually, the limitation $M_\alpha^2 - 1 \ll 1$ on the linearized analysis that leads to equation (9.4) is more restrictive than the limitations on the general Navier-Stokes solution. In fact, comparisons with experiment show that the Navier-Stokes solution is accurate for larger values of M_α then might be expected from purely theoretical considerations.[14]

To illustrate the Navier-Stokes solution for shock waves of moderate strength we shall make use of a simplified solution that is possible for the special value $\Pr = \frac{3}{4}$. Within our context this must be viewed as an

[14] It is sometimes said that the test of a good theory is whether its usefulness exceeds its expected range of validity; the Navier-Stokes equations amply satisfy this criterion.

approximation, since for monatomic gases $\Pr \simeq \frac{2}{3}$. Liepmann, Narasimha, and Chahine (1962) have shown, however, that the Navier-Stokes solution is not particularly sensitive to small changes in the Prandtl number. When $\Pr = \frac{3}{4}$, it follows from equations (9.1) and (9.3) (Exercise 9.1) that the stagnation enthalpy $h + u^2/2$ is constant throughout the wave and $u\tau_{xx} = q_x$. In this case it can be shown that equation (9.6) remains valid for any value of M_α if $\bar{C}_\alpha\lambda_\alpha$ is replaced by the local value $\bar{C}\lambda$ and $\Pr = \frac{3}{4}$ is used in G. We then have

$$\frac{du}{dx} = -G\frac{(u_\alpha - u)(u - u_\beta)}{\bar{C}\lambda}. \tag{9.10}$$

Since by equation (9.5) $\bar{C}\lambda = 2\mu/\rho = 2u\mu/\rho_\alpha u_\alpha$, equation (9.10) cannot be integrated without a knowledge of the dependence of the coefficient of viscosity on temperature.[15] Such an integration will not be required for the observations that we wish to make.

We can see from equation (9.10) that, since a shock wave causes an increase in temperature in the direction of flow and since μ increases with increasing temperature, neglect of the temperature dependence of μ would underestimate the thickness of the wave. With equation (9.10) we can also assess the limitations on the applicability of the Navier-Stokes equations. From the Chapman-Enskog distribution function (7.5), the first-order correction to the Maxwellian distribution function contains in the present case the terms τ_{xx}/p and $q_x/p\bar{C}$. Since in the present approximation $q_x = u\tau_{xx}$, the local values of both τ_{xx}/p and $u\tau_{xx}/p\bar{C}$ must be everywhere small for the Navier-Stokes equations to be expected to apply. The restriction on $u\tau_{xx}/p\bar{C}$ is, however, the most severe. From equation (9.10) and the relation $h + u^2/2 = h_\alpha + u_\alpha^2/2$, we have for a monatomic gas

$$\left|\frac{q_x}{p\bar{C}}\right| = \left|\frac{u\tau_{xx}}{p\bar{C}}\right| = \frac{16u\lambda}{3\pi\bar{C}^2}\frac{du}{dx} = G\frac{5}{3}\sqrt{\frac{5\pi}{16}}\frac{(u_\alpha - u)(u - u_\beta)u}{(h_\alpha + u_\alpha^2/2 - u^2/2)^{3/2}}. \tag{9.11}$$

In Fig. 1, $|q_x/p\bar{C}|$ from equation (9.11) is shown as a function of u for several values of M_α. From this result we can infer that deviations from the Navier-Stokes solution are to be expected as M_α becomes large, particularly in the upstream portion of the shock wave.

To improve the theory and extend its validity to higher values of M_α, the next logical step would be to include terms of order ξ^2 in the Chapman-Enskog expansion. This leads to the so-called Burnett equations (see Burnett, 1936, and Chapman and Cowling, 1952, Chapter 15). Talbot and Sherman (1959) have made detailed measurements of the structure

[15] Temperature and velocity are related by means of $h + u^2/2 = h_\alpha + u_\alpha^2/2$.

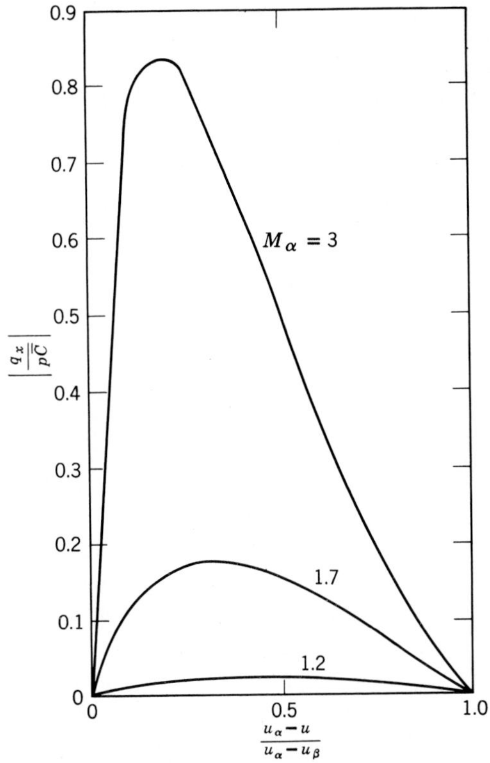

Fig. 1. Magnitude of the heat-flux contribution to the Chapman-Enskog correction to f_0 within a shock wave (monatomic gas).

of shock waves in argon at moderate values of M_α and for these cases have integrated the Burnett equations numerically. The experimental data were obtained by means of a wire probe, where the ratio of wire diameter to mean free path within the shock wave was small. This "free-molecule probe" was oriented perpendicularly to the direction of flow and was traversed through the wave. A comparison between the measured profile of the steady-state wire temperature and the values predicted by the Navier-Stokes and Burnett solutions is shown in Fig. 2 for argon at $M_\alpha = 1.7$. It can be seen that the difference between the two theoretical profiles is small and that both calculations are in excellent agreement with experiment. The Burnett solution thus gives little, if any, improvement over the Navier-Stokes solution. This is representative of the general situation with regard to the Burnett equations, where at present there is no conclusive evidence to indicate that these equations are to be preferred to the Navier-Stokes equations. Grad (1963) has suggested that this may

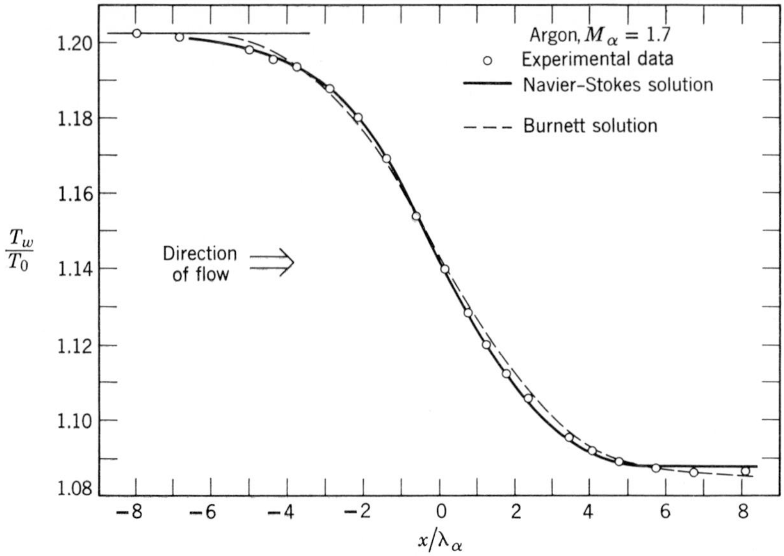

Fig. 2. Shock-wave structure in argon (after Talbot and Sherman, 1959). T_w/T_0 is the ratio of the temperature of the wire probe to the stagnation temperature.

be indicative of an asymptotic convergence of the Chapman-Enskog expansion such that only terms of order ξ or perhaps ξ^2 should be retained.

Since we do not expect the Navier-Stokes solution to be valid for strong shock waves and since the Boltzmann equation itself has not been solved directly, there exists a strong motivation to find approximate solutions of the Boltzmann equation that may be useful for large M_α. The shock-wave solution proposed by Mott-Smith (1951) together with the method of Lees that is discussed in the next section serve to illustrate an interesting intuitive approach to problems of translational nonequilibrium. In this approach an approximate functional form of the distribution function is selected in terms of one or more unspecified parameters. These parameters are then found in such a way that a certain number of moments of the Boltzmann equation are satisfied.

Mott-Smith reasoned that for large values of M_α, where the shock wave is thin, the distribution function can be represented as a weighted sum of the upstream and downstream distributions functions. He assumed, with no rigorous justification, the *bimodal* distribution function

$$nf = N_\alpha f_0(u_\alpha, T_\alpha) + N_\beta f_0(u_\beta, T_\beta), \tag{9.12}$$

where

$$f_0(u_\alpha, T_\alpha) = \left(\frac{m}{2\pi k T_\alpha}\right)^{3/2} \exp\left\{-[(c_x - u_\alpha)^2 + c_y^2 + c_z^2]\frac{m}{2kT_\alpha}\right\}$$

and

$$f_0(u_\beta, T_\beta) = \left(\frac{m}{2\pi k T_\beta}\right)^{3/2} \exp\left\{-[(c_x - u_\beta)^2 + c_y^2 + c_z^2]\frac{m}{2kT_\beta}\right\}.$$

The unspecified weighting functions N_α and N_β must vary with position so that upstream $N_\alpha = n_\alpha$, $N_\beta = 0$ and downstream $N_\alpha = 0$, $N_\beta = n_\beta$. Since

$$\int_{-\infty}^{\infty} nf\, dV_c = n,$$

we also must have

$$n = N_\alpha + N_\beta. \tag{9.13}$$

The three conservation equations (9.1) express the fact that the fluxes of mass, momentum, and energy are constant within the shock wave. With the distribution function (9.12) these conditions can be written directly without recourse to equations (9.1) as

$$N_\alpha u_\alpha + N_\beta u_\beta = n_\alpha u_\alpha, \tag{9.14a}$$

$$N_\alpha(kT_\alpha + mu_\alpha^2) + N_\beta(kT_\beta + mu_\beta^2) = p_\alpha + \rho_\alpha u_\alpha^2, \tag{9.14b}$$

$$N_\alpha u_\alpha\left(\frac{5kT_\alpha}{2} + \frac{mu_\alpha^2}{2}\right) + N_\beta u_\beta\left(\frac{5kT_\beta}{2} + \frac{mu_\beta^2}{2}\right) = \rho_\alpha u_\alpha\left(h_\alpha + \frac{u_\alpha^2}{2}\right). \tag{9.14c}$$

Equations (9.14) are then equivalent, for this case, to equations (9.1). Since u_α, T_α, u_β, and T_β must satisfy equations (9.2), the conservation equations (9.14) are doubly redundant and can be replaced by the single equation (9.14a). Equations (9.13), (9.14a), and (9.1a) can then be used to express the two variables N_α and N_β in terms of the local velocity u as follows:

$$\begin{aligned} N_\alpha &= \frac{n_\alpha u_\alpha - nu_\beta}{u_\alpha - u_\beta} = \frac{n_\alpha u_\alpha}{u_\alpha - u_\beta}\left(1 - \frac{u_\beta}{u}\right), \\ N_\beta &= \frac{nu_\alpha - n_\alpha u_\alpha}{u_\alpha - u_\beta} = \frac{n_\alpha u_\alpha}{u_\alpha - u_\beta}\left(\frac{u_\alpha}{u} - 1\right). \end{aligned} \tag{9.15}$$

One additional condition is required to find u as a function of position. For this condition Mott-Smith used the equation of transfer (IX 6.2), with alternatively $Q = c_x^2$ and $Q = c_x^3$. It can be shown that the form of the predicted shock-wave structure is independent of both the intermolecular potential and the choice of Q. We shall consider the simplest case of Maxwellian molecules and $Q = c_x^2$. The equation of transfer [equations (IX 6.2) and (IX 9.6)] is then

$$\frac{d}{dx}(n\overline{c_x^3}) = 3\pi A_2(4)\sqrt{2a/m}\,\frac{n}{m}\,\tau_{xx}. \tag{9.16}$$

With equation (9.5) for λ in terms of μ and equation (7.4b) for μ, the upstream mean free path

$$\lambda_\alpha = \frac{5}{16A_2(4)} \frac{1}{n_\alpha} \sqrt{kT_\alpha/a} \frac{1}{\Gamma(\frac{7}{2})} \tag{9.17}$$

can be used in equation (9.16) in place of the intermolecular-potential constants $A_2(4)$ and a. With Mott-Smith's assumed distribution function (9.12), the equation of transfer (9.16) then becomes

$$\frac{d}{dx}\left[N_\alpha u_\alpha\left(\frac{3kT_\alpha}{m} + u_\alpha^2\right) + N_\beta u_\beta\left(\frac{3kT_\beta}{m} + u_\beta^2\right)\right]$$
$$= -\frac{1}{m\lambda_\alpha}\sqrt{\pi kT_\alpha/2m}\,\frac{u_\alpha}{u}\,[p_\alpha + \rho_\alpha u_\alpha^2 - \rho_\alpha u_\alpha u - (N_\alpha kT_\alpha + N_\beta kT_\beta)]. \tag{9.18}$$

Here τ_{xx} in equation (9.16) has been evaluated by means of the momentum equation (9.1b). Equation (9.18) can be simplified further by the use of the conservation equations (9.14), and with equations (9.15), N_α and N_β can be eliminated in favor of u. After some manipulation we obtain

$$A\frac{\lambda_\alpha}{u_\alpha}\frac{du}{dx} = -\left(1 - \frac{u}{u_\alpha}\right)\left(\frac{u}{u_\alpha} - \frac{u_\beta}{u_\alpha}\right), \tag{9.19}$$

where

$$A \equiv \sqrt{\frac{2}{\pi}}\frac{u_\beta}{u_\alpha}\left[\left(1 - 2\frac{u_\beta}{u_\alpha}\right)\frac{u_\alpha}{\sqrt{kT_\alpha/m}} + 3\frac{\sqrt{kT_\alpha/m}}{u_\alpha}\right].$$

Equation (9.19) can be integrated directly to give the symmetric profile

$$\boxed{\frac{u - u_\beta}{u_\alpha - u_\beta} = \left\{1 + \exp\left[\left(1 - \frac{u_\beta}{u_\alpha}\right)\frac{x}{A\lambda_\alpha}\right]\right\}^{-1}}. \tag{9.20}$$

The Mott-Smith solution (9.20) satisfies only one equation of transfer [equation (9.16)] in addition to the conservation equations and is only approximate; an exact solution of the Boltzmann equation would satisfy any and all equations of transfer.

From the equation (9.19), the velocity-gradient thickness $\bar{x}_u$ is given by

$$\frac{\bar{x}_u}{\lambda_\alpha} = \frac{4A}{1 - (u_\beta/u_\alpha)}. \tag{9.21}$$

For weak shock waves, the Mott-Smith solution is found to differ significantly from the Navier-Stokes solution and from detailed measurements of the shock-wave structure (Sherman, 1955), the Mott-Smith thickness (9.21) being somewhat greater than that observed. For strong shock waves, for which the Mott-Smith theory was intended, detailed measurements of the structure are not yet available. Measurements of the density-gradient thickness

$$\bar{x}_\rho \equiv \frac{\rho_\beta - \rho_\alpha}{(d\rho/dx)_{\max}} \tag{9.22}$$

for values of M_α up to about 5 have been made, however, by means of an optical-reflectivity technique (Linzer and Hornig, 1963, and Hansen and

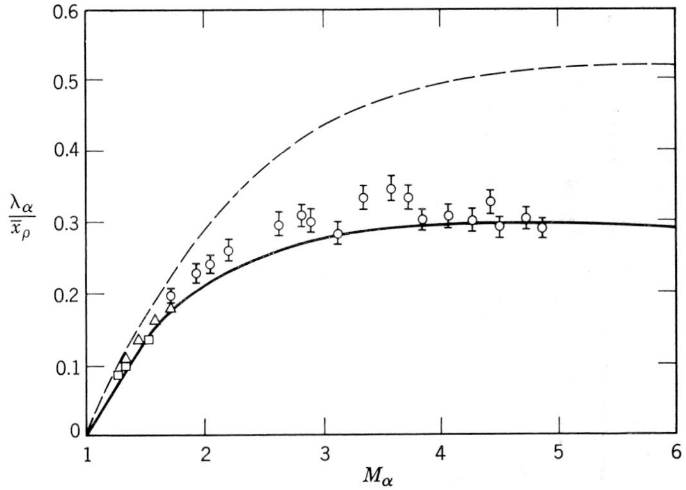

Fig. 3. The reciprocal of the density-gradient shock-wave thickness in argon; comparison of the Mott-Smith (———) and Navier-Stokes (- - - -) theories for the Lennard-Jones 6-12 potential with the experiments of Linzer and Hornig (1963) (⚲), Talbot and Sherman (1959) (△), and Hansen and Hornig (1960) (□) (after Linzer and Hornig, 1963).

and Hornig, 1960). A symmetric density profile of the Mott-Smith form [see equation (9.20)] was used to infer $\bar{x}_\rho$ from the experimental data, and for this reason the interpretation of the experiments is open to some question. In Fig. 3 a comparison is made between the results of these experiments and the shock-wave thickness $\bar{x}_\rho$ in argon as predicted by the Mott-Smith and Navier-Stokes theories. [For the Mott-Smith solution (9.20), $\bar{x}_\rho$ is equal to $\bar{x}_u$; this is not true for the Navier-Stokes solution. See Schwartz and Hornig, 1963.] The Mott-Smith thickness, as calculated by Muckenfuss (1962) for realistic (Lennard-Jones 6-12) intermolecular potentials and $Q = c_x^2$, is seen to compare favorably with the experimental results. The importance of the influence of the intermolecular potential on calculated shock-wave thicknesses and on comparisons with experiment is emphasized by the calculations of Muckenfuss and of Schwartz and Hornig. Preliminary values of the shock-wave thickness from the electron-beam measurements of Camac (1964) for M_α up to 10 are also in better agreement with the Mott-Smith solution than with the Navier-Stokes solution.[16]

A more recent approach to the problem of shock-wave structure (Liepmann, Narasimha, and Chahine, 1962) employs the B-G-K model (see Sec. 2). This approach is more in the spirit of an exact solution in that the distribution function is an outcome rather than a postulate of the analysis. If the collision integral in the Boltzmann equation is replaced by the B-G-K model, we have for the one-dimensional flow through a shock wave

$$c_x \frac{d}{dx}(nf) = n\nu(f_0 - f). \tag{9.23}$$

Since $f_0 = f_0(u, T)$, where u and T are defined as moments of the distribution function, equation (9.23) is an integro-differential equation. Formal integration of this equation yields

$$nf = \int_{-\infty}^{x} \frac{n\nu f_0(x')}{c_x}\left[\exp\left(-\int_{x'}^{x}\frac{\nu\, dx''}{c_x}\right)\right] dx' \qquad (c_x > 0) \tag{9.24a}$$

and

$$nf = \int_{\infty}^{x} \frac{n\nu f_0(x')}{c_x}\left[\exp\left(-\int_{x'}^{x}\frac{\nu\, dx''}{c_x}\right)\right] dx' \qquad (c_x < 0), \tag{9.24b}$$

where x' and x'' are dummy variables for x. As a result of the boundary conditions at $x = \pm\infty$, the distribution function (9.24a) for $c_x > 0$ differs from the function (9.24b) for $c_x < 0$. Equations (9.24) were solved by numerical iteration by Liepmann, Narasimha, and Chahine. The collision frequency ν was evaluated from $\nu = nkT/\mu$ [equation (3.13)] with the viscosity of argon as given by the Sutherland model, and the B-G-K value of the Prandtl number Pr = 1 was used to find the corresponding Navier-Stokes solution. At $M_\alpha = 1.5$ the B-G-K and Navier-Stokes solutions were found to be in almost exact agreement. The calculated velocity profiles at $M_\alpha = 5.0$ are shown in Fig. 4. In this case the B-G-K solution follows closely the Navier-Stokes solution in the downstream, high-density portion of the shock wave but gives a much

[16] Additional measurements of shock-wave thicknesses were reported at the Fourth International Symposium on Rarefied Gas Dynamics held in Toronto in 1964. The proceedings of this symposium will be published in 1965.

thicker profile in the upstream region. This is in accord with our observations from Fig. 1 with regard to expected deviations from the Navier-Stokes solution in the upstream region. It can be seen also from Fig. 4 that the B-G-K and Navier-Stokes velocity-gradient thicknesses [equation (9.8)] are nearly the same, although the profiles differ significantly.

In summary, the investigation of the structure of shock waves has been both theoretically and experimentally successful for values of M_α somewhat less than 2. Under these conditions the Navier-Stokes solution is in

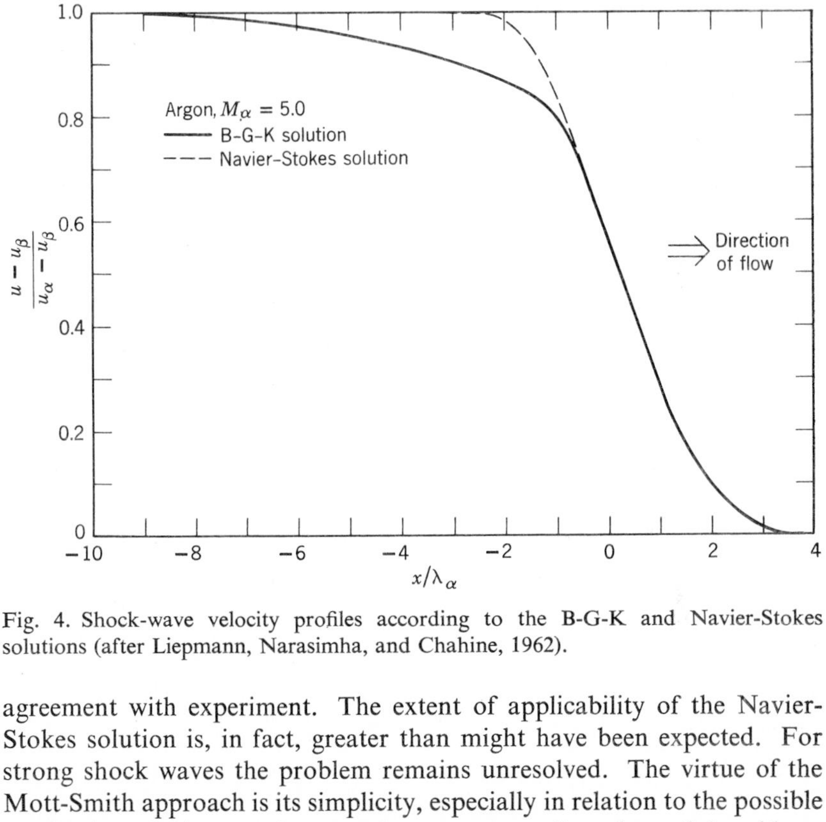

Fig. 4. Shock-wave velocity profiles according to the B-G-K and Navier-Stokes solutions (after Liepmann, Narasimha, and Chahine, 1962).

agreement with experiment. The extent of applicability of the Navier-Stokes solution is, in fact, greater than might have been expected. For strong shock waves the problem remains unresolved. The virtue of the Mott-Smith approach is its simplicity, especially in relation to the possible application of this general approach to more complicated practial problems for which more exact solutions are not feasible. The success of the Mott-Smith solution in comparison with shock-thickness experiments is certainly encouraging in this regard. This may indicate that the specification of the form of the distribution function is not critical if only gross quantities such as some measure of the shock-wave thickness are of interest. The B-G-K model, on the other hand, provides a means whereby the distribution function itself can be found. Although the limitations and accuracy of this method have not yet been definitely established, it is reasonable to suppose that the shock-wave profiles that result from the B-G-K solution are at least qualitatively correct.

Exercise 9.1. When Pr $= \frac{3}{4}$, show for the Navier-Stokes solution that $h + u^2/2$ is constant throughout the shock wave and that $u\tau_{xx} = q_x$. Show that these conditions result in equation (9.10).

Exercise 9.2. Verify that with the Mott-Smith distribution function the equation of transfer (9.16) is given by equation (9.18).

Exercise 9.3. Apply the Mott-Smith method to the Krook equation (9.23). Calculate first $\Delta[c_x^2]$ for the B-G-K model and compare your result with equation (9.16). This should enable you to compare the Mott-Smith solutions of the Krook and Boltzmann equations.

Exercise 9.4. Prove that according to the Mott-Smith solution the kinetic temperature $m\overline{C^2}/3k$ is a monotonic function of position throughout the shock wave. (As a result of an error in calculation, a statement to the contrary appears in Mott-Smith, 1951.)

10 LINEARIZED COUETTE FLOW

Couette flow results from the steady, relative motion of two parallel surfaces that are separated by a fluid (Fig. 5). It is the simplest situation that involves the interaction between a flow and solid surfaces. We shall discuss the case for which the temperatures of the two surfaces are equal and restrict our attention to linearized Couette flow, where the kinetic energy of mean motion is assumed to be small compared with the kinetic energy of thermal motion. In this situation, which prevails when

$$\frac{u_0^2}{2kT_0/m} \ll 1, \tag{10.1}$$

the temperature T_0 can be assumed uniform throughout the gas and the energy equation, which then involves only second-order quantities, need not be considered. Since the flow is steady and the gradients in the x- and z-direction and the velocity in the z-direction are zero, the relevant conservation equations, which are obtained from equations (IX 2.22) and (IX 2.23), are

Mass: $$\frac{d}{dy}(\rho v) = 0, \tag{10.2a}$$

y-momentum: $$\frac{d}{dy}(p - \tau_{yy}) = 0, \tag{10.2b}$$

x-momentum: $$\frac{d}{dy}(\tau_{xy}) = 0. \tag{10.2c}$$

The continuity equation (10.2a) requires that the net mass flux ρv in the y-direction be constant. Since there is no mass flux through the surfaces at $y = \pm d/2$, the velocity v in the y-direction must therefore be everywhere

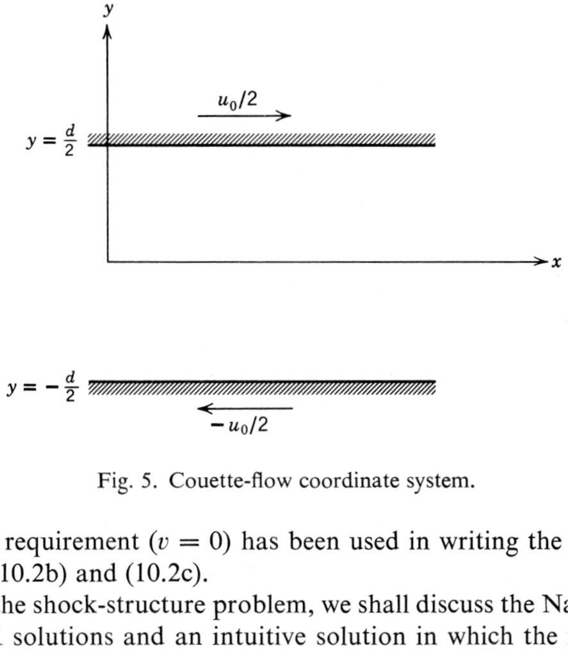

Fig. 5. Couette-flow coordinate system.

zero. This requirement ($v = 0$) has been used in writing the momentum equations (10.2b) and (10.2c).

As with the shock-structure problem, we shall discuss the Navier-Stokes and B-G-K solutions and an intuitive solution in which the form of the distribution function is assumed *a priori*. In this problem, there is also a free-molecule solution that applies when the Knudsen number λ/d is large. In addition, we now have the complication of the molecular boundary condition at the solid surfaces.

The Navier-Stokes (Chapman-Enskog) solution follows from the relations

$$\tau_{yy} = 0 \tag{10.3a}$$

and

$$\tau_{xy} = \mu \frac{du}{dy} \tag{10.3b}$$

as given by equations (5.1) and (5.4). With $T = T_0$, the conservations equations then require that p and n are constant and that

$$\mu \frac{du}{dy} = \tau_0, \tag{10.4}$$

where τ_0 is the shear stress at the solid surfaces. Since μ depends only on temperature and hence is constant, equation (10.4) can be integrated directly to give

$$u = (u_0 - 2u_s)\frac{y}{d} \tag{10.5a}$$

and

$$\tau_0 = \frac{\mu}{d}(u_0 - 2u_s). \tag{10.5b}$$

Here the *slip velocity* u_s enters the solutions (10.5) through the as yet unspecified boundary condition at $y = \pm d/2$. The slip velocity is defined as the difference between the velocity $\pm u_0/2$ of the surfaces and the velocity $u(y = \pm d/2)$ that is obtained from the extrapolation to the surfaces of the velocity profile in the gas (see Fig. 6). The actual mean velocity of the gas at the surface, however, is not necessarily equal to the extrapolated velocity nor can the Navier-Stokes equations be expected to apply rigorously in the immediate vicinity of the surface. In the familiar continuum solution, u_s is set equal to zero on the basis of experiment.

To find the boundary condition u_s we must consider the interaction between the gas molecules and the solid surfaces. This is a difficult problem and the situation is not well understood from a fundamental viewpoint. Fortunately, for practical engineering surfaces at moderate temperatures and gas velocities, available experimental data (for example, Hurlbut, 1957) are reasonably correlated by the assumption of *diffuse reflection* (or re-emission). In this approximation gas molecules that strike the surface are assumed to leave it with a Maxwellian velocity distribution in which the mean velocity relative to the surface is zero. That is, the velocity distribution of reflected molecules is as if they came from an imaginary gas behind the surface and at rest, in the mean, with respect to the surface. The temperature of this imaginary gas is specified by an energy accommodation coefficient that is unity if the gas temperature and surface temperature are equal. Although we do not wish to go into the details of such models here, the accommodation of energy is found experimentally to be less complete than the accommodation of momentum. In our problem the temperature of the reflected gas is, in any case, equal to the surface temperature T_0.

Even with this simple model of the surface interaction a general and completely rigorous solution for the slip velocity u_s has not been found. A useful approximation can be obtained, however, by means of an argument that is similar to the mean-free-path discussion of viscous stress

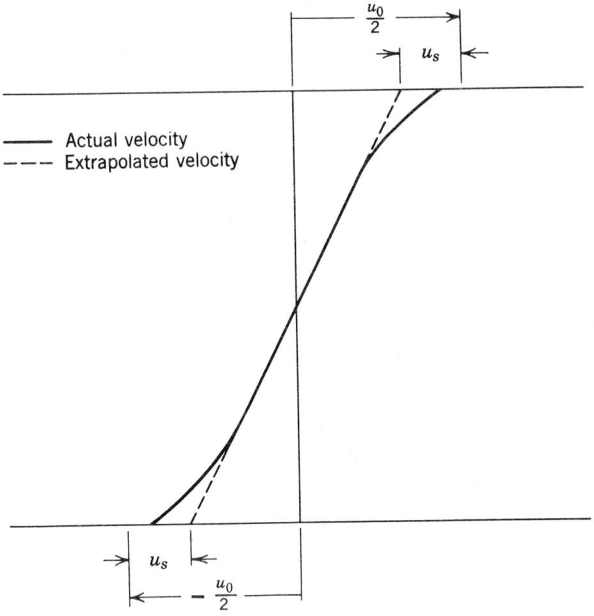

Fig. 6. Couette-flow velocity profile.

in Chapter I, Sec. 5. It is convenient to adopt temporarily a coordinate system that is fixed with respect to one of the surfaces. By equation (10.2c) the shear stress τ_{xy} is everywhere constant and equal to the value τ_0 at the surface. With diffuse reflection, the reflected molecules make no contribution to τ_{xy} at the surface since, relative to the surface, they carry equal amounts of momentum in the positive and negative x-directions. If the molecules that are incident on the surface carry on the average the x-momentum per molecule that prevails at a distance of one mean free path from the surface, τ_0 will be given by the product of the flux of incident molecules and $m(u_s + \lambda\, du/dy)$. It is easily shown that the flux of incident molecules is $\frac{1}{4}n\bar{C}$, so that the shear stress at the surface is

$$\tau_0 = \tfrac{1}{4}\rho\bar{C}\left(u_s + \lambda\frac{du}{dy}\right).$$

In the present linearized solution ρ, $\bar{C}$, and λ are all constant. With $\mu = \frac{1}{2}\rho\bar{C}\lambda$ and $\tau_0 = \mu\, du/dy$ from equation (10.4), evaluated at a sufficient distance from the surface that the Navier-Stokes equations apply, we then have

$$u_s = \lambda \frac{du}{dy}. \tag{10.6}$$

More general expressions for the slip velocity and for an analogous quantity, the temperature jump (see Exercise 10.1), are given by Schaaf and Chambré (1958).

Since by equation (3.2) the expansion parameter ξ in the Chapman-Enskog expansion is of the order of λ/d, we would expect that under conditions for which the Navier-Stokes equations are applicable, $\lambda\, du/dy$ in equation (10.6) would be small compared with u_0. Our rough analysis then supports by molecular arguments the familiar zero-slip boundary conditions. Nonetheless, it has been suggested (Schaaf, 1960) that the range of applicability of the Navier-Stokes equations can be extended somewhat by the retention of slip and temperature-jump boundary conditions. With the slip velocity (10.6), the Navier-Stokes solution (10.5) can be written

$$\boxed{\begin{aligned} \frac{u}{u_0} &= \frac{y}{d}\left(\frac{1}{1 + 2\lambda/d}\right) \\ \tau_0 &= \mu \frac{u_0}{d}\left(\frac{1}{1 + 2\lambda/d}\right) \end{aligned}} \qquad \begin{aligned} &(10.7\text{a}) \\ &(10.7\text{b}) \end{aligned}$$

In any case, however, the Navier-Stokes solution will fail as λ/d becomes large. (For fixed d this would occur as the density is decreased.)

As the Navier-Stokes solution is exact for $\lambda/d \ll 1$, the *free-molecule* solution is exact for $\lambda/d \gg 1$. Under these conditions molecular collisions can be neglected in the region between the solid surfaces, and the distribution function can be found from a knowledge of the nature of the molecular interaction with the surfaces. With no molecular collisions and the Couette-flow geometry, at any position y the distribution function for molecules with velocities that are directed away from a given surface is unchanged from the distribution function of such molecules at the surface. For diffuse reflection we then have

$$\begin{aligned} f &= \left(\frac{m}{2\pi k T_0}\right)^{3/2} \exp\left\{-\left[\left(c_x - \frac{u_0}{2}\right)^2 + c_y^2 + c_z^2\right]\frac{m}{2kT_0}\right\} \\ &= f_0(u_0/2, T_0) \qquad (c_y < 0) \end{aligned} \tag{10.8}$$

and

$$f = \left(\frac{m}{2\pi kT_0}\right)^{3/2} \exp\left\{-\left[\left(c_x + \frac{u_0}{2}\right)^2 + c_y^2 + c_z^2\right]\frac{m}{2kT_0}\right\}$$
$$= f_0(-u_0/2, T_0) \qquad (c_y > 0).$$

By the same argument the density must be uniform. Inspection of equations (10.8) shows that the mean velocity u is zero everywhere as is v, so that $C_x = c_x$ and $C_y = c_y$. The shear stress τ_{xy} can be found by direct integration as follows:

$$\tau_{xy} \equiv -\rho\overline{C_x C_y} = -\rho\overline{c_x c_y} = -\rho\left[\int_{-\infty}^{\infty}\int_{-\infty}^{0}\int_{-\infty}^{\infty} c_x c_y f_0\left(\frac{u_0}{2}, T_0\right) dc_x\, dc_y\, dc_z + \int_{-\infty}^{\infty}\int_{0}^{\infty}\int_{-\infty}^{\infty} c_x c_y f_0\left(-\frac{u_0}{2}, T_0\right) dc_x\, dc_y\, dc_z\right].$$

The integration is straightforward and gives

$$\boxed{\tau_{xy} = \tau_0 = \tfrac{1}{2}\rho u_0\sqrt{\frac{2kT_0}{\pi m}}}\,. \tag{10.9}$$

The shear stress (10.9) in the free-molecule limit $\lambda/d \gg 1$ is proportional to density, in contrast to the Navier-Stokes shear stress (10.7b) in the continuum limit $\lambda/d \ll 1$, which is independent of density if the slip-velocity correction is neglected.

Lees (1959) has proposed an approximate method for problems of rarefied-gas dynamics that is similar in spirit to the Mott-Smith approach to shock-wave structure (see also Liu and Lees, 1961, and Lees and Liu, 1962). As the Knudsen number λ/d increases, the Lees solution for linearized Couette flow goes smoothly from the Navier-Stokes solution (10.7) to the free-molecule solution (10.9). Lees assumed a "two-stream Maxwellian" distribution function of the form

$$\begin{aligned} nf &= N_\alpha\left(\frac{m}{2\pi kT_0}\right)^{3/2} \exp\left\{-\left[(c_x - u_\alpha)^2 + c_y^2 + c_z^2\right]\frac{m}{2kT_0}\right\} \\ &= N_\alpha f_0(u_\alpha, T_0) \qquad (c_y < 0) \\ nf &= N_\beta\left(\frac{m}{2\pi kT_0}\right)^{3/2} \exp\left\{-\left[(c_x - u_\beta)^2 + c_y^2 + c_z^2\right]\frac{m}{2kT_0}\right\} \\ &= N_\beta f_0(u_\beta, T_0) \qquad (c_y > 0). \end{aligned} \tag{10.10}$$

Here N_α, N_β, u_α, and u_β are functions of y that are to be found from the conservation equations and the required number of additional moments of the Boltzmann equation. The distribution function (10.10) is particularly suited to the assumption of diffuse reflection, for which we have the boundary conditions [see equations (10.8)]

$$u_\alpha = \frac{u_0}{2} \quad \text{at} \quad y = \frac{d}{2} \tag{10.11a}$$

and

$$u_\beta = -\frac{u_0}{2} \quad \text{at} \quad y = -\frac{d}{2}. \tag{10.11b}$$

With the distribution function (10.10), the conditions $\int_{-\infty}^{\infty} nf\, dV_c = n$ and $v = 0$ and the momentum equations (10.2b) and (10.2c) become, after evaluation of the moments that appear therein,

$$\tfrac{1}{2}(N_\alpha + N_\beta) = n, \tag{10.12a}$$

$$\sqrt{kT_0/2\pi m}\,(N_\alpha - N_\beta) = 0, \tag{10.12b}$$

$$\frac{d}{dy}\,[\tfrac{1}{2}kT_0(N_\alpha + N_\beta)] = 0, \tag{10.12c}$$

and

$$m\sqrt{kT_0/2\pi m}\,(N_\alpha u_\alpha - N_\beta u_\beta) = \tau_0. \tag{10.12d}$$

In addition, the definition $nu \equiv \int_{-\infty}^{\infty} c_x nf\, dV_c$ of the mean velocity u gives

$$nu = \tfrac{1}{2}(N_\alpha u_\alpha + N_\beta u_\beta). \tag{10.12e}$$

The first three of these equations require that

$$N_\alpha = N_\beta = n, \qquad \text{a constant.} \tag{10.13}$$

The fourth equation provides a single relation between the two remaining variables u_α and u_β and the constant shear stress τ_0. For the additional relation Lees chose the equation of transfer with $Q = c_x c_y$, evaluated for Maxwellian molecules. Although the method can be extended to other intermolecular potentials, we shall follow Lees' example for simplicity. The equation of transfer for $Q = c_x c_y$ [equations (IX 6.2) and (IX 9.7a)] is then

$$\frac{d}{dy}\,(n\overline{c_x c_y^2}) = 3\pi A_2(4)\sqrt{2a/m}\,\frac{n}{m}\,\tau_{xy}. \tag{10.14}$$

With the two-stream Maxwellian distribution function (10.10), we have

$$n\overline{c_x c_y^2} = \frac{kT_0}{2m}\,(N_\alpha u_\alpha + N_\beta u_\beta),$$

so that with equations (10.12d,e) and (10.13), equation (10.14) becomes

$$kT_0 \frac{du}{dy} = 3\pi A_2(4)\sqrt{2a/m}\,\tau_0. \tag{10.15}$$

When equation (10.15) is integrated, the constant of integration and the as yet unknown shear stress τ_0 are fixed by equations (10.12d,e) and the boundary conditions (10.11) for diffuse reflection. (Equivalently, we could use only one of these conditions and a symmetry argument.) The final results, written in terms of the mean free path (9.17), are

$$\frac{u_\alpha}{u_0} = \left(\frac{y}{d} + \frac{\lambda}{d}\right)\left(\frac{1}{1 + 2\lambda/d}\right), \tag{10.16a}$$

$$\frac{u_\beta}{u_0} = \left(\frac{y}{d} - \frac{\lambda}{d}\right)\left(\frac{1}{1 + 2\lambda/d}\right), \tag{10.16b}$$

and

$$\boxed{\frac{u}{u_0} = \frac{y}{d}\left(\frac{1}{1 + 2\lambda/d}\right)} \tag{10.16c}$$

$$\boxed{\tau_0 = \tfrac{1}{2}\rho u_0\sqrt{2kT_0/\pi m}\left(\frac{2\lambda/d}{1 + 2\lambda/d}\right)}\,. \tag{10.16d}$$

As we indicated previously, it can be seen (with $\mu = \frac{1}{2}\rho\bar{C}\lambda$) that the Lees shear-stress (10.16d) reduces to the Navier-Stokes value (10.7b) and to the free-molecule value (10.9) when $\lambda/d \ll 1$ and $\lambda/d \gg 1$, respectively. In fact, although the slip-velocity correction would be expected to apply only for small λ/d, for the present problem the Navier-Stokes expressions (10.7) with the slip-velocity correction are formally identical, over the entire range of λ/d, to the Lees results (10.16c,d). It is also interesting to note from equations (10.16a,b,c) that

$$u_\alpha(y) = u(y + \lambda)$$

and

$$u_\beta(y) = u(y - \lambda),$$

so that the Lees solution is equivalent in the Navier-Stokes limit $\lambda/d \ll 1$ to the mean-free-path interpretation of viscous stress (Chapter I, Sec. 5).

The problem of linearized Couette flow for values of λ/d intermediate between the Navier-Stokes and free-molecule limits has been studied by several investigators. We shall complete our discussion, as with the shock-structure problem in Sec. 9, by reference to the B-G-K solution.

The appropriate form of the Krook equation in the present case is

$$c_y \frac{d}{dy}(nf) = n\nu(f_0 - f), \tag{10.17}$$

with

$$f_0 = f_0(u, T_0).$$

As before, this equation can be integrated formally. With the boundary conditions of diffuse reflection we have

$$f = e^{-\nu y/c_y}\left[\int_{-d/2}^{y} \frac{\nu f_0(y')}{c_y} e^{\nu y'/c_y}\, dy' + f_0\left(-\frac{u_0}{2}, T_0\right) e^{\nu d/2c_y}\right] \qquad (c_y > 0)$$

and

$$f = e^{-\nu y/c_y}\left[\int_{d/2}^{y} \frac{\nu f_0(y')}{c_y} e^{\nu y'/c_y}\, dy' + f_0\left(\frac{u_0}{2}, T_0\right) e^{\nu d/2c_y}\right] \qquad (c_y < 0),$$

$$\tag{10.18}$$

where in accord with the previous solutions of this section we have taken n and $\nu(n, T_0)$ as independent of y. In contrast to the shock-structure problem [cf. equations (9.24)], the integrals in equations (10.18) involve, through $f_0 = f_0(u)$, only the single unknown macroscopic variable u. By means of the definition $u = \int_{-\infty}^{\infty} c_x f\, dV_c$, an integral equation for the velocity u then can be obtained from equation (10.18). This integral equation for u, with simplifications appropriate to the linearization and with $\nu = nkT/\mu$ [equation (3.13a)], was solved numerically by Willis (1962).

The velocity profiles found by Willis are of the form shown previously in Fig. 6. In qualitative agreement with the results obtained by other methods (but not with the Lees solution), the calculated slip velocity for small values of λ/d is somewhat in excess (by a factor of 1.14) of the value given by equation (10.6). Once the velocity $u(y)$ is known, the distribution function is given by equations (10.18), and the shear stress $\tau_{xy} = \tau_0$ can be found from its definition as $-\rho\overline{c_x c_y}$, In Fig. 7 the shear stress as calculated for the B-G-K solution by Willis is compared with that given by the Lees solution (10.16). It can be seen that the two methods are in reasonable agreement in their prediction of the gross quantity τ_0. Larger relative differences are found in the more detailed comparisons given by Willis (1962).

The results of Couette-flow experiments, made with a concentric-cylinder geometry, confirm the smooth transition between the Navier-Stokes and free-molecule limits that is predicted by the Lees and B-G-K solutions and shown in Fig. 7. Because of a non-negligible curvature correction, however, these measurements do not provide a discriminating check on the relative merits of the various theories (see Sherman and Talbot, 1960).

The Couette-flow problem is representative of the field of rarefied-gas dynamics in that further theoretical and experimental progress will be required for a more complete understanding of flows with intermediate values of the Knudsen number.

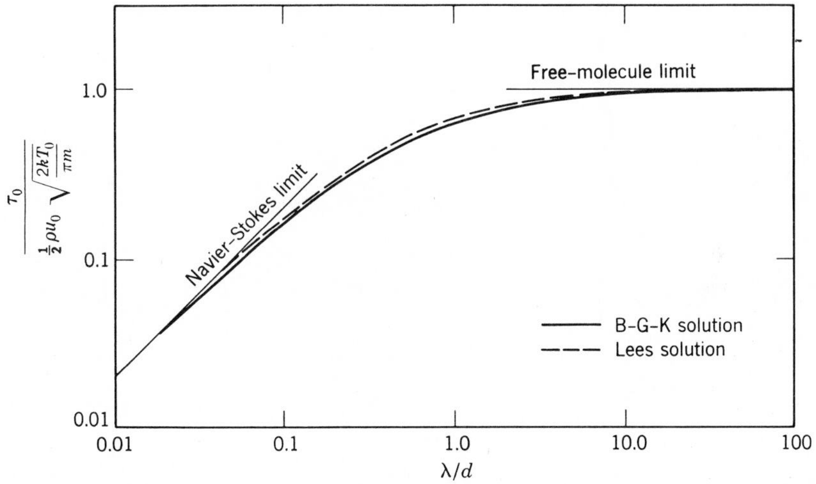

Fig. 7. Shear stress in linearized Couette flow. Comparison of the B-G-K (Willis, 1962) and Lees (1959) solutions.

Exercise 10.1. Show that for diffuse reflection with an accommodation coefficient of unity the difference between the extrapolated gas temperature at a surface and the surface temperature—the temperature jump—is of the order of $\lambda\, dT/dy$.

Exercise 10.2. Find the Navier-Stokes and free-molecule heat-transfer rates between two stationary parallel surfaces, at temperatures T_1 and T_2, that are separated by a gas with no mean motion. Assume diffuse reflection and an accommodation coefficient of unity.

Exercise 10.3. Apply the method of Lees to the heat-transfer problem in Exercise 10.2. Take N_α, N_β, T_α, T_β as functions of y and $u_\alpha = u_\beta = 0$. In addition to the conservation equations, use the equation of transfer with $Q = c_y c^2$ and $\Delta[Q]$ for Maxwellian molecules (cf. Liu and Lees, 1961, Sec. II B).

Exercise 10.4. Apply the method of Lees, for linearized Couette flow, to the Krook equation (10.17) (see Exercise 9.3).

References

Bhatnagar, P. L., E. P. Gross, and M. Krook, 1954, A Model for Collision Processes in Gases. I. Small Amplitude Processes in Charged and Neutral One-Component Systems. *Phys. Rev.*, vol. 94, no. 3, p. 511.

Burnett, D., 1935, The Distribution of Velocities in a Slightly Non-uniform Gas. *Proc. Lond. Math. Soc.*, Ser. 2, vol. 39, p. 385.

Burnett, D., 1936, The Distribution of Molecular Velocities and the Mean Motion in a Non-uniform Gas. *Proc. Lond. Math. Soc.*, Ser. 2, vol. 40, p. 382.

Camac, M., 1964, Argon Shock Thickness. *Phys. Fluids*, vol. 7, no. 7, p. 1076; Argon and Nitrogen Shock Thicknesses. *AIAA*, Preprint no. 64–35.

Chapman, S., 1916, On the Law of the Distribution of Molecular Velocities, and on the Theory of Viscosity and Thermal Conduction, in a Non-uniform Simple Monatomic Gas. *Phil. Trans. Roy. Soc.* A, vol. 216, p. 279.

Chapman, S., and T. G. Cowling, 1952, *The Mathematical Theory of Non-Uniform Gases*, 2nd ed., Cambridge.

De Groot, S. R., and P. Mazur, 1962, *Non-equilibrium Thermodynamics*, Interscience.

Gilbarg, D., and D. Paolucci, 1953, The Structure of Shock Waves in the Continuum Theory of Fluids. *J. Rat. Mech. Anal.*, vol. 2, no. 4, p. 617.

Grad, H., 1958, Principles of the Kinetic Theory of Gases; in *Handbuch der Physik*, vol. 12, p. 205, Springer.

Grad, H., 1963, Asymptotic Theory of the Boltzmann Equation. *Phys. Fluids*, vol. 6, no. 2, p. 147.

Gross, E. P., and E. A. Jackson, 1959, Kinetic Models and the Linearized Boltzmann Equation. *Phys. Fluids*, vol. 2, no. 4, p. 432.

Hansen, K., and D. F. Hornig, 1960, Thickness of Shock Fronts in Argon. *J. Chem. Phys.*, vol. 33, no. 3, p. 913.

Hirschfelder, J. O., C. F. Curtiss, and R. B. Bird, 1964, *Molecular Theory of Gases and Liquids*, corrected printing, Wiley.

Hirschfelder, J. O., C. F. Curtiss, R. B. Bird, and E. L. Spotz, 1955, The Transport Properties of Gases and Gaseous Mixtures; in *High Speed Aerodynamics and Jet Propulsion*, vol. I, Sec. D., p. 339, Princeton.

Hurlbut, F. C., 1957, Studies of Molecular Scattering at the Solid Surface. *J. Appl. Phys.*, vol. 28, no. 8, p. 844.

Lees, L., 1959, A Kinetic Theory Description of Rarefied Gas Flows. *GALCIT Hypersonic Research Project*, Memo. no. 51.

Lees, L., and C. Liu, 1962, Kinetic Theory Description of Conductive Heat Transfer from a Fine Wire. *Phys. Fluids*, vol. 5, no. 10, p. 1137.

Liepmann, H. W., R. Narasimha, and M. T. Chahine, 1962, Structure of a Plane Shock Layer. *Phys. Fluids*, vol. 5, no. 11, p. 1313. (For errata, see also vol. 8, no. 3, p. 551.)

Lighthill, M. J., 1956, Viscosity Effects in Sound Waves of Finite Amptitude; in *Surveys in Mechanics*, p. 250, Cambridge.

Linzer, M., and D. F. Hornig, 1963, Structure of Shock Fronts in Argon and Nitrogen. *Phys. Fluids*, vol. 6, no. 12, p. 1661.

Liu, C., and L. Lees, 1961, Kinetic Theory Description of Plane Compressible Couette Flow; in *Advances in Applied Mechanics, Supplement* **1**, *Rarefied Gas Dynamics*, p. 391, Academic.

Mason, E. A., and L. Monchick, 1962, Heat Conductivity of Polyatomic and Polar Gases. *J. Chem. Phys.*, vol. 36, no. 6, p. 1622.

Mott-Smith, H. M., 1951, The Solution of the Boltzmann Equation for a Shock Wave. *Phys. Rev.*, vol. 82, no. 11, p. 885.

Muckenfuss, C., 1962, Some Aspects of Shock Structure According to the Bimodal Model. *Phys. Fluids*, vol. 5, no. 11, p. 1325.

Schaaf, S. A., 1960, Recent Progess in Rarefied Gasdynamics. *ARS J.*, vol. 30, no. 5, p. 443.

Schaaf, S. A., and P. L. Chambré, 1958, Flow of Rarefied Gases; in *High Speed Aerodynamics and Jet Propulsion*, vol. III, Sec. H, p. 687, Princeton.

Schwartz, L. M., and D. F. Hornig, 1963, Navier-Stokes Calculations of Argon Shock Wave Structure. *Phys. Fluids*, vol. 6, no. 12, p. 1669.

Serrin, J., 1959, Mathematical Principles of Classical Fluid Mechanics; in *Handbuch der Physik*, vol. VIII/1, p. 125, Springer.

Sherman, F. S., 1955, A Low-Density Wind-Tunnel Study of Shock-Wave Structure and Relaxation Phenomena in Gases. *NACA*, TN 3298.

Sherman, F. S., and L. Talbot, 1960, Experiment versus Kinetic Theory for Rarefied Gases; in *Rarefied Gas Dynamics*, p. 161, Pergamon.

Talbot, L., and S. M. Scala, 1961, Shock Wave Structure in a Relaxing Diatomic Gas; in *Advances in Applied Mechanics, Supplement* **1**, *Rarefied Gas Dynamics*, p. 603, Academic.

Talbot, L., and F. S. Sherman, 1959, Structure of Weak Shock Waves in a Monatomic Gas. *NASA*, Memo. 12-14-58W.

Uhlenbeck, G. E., and G. W. Ford, 1953, *Lectures in Statistical Mechanics*, American Mathematical Society.

Wang Chang, C. S., and G. E. Uhlenbeck, 1951, Transport Phenomena in Polyatomic Gases. *Univ. Mich., Engr. Res. Inst.*, Report no. CM-681, Project M604-6.

Wang Chang, C. S., and G. E. Uhlenbeck, 1952, On the Propagation of Sound in Monatomic Gases. *Univ. Mich., Engr. Res. Inst.*, Project M999.

Willis, D. R., 1962, Comparison of Kinetic Theory Analysis of Linearized Couette Flow. *Phys. Fluids*, vol. 5, no. 2, p. 127.

Chapter XI

Radiative Transfer in Gases

1 INTRODUCTION

It is a matter of observation that at high temperatures gases, in common with other substances, emit energy in the form of electromagnetic radiation. This emission of thermal radiation results in part from transitions from upper to lower energy levels of the atoms or molecules of the gas and in part from transitions that involve free electrons. The reverse process also occurs and is present when the gas absorbs radiant energy, either that emitted by other portions of the gas or that incident on it from an external source. Under equilibrium conditions the energy absorbed by the gas at every frequency will just balance that emitted. When this is not the case, the gas is in radiative nonequilibrium. The radiation of energy by a gas can be important not only through the resultant heat transfer to the surroundings of the gas, but also as the nonadiabatic process of radiation affects the motion of the gas itself.

Although the subject of electromagnetic radiation is a broad one that encompasses many disciplines, such as radio communications, we shall adopt a rather specialized point of view. Specifically, we shall be interested in the transfer of energy in a gas by thermal radiation. In contrast to the emphasis that is appropriate in certain applications, we shall give scant attention to the radiative properties of surfaces and shall avoid problems that are largely geometrical in character. Our concern is with the coupling between radiative transfer and gas dynamics. Thus, without denying the importance of the aspect of the formulation that enables the calculation of the radiant transfer from a gas to its surroundings, we emphasize the interaction between the emitted and absorbed radiation and the motion of the gas.

In Chapter VII we used microscopic considerations to formulate certain terms, relevant to vibrational and chemical nonequilibrium, which were then incorporated into the otherwise macroscopic equations of gas dynamics. These macroscopic equations were used in Chapter VIII to solve flow problems. In Chapters IX and X on the other hand we used a microscopic description of translational nonequilibrium both to derive the macroscopic equations of gas dynamics and to find direct solutions of problems of nonequilibrium flow. In the case of radiative nonequilibrium we shall return, in this and the following chapter, to the former point of view.

In the present chapter we shall formulate a description of radiative transfer in a gas and, with the methods of statistical mechanics from Chapter IV, obtain the equilibrium values for certain quantities of interest. In this formulation we shall emphasize the particle rather than wave aspect of radiation. In analogy with the methods of kinetic theory, we consider the radiation field within the gas as composed of photons moving in varying directions with a fixed speed. In Chapter XII we shall incorporate this formulation into the equations of motion of a gas in radiative nonequilibrium and apply these equations, with certain simplifications, to problems of radiative flow.

2 ENERGY TRANSFER BY RADIATION

In our consideration of the interaction between radiation and the motion of gases we shall be concerned with the rate at which energy is transferred by radiation. We can express this energy transfer quantitatively by means of the *radiant heat-flux vector* q_i^R. A typical component of q_i^R, for example q_1^R, is the net rate per unit area at which radiant energy is transferred across a real or imagined surface perpendicular to the x_1-coordinate axis. As we shall discuss, the flux of radiant energy involves both the direction and the frequency of the radiation. For this reason the value of q_i^R itself must be found from a more detailed description of the radiation field.

To formulate this description we emphasize the particle aspect of radiation and consider the radiation field to be composed of n^R photons per unit volume. In general, n^R is a function of position and time. Unlike gas molecules, all photons move with the same speed, the speed of light c. There is, on the other hand, a distribution of energy among the photons, the photon energy being equal to $h\nu$, where h is Planck's constant and ν is the frequency of the radiation. The magnitude of the momentum per photon is known to be $h\nu/c$.

In addition to the distribution of energy and momentum, there is also a distribution in the direction of propagation of the photons. The direction of propagation can be specified by a unit vector l_i through the position x_i (see Fig. 1). The components of l_i are the direction cosines of the direction of propagation with respect to the positive coordinate axes. Since $l_1^2 + l_2^2 + l_3^2 = 1$, only two components of l_i are independent. We now fix our attention on those photons having a direction of propagation within the differential solid angle $d\Omega$ centered about the direction l_i and a frequency within the range ν to $\nu + d\nu$. A photon distribution function

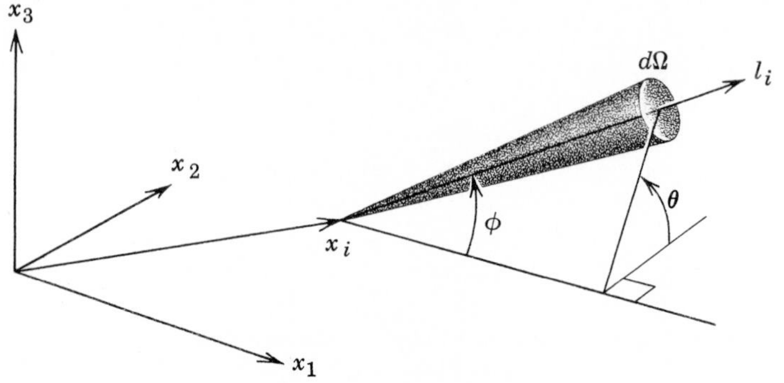

Fig. 1. Coordinate system.

f^R can be defined in terms of the number of such photons per unit volume, so that this number is given by

$$n^R f^R \, d\Omega \, d\nu.$$

In general f^R is a function of position, time, direction, and frequency, that is,

$$f^R = f^R(x_i, t, l_i, \nu).$$

Here the quantities l_i and ν specify the direction of propagation and the energy of the photons. Therefore, in a comparison of f^R with the molecular distribution function f of Chapter IX, Sec. 2, l_i and ν together are analogous to the molecular velocity c_i.

It is conventional to replace $n^R f^R$ as the fundamental quantity that describes the radiation field by the *specific intensity* I_ν, where I_ν is related to $n^R f^R$ by

$$I_\nu \equiv ch\nu n^R f^R. \tag{2.1}$$

The factor $ch\nu$ that appears in equation (2.1) enters from a more usual definition of I_ν in terms of the energy flux across a surface, which we shall note below. Since by definition

$$\int_0^\infty \int_0^{4\pi} n^R f^R \, d\Omega \, d\nu = n^R,$$

it follows from equation (2.1) that the photon number density at a point is given in terms of I_ν by

$$n^R = \frac{1}{c}\int_0^\infty \int_0^{4\pi} \frac{I_\nu}{h\nu}\, d\Omega\, d\nu. \tag{2.2}$$

Here the integration over solid angle accounts for all possible directions of propagation, that is, includes 4π steradians, and the integration over ν accounts for all possible radiation frequencies. Similarly the radiant energy density at frequency ν, denoted by u_ν^R (energy per unit volume per unit frequency), is given by the integration over solid angle of $h\nu n^R f^R$. In terms of I_ν this is

$$u_\nu^R = \frac{1}{c}\int_0^{4\pi} I_\nu\, d\Omega. \tag{2.3a}$$

The total *radiant energy density* u^R at a point is then

$$\boxed{u^R = \frac{1}{c}\int_0^\infty \int_0^{4\pi} I_\nu\, d\Omega\, d\nu}\,. \tag{2.3b}$$

We now express the fluxes of radiant energy and momentum across a real or imagined elementary surface in terms of integrals (moments) of the specific intensity. Since the photon velocity vector is cl_i, it follows from the argument used in the derivation of the molecular flux expressions [equations (IX 2.6)] that the differential photon number flux is

$$cl_j n^R f^R\, d\Omega\, d\nu. \tag{2.4a}$$

With the use of equation (2.1) this can be written in terms of I_ν as

$$\frac{l_j I_\nu}{h\nu}\, d\Omega\, d\nu. \tag{2.4b}$$

Each component of the flux vector (2.4a) or (2.4b) gives the differential photon number flux across a surface that is perpendicular to the corresponding coordinate axis. Since the energy per photon is $h\nu$ and the momentum per photon is $(h\nu/c)l_i$, the differential fluxes of radiant energy and momentum can be obtained directly from equation (2.4b) and are

$$l_j I_\nu\, d\Omega\, d\nu \tag{2.5}$$

and

$$\frac{1}{c}\, l_i l_j I_\nu\, d\Omega\, d\nu. \tag{2.6}$$

A more usual and completely equivalent definition of the specific intensity I_ν can be obtained from the expression (2.5) for the differential radiant energy flux. The specific intensity I_ν is then *defined* such that the differential flux of radiant energy across a surface is

$$\cos \Phi \, I_\nu \, d\Omega \, d\nu, \tag{2.7}$$

where Φ is the angle between the direction of propagation and the normal to the surface. Expression (2.7) follows from expression (2.5) if one axis is taken temporarily to coincide with the normal to the surface in question

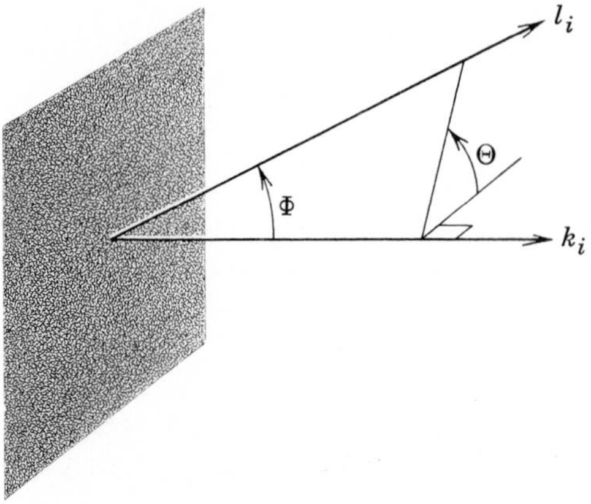

Fig. 2. Coordinate system at a surface.

and the component of the radiant energy flux (2.5) along that axis is considered. From the expression (2.7) it follows that for a given direction of propagation the specific intensity is equal to the radiant energy flux, per unit solid angle per unit frequency, across a surface that is normal to the direction of propagation.

By integration of expression (2.5) the radiant heat-flux vector q_i^R, which was introduced at the beginning of the section, is given by

$$\boxed{q_i^R = \int_0^\infty \int_0^{4\pi} I_\nu l_i \, d\Omega \, d\nu}\,, \tag{2.8}$$

where we have replaced the subscript j in equation (2.5) by i.

We shall also need to calculate Q^R, the net radiant heat flux across a surface that is arbitrarily oriented with respect to the coordinate axes. From the expression (2.7) for the differential heat flux and with reference to Fig. 2, where we take the angles Φ and Θ as measured in relation to the normal to the surface, we have

$$Q^R = \int_0^\infty \int_0^{4\pi} I_\nu(\Phi, \Theta) \cos \Phi \, d\Omega \, d\nu$$

$$= \int_0^\infty \int_0^{2\pi} \int_0^{\pi} I_\nu(\Phi, \Theta) \cos \Phi \sin \Phi \, d\Phi \, d\Theta \, d\nu. \qquad (2.9)$$

Here we have used the relation

$$d\Omega = \sin \Phi \, d\Phi \, d\Theta.$$

In terms of k_i, the unit vector normal to the surface in question (see Fig. 2), we can write $\cos \Phi = l_j k_j$ and equation (2.9) becomes

$$Q^R = \int_0^\infty \int_0^{4\pi} I_\nu l_j k_j \, d\Omega \, d\nu = q_j^R k_j. \qquad (2.10)$$

When the surface is taken normal to a coordinate axis, Q^R as given by equation (2.10) reduces to the corresponding component of the heat-flux vector (2.8).

In certain situations it is useful to separate Q^R into two parts: Q_+^R, the heat flux in the direction of k_i, and Q_-^R, the heat flux in the direction opposite to k_i. We first calculate $Q_{\nu+}^R$ and $Q_{\nu-}^R$, the one-sided fluxes per unit frequency at frequency ν. Subdividing the range of integration with respect to Φ in equation (2.9), we have

$$Q_{\nu+}^R = \int_0^{2\pi} \int_0^{\pi/2} I_\nu(\Phi, \Theta) \cos \Phi \sin \Phi \, d\Phi \, d\Theta \qquad (2.11a)$$

and

$$Q_{\nu-}^R = \int_0^{2\pi} \int_{\pi/2}^{\pi} I_\nu(\Phi, \Theta)(-\cos \Phi) \sin \Phi \, d\Phi \, d\Theta. \qquad (2.11b)$$

The minus sign in equation (2.11b) arises from the fact that by definition $Q_{\nu+}^R$ and $Q_{\nu-}^R$ are both *positive* quantities. The corresponding frequency-integrated fluxes are then

$$Q_+^R = \int_0^\infty Q_{\nu+}^R \, d\nu \qquad (2.12a)$$

and

$$Q_-^R = \int_0^\infty Q_{\nu-}^R \, d\nu. \qquad (2.12b)$$

The net radiant heat flux

$$Q^R = Q_+^R - Q_-^R$$

from equations (2.12) is, of course, identical to that given by equation (2.9).

The *radiant pressure tensor* p_{ij}^R, which is analogous to the molecular pressure tensor given by equation (X 8.3), can be obtained from expression

(2.6) and is

$$\boxed{p_{ij}^R = \frac{1}{c}\int_0^\infty \int_0^{4\pi} I_\nu l_i l_j \, d\Omega \, d\nu} \, . \tag{2.13}$$

A typical component of p_{ij}^R, for example p_{12}^R, gives the net flux of the x_1-component of radiant momentum across a surface that is perpendicular to the x_2-coordinate axis.

When the radiation field is *isotropic*, that is, when I_ν is independent of the direction of propagation, the radiant pressure tensor (2.13) can be written

$$p_{ij}^{R^0} = \frac{1}{c}\int_0^\infty I_\nu \, d\nu \int_0^{4\pi} l_i l_j \, d\Omega.$$

To evaluate the integral over solid angle, we use the angular coordinates shown in Fig. 1 and write

$$d\Omega = \sin\phi \, d\phi \, d\theta$$

and

$$l_1 = \cos\phi,$$
$$l_2 = \sin\phi \, \cos\theta,$$
$$l_3 = \sin\phi \, \sin\theta.$$

We then have

$$\int_0^{4\pi} l_i l_j \, d\Omega = \int_0^{2\pi}\int_0^{\pi} l_i l_j \sin\phi \, d\phi \, d\theta$$
$$= \begin{cases} 4\pi/3 & \text{for } i = j \\ 0 & \text{for } i \neq j. \end{cases}$$

Thus the radiant pressure tensor $p_{ij}^{R^0}$ for an isotropic field is

$$p_{ij}^{R^0} = \frac{4\pi}{3c}\int_0^\infty I_\nu \, d\nu \, \delta_{ij}.$$

Since from equation (2.3b) the radiant energy density under these conditions is

$$u^{R^0} = \frac{4\pi}{c}\int_0^\infty I_\nu \, d\nu,$$

we find that for the special case of an isotropic field the pressure tensor and energy density are related by

$$p_{ij}^{R^0} = \tfrac{1}{3} u^{R^0} \, \delta_{ij}. \tag{2.14}$$

We shall make use of this relation in Chapter XII.

Having formulated a description of the quantities of interest in terms of the specific intensity, we next derive the equation that governs I_ν. This *equation of radiative transfer* is analogous to the Boltzmann equation of kinetic theory.

Exercise 2.1. Find an expression for the radiant momentum density in terms of I_ν.

Exercise 2.2. Find the value of q_i^R that applies when the radiation field is isotropic.

3 THE EQUATION OF RADIATIVE TRANSFER

The equation of radiative transfer is a continuity equation for the number of photons having a direction of propagation in the solid angle $d\Omega$ and a frequency in the range ν to $\nu + d\nu$. If we consider an elementary volume element dV_x, the number of such photons in dV_x is

$$n^R f^R \, d\Omega \, d\nu \, dV_x,$$

where we have used the photon distribution function f^R to emphasize the analogy with the derivation of the Boltzmann equation in Chapter IX, Sec. 3.

The time rate of change of the number of photons under consideration will result from the convection (flux) of photons through the surfaces of dV_x and from the absorption, emission, and scattering of photons by the gas molecules within dV_x. As in Chapter IX, Sec. 3, the contribution of the convection is proportional to the divergence of the number flux and, with equation (2.4a) for the flux, is [cf. equation (IX 3.1b)]

$$-\frac{\partial}{\partial x_j}(c l_j n^R f^R \, d\Omega \, d\nu) \, dV_x = -c l_j \frac{\partial}{\partial x_j}(n^R f^R) \, d\Omega \, d\nu \, dV_x.$$

If we write

$$\left[\frac{\partial}{\partial t}(n^R f^R)\right]_{e,a,s} d\Omega \, d\nu \, dV_x$$

for the net rate of increase of the photons under consideration as a result of emission, absorption, and scattering in dV_x, the equation of radiative transfer is then

$$\frac{\partial}{\partial t}(n^R f^R) + c l_j \frac{\partial}{\partial x_j}(n^R f^R) = \left[\frac{\partial}{\partial t}(n^R f^R)\right]_{e,a,s}. \tag{3.1}$$

The right-hand side of equation (3.1) can be written in terms of properties of the gas through the definition of coefficients of emission and absorption. The *mass emission coefficient* j_ν is defined such that the rate per unit volume at which the gas emits radiant energy in the frequency range ν to $\nu + d\nu$ and in the solid angle $d\Omega$ is given by

$$\rho j_\nu \, d\Omega \, d\nu, \tag{3.2a}$$

where ρ is the mass density of the gas. The rate of absorption of radiant energy by a gas is found to be proportional to the specific intensity I_ν. The *mass absorption coefficient* κ_ν is accordingly defined such that the rate per unit volume at which radiant energy in $d\nu$ and $d\Omega$ is absorbed by the gas is given by

$$\rho \kappa_\nu I_\nu \, d\Omega \, d\nu. \tag{3.2b}$$

The emission and absorption terms thus introduced into the equation of radiative transfer are analogous to the chemical-reaction terms that would appear in a generalized Boltzmann equation to account for the rates of formation and depletion of molecules of a particular chemical species.[1]

Photons may also be scattered by the molecules of a gas. That is to say, the direction of propagation of the photons may be changed as a result of their interaction with gas molecules. We refer, in our use of the term scattering, only to those interactions where the energy of the photons is unchanged, in contrast to emission and absorption where there is an energy exchange between the gas molecules and the radiation. The scattering of photons is roughly analogous to elastic collisions between molecules of one chemical species and molecules of other species, as in the Boltzmann equation for a given species in a gas mixture [equation (IX 10.12)]. In most situations of interest in gas dynamics, scattering is unimportant (see Lighthill, 1960). For simplicity, we therefore omit this effect in the discussion that follows. Thus, although in the more general case the definitions of the absorption and emission coefficients can be modified to allow for scattering (see Chandrasekhar, 1950), our use of these quantities includes only "true" emission and absorption.

We can complete our analogy between the equation of radiative transfer and the Boltzmann equation by observing that the analog of the collision term of primary interest in Chapters IX and X—that resulting from collisions between like molecules [equation (IX 7.10)]—is absent here. This results from the fact that photons do not interact with other photons.

[1] If the speed of the fluid is small when compared to the speed of light, it is unimportant whether the quantities that appear in equations (3.1) and (3.2) are defined in a laboratory coordinate system or in a coordinate system for which the mean velocity of the gas is zero. To obtain a relativistically correct formulation, however, this distinction must be taken into account.

From the definitions (3.2) of the absorption and emission coefficients and the fact that the energy of a photon is $h\nu$, the right-hand side of equation (3.1) can be written

$$\left[\frac{\partial}{\partial t}(n^R f^R)\right]_{e,a} = \frac{1}{h\nu}\rho(j_\nu - \kappa_\nu I_\nu). \tag{3.3}$$

The equation of radiative transfer, in the absence of scattering, is then

$$\boxed{\frac{1}{c}\frac{\partial I_\nu}{\partial t} + l_j \frac{\partial I_\nu}{\partial x_j} = \rho(j_\nu - \kappa_\nu I_\nu)}, \tag{3.4}$$

where we have used equation (2.1) to replace $n^R f^R$ on the left-hand side in favor of I_ν. Implicit in our derivation of equation (3.4) is the assumption that the index of refraction is unity, as is very nearly the case for gases. Thus c is taken as the speed of light in a vacuum.

The quantity $\rho\kappa_\nu/n$, which has the dimensions of an area, can be interpreted as a *cross-section* σ_ν for the absorption of radiation of frequency ν by gas molecules of number density n. The correspondence of $\rho\kappa_\nu/n$ to the kinetic-theory cross-section of Chapter IX, Sec. 7 follows from expression (3.2b) if we eliminate I_ν in favor of the photon distribution function by means of equation (2.1). With

$$n\sigma_\nu = \rho\kappa_\nu,$$

expression (3.2b) then becomes

$$h\nu(ncn^R f^R \sigma_\nu \, d\Omega \, d\nu).$$

Here the quantity in parentheses can be interpreted as the collision rate per unit volume between photons, in the frequency range $d\nu$ and the solid angle $d\Omega$, and gas molecules. This expression is of a similar form to the kinetic-theory collision rate that applies to collisions between molecules of one species in a particular velocity class and approximately stationary molecules of another species, as for example in the case of collisions between electrons and atoms. (In a more detailed description $n\sigma_\nu$ would be replaced by a summation of similar quantities over all species that can absorb radiation of frequency ν.) We note, however, that because κ_ν depends in general on the state of the gas, σ_ν is not a molecular constant. By an extension of the foregoing reasoning, the radiative mean free path λ_ν for photons of frequency ν is given by

$$\lambda_\nu = \frac{1}{n\sigma_\nu} = \frac{1}{\rho\kappa_\nu}. \tag{3.5}$$

The presence of radiation adds a new quantity, the specific intensity, to the formulation of problems of gas dynamics. The equation of transfer (3.4) provides the corresponding equation. Although this equation avoids the details of the electromagnetic-wave aspect of radiation, it gives for our purposes a complete description of radiant energy transfer. In a nonequilibrium situation a knowledge of the emission and absorption coefficients is a requisite to a useful solution of the equation of transfer. We shall discuss the physical mechanisms of emission and absorption in Sec. 6. First, however, we turn our attention to the important special case of radiative equilibrium.

Exercise 3.1. Let s be the scalar distance in a given direction l_i through a given point. Assuming a steady state and *no emission*, obtain the following equation for the specific intensity in the given direction:

$$\frac{\partial I_\nu}{\partial s} = -\rho\kappa_\nu I_\nu.$$

An alternate (and equivalent) definition of the absorption coefficient can be based on this expression, that is, $\rho\kappa_\nu$ can be defined as the fractional reduction of intensity per unit length along the direction of propagation as the result of absorption.

4 RADIATIVE EQUILIBRIUM

For a system composed of a gas and its associated radiation field, enclosed by a container and isolated from its environment, the specific intensity of the radiation will eventually reach an equilibrium value that is uniform in space and constant in time. A necessary condition for such a system to be isolated is that no radiation be transmitted through the walls of the container, that is, that the walls be opaque. In this section we shall indicate how the methods of statistical mechanics can be used to calculate the specific intensity and other quantities of interest under such conditions of radiative equilibrium. Our purpose here is to make plausible certain important results, rather than to present a complete and rigorous derivation. We shall find that the equilibrium radiation field depends only on the temperature of the system and not on its nature. In fact, in the example mentioned above, the radiation field at equilibrium is the same whether a gas is present or not. This latter case of an evacuated space surrounded by opaque walls and containing an equilibrium radiation field is called a *hohlraum*. It is important to note, however, that for the existence of radiative equilibrium the radiation field must interact with some matter, such as the walls of the container, which itself is in equilibrium.

Although we do not wish to go into the quantum-mechanical reasons here, it is known that photons obey Bose-Einstein statistics. The analysis of radiative equilibrium is, in fact, one of the most important applications of this form of quantum statistics. In Chapter IV we found that for a Bose-Einstein system the equilibrium distribution of particles among the groups of quantum states is given by [see equation (IV 5.8)]

$$\frac{N_j^*}{C_j} = \frac{1}{e^{\alpha+\beta\epsilon_j} - 1}. \tag{4.1}$$

Here N_j^* and C_j are, respectively, the number of particles and number of states in the jth group of quantum states, and ϵ_j is the value of the energy for this group of states. To apply equation (4.1) to an equilibrium radiation field we set C_j equal to the number of quantum states for radiation in the frequency interval $d\nu$. This number can be obtained from an analysis based on the electromagnetic boundary conditions at the walls of of the container. It is found that the quantum condition on the wavelength is the same as that given by equation (IV 3.13) for the translational motion of material particles. By arguments similar to those used in Chapter IV to find C_j for the translational energy states allowable in a volume V [see equation (IV 9.10)], C_j is then found to be (see Mayer and Mayer, 1940, Sec. 16c, or Davidson, 1962, Sec. 12-2)

$$C_j = 8\pi V \frac{\nu^2}{c^3}\, d\nu. \tag{4.2}$$

At equilibrium the radiation field can have no preferred direction, so that the specific intensity I_ν must be isotropic. It then follows from their definitions [see equation (2.1)] that N_j^*, $n^{R*}f^{R*}$, and I_ν^* are related by

$$N_j^* = 4\pi V n^{R*} f^{R*}\, d\nu = \frac{4\pi V}{ch\nu} I_\nu^*\, d\nu. \tag{4.3}$$

Putting expressions (4.2) and (4.3) into the Bose-Einstein distribution (4.1) and setting $\epsilon_j = h\nu$, we obtain the equilibrium specific intensity as

$$I_\nu^* = \frac{2h\nu^3/c^2}{e^{\alpha+\beta h\nu} - 1}. \tag{4.4}$$

The constants α and β that appear in this equation entered the theory of Chapter IV, Sec. 5 as a result respectively of the requirements that the total number of particles and the total energy of the particles be fixed. For radiation it turns out that the restriction on the number of particles is not appropriate. This follows from the fact that there is no conservation requirement on the number of photons in a system. As a consequence, in contrast to the situation with a molecular system, a single intensive variable (conveniently taken as the temperature) is sufficient to specify the state of an equilibrium radiation field. For example, consider a system composed of a gas, its associated radiation field, and an opaque container. It is operationally possible to adjust the energy of the system at will, and once the system is isolated the energy will remain constant. On the other hand, the number of photons cannot be set *a priori*. In fact, if the system is initially not in equilibrium and is then isolated, the number of photons will change as a result of an imbalance between absorption and emission. The consequence of the fact that the number of photons is not fixed can be found by a revision of the analysis that led to equation (4.1) and is that $\alpha = 0$ (see Mayer and Mayer, 1940, Sec. 16d). In other words, since the first of equations (IV 5.6) (i.e., $\sum_j \delta N_j = 0$) is not applicable, the Lagrange multiplier α does not enter at all into the calculation. As before, the quantity β is most easily evaluated by means of the thermodynamic definition of temperature: $1/T \equiv (\partial S/\partial E)_V$, where S and E are the entropy and energy of the radiation field. We thus obtain (Exercise 4.1) the familiar result $\beta = 1/kT$. It can be shown that the temperature of the radiation field, defined in this way, must be equal to the temperature of the matter with which the radiation is in equilibrium.

With $\alpha = 0$ and $\beta = 1/kT$, the equilibrium specific intensity (4.4)—called the *Planck function* and given the special symbol $B_\nu(T)$—is then

$$\boxed{B_\nu(T) \equiv I_\nu^* = \frac{2h\nu^3/c^2}{e^{h\nu/kT} - 1}}\,. \tag{4.5}$$

An examination of the form of this result shows that regardless of the temperature the term -1 that appears in the denominator as a result of the use of Bose-Einstein statistics is important for frequencies such that $\exp(h\nu/kT)$ is not much greater than unity.

The equilibrium values of the radiant energy density, heat-flux vector, and other quantities introduced in Sec. 2 are easily found from the equilibrium specific intensity B_ν. The evaluation of these quantities involves the integral[2]

$$\int_0^\infty B_\nu\, d\nu = \frac{2k^4T^4}{h^3c^2}\int_0^\infty \frac{(h\nu/kT)^3}{e^{h\nu/kT} - 1}\, d\left(\frac{h\nu}{kT}\right)$$

$$= \frac{2\pi^4k^4T^4}{15h^3c^2}\,.$$

[2] The definite integral $\int_0^\infty \frac{x^3}{e^x - 1}\,dx$ has the value $\pi^4/15$ (see Mayer and Mayer, 1940, p. 372).

Here the coefficient of T^4 differs by a factor of π from the *Stefan-Boltzmann constant*[3]

$$\sigma \equiv \frac{2\pi^5 k^4}{15h^3c^2}, \tag{4.6}$$

which has the numerical value $\sigma = 5.6697 \times 10^{-5}$ erg/cm² °K⁴ sec. We thus obtain the important formula

$$\boxed{\int_0^\infty B_\nu \, d\nu = \frac{\sigma}{\pi} T^4}. \tag{4.7}$$

With this result and the definitions of Sec. 2 we have

$$u_\nu^{R*} = \frac{4\pi}{c} B_\nu(T), \tag{4.8a}$$

$$u^{R*} = \frac{4}{c} \sigma T^4, \tag{4.8b}$$

$$q_i^{R*} = 0, \tag{4.8c}$$

and

$$p_{ij}^{R*} = \frac{4}{3c} \sigma T^4 \, \delta_{ij}. \tag{4.8d}$$

The equilibrium energy density and pressure tensor are related by

$$p_{ij}^{R*} = \tfrac{1}{3} u^{R*} \, \delta_{ij}, \tag{4.8e}$$

in agreement with equation (2.14) which applies whenever the specific intensity is isotropic. In addition, as in kinetic theory, we find from equation (4.8d) that the scalar pressure $p^{R*} \equiv \frac{1}{3} p_{jj}^{R*}$ calculated as a momentum flux is equal to the value given by the thermodynamic definition $p \equiv T(\partial S/\partial V)_E$ (Exercise 4.1).

Under conditions of radiative equilibrium the one-sided heat fluxes [equations (2.11) and (2.12)] are independent of the orientation of the surface in question and are

$$Q_{\nu+}^{R*} = Q_{\nu-}^{R*} = \pi B_\nu(T), \tag{4.9a}$$

$$Q_+^{R*} = Q_-^{R*} = \sigma T^4. \tag{4.9b}$$

From the expression (2.7) and the fact that B_ν is independent of the direction of propagation, we obtain also the "cosine-law" angular distribution of the one-sided differential heat flux at equilibrium. That is, the flux in the frequency interval $d\nu$ and solid angle $d\Omega$ varies directly as

[3] The quantity $a \equiv 4\sigma/c$ is called the Stefan-Boltzmann constant by some authors.

the cosine of the angle Φ according to the relation

$$B_\nu(T)\,|\cos\Phi|\,d\nu\,d\Omega, \tag{4.10}$$

where Φ is the angle measured from the normal to the surface.

The foregoing results are called the *black-body* values of the radiant heat flux, energy density, and pressure. The term "black body" is applied also to any object or surface that emits a radiant heat flux in the amount σT^4. Such body can be constructed, at least in thought, if a hole is opened in the wall that surrounds a hohlraum. If the hole is sufficiently small that the equilibrium radiation field within the hohlraum is not disturbed, the outward radiant energy flux through the hole is then σT^4 [see equation (4.9b)]. It follows also that any radiation that might be incident upon the hole from an external source will be absorbed (trapped) by the hohlraum. Thus a hohlraum is not only a perfect, black-body emitter of radiation but is also a perfect absorber. This is a special case of Kirchhoff's law, which we shall discuss further in the next section.

Exercise 4.1. From the thermodynamic definition $1/T \equiv (\partial S/\partial E)_V$, show that the constant β that appears in the equilibrium specific intensity (4.4) is equal to $1/kT$ (cf. Chapter IV, Sec. 8). Here S and E are the entropy and energy of the radiation field. Show that the thermodynamic pressure $p \equiv T(\partial S/\partial V)_E$ is in agreement with the equilibrium scalar pressure $p^{R*} = \frac{1}{3}p_{jj}^{R*}$ obtained from equation (4.8d).

Exercise 4.2. Find N^{R*}, the number of photons in a given volume V at equilibrium. Note that N^{R*} is not an independent variable, but depends on the temperature and hence on the energy of the radiation field. (You may express your result in terms of a definite integral.)

Exercise 4.3. Derive equations (4.8) and (4.9).

5 THE INTERACTION OF RADIATION WITH SOLID SURFACES

Before we proceed to a more detailed discussion of the processes of emission and absorption of radiation in a gas, it is convenient to set down certain definitions that apply to the interaction of radiation with solid surfaces. This topic is one of some complexity and is the subject of specialized study. We shall be interested here, however, only in a gross description sufficient to provide simplified boundary conditions for the treatment of the influence of radiation in gas dynamics.

We consider first radiation that is incident upon a surface from a gas. Such radiation can be either absorbed, reflected, or transmitted through the surface. The description of this situation involves in general not only the frequency and angle with respect to the surface of the incident radiation, but also the angular distribution of the reflected and transmitted energy. Although the definition (but not the determination) of coefficients to account for the angular dependence of the surface interaction is straightforward (see, for example, Goulard, 1964, or Eckert and Drake, 1959), we shall avoid such details and consider only quantities that are integrated over direction. We restrict our attention, furthermore, to surfaces such that the absorbed, reflected, and transmitted radiation is of the same frequency as the incident radiation. The hemispheric surface absorptivity a_ν, reflectivity r_ν, and transmissivity t_ν are defined, respectively, as the fractions of the total incident radiant energy flux at frequency ν that are absorbed, reflected, and transmitted by the surface. Since the energy of the incident radiation is conserved, we must have

$$a_\nu + r_\nu + t_\nu = 1. \tag{5.1a}$$

In the material that follows we shall consider only opaque surfaces, in which case $t_\nu = 0$. We then have

$$a_\nu + r_\nu = 1. \tag{5.1b}$$

We now turn our attention to the emitted radiation. To account for the emission by a solid surface at temperature T_w, the hemispheric surface emissivity e_ν is *defined* as the ratio of the energy flux at frequency ν actually emitted by the surface to the energy flux that would be emitted by a black body at the same temperature. The emitted energy flux is thus [see equation (4.9a)]

$$(Q^R_{\nu+})_e = e_\nu \pi B_\nu(T_w). \tag{5.2}$$

Here we have taken k_i in Fig. 2 as the outward-drawn normal from the solid surface. If equation (5.2) is used in the left-hand side of equation (2.11a) when that equation is applied to the emitted radiation, and if it is *assumed* that the specific emitted intensity $(I_{\nu_w})_e$ from the walls is independent of direction as for a black body, we obtain

$$(I_{\nu_w})_e = e_\nu B_\nu(T_w). \tag{5.3}$$

Note, however, that equation (5.2), which defines e_ν, does not depend on the condition (5.3). It is also important to note that the assumption that I_ν is independent of direction is permissible only for the emitted radiation (and for the reflected radiation if the reflection is completely diffuse). The intensity of the incident radiation in a nonequilibrium situation will be in general a function of direction.

The experimental determination of the coefficients of absorptivity and emissivity (or their directionally dependent counterparts) for surfaces of engineering interest is not an easy matter. There is, however, one simplification of the formulation that can be made on theoretical grounds. If we consider a radiation field that is in equilibrium with and contained within an opaque surface at a given temperature T, the radiant energy flux incident on the surface will be

$$Q_{\nu-}^{R*} = \pi B_\nu(T).$$

The radiant energy flux away from the surface is made up of emitted and reflected radiation and is given according to our definitions by

$$Q_{\nu+}^{R*} = e_\nu^* \pi B_\nu(T) + r_\nu^* Q_{\nu-}^{R*} = (e_\nu^* + r_\nu^*)\pi B_\nu(T).$$

Since at equilibrium the net flux $Q_{\nu+}^{R*} - Q_{\nu-}^{R*}$ at each frequency must be zero everywhere [see equation (4.9a)], it follows that

$$e_\nu^* + r_\nu^* = 1,$$

or, in view of equation (5.1b),

$$e_\nu^* = a_\nu^*.$$

Now, if we make the quasi-equilibrium asumption that the coefficients e_ν and a_ν for a given value of ν depend only on the nature of the surface and the surface temperature, but not on the radiation field to which the surface is exposed, the foregoing result holds whether or not the incident intensity is equal to the equilibrium value B_ν. We thus write, removing the equilibrium asterisks,

$$\boxed{e_\nu = a_\nu}\,. \tag{5.4}$$

This result is one form of *Kirchhoff's law*. As noted before, it shows that a perfectly absorbing surface ($a_\nu = 1$ for all ν) is also a perfect, black-body emitter.

Exercise 5.1. Find expressions for the total (frequency-integrated) energy flux (a) absorbed, (b) emitted, and (c) reflected in a nonequilibrium situation by an opaque surface at temperature T_w in terms of e_ν and the incident specific intensity $(I_\nu)_{\text{in}}$. What is the net flux?

6 EMISSION AND ABSORPTION OF RADIATION

There are in a gas a variety of mechanisms for emission and absorption of radiation. Although we are unable in our brief treatment to discuss these mechanisms in detail, it will be useful to consider the types of processes that can occur. The emission or absorption of radiation is associated with transitions between energy levels of the gas atoms or molecules. In these transitions the change in internal energy of the atom or molecule is equal to the radiant energy absorbed or emitted. The types of transitions that are important in a given situation depend on the nature of the gas, on the relative number of gas molecules in the various energy levels, and on the radiation frequency range of interest. Distinction can be made, as we shall discuss later, between atomic and molecular transitions, and between bound-bound, bound-free, and free-free transitions.

A radiative transition between two quantized energy levels of an atom or molecule is called a *bound-bound transition*. Since the energy difference between two quantized energy levels is established by internal molecular structure, the emission and absorption coefficients for bound-bound transitions are sharply peaked functions of the photon energy $h\nu$. For this reason bound-bound transitions are said to give rise to emission and absorption *lines*. In a monatomic gas such transitions are between electronic states of the atom. In a molecular gas rotational, vibrational, and electronic states of the molecule can all change in a bound-bound transition. Molecular emission and the corresponding absorption spectra consist of systems composed of bands of lines.

Actually, bound-bound transitions and the resulting individual lines do not occur for a precise value of the photon energy $h\nu$ but exhibit a distribution of emitted and absorbed energy over a more or less narrow range of $h\nu$. The extent of this distribution is described as the width of the line. There is a natural line width that can be understood in terms of the uncertainty principle. In addition, line broadening can result from several effects, such as Doppler broadening and collision broadening, that depend on the state of the gas. Doppler broadening occurs as a result of the frequency shift that is associated with the thermal motion of the emitting and absorbing atom or molecule relative to the observer. Collision broadening occurs when a radiating atom is perturbed as a result of its interaction with other atoms, ions, or electrons.

The interaction of radiation with an atom or molecule can also result in photoionization or photodissociation of the atom or molecule. Such transitions give rise to *bound-free* absorption. The reverse emission process is called radiative recombination. Consider in particular the removal of an electron from an atom by photoionization. The electron is initially in a bound, quantized state. After ionization, however, the electron is free to take on any value of kinetic energy. Therefore the absorption coefficient for photionization is a continuous non-zero function of photon energy for all values of photon energy that exceed the ionization potential of the atom. Similar remarks apply to photodissociation and radiative recombination.

Continuous emission and absorption also result from *free-free transitions*, which are associated with changes in energy of the free electrons in a gas. This process, called Bremsstrahlung, occurs when a free electron interacts with the field of a positive ion.

The radiation field associated with continuous absorption and emission, whether it be from bound-free or free-free transitions, is said to be a continuum, or continuum radiation.

We shall now give an indication of how the emission and absorption processes that we have discussed can be described quantitatively in terms of molecular constants. By molecular constants we refer to microscopic quantities, such as the collision cross-section in kinetic theory, that are characteristic of the atoms or molecules that make up the gas. These are in contrast to macroscopic quantities such as pressure and temperature that are characteristic of the gas as a whole. A complete microscopic description of the various mechanisms of radiative transition would be lengthy and involved. In accord with the introductory nature of our treatment, we shall be content to discuss only the specific example of atomic line absorption and emission (bound-bound transitions), and shall not attempt an exposition of the role of line broadening. The interested reader will find more complete treatments in various books on astrophysics and quantum mechanics as, for example, Aller (1963), Griem (1964), and for an advanced treatment, Heitler (1954).

Fundamental quantities of atomic line radiation are the *Einstein coefficients*. Let us consider transitions from an upper electronic state n to a lower state m such that the difference in energy between the two states is $h\nu$. As a result of line broadening, the energy and hence frequency of the radiation associated with transitions between the two states n and m will not be precise but will vary somewhat over the line width. For simplicity, however, we assume that the line width and the radiation field are such that both I_ν and $h\nu$ can be taken as approximately constant over the line in our calculations. The Einstein coefficient A_{nm} for spontaneous emission is defined such that the atomic probability per unit time for a spontaneous transition from state n to state m with the emission of radiant energy in the solid angle $d\Omega$ is[4]

$$A_{nm}\, d\Omega.$$

The total number of spontaneous radiative transitions from n to m per unit volume per unit time is then

$$-\left(\frac{dn_n}{dt}\right)_{n\to m} = n_n 4\pi A_{nm}, \tag{6.1}$$

[4] The reader is warned that different forms for the definitions of the Einstein coefficients are used by different authors.

where n_n is the number of atoms per unit volume in the electronic state n. Since spontaneous emission is isotropic in direction, we have set $\int d\Omega = 4\pi$ in this equation. The quantity $(4\pi A_{nm})^{-1}$ that enters equation (6.1) as a radiative relaxation time is known as the radiative lifetime of state n for transitions to state m. The total rate of change of n_n as a result of spontaneous emission is obtained by summing equation (6.1) over all lower states m for which such transitions are allowed.

The Einstein coefficient B_{mn} for absorption is defined such that, in a radiation field of specific intensity I_ν, the atomic probability per unit time for a transition from state m to state n with the absorption of radiant energy from the solid angle $d\Omega$ is

$$B_{mn} I_\nu \, d\Omega.$$

The number of transitions from m to n per unit volume per unit time as a result of absorption is then

$$\left(\frac{dn_n}{dt}\right)_{m\to n} = n_m B_{mn} \int_0^{4\pi} I_\nu \, d\Omega. \tag{6.2}$$

The interaction of a radiation field with the atoms of a gas results not only in absorption but also in *induced emission* (known also as negative absorption). That is to say, in the presence of radiation the emission in a given direction and of a given frequency is increased above and beyond that resulting purely from spontaneous emission. The amount of such induced emission is proportional to the value of I_ν in the given direction and at the frequency in question. Induced emission has a classical counterpart in that a material oscillator acted on by a force imposed by an alternating electromagnetic field can either absorb energy from the field or add energy to the field, depending on the phase of the oscillator with respect to the phase of the field. The atomic probability per unit time for a transition from state n to state m as a result of induced emission in the solid angle $d\Omega$ is

$$B_{nm} I_\nu \, d\Omega,$$

where B_{nm} is the Einstein coefficient for induced emission. To take account of such transitions, equation (6.1) must be replaced by

$$\left(\frac{dn_n}{dt}\right)_{n\to m} = -n_n\left(4\pi A_{nm} + B_{nm} \int_0^{4\pi} I_\nu \, d\Omega\right). \tag{6.3}$$

Although we shall discuss induced emission in the context of bound-bound transitions, this phenomenon occurs for continuum radiation as well. Mention of induced emission was avoided in our introduction of the equation of radiative transfer (3.4). That equation remains valid, however, if the emission coefficient j_ν is interpreted to include both induced and spontaneous emission. As we shall see in the next section, it is convenient to rewrite the equation of transfer in a form where the absorption and induced-emission terms, which are both proportional to I_ν, are grouped together.

The three Einstein coefficients A_{nm}, B_{mn}, and B_{nm} are not independent but can be related by consideration of the special case of radiative equilibrium. A condition for equilibrium is that the number density of atoms in the states n and m be independent of time. Equations (6.2) and (6.3) give the rate of change of n_n as a result of radiative transitions from and to state m. Now, in general, a variety of mechanisms in addition to radiative transitions from and to state m, for example, atomic collisions and other radiative transitions, can change the number of atoms in state n. According to the principle of detailed balancing discussed in Chapter II, Sec. 4, however, the rate of radiative transitions upward from m to n is at equilibrium equal to the rate of radiative transitions downward from n to m. The situation is similar to that of vibrational nonequilibrium discussed in Chapter VII, Sec. 2. Applying the principle of detailed balancing we have, from equations (6.3) and (6.2),

$$\left(\frac{dn_n}{dt}\right)^*_{n\to m} + \left(\frac{dn_n}{dt}\right)^*_{m\to n}$$

$$= -n_n^*[4\pi A_{nm} + 4\pi B_{nm}B_\nu(T)] + n_m^* 4\pi B_{mn}B_\nu(T) = 0, \tag{6.4}$$

where we have set $I_\nu = I_\nu^* = B_\nu(T)$ and have used the symbols n_n^* and n_m^* that are appropriate to equilibrium. Solving equation (6.4) for $B_\nu(T)$ and equating the result to the expression (4.5) for $B_\nu(T)$, we obtain

$$\frac{2h\nu^3/c^2}{e^{h\nu/kT} - 1} = \frac{A_{nm}/B_{nm}}{(n_m^*/n_n^*)(B_{mn}/B_{nm}) - 1}. \tag{6.5}$$

The equilibrium populations of the states n and m are related by the Boltzmann distribution [equation (IV 6.12)] as follows:

$$\frac{n_m^*}{n_n^*} = e^{-(\epsilon_m - \epsilon_n)/kT} = e^{h\nu/kT}. \tag{6.6}$$

Since the Einstein coefficients are molecular constants and independent of the macroscopic quantity temperature, equation (6.5) with n_m^*/n_n^* given by equation (6.6) will be satisfied only if

$$\frac{A_{nm}}{B_{nm}} = \frac{2h\nu^3}{c^2}, \qquad \frac{B_{mn}}{B_{nm}} = 1. \tag{6.7}$$

Thus by the use of the equilibrium condition of detailed balancing, the number of independent Einstein coefficients for the $n \rightleftharpoons m$ transitions is reduced from three to one. Similar expressions can be written for the Einstein coefficients A_{NM}, B_{NM}, and B_{MN} that give the overall probabilities of radiative transitions between degenerate energy *levels* N and M rather than individual *states* n and m as we have here (see Exercise 6.1).

We are now in a position to relate the macroscopic emission and absorption coefficients that were introduced in Sec. 3 to the microscopic transition probabilities. We shall continue to restrict our consideration to atomic line radiation, although other forms of radiative transitions can be included in a similar treatment. In addition, we confine our attention for simplicity to the quantities

$$\bar{j}_{nm} \equiv \int j_{\nu_{nm}}\, d\nu \tag{6.8a}$$

and

$$\bar{\kappa}_{nm} \equiv \int \kappa_{\nu_{nm}}\, d\nu. \tag{6.8b}$$

These quantities are obtained by integrating over the line width the contributions to j_ν and κ_ν that result from $n \rightleftharpoons m$ transitions. As before we assume that conditions are such that I_ν and $h\nu$ can be taken as constant over the line.

In view of the definitions of j_ν and κ_ν in Sec. 3, the quantities $\rho\bar{j}_{nm}$ and $\rho\bar{\kappa}_{nm}I_\nu$ give the rates per unit volume *per unit solid angle* at which radiant energy is emitted and absorbed as a result of transitions between the particular states n and m. These rates are simply related to the Einstein coefficients and, with the use of equations (6.3) and (6.2), are

$$\rho\bar{j}_{nm} = n_n(A_{nm} + B_{nm}I_\nu)h\nu \tag{6.9a}$$

and

$$\rho\bar{\kappa}_{nm}I_\nu = n_m B_{mn} I_\nu h\nu. \tag{6.9b}$$

From equations (6.9) we see that the absorption coefficient is independent of the radiation field (as given by I_ν) whereas the emission coefficient is not. If we now use equations (6.7) to eliminate the two Einstein coefficients in equation (6.9a) in favor of the single quantity B_{mn}, we obtain

$$\rho\bar{j}_{nm} = n_n B_{mn}\left(\frac{2h\nu^3}{c^2} + I_\nu\right)h\nu. \tag{6.9c}$$

With these results we find finally that the net rate, per unit volume per unit solid angle, of energy exchange between the gas and the radiation field as a result of $n \rightleftharpoons m$ transitions is

$$\rho(\bar{j}_{nm} - \bar{\kappa}_{nm}I_\nu) = n_m B_{mn} h\nu\left[\frac{n_n}{n_m}\frac{2h\nu^3}{c^2} + \left(\frac{n_n}{n_m} - 1\right)I_\nu\right]. \tag{6.10}$$

This expression gives the contribution of $n \rightleftharpoons m$ transitions, integrated over the line width, to the right-hand side of the equation of radiative transfer (3.4). Thus in this example a knowledge of the populations of the emitting and absorbing electronic states together with the single transition probability B_{mn} is sufficient to evaluate the macroscopic radiative properties of the gas that appear in the equation of radiative transfer.[5] In the next section we shall see how the situation can be further simplified by means of an assumption regarding the populations of the electronic states.

Exercise 6.1. (a) Obtain the relations

$$\frac{A_{NM}}{B_{NM}} = \frac{2h\nu^3}{c^2}, \qquad \frac{B_{MN}}{B_{NM}} = \frac{g_N}{g_M},$$

which correspond to equations (6.7) but apply to the Einstein coefficients A_{NM}, B_{NM}, and B_{MN} that give the overall probabilities of radiative transitions between degenerate energy *levels* N and M containing respectively the typical individual quantum *states* n and m. This can be done by expressing A_{NM}, B_{NM}, and B_{MN} in terms of A_{nm}, B_{nm}, and B_{mn} by means of summations over all states n and m in the levels N and M and then applying equations (6.7). Assume that $n_N = g_N n_n$ and $n_M = g_M n_m$, where g_N and g_M are the degeneracies of the levels. (In general, the individual coefficients such as A_{nm} are different for each of the separate $n \rightleftharpoons m$ transitions between the levels N and M.)

(b) Show that the net rate $\rho(\bar{j}_{NM} - \bar{\kappa}_{NM} I_\nu)$, per unit volume per unit solid angle, of energy exchange between the gas and the radiation field as a result of all transitions between the energy levels N and M is given by

$$\rho(\bar{j}_{NM} - \bar{\kappa}_{NM} I_\nu) = n_M B_{MN} h\nu \left[\frac{g_M n_N}{g_N n_M} \frac{2h\nu^3}{c^2} + \left(\frac{g_M n_N}{g_N n_M} - 1 \right) I_\nu \right].$$

Again, assume that $n_N = g_N n_n$ and $n_M = g_M n_m$.

Exercise 6.2. Indicate which terms on the right-hand side of equation (6.10) arise respectively from spontaneous emission, absorption, and induced emission.

7 QUASI-EQUILIBRIUM HYPOTHESIS

The equation of radiative transfer is simplified significantly if it is assumed that the populations of the various atomic and molecular states that participate in the emission and absorption of radiation are given by their equilibrium values. This assumption is valid over a wide range of

[5] An alternative formulation can be given in terms of a quantity called the oscillator strength or the *f*-number of a given transition. This quantity is proportional to the Einstein coefficients.

conditions in gas dynamics, where the excitation and de-excitation of these states occurs predominantly as a result of molecular collisions rather than through radiative emission and absorption. In this situation the populations of the states, and hence of the energy levels, will be controlled by collisional processes. As a result, if the collisional processes themselves are in quasi-equilibrium, the distribution of molecules among the energy levels corresponds closely to the Boltzmann value for the gas temperature, even though the radiant energy density differs from its equilibrium value. The assumption that this condition holds is known as the *quasi-equilibrium hypothesis* or the *assumption of local thermodynamic equilibrium.* It is important to realize that the assumed equilibrium is among the gas molecules rather than between the gas and the radiation. We note also that *small* departures from molecular equilibrium have little effect on the emission and absorption of radiation and hence will not affect the applicability of the quasi-equilibrium hypothesis.

In the preceding section we found that the Einstein coefficients for line emission, absorption, and induced emission can be related through considerations of detailed balancing. We emphasize that although detailed balancing actually occurs only under conditions of radiative equilibrium, the relations (6.7) between the Einstein coefficients are generally valid. We now assume in addition that the relative populations of the electronic states under consideration is given by the equilibrium relation (6.6). Equations (6.9b) and (6.9c) give the absorption and emission coefficients integrated over a spectral line for which I_ν is assumed approximately constant. With equation (6.6), together with equation (4.5) for $B_\nu(T)$, these equations now become

$$\rho\bar{\kappa}_{nm} = n_m^* B_{mn} h\nu \tag{7.1a}$$

and

$$\begin{aligned}\rho\bar{j}_{nm} &= n_m^* B_{mn} h\nu\left[\frac{n_n^*}{n_m^*}\left(\frac{2h\nu^3}{c^2} + I_\nu\right)\right] \\ &= n_m^* B_{mn} h\nu\left[e^{-h\nu/kT}\left(\frac{2h\nu^3}{c^2} + I_\nu\right)\right] \\ &= n_m^* B_{mn} h\nu[(1 - e^{-h\nu/kT})B_\nu(T) + e^{-h\nu/kT} I_\nu].\end{aligned} \tag{7.1b}$$

The contribution of $n \rightleftharpoons m$ transitions to the right-hand side of the equation of radiative transfer, integrated over the line width, is then [cf. equation (6.10)]

$$\begin{aligned}\rho(\bar{j}_{nm} - \bar{\kappa}_{nm} I_\nu) &= \rho\bar{\kappa}_{nm}[(1 - e^{-h\nu/kT})B_\nu(T) + e^{-h\nu/kT} I_\nu - I_\nu] \\ &= \rho\bar{\kappa}'_{nm}[B_\nu(T) - I_\nu],\end{aligned} \tag{7.2}$$

where

$$\bar{\kappa}'_{nm} \equiv \bar{\kappa}_{nm}(1 - e^{-h\nu/kT}).$$

Although equation (7.2) was derived for line emission and absorption, relations of the same form hold for other radiative processes. Consistent with equation (7.2) it is found that under quasi-equilibrium conditions the general equation of radiative transfer (3.4) becomes

$$\frac{1}{c}\frac{\partial I_\nu}{\partial t} + l_j \frac{\partial I_\nu}{\partial x_j} = \rho\kappa'_\nu[B_\nu(T) - I_\nu], \tag{7.3a}$$

where

$$\kappa'_\nu \equiv \kappa_\nu(1 - e^{-h\nu/kT}). \tag{7.3b}$$

Here the absorption coefficient κ_ν retains the definition given in Sec. 3 and includes the effects of bound-bound, bound-free, and free-free transitions.

The bound-bound contribution to $\rho\kappa_\nu$ associated with a given frequency *within* a single line is given by

$$\rho\kappa_{\nu_{nm}} = n^*_m B_{mn} h\nu\phi(\nu).$$

This expression differs from equation (7.1a) by the line-shape factor $\phi(\nu)$. The factor $\phi(\nu)$ is a kind of distribution function such that in an integration over the line we have $\int\phi(\nu)\,d\nu = 1$. To recover equation (7.1a) we write

$$\int \rho\kappa_{\nu_{nm}} I_\nu \,d\nu = \int n^*_m B_{mn} h\nu\phi(\nu) I_\nu \,d\nu.$$

When I_ν and $h\nu$ are approximately constant over the line, this becomes

$$\rho\bar{\kappa}_{nm} I_\nu = n^*_m B_{mn} h\nu I_\nu \int \phi(\nu)\,d\nu = n^*_m B_{mn} h\nu I_\nu,$$

which is equivalent to equation (7.1a). The simplifying restriction in Sec. 6 and the present section that I_ν be approximately constant over a line can be removed in these sections by similar modifications involving $\phi(\nu)$.

The arguments preceding equations (7.3) and based on the example of line emission and absorption do not, of course, constitute a derivation of these equations. They are meant to make the equations plausible and at the same time to illustrate the connection between macroscopic emission and absorption coefficients and microscopic transition probabilities, such as the Einstein coefficients.

The right-hand side of the equation of transfer (7.3a) contains the sum of the contributions

$$\rho\kappa'_\nu B_\nu(T) = \rho\kappa_\nu(1 - e^{-h\nu/kT}) B_\nu(T) \tag{7.4a}$$

from spontaneous emission,

$$\rho\kappa_\nu e^{-h\nu/kT} I_\nu \tag{7.4b}$$

from induced emission, and

$$-\rho\kappa_\nu I_\nu \tag{7.4c}$$

from absorption. The factor $1 - e^{-h\nu/kT}$ in κ_ν' [equation (7.3b)] is called, somewhat misleadingly, the correction to κ_ν that results from induced emission. In any event, it is important to understand the distinction between the three terms (7.4) that contribute to the right-hand side of equation (7.3a).

Under quasi-equilibrium conditions the absorption coefficient κ_ν is an equilibrium property of the gas. The value of κ_ν can be found by experiment or calculated from quantum mechanics. Such calculations, a

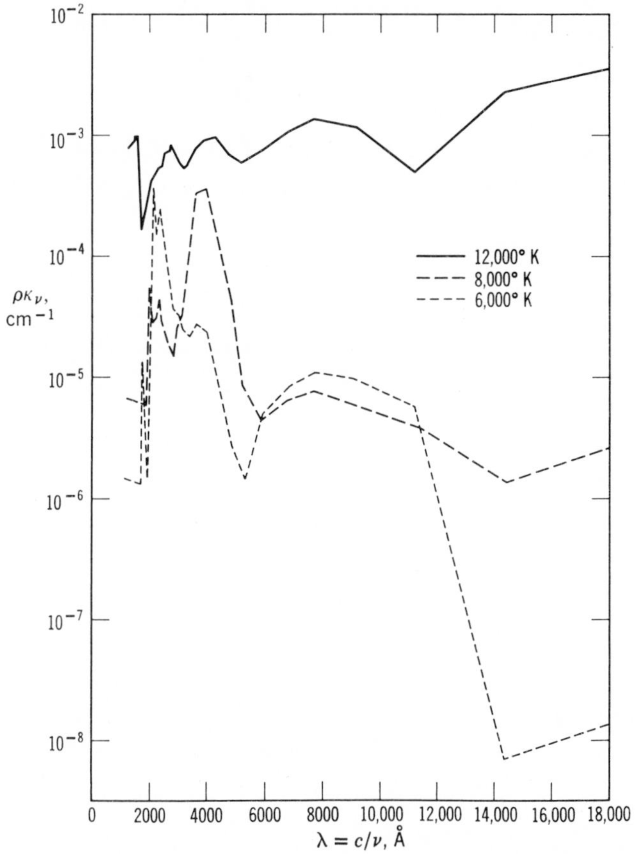

Fig. 3. Absorption coefficients for air at $\rho/\rho_0 = 0.01$, where ρ_0 is the standard density 1.29×10^{-3} gm/cm³ (after Meyerott, Sokoloff, and Nicholls, 1960).

detailed discussion of which is outside the scope of this book, involve two distinct parts. The first, a problem in statistical mechanics that is independent of considerations of radiation, is the calculation of the populations of the various atomic and molecular energy levels. This includes the calculation of the degree of dissociation and the degree of ionization. The second, a problem in the quantum theory of radiation, is the calculation of the radiative transition probabilities for transitions between the levels. The nature of these calculations for different ranges of temperature and density depends strongly on such factors as the importance of molecular transitions and the number of free electrons. These in turn depend on the degree of dissociation and the degree of ionization. A summary of such calculations for air is given by Armstrong et al. (1961). The work of Kivel and Bailey (1957) also applies to air. In Fig. 3 calculated values of $\rho\kappa_\nu$ for air, from Meyerott, Sokoloff, and Nicholls (1960), are shown as a function of wavelength λ for several temperatures. At the lowest temperature shown (6000°K), the contribution of molecular lines is found to be the most important, except for the smaller values of λ, whereas at 12,000 °K bound-free transitions tend to dominate.[6]

Assuming, as with the transport properties in continuum fluid mechanics, that κ_ν can be found for the gas in question, we return to the consideration of the equation of radiative transfer.

Exercise 7.1. On the basis of the quasi-equilibrium hypothesis, for what gas temperatures will the rate of induced emission be at least 10% as large as the rate of absorption throughout the visible spectrum ($4000 < \lambda < 7000$ Å)?

Ans: $T > 15{,}600$°K.

8 FORMAL SOLUTION OF THE EQUATION OF RADIATIVE TRANSFER

A formal solution to the quasi-equilibrium equation of radiative transfer (7.3a) can be obtained if the time derivative is neglected. The ratio of the time-derivative term to the spatial-gradient term in equation (7.3a) is of the order of the ratio of a speed formed from a characteristic distance and a characteristic time to the speed of light. Since this ratio is small for almost all problems of gas dynamics, we can safely omit the time-derivative term.

[6] In these calculations the contribution of molecular lines to κ_ν has been averaged over a frequency interval that includes many lines, hence the continuous nature of the resulting curves.

Before writing the solution we shall make two changes of variable. If the coordinate r is measured from the point in question in the direction *opposite* to the direction of propagation of the radiation, we have from the definition of the direction cosines l_1, l_2, l_3 and the usual expression for a directional derivative

$$l_j \frac{\partial I_\nu}{\partial x_j} = -\frac{\partial x_1}{\partial r}\frac{\partial I_\nu}{\partial x_1} - \frac{\partial x_2}{\partial r}\frac{\partial I_\nu}{\partial x_2} - \frac{\partial x_3}{\partial r}\frac{\partial I_\nu}{\partial x_3} = -\frac{\partial I_\nu}{\partial r}. \tag{8.1}$$

We also define a *volumetric absorption coefficient* α_ν by

$$\alpha_\nu \equiv \rho\kappa_\nu',$$

which has the dimensions of inverse length. The equation of radiative transfer (7.3a) with the time-derivative term omitted then becomes

$$\boxed{\frac{\partial I_\nu}{\partial r} = \alpha_\nu(I_\nu - B_\nu)}\,. \tag{8.2}$$

For the solution of equation (8.2) we take $r = 0$ at the point in question and suppose that I_ν has a known boundary value

$$I_\nu = I_\nu(r_s) \qquad \text{at} \quad r = r_s. \tag{8.3}$$

We shall ordinarily take $r = r_s$ to be the location of a solid wall, although the boundary condition (8.3) can be applied at any point where I_ν is known. If the surface at r_s radiates as a black body ($e_\nu = 1$) at a temperature T_w, then we have in particular from Sec. 5

$$I_\nu(r_s) = B_\nu(T_w). \tag{8.4}$$

With the condition (8.3) the formal solution to the equation of transfer (8.2) can be obtained by standard methods [see equation (VIII 13.5)] and is

$$\boxed{I_\nu = I_\nu(r_s)e^{-\tau_\nu(r_s)} + \int_0^{r_s} \alpha_\nu B_\nu(r)e^{-\tau_\nu(r)}\,dr}\,. \tag{8.5a}$$

Here the *optical depth* τ_ν at frequency ν is defined by

$$\tau_\nu(r) \equiv \int_0^r \alpha_\nu\,d\hat{r}, \tag{8.6}$$

where $\hat{r}$ is a running variable of integration. Equation (8.5a) is a formal solution in the sense that I_ν is expressed in terms of integrals that can be evaluated only if the state of the gas is known. Since $\alpha_\nu\,dr = d\tau_\nu$, the solution (8.5a) can also be written as

$$I_\nu = I_\nu(r_s)e^{-\tau_\nu(r_s)} + \int_0^{\tau_\nu(r_s)} B_\nu(\tau_\nu)e^{-\tau_\nu}\, d\tau_\nu. \tag{8.5b}$$

In the special case where α_ν is independent of r and if spontaneous emission from the gas is neglected for $0 < r < r_s$, the solution (8.5) becomes

$$I_\nu = I_\nu(r_s)e^{-\alpha_\nu r_s}.$$

The quantity $1/\alpha_\nu$ can thus be interpreted as a relaxation length for radiation of frequency ν.

Equations (8.5) give the specific intensity of radiation at a point and in a given direction as the sum of two parts: (1) the contribution of the boundary located at a given distance r_s in the direction opposite to the direction of propagation of the radiation, attenuated by the factor $\exp[-\tau_\nu(r_s)]$ to account for absorption and induced emission in the intervening gas, and (2) the contribution of the spontaneous emission from elements of the gas at a varying distance r, each elementary contribution attentuated by the factor $\exp[-\tau_\nu(r)]$, and the whole summed over all elements between the point in question and the boundary. With the solution (8.5), the radiant heat-flux vector (2.8) can be calculated by direct integration if the temperature and absorption coefficient of the gas are everywhere known.

The radiation term that will appear in the energy equation of gas dynamics represents the volumetric rate of energy addition to the gas by radiation. This quantity is easily shown to be equal to $-\partial q_j^R/\partial x_j$, the negative of the divergence of the heat-flux vector (see Chapter XII, Sec. 2). From the definition (2.8) of q_i^R and equation (8.1) we obtain

$$-\frac{\partial q_j^R}{\partial x_j} = -\int_0^\infty \int_0^{4\pi} l_j \frac{\partial I_\nu}{\partial x_j}\, d\Omega\, d\nu = \int_0^\infty \int_0^{4\pi} \frac{\partial I_\nu}{\partial r}\, d\Omega\, d\nu. \tag{8.7a}$$

With the equation of radiative transfer (8.2), this can be written

$$\boxed{-\frac{\partial q_j^R}{\partial x_j} = \int_0^\infty \alpha_\nu\left(\int_0^{4\pi} I_\nu\, d\Omega - 4\pi B_\nu\right) d\nu}\ . \tag{8.7b}$$

The two terms on the right-hand side of equation (8.7b) account for the energy addition to the gas that results, respectively, from absorption less induced emission, and from spontaneous emission.

Exercise 8.1. Find the specific intensity at a point under the following conditions: (a) if $\tau_\nu(r_s) \ll 1$; (b) if B_ν and α_ν are uniform throughout the gas.

In case (b), if $I_\nu(r_s) = 0$ what is the maximum possible value of I_ν? Again in case (b), if the gas is completely surrounded by an opaque (but not necessarily black) wall with T_w equal to the gas temperature, what will be the value of I_ν?

9 SIMPLIFICATIONS AND APPROXIMATIONS

With the equation of radiative transfer in the form (8.2) we could proceed to a discussion of the equations that govern the coupling of radiative nonequilibrium and gas dynamics. As we shall see in the next chapter, however, this system of equations is of such a nature that some simplification is required for the solution of most physical problems. It is of interest therefore to examine first, in general terms, certain limiting cases and approximations.

In many physical situations the specific intensity I_ν is much less than its equilibrium value $B_\nu(T)$, that is,

$$I_\nu \ll B_\nu \tag{9.1}$$

for some range of frequency and some region in space. When I_ν is negligible compared with B_ν, the time-independent equation of radiative transfer (8.2) reduces to

$$\frac{\partial I_\nu}{\partial r} = -\alpha_\nu B_\nu. \tag{9.2a}$$

In a region $0 \le r \le r_s$ for which the inequality (9.1) holds, the formal solution of the equation of transfer is then

$$I_\nu = I_\nu(r_s) + \int_0^{r_s} \alpha_\nu B_\nu \, dr, \tag{9.2b}$$

where r and r_s have the same meaning as in the preceding section.

The physical interpretation of this situation is as follows: Under conditions of local thermodynamic equilibrium, the rate at which radiant energy in the frequency range $d\nu$ and the solid angle $d\Omega$ is spontaneously emitted per unit volume of the gas is

$$\alpha_\nu B_\nu \, d\Omega \, d\nu, \tag{9.3a}$$

whereas the rate of energy absorption minus the rate of induced emission is

$$\alpha_\nu I_\nu \, d\Omega \, d\nu. \tag{9.3b}$$

Thus the assumption (9.1) that $I_\nu \ll B_\nu$ is equivalent to the assumption that spontaneous emission processes dominate over absorption and induced emission processes. We describe the gas-radiation interaction as being *emission-dominated* in a certain frequency range when the inequality (9.1) holds for all directions of propagation.

For an emission-dominated interaction many practical problems can be solved most conveniently in a direct way rather than through the formal equations such as (9.2a). For example, when absorption and induced emission are negligible throughout a gas, the radiant heat flux from the gas to any elementary surface can be obtained from an integration over the volume of radiating gas of the rate of spontaneous emission in the direction of the surface (see Exercise 9.1). Similarly, the radiation term $\partial q_j^R/\partial x_j$ in the energy equation of gas dynamics can be calculated directly. That is to say, if the condition (9.1) is assumed to hold for all frequencies and directions of propagation, $\partial q_j^R/\partial x_j$ will depend only on local conditions and will be equal to the total volumetric rate of spontaneous emission. From equation (9.3a) and the fact that spontaneous emission is isotropic, we therefore have [cf. equation (8.7b)]

$$-\frac{\partial q_j^R}{\partial x_j} = -4\pi \int_0^\infty \alpha_\nu B_\nu \, d\nu. \tag{9.4a}$$

It is conventional to write this relation in the form

$$-\frac{\partial q_j^R}{\partial x_j} = -4\alpha_P \sigma T^4, \tag{9.4b}$$

where the *Planck* or *emission mean absorption coefficient* α_P is defined by

$$\alpha_P \equiv \frac{\displaystyle\int_0^\infty \alpha_\nu B_\nu \, d\nu}{\displaystyle\int_0^\infty B_\nu \, d\nu} = \frac{\pi}{\sigma T^4} \int_0^\infty \alpha_\nu B_\nu \, d\nu. \tag{9.5a}$$

This can be written with the aid of equations (4.5) and (4.6) as

$$\alpha_P = \frac{15}{\pi^4} \int_0^\infty \alpha_\nu \frac{(h\nu/kT)^3}{e^{h\nu/kT} - 1} \, d\left(\frac{h\nu}{kT}\right). \tag{9.5b}$$

The inverse quantity $\lambda_P \equiv \alpha_P^{-1}$ is called the *Planck mean free path* for radiation [cf. equation (3.5)]. We note that the right-hand side of equations (9.4) is a general expression for the negative of the volumetric rate of spontaneous emission under conditions of local thermodynamic equilibrium. It is equal to $-\partial q_j^R/\partial x_j$, however, only if the condition (9.1) is satisfied.

A comparison of the solution (9.2b) for I_ν when $I_\nu \ll B_\nu$ and the general solution (8.5a) shows that the two agree if we take

$$\tau_\nu(r_s) \equiv \int_0^{r_s} \alpha_\nu \, dr \ll 1. \tag{9.6}$$

For the frequency range and directions of propagation for which the condition (9.6) holds, the gas is said to be *optically thin.* When the gas is thin the specific intensity at a given point is the unattenuated sum of the contribution $\int_0^{r_s} \alpha_\nu B_\nu \, dr$ that results from spontaneous emission at all points along the direction of propagation plus the specific intensity $I_\nu(r_s)$ at the boundary. Although the condition (9.6) for a thin gas is not a sufficient condition for the interaction to be emission-dominated (i.e., for $I_\nu \ll B_\nu$), the two conditions are closely related. From the solution (8.5b) for I_ν we see that, if $\tau_\nu(r_s) \ll 1$, then we must also have

$$I_\nu(r_s) + \int_0^{\tau_\nu(r_s)} B_\nu \, d\tau_\nu \ll B_\nu \tag{9.7}$$

in order to have $I_\nu \ll B_\nu$. If $B_\nu(r)$ does not differ greatly from B_ν at the point in question, $\tau_\nu(r_s) \ll 1$ will ensure that the second term on the left in the inequality (9.7) is small, so that (9.7) reduces to a condition on $I_\nu(r_s)$. When a gas is optically thin *and* the inequality (9.7) applies for all frequencies and directions of propagation, then the situation is emission-dominated and $-\partial q_j^R/\partial x_j$ is given by equations (9.4) as before.

With reference to equation (9.6), if $\tau_\nu(r_s) \ll 1$ for all ν, the mean optical depth based on the Planck mean absorption coefficient will also be small, that is,

$$\tau_P(r_s) \equiv \int_0^{r_s} \alpha_P \, d\hat{r} \ll 1. \tag{9.8}$$

Although the converse is *not* necessarily true, that is, equation (9.6) does not follow from equation (9.8), the condition (9.8) is often used as an approximate criterion for thinness.

The situation in which I_ν differs by only a small amount from B_ν, which will exist when the departure from radiative equilibrium is small, can be considered as the opposite case from an emission-dominated interaction. For reasons that we shall see presently a gas in these conditions is said to be *optically thick.*

For simplicity we limit our discussion to the consideration of a point in the gas that is sufficiently far from any boundary, in terms of optical depth, that the contribution to I_ν from $I_\nu(r_s)$ is negligible. The formal solutions (8.5) to the equation of radiative transfer then become

$$I_\nu(r = 0) = \int_0^{r_s} \alpha_\nu(r) B_\nu(r) e^{-\tau_\nu(r)}\, dr \tag{9.9a}$$

or

$$I_\nu(r = 0) = \int_0^{\infty} B_\nu[\tau_\nu(r)] e^{-\tau_\nu(r)}\, d\tau_\nu. \tag{9.9b}$$

Here we have indicated explicitly the dependence of α_ν and B_ν on the distance r in the direction opposite to the propagation of the radiation and equivalently in the second form have written B_ν as a function of $\tau_\nu(r)$. Equations (9.9) apply for a fixed frequency and direction of propagation. We now assume that away from the point in question the optical depth $\tau_\nu(r)$ becomes large at values of r for which $B_\nu(r)$ has changed by only a relatively small amount from $B_\nu(r = 0)$. That is, we assume that

$$\frac{1}{B_{\nu 0}} \left| \left(\frac{\partial B_\nu}{\partial \tau_\nu}\right)_0 \right| \ll 1, \tag{9.10}$$

where the subscript 0 denotes quantities evaluated at the point in question ($r = 0$). Remembering that for fixed ν, B_ν is a function only of temperature, we can write for B_ν the Taylor's-series expansion

$$\begin{aligned} B_\nu[\tau_\nu(r)] &= B_{\nu_0} + \left(\frac{\partial B_\nu}{\partial \tau_\nu}\right)_0 \tau_\nu + \cdots \\ &= B_{\nu_0}\left[1 + \frac{1}{B_{\nu_0}}\left(\frac{\partial B_\nu}{\partial \tau_\nu}\right)_0 \tau_\nu + \cdots\right]. \end{aligned}$$

In accord with the condition (9.10) we retain only the first two terms in this series. With

$$\frac{1}{B_{\nu_0}}\left(\frac{\partial B_\nu}{\partial \tau_\nu}\right)_0 \tau_\nu = \frac{1}{B_{\nu_0}}\left(\frac{dB_\nu}{dT}\right)_0 \left(\frac{\partial T}{\partial \tau_\nu}\right)_0 \tau_\nu = \frac{1}{\alpha_{\nu_0} B_{\nu_0}}\left(\frac{dB_\nu}{dT}\right)_0 \left(\frac{\partial T}{\partial r}\right)_0 \tau_\nu,$$

equation (9.9b) for I_ν then becomes

$$\begin{aligned} I_{\nu_0} &= B_{\nu_0}\left[\int_0^\infty e^{-\tau_\nu}\, d\tau_\nu + \frac{1}{\alpha_{\nu_0} B_{\nu_0}}\left(\frac{dB_\nu}{dT}\right)_0 \left(\frac{\partial T}{\partial r}\right)_0 \int_0^\infty \tau_\nu e^{-\tau_\nu}\, d\tau_\nu\right] \\ &= B_{\nu_0}\left[1 + \frac{1}{\alpha_{\nu_0} B_{\nu_0}}\left(\frac{dB_\nu}{dT}\right)_0 \left(\frac{\partial T}{\partial r}\right)_0\right]. \end{aligned} \tag{9.11}$$

From equation (9.11) we see that the condition (9.10) yields the result that I_ν differs by only a small amount from B_ν. The gas is optically thick in the sense that only radiation from points for which $B_\nu[\tau_\nu(r)]$ is nearly equal to B_{ν_0} reaches the point $r = 0$.

If equation (9.11) for I_ν holds for all frequencies and all directions of propagation, the radiant heat-flux vector [equation (2.8)] is

$$q_i^R = \int_0^\infty \left[B_\nu \int_0^{4\pi} l_i \, d\Omega + \frac{1}{\alpha_\nu} \frac{\partial B_\nu}{\partial T} \int_0^{4\pi} l_i \left(\frac{\partial T}{\partial r} \right) d\Omega \right] d\nu,$$

where we have now dropped the subscript 0. As in equation (4.8c), the first term on the right integrates to zero. Detailed consideration of the integral over solid angle in the second term yields (Exercise 9.2)

$$\int_0^{4\pi} l_i \left(\frac{\partial T}{\partial r} \right) d\Omega = - \frac{4\pi}{3} \frac{\partial T}{\partial x_i}.$$

We thus obtain

$$q_i^R = - \frac{4\pi}{3} \left(\int_0^\infty \frac{1}{\alpha_\nu} \frac{dB_\nu}{dT} \, d\nu \right) \frac{\partial T}{\partial x_i}. \tag{9.12}$$

The condition (9.10), on which this result depends, is analogous to the condition associated with the Chapman-Enskog expansion in kinetic theory (Chapter X, Sec. 4). In both cases the heat flux depends only on the local value of the first derivative of the temperature. We note from equation (9.12) that a relation of this kind applies separately at any frequency for which equation (9.11) holds in all directions of propagation.

A straightforward calculation (Exercise 9.2) based on equation (4.5) for $B_\nu(T)$ shows that the integral in equation (9.12) can be written

$$\int_0^\infty \frac{1}{\alpha_\nu} \frac{dB_\nu}{dT} \, d\nu = \frac{4}{\pi} \frac{\sigma T^3}{\alpha_R},$$

where α_R is defined by

$$\frac{1}{\alpha_R} \equiv \frac{\displaystyle\int_0^\infty \frac{1}{\alpha_\nu} \frac{dB_\nu}{dT} \, d\nu}{\displaystyle\int_0^\infty \frac{dB_\nu}{dT} \, d\nu} = \frac{\pi}{4\sigma T^3} \int_0^\infty \frac{1}{\alpha_\nu} \frac{dB_\nu}{dT} \, d\nu. \tag{9.13a}$$

With the aid of equations (4.5) and (4.6) this can be written

$$\frac{1}{\alpha_R} = \frac{15}{4\pi^4} \int_0^\infty \frac{1}{\alpha_\nu} \frac{e^{h\nu/kT}}{(e^{h\nu/kT} - 1)^2} \left(\frac{h\nu}{kT} \right)^4 d\left(\frac{h\nu}{kT} \right). \tag{9.13b}$$

The quantity α_R is called the *Rosseland mean absorption coefficient*, and $\lambda_R \equiv \alpha_R^{-1}$ is the *Rosseland mean free path* for radiation. In terms of α_R, equation (9.12) for q_i^R is

$$q_i^R = - \frac{16\sigma T^3}{3\alpha_R} \frac{\partial T}{\partial x_i}. \tag{9.14}$$

This result is known as the *Rosseland diffusion approximation*.

Values of λ_R and λ_P for air over a wide range of conditions are given by Armstrong et al. (1961). Under some conditions the two mean free paths can differ by several orders of magnitude. It has been shown, however, that λ_R forms an effective upper bound to the value of λ_P (see Armstrong, 1962). More precisely, it is found that in general $\lambda_P \leqq 1.053\ \lambda_R$.

The Planck and Rosseland mean absorption coefficients [equations (9.5) and (9.13)] prescribe, for two asymptotic cases, an average of α_ν over frequency. Insofar as such averages can be applied to a more general situation, the analysis of radiative transfer can be simplified significantly. Thus, if we replace α_ν by some average value α that is independent of frequency, the equation of radiative transfer (8.2) and its formal solution (8.5a) can be integrated over ν to yield

$$\frac{\partial I}{\partial r} = \alpha\left(I - \frac{\sigma}{\pi} T^4\right) \tag{9.15}$$

and

$$I = I(r_s)e^{-\tau(r_s)} + \frac{\sigma}{\pi}\int_0^{r_s} \alpha T^4 e^{-\tau(r)}\, dr. \tag{9.16}$$

Here we have written

$$I \equiv \int_0^\infty I_\nu\, d\nu, \tag{9.17a}$$

$$\tau(r) \equiv \int_0^r \alpha\, d\hat{r}, \tag{9.17b}$$

and have used, from equation (4.7),

$$\int_0^\infty B_\nu\, d\nu = \frac{\sigma}{\pi} T^4.$$

The quantity I is called the *integrated intensity*. From equations (2.8) and (8.7) the heat-flux vector and its divergence are then

$$q_i^R = \int_0^{4\pi} I\, l_i\, d\Omega, \tag{9.18}$$

and

$$-\frac{\partial q_j^R}{\partial x_j} = \int_0^{4\pi} \frac{\partial I}{\partial r}\, d\Omega = \alpha\left[\int_0^{4\pi} I\, d\Omega - 4\sigma T^4\right]. \tag{9.19}$$

The foregoing formulation is called the *grey-gas approximation*, and α is known as a grey absorption coefficient. It is important to realize that these results are obtained by arbitrarily replacing α_ν with a frequency-independent value α. Thus the grey-gas approximation does *not* proceed from a rigorous average of the equations over frequency. The specification of α in a given situation then involves both judgment and a certain arbitrariness. The Planck and Rosseland mean absorption coefficients as well as other averages of α_ν have been used. It should be noted, however, that α_P and α_R enter the two previously discussed asymptotic cases independently of the grey-gas approximation.

At this point the reader who is unfamiliar with the solution of problems of radiative transfer may be impressed more with the *ad hoc* nature of the grey-gas approximation than with the motivation for this simplification. This is particularly true in view of the strong frequency dependence of actual absorption coefficients (Fig. 3). As we shall see, however, in many problems involving the coupling of radiative transfer and gas dynamics, a strong motivation for simplification arises from the complexity associated with the general frequency-dependent formulation.

Exercise 9.1. Consider a hemispheric volume of gas at uniform temperature and density, so that α_ν is uniform. Assume that the gas is surrounded by a transparent surface outside of which there is a vacuum. If absorption and induced emission are negligible within the gas, calculate the radiant heat flux incident upon a surface element on the base of the hemisphere at its center (a) by a direct integration based on the intercepted portion of the energy $4\pi \int_0^\infty \alpha_\nu B_\nu \, d\nu$ emitted by each elementary volume of the gas, and (b) by means of the specific intensity through equations (9.2b) and (2.12). Repeat the latter calculation for the situation where the spherical surface has an emissivity of unity and a temperature T_w. Do your results apply when the gas-radiation interaction is emission-dominated or when the gas is optically thin or in both cases?

Exercise 9.2. (a) Show that $\int_0^{4\pi} l_i(\partial T/\partial r)\, d\Omega = -(4\pi/3)\, \partial T/\partial x_i$ to give equation (9.12) [see equation (8.1)]. (b) Derive equations (9.13).

Exercise 9.3. Under what circumstances will the grey-gas formulas (i) equation (9.16) for I and (ii) equation (9.19) for $-\partial q_j^R/\partial x_j$ be exact with the Planck mean absorption coefficient α_P used for α?

References

Aller, L. H., 1963, *The Atmospheres of the Sun and the Stars*, 2nd. ed., Ronald.

Armstrong, B. H., 1962, A Maximum Opacity Theorem. *Astrophys. J.*, vol. 136, no. 1, p. 309.

Armstrong, B. H., J. Sokoloff, R. W. Nicholls, D. H. Holland, and R. E. Meyerott, 1961, Radiative Properties of High Temperature Air. *J. Quant. Spect. Rad. Trans.*, vol. 1, no. 2, p. 143.

Chandrasekhar, S., 1950, *Radiative Transfer*, Oxford. (Also Dover, 1960.)

Davidson, N., 1962, *Statistical Mechanics*, McGraw-Hill.

Eckert, E. R. G., and R. M. Drake, 1959, *Heat and Mass Transfer*, 2nd. ed., McGraw-Hill.

Goulard, R., 1964, Fundamental Equations of Radiation Gas Dynamics; in *The High Temperature Aspects of Hypersonic Flow*, ed. by W. C. Nelson, Macmillan.

Griem, H. R., 1964, *Plasma Spectroscopy*, McGraw-Hill.

Heitler, W., 1954, *The Quantum Theory of Radiation*, 3rd. ed., Oxford.

Kivel, B., and K. Bailey, 1957, Tables of Radiation from High Temperature Air. *AVCO*, Res. Rep. No. 21.

Lighthill, M. J., 1960, Dynamics of a Dissociating Gas. Part 2. Quasi-equilibrium Transfer Theory. *J. Fluid Mech.*, vol. 8, pt. 2, p. 161.

Mayer, J. E., and M. G. Mayer, 1940, *Statistical Mechanics*, Wiley.

Meyerott, R. E., J. Sokoloff, and R. W. Nicholls, 1960, Absorption Coefficients of Air. *Air Force Cambridge Research Center*, Geophysics Research Paper No. 68.

Chapter XII

Flow with Radiative Nonequilibrium

1 INTRODUCTION

We are now ready to discuss the interaction of radiative transfer and gas motion. Since energy transfer by radiation is a nonequilibrium process, we here encounter another type of nonequilibrium flow. The subject is in some respects similar to the flow with translational nonequilibrium discussed in Chapter X; in others it resembles the flow with vibrational or chemical nonequilibrium treated in Chapter VIII.

To simplify the problem and isolate the influence of radiation, we assume that nonequilibrium effects from other sources, such as molecular transport, dissociation, and vibration, are negligible. We also assume that the gas is perfect. These assumption are unrealistic, since at the temperatures required for radiation the gas would also be expected to ionize, dissociate, etc., with attendant nonperfect, nonequilibrium phenomena. The theory is sufficiently complicated without these effects, however, and for an initial study it is advisable to omit them. In setting down the conservation equations we shall also neglect the contributions of radiative pressure, energy density, and scattering. For most nonastrophysical applications, these quantities are in fact insignificant (see Lighthill, 1960, and Zhigulev et al., 1963). The radiative effects will be taken into account on the basis of the quasi-equilibrium hypothesis, and in most of the analysis the grey-gas assumption will be used. For many situations these assumptions will certainly be quantitatively inaccurate. Again, however, the complexity of the problem makes them warranted in a study in which qualitative understanding is the primary object. In this, more than any other section of the book, the subject is in a state of development, and the accuracy of many of the assumptions is not yet known.

As mentioned in the introduction to the preceding chapter, we return in the formulation of our conservation equations to the macroscopic point of view used for the discussion of reactive nonequilibrium in Chapter VIII. The structure of our treatment, in fact, will parallel that of Chapter VIII in many respects.

2 BASIC NONLINEAR EQUATIONS

We begin by writing the equations for three-dimensional time-dependent flow under the assumptions just outlined.

The conservation equations of mass, momentum, and energy are the same as in Chapter VIII, Sec. 2, except for the inclusion of a radiative-transfer term in the energy equation. Using the same notation as before, we write for the conservation of mass, as in equation (VIII 2.2b),

$$\frac{D\rho}{Dt} + \rho \frac{\partial u_j}{\partial x_j} = 0. \tag{2.1}$$

If radiation pressure is neglected, the conservation of momentum in the absence of translational nonequilibrium and with body forces omitted is given again by equation (VIII 2.3):

$$\rho \frac{Du_i}{Dt} + \frac{\partial p}{\partial x_i} = 0. \tag{2.2}$$

With the present assumptions, the conservation of energy is again expressed most readily by applying the first law of thermodynamics to a unit mass moving with the fluid. Here, however, the process is non-adiabatic as a result of the nonequilibrium radiative transfer. By the usual considerations with reference to a differential element, the net rate of heat addition per unit mass as a result of radiation can be represented by $-(1/\rho)\, \partial q_j^R/\partial x_j$, where q_i^R is the radiant heat-flux vector introduced in Chapter XI, Sec. 2. Applying the first law in the form (III 4.8), we thus obtain for the energy equation

$$\rho \frac{Dh}{Dt} - \frac{Dp}{Dt} = -\frac{\partial q_j^R}{\partial x_j}. \tag{2.3}$$

The enthalpy h and pressure p here are the values associated with the material particles of the gas; in accord with our assumptions, they contain no contribution from the radiative energy density or radiative pressure. The term that appears on the right-hand side of (2.3) is identical in form to the heat-transfer term that appears in equation (IX 2.24b) as the result of molecular transport effects. This term constitutes the only difference between the foregoing equations and the corresponding equations of Chapter VIII.

In line with the neglect of all effects of vibrational and chemical non-equilibrium, the enthalpy is related to the other state variables by an equilibrium thermodynamic relation of the form $h = h(p, \rho)$. In particular, for the perfect gas assumed here we have

$$h = \frac{\gamma}{(\gamma - 1)} \frac{p}{\rho}, \tag{2.4}$$

where γ is the (constant) ratio of specific heats. The temperature T, which must enter in the evaluation of the radiative transfer, is given correspondingly by

$$T = \frac{p}{R\rho}, \tag{2.5}$$

where R is the ordinary gas constant.

The heat-flux vector q_i^R in equation (2.3) is expressed in terms of the frequency-dependent specific intensity I_ν by equation (XI 2.8):

$$q_i^R = \int_0^\infty \int_0^{4\pi} I_\nu l_i \, d\Omega \, d\nu, \tag{2.6a}$$

where l_i is the unit vector specifying the direction of propagation of I_ν and l_1, l_2, l_3 are the associated direction cosines. To evaluate I_ν we use the equation of radiative transfer in the quasi-equilibrium form introduced in Chapter XI, Secs. 7 and 8. For reasons that were explained at the beginning of Sec. 8, the time-derivative term is omitted. This is formally equivalent to taking the speed of light as infinite in the left-hand side of the equation. We have then from equations (XI 8.1) and (XI 8.2)

$$l_j \frac{\partial I_\nu}{\partial x_j} = -\alpha_\nu (I_\nu - B_\nu), \tag{2.7a}$$

where α_ν is the frequency-dependent absorption coefficient. The Planck function $B_\nu = B_\nu(T)$ that appears here is given by equation (XI 4.5) as

$$B_\nu(T) = \frac{2h\nu^3/c^2}{e^{h\nu/kT} - 1}.$$

By differentiation of (2.6a) and substitution from (2.7a), the heat-addition term in equation (2.3) can also be written directly as follows [see equation (XI 8.7b)]:

$$-\frac{\partial q_j^R}{\partial x_j} = \int_0^\infty \alpha_\nu \left(\int_0^{4\pi} I_\nu \, d\Omega - 4\pi B_\nu \right) d\nu. \tag{2.6b}$$

If desired, I_ν in equations (2.6a or b) can be written alternatively in terms of the formal solution of equation (2.7a), given as follows by equation (XI 8.5a):

$$I_\nu = I_\nu(r_s)e^{-\tau_\nu(r_s)} + \int_0^{r_s} \alpha_\nu B_\nu e^{-\tau_\nu(r)}\, dr. \tag{2.7b}$$

In applying this formula it must be recalled that r is the radial distance measured from the point x_i at which I_ν is evaluated and in the direction *opposite* to l_i; r_s is the particular value of r at some distance at which I_ν along the direction in question has the boundary value $I_\nu(r_s)$. The optical depth $\tau_\nu(r)$ of the gas over the radial distance from the point x_i is given by equation (XI 8.6) as

$$\tau_\nu(r) = \int_0^r \alpha_\nu\, d\hat{r}. \tag{2.8}$$

Equations (2.1) through (2.7) provide a set of eleven scalar equations for the eleven unknowns ρ, u_i, p, h, q_i^R, T, and I_ν. The set is obviously nonlinear. Because of the coupling of the transfer equation (2.7a) with the flow equations through the integral relation (2.6a or b), it is also of a complicated integro-differential form.

The foregoing equations are written on the assumption that radiative pressure and radiative energy density are negligible. How they should be modified to include these quantities is not obvious on simple physical grounds. This is because of the relativistic character of radiation, which must be taken into account in any transformations from a fixed frame of reference to a frame of reference moving with the fluid. For a treatment of this question, see Prokof'ev (1962) and Simon (1963). Many of the earlier treatments of this question are now known to be in error.

Exercise 2.1. Carry through the details to show that the net rate of heat addition per unit mass due to radiation is given by $-(1/\rho)\, \partial q_j^R / \partial x_j$.

3 ASYMPTOTIC SITUATIONS; GREY-GAS APPROXIMATION

Owing to their nonlinear integro-differential form, the equations as they stand are hardly susceptible of analytical solution. We therefore look for useful approximations. As we have seen in Sec. 9 of the preceding chapter, the emission-dominated and thick-gas approximations offer considerable simplification in the radiative transfer. We examine these first.

A condition for the applicability of the emission-dominated approximation, as given by the inequality (XI 9.6), is that

$$\tau_\nu(r_s) \equiv \int_0^{r_s} \alpha_\nu \, d\hat{r} \ll 1 \tag{3.1}$$

for all frequencies and all points in the field. To express this in another way, let L be a characteristic length in the flow field and α_{ν_r} a characteristic, reference value of α_ν. Dimensionless quantities, assumed to be of order one, are then defined by $\bar{r} = r/L$ and $\bar{\alpha}_\nu = \alpha_\nu/\alpha_{\nu_r}$. In terms of these quantities, condition (3.1) becomes

$$\alpha_{\nu_r} L \int_0^{\bar{r}_s} \bar{\alpha}_\nu \, d\hat{\bar{r}} \ll 1,$$

which requires that $\alpha_{\nu_r} L \ll 1$ or finally that

$$\frac{1}{\alpha_{\nu_r}} \gg L. \tag{3.2}$$

The emission-dominated approximation thus requires the characteristic radiative mean free path for all frequencies to be much larger than the dimensions of the region in question. If this is so and if the values of $I_\nu(r_s)$ and $B_\nu(r)$ are such that the condition (XI 9.7) is satisfied, then the situation is emission-dominated and the term on the right-hand side of equation (2.3) is given by equation (XI 9.4b) as

$$-\frac{\partial q_j^R}{\partial x_j} = -4\alpha_P \sigma T^4, \tag{3.3}$$

where α_P is the Planck mean absorption coefficient given by equations (XI 9.5). With this expression the governing equations (2.1) to (2.3) become purely differential equations. The effect of radiation is then formally the same as a system of heat sinks distributed throughout the flow field, the strength of the sinks being a known function of density and temperature. For an example of the treatment of a problem by this method, see Zhigulev et al. (1963), who discuss the radiating hypersonic flow past a wedge.

The condition for the thick-gas approximation, as given by the inequality (XI 9.10), is that

$$\frac{1}{B_\nu}\left|\frac{\partial B_\nu}{\partial \tau_\nu}\right| = \frac{1}{B_\nu \alpha_\nu}\left|\frac{\partial B_\nu}{\partial r}\right| \ll 1.$$

Writing this in terms of the dimensionless quantities as before and introducing the additional quantity $\bar{B}_\nu = B_\nu/B_{\nu_r}$, we obtain

$$\frac{1}{\alpha_{\nu_r} L \bar{B}_\nu \bar{\alpha}_\nu}\left|\frac{\partial \bar{B}_\nu}{\partial \bar{r}}\right| \ll 1.$$

This requires that $\alpha_{\nu_r} L \gg 1$ or

$$\frac{1}{\alpha_{\nu_r}} \ll L. \tag{3.4}$$

The thick-gas approximation thus applies when the radiative mean free path is much smaller than the characteristic dimension of the field. Under these conditions the heat flux in equation (2.3) is given by equation (XI 9.14) as

$$q_i^R = -\frac{16\sigma T^3}{3\alpha_R}\frac{\partial T}{\partial x_i}, \tag{3.5}$$

where α_R is the Rosseland mean absorption coefficient given by equations (XI 9.13). Thus again the governing equations (2.1) to (2.3) become purely differential equations. The effect of radiation is now formally the same as a molecular heat conduction with a density- and temperature-dependent thermal conductivity given by $K^R = 16\sigma T^3/3\alpha_R$. An example of a problem treated by this method is given by Sen and Guess (1957) in their discussion of the effect of radiation on shock-wave structure. Difficulties arise, however, in the application of the thick-gas approximation near solid boundaries, since the series expansion leading to equation (3.5) is not valid in the immediate vicinity of the boundary. This is a matter of current research.

The emission-dominated and thick-gas approximations simplify the problem by replacing the integro-differential equations with a system of purely differential form. They are open to the objection, however, that they apply asymptotically in only limited ranges of α_ν. For this reason we do not pursue them further. Instead we look for methods that apply over the full range of the absorption coefficient.

The first step in this direction is to adopt the grey-gas approximation, which assumes a constant absorption coefficient α independent of ν. As we have seen in Chapter XI, Sec.9, this simplifies the radiation equations by allowing us to perform at once the necessary integrations with respect to ν. In terms of the integrated intensity $I \equiv \int_0^\infty I_\nu \, d\nu$, equations (2.6) to (2.8) then assume the following forms, which we take over from Chapter XI, Sec. 9:

$$q_i^R = \int_0^{4\pi} I l_i \, d\Omega, \tag{3.6a}$$

$$l_j \frac{\partial I}{\partial x_j} = -\alpha\left(I - \frac{\sigma}{\pi} T^4\right), \tag{3.7a}$$

$$-\frac{\partial q_j^R}{\partial x_j} = \alpha\left(\int_0^{4\pi} I \, d\Omega - 4\sigma T^4\right), \tag{3.6b}$$

$$I = I(\tau_s)e^{-\tau_s} + \frac{\sigma}{\pi}\int_0^{\tau_s} T^4 e^{-\tau} \, d\tau. \tag{3.7b}$$

In the last of these, as in equation (XI 8.5b), the variable of integration has been changed from r to τ in accord with the equation

$$\tau(r) \equiv \int_0^r \alpha \, d\hat{r}. \tag{3.8}$$

We have also introduced the notation $\tau_s = \tau(r_s)$.

All our subsequent work will be based on the grey-gas approximation. It is not to be expected that the resulting theory will give accurate quantitative predictions in all situations. We may hope, however, that it will give many of the essential qualitative features. Although the approximation does avoid the integration with respect to ν, the resulting flow equations are still of integro-differential form.

4 ONE-DIMENSIONAL EQUATIONS

Since the equations for the grey-gas approximation remain of integro-differential form, their solution is still difficult. It is therefore of interest to see if any simplification can be attained by restricting ourselves to problems where quantities are a function of only a single space variable, say x. The simplification of the fluid-flow terms on the left in equations (2.1), (2.2), and (2.3) is obvious, and the results need not be set down. We merely retain spatial derivatives in the one variable x and write the substantial derivatives as $D(\ )/Dt = \partial(\ )/\partial t + u\partial(\ )/\partial x$, where u is the magnitude of the velocity, which is now entirely in the x-direction. The heat-addition term on the right in equation (2.3), however, requires special attention.

To fix our ideas, let us consider the particular case of a semi-infinite expanse of gas on one side of a solid wall perpendicular to the x-axis and located at the given value $x = x_w$ (see Fig. 1). Later x_w will be regarded as a function of time; since the speed of light has been effectively taken as infinite, however, this is of no relevance in the treatment of the radiative transfer itself. To evaluate the heat-addition term as a function of x we employ equation (3.6b). To carry out the integration over solid angle in that equation, consider the intensity I at a given (i.e., temporarily fixed) location x and in a direction defined by the angles ϕ and θ as shown in the figure. The element of solid angle $d\Omega$ in equation (3.6b) is then $d\Omega = \sin\phi \, d\phi \, d\theta$. This can also be written $d\Omega = -dl \, d\theta$, where $l = \cos\phi$ is here the direction cosine, relative to the x-axis, of the direction of propagation in question. In the one-dimensional case, the value of I at a given point x and in a given direction ϕ is independent of θ. We can therefore write

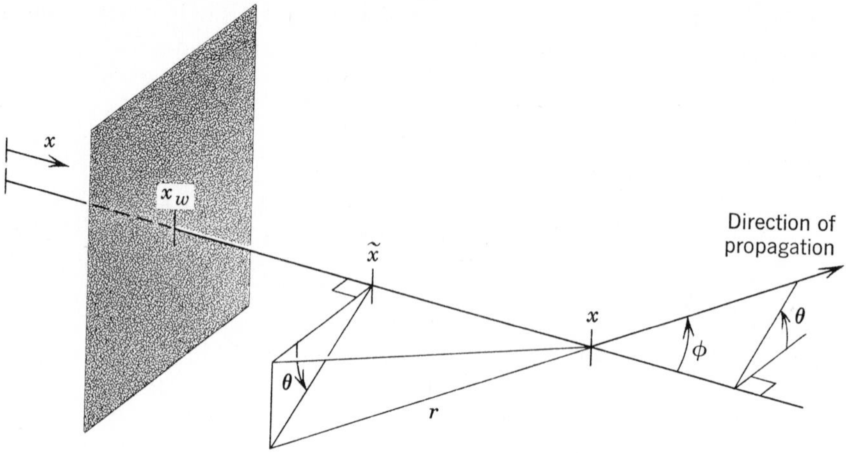

Fig. 1. Coordinate system.

$I = I(x, \phi) = I(x, l)$. Putting these expressions into equation (3.6b) and carrying out the integration with respect to θ from 0 to 2π, we obtain

$$-\frac{\partial q^R}{\partial x} = \alpha\left(-2\pi\int_{+1}^{-1} I(x, l)\, dl - 4\sigma T^4\right), \tag{4.1}$$

where q^R is the magnitude of the heat-flux vector, which is now entirely in the x-direction.

To obtain $I(x, l)$, we utilize equation (3.7b) and express the radial variable r as a function of x and l. If $\tilde{x}$ is a running (i.e., variable) value of x, the necessary relation is given by $r = (x - \tilde{x})/\cos\phi = (x - \tilde{x})/l$. The element of length dr along the direction specified by a given value of l can thus be expressed in terms of $d\tilde{x}$ by $dr = -d\tilde{x}/l$. The optical depth τ is then given by equation (3.8) as

$$\tau = \int_x^{\tilde{x}} \alpha\left(-\frac{d\hat{x}}{l}\right) = \frac{1}{l}\int_{\tilde{x}}^{x} \alpha\, d\hat{x}.$$

Since α in the one-dimensional situation is a function of the single position variable x, it is convenient to define a quantity

$$\eta = \eta(x) \equiv \int_{x_w}^{x} \alpha\, d\hat{x}. \tag{4.2}$$

This is called the *optical thickness* of the gas between the wall and the station x. In terms of this quantity, the optical depth τ can here be written

$$\tau = \frac{\eta(x) - \eta(\tilde{x})}{l} = \frac{\eta - \tilde{\eta}}{l}. \tag{4.3}$$

The optical depth of a point along the direction opposite to that of I, as measured from the specified point at x, is thus a function of η, $\tilde{\eta}$, and l. In terms of these quantities equation (3.7b) gives $I(x, l)$, or alternatively $I(\eta, l)$, as follows:

$$I(\eta, l) = I(\eta_s, l) \exp\left(-\frac{\eta - \eta_s}{l}\right) - \frac{\sigma}{\pi}\int_{\eta}^{\eta_s} T^4 \exp\left(-\frac{\eta - \tilde{\eta}}{l}\right)\frac{d\tilde{\eta}}{l}, \tag{4.4}$$

where η_s is the particular value of η at which I in the direction l has the boundary value $I(\eta_s, l)$.

We have now merely to substitute expression (4.4) into equation (4.1) and carry out the integration formally with respect to l. Here a distinction is necessary between the ranges $+1$ to 0 and 0 to -1. In the former range the boundary value $I(\eta_s, l)$ is given by the value $I_w(l)$ at the wall, which is located at $\eta_s = \eta(x_w) = 0$. In the latter range, the boundary is at infinity, where $\eta_s = \eta(\infty) = \infty$. We assume here that $I(\infty, l)$ is finite. Dividing the integration into these two ranges, we thus write for the integral in (4.1)

$$\int_{+1}^{-1} I(\eta, l)\, dl = \int_{+1}^{0}\left[I_w(l)e^{-\eta/l} - \frac{\sigma}{\pi}\int_{\eta}^{0} T^4 \exp\left(-\frac{\eta - \tilde{\eta}}{l}\right)\frac{d\tilde{\eta}}{l}\right] dl$$
$$+ \int_{0}^{-1}\left[-\frac{\sigma}{\pi}\int_{\eta}^{\infty} T^4 \exp\left(-\frac{\eta - \tilde{\eta}}{l}\right)\frac{d\tilde{\eta}}{l}\right] dl.$$

For simplicity we assume that the wall radiates as a black body, that is, that it has an emissivity $e_\nu = 1$ (see Chapter XI, Sec. 5). With this assumption I_w is independent of l and has the value $I_w = \int_0^\infty B_\nu(T_w)\, d\nu = (\sigma/\pi)T_w^4$, where T_w is the temperature of the wall.[1] Introducing this value into the foregoing relation and interchanging the order of integration in the double integrals, as well as the order of the limits in certain of the integrals themselves, we write

$$\int_{+1}^{-1} I(\eta, l)\, dl = -\frac{\sigma}{\pi}\left[T_w^4 \int_0^1 e^{-\eta/l}\, dl + \int_0^\eta T^4 \int_0^1 \exp\left(-\frac{\eta - \tilde{\eta}}{l}\right)\frac{dl}{l}\, d\tilde{\eta}\right.$$
$$\left. + \int_\eta^\infty T^4 \int_0^1 \exp\left(-\frac{\tilde{\eta} - \eta}{l}\right)\frac{dl}{l}\, d\tilde{\eta}\right]. \tag{4.5}$$

To arrive at the third term on the right, the previous integration over l from 0 to -1 has also been replaced formally by an integration over $-l$ from 0 to $+1$. In the present one-dimensional case, T is a function only of x or alternatively of η, which is the reason it can be taken outside the integration with respect to l. The integrals with respect to l that appear

[1] For treatment of a nonblack body, including the effects of both diffuse and specular reflection, see Vincenti and Baldwin (1962).

here are particular cases of the *integro-exponential function* of order n, denoted by $E_n(z)$ and defined by

$$E_n(z) = \int_0^1 l^{n-2} e^{-z/l}\, dl. \tag{4.6}$$

We are concerned here only with positive integral values of n. The properties of this function are discussed and numerical values tabulated in the book by Kourganoff (1952), pp. 253 and 266. The following properties will be of use to us:

$$\frac{dE_n(z)}{dz} = -E_{n-1}(z), \tag{4.7a}$$

$$E_n(0) = \frac{1}{n-1}, \tag{4.7b}$$

$$E_n(z) \to e^{-z}/z \qquad \text{as } z \to \infty. \tag{4.7c}$$

Writing expression (4.5) in terms of the E_n-functions and putting the result into (4.1), we obtain finally the following expression for the radiant-transfer term in equation (2.3) in the one-dimensional case:

$$-\frac{\partial q^R}{\partial x} = 2\alpha\sigma\left[T_w^4 E_2(\eta) + \int_0^\eta T^4 E_1(\eta - \tilde{\eta})\, d\tilde{\eta} + \int_\eta^\infty T^4 E_1(\tilde{\eta} - \eta)\, d\tilde{\eta} - 2T^4\right]. \tag{4.8a}$$

The first term on the right represents the net energy input to an element of gas at a point η as a result of emission from the wall; the next two terms represent the net input as a result of radiation from other elements of the gas, the first from elements to the left of η and the second from elements to the right; the last term represents the heat loss from the element by spontaneously emitted radiation.

Relation (4.8a) can be put into a useful alternative form by integrating each integral once by parts. In this, use is made of the properties (4.7) of the E_n-functions. Under the condition that T is a continuous function of η (cf. Exercise 4.2), we obtain finally

$$-\frac{\partial q^R}{\partial x} = 2\alpha\sigma\left[(T_w^4 - T_{x=x_w}^4)E_2(\eta) - 4\int_0^\eta T^3 \left.\frac{\partial T}{\partial \eta}\right|_{\tilde{\eta}} E_2(\eta - \tilde{\eta})\, d\tilde{\eta} \right.$$
$$\left. + 4\int_\eta^\infty T^3 \left.\frac{\partial T}{\partial \eta}\right|_{\tilde{\eta}} E_2(\tilde{\eta} - \eta)\, d\tilde{\eta}\right]. \tag{4.8b}$$

Here $T_{x=x_w}$ is the temperature of the gas immediately adjacent to the wall, *which need not be the same as that of the wall in the assumed absence of molecular transport processes.* The partial derivative $\partial T/\partial\eta|_{\tilde{\eta}}$ is used instead of an ordinary derivative in the integral terms to indicate that the time is held fixed in taking the derivative; the subscript $\tilde{\eta}$ is added to emphasize that the derivative is to be evaluated as a function of the running variable of integration $\tilde{\eta}$. Equation (4.8b) shows clearly that the net radiative transfer vanishes when T is everywhere the same.

The expressions (4.8) are considerably simpler than the more general equations (3.6b) and (3.7b) that apply to a multidimensional problem. Expressions of this type have, in fact, been applied extensively in astrophysics in the study of the classical problem of the plane-parallel stellar atmosphere. Even in the one-dimensional form, however, the expressions are far from simple, involving as they do integrals over the entire field. Their successful use in astrophysics stems in part from the fact that in such application the motion of the gas may be neglected, and solid boundaries do not enter the problem. In gas dynamics, however, the coupling with the fluid motion as represented by the differential terms in the conservation equations is essential, and the presence of solid boundaries is unavoidable. For these reasons, further simplification is desirable even in the one-dimensional case.

A simplification that has been used to good advantage is the so-called "exponential approximation." This consists in arbitrarily replacing the exponential integral $E_2(z)$ by a purely exponential function of the form

$$E_2(z) \cong ae^{-bz}. \tag{4.9}$$

The other E_n-functions are taken correspondingly so as to satisfy the differentiation formula (4.7a), for example, $E_1(z) \cong abe^{-bz}$. The values for the constants a and b are a matter of choice. Matching the area and first moment of the exact and approximate expressions for $E_2(z)$ gives $a = \frac{3}{4}$ and $b = \frac{3}{2}$. Another method that we shall encounter later leads to $a = 1$ and $b = \sqrt{3}$, and still other methods are possible (see Vincenti and Baldwin, 1962). The two foregoing approximations are compared with the exact relation in Fig. 2. We note in passing that the exponential approximation (4.9) does not reproduce the correct infinite derivative of $E_2(z)$ at $z = 0$.

On the basis of the exponential approximation and with certain mathematical manipulations, the integral terms can sometimes be removed entirely in the one-dimensional problem. This reduces the governing equations to purely differential equations, with attendant reduction in the difficulties of solution. The method has the advantage, compared with the emission-dominated and thick-gas approximations of Sec. 3, of being valid for all values of the absorption coefficient α. An example of the technique is given in Sec. 5 in connection with the linearized equations for one-dimensional flow. The method is not limited, however, to linearized

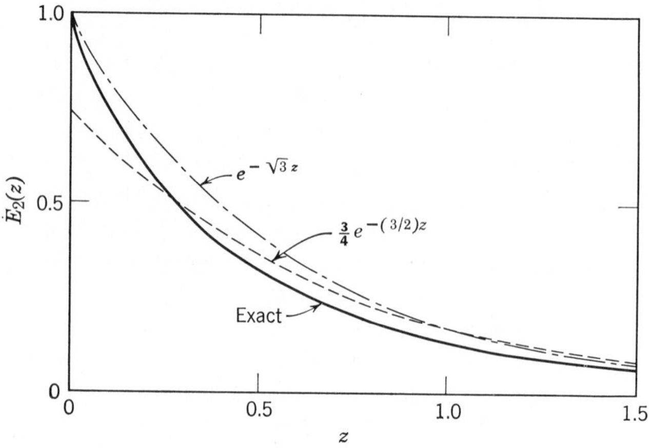

Fig. 2. Comparison of exact and approximate values for $E_2(z)$.

problems. An illustration of its application to the nonlinear equations of the present section is given by Heaslet and Baldwin (1963) in their study of shock structure in the presence of radiation.

Exercise 4.1. Starting from equation (3.6a) and using procedures similar to those of this section, obtain the following expression for the heat-flux vector in the one-dimensional case:

$$q^R = 2\sigma\left[T_w^4 E_3(\eta) + \int_0^\eta T^4 E_2(\eta - \tilde{\eta})\, d\tilde{\eta} - \int_\eta^\infty T^4 E_2(\tilde{\eta} - \eta)\, d\tilde{\eta}\right].$$

Show that differentiation of this formula leads again to the expression (4.8a). [This expression shows, by virtue of the integration, that q^R is a continuous function, even if discontinuities are present in T. Because of the final T^4-term in (4.8a), however, the derivative $\partial q^R/\partial x$ must be discontinuous when T has a discontinuity.]

Exercise 4.2. By suitable division of the range of integration in equation (4.8a), show that when a discontinuity in T (as, for instance, through a shock wave) exists at a location η_D, then equation (4.8b) is replaced by

$$-\frac{\partial q^R}{\partial x} = 2\alpha\sigma\left[(T_w^4 - T_{x=x_w}^4)E_2(\eta) + 4\int_0^\infty T^3 \frac{\partial T}{\partial \eta}\bigg|_{\tilde{\eta}} \operatorname{sgn}(\tilde{\eta} - \eta) E_2(|\tilde{\eta} - \eta|)\, d\tilde{\eta} \right.$$
$$\left. + (T_{\eta=\eta_D+}^4 - T_{\eta=\eta_D-}^4)\operatorname{sgn}(\eta_D - \eta)E_2(|\eta_D - \eta|)\right].$$

Here we have for convenience combined the two integrals into a single term by use of the function sgn(z), which is equal to $+1$ when z is positive and -1 when z is negative. The quantities $T_{\eta=\eta_D+}$ and $T_{\eta=\eta_D-}$ are the temperatures on the positive- and negative-facing sides of the discontinuity, respectively. Show as part of your derivation that this expression is valid for a discontinuity either to the right or left of the location η at which $-\partial q^R/\partial x$ is being evaluated.

5 LINEARIZED ONE-DIMENSIONAL EQUATIONS

Without the exponential approximation, the one-dimensional equations of the preceding section are still of a complicated integro-differential form. It is therefore of interest to see what can be gained by linearization, that is, by going to the corresponding acoustic approximation.

To do this we proceed as in Chapter VIII, Sec. 4. We consider, in particular, disturbances on a uniform gas at rest and write $u = u'$, $p = p_0 + p'$, $T = T_0 + T'$, etc., where subscript 0 denotes the uniform conditions in the undisturbed gas and the prime denotes the disturbances therefrom, which are regarded as small. For the undisturbed gas to be uniform, it must be in radiative equilibrium. This means that the undisturbed heat-flux vector q_0^R must be zero, so that we write $q^R = q^{R'}$. The undisturbed wall temperature T_{w_0} must also be equal to the uniform undisturbed gas temperature T_0.

The linearization of the terms on the left-hand side of the conservation equations proceeds straightforwardly as in Chapter VIII, Sec. 4. We thus find for the linearized equivalent of equations (2.1), (2.2), and (2.3) in the present one-dimensional situation

$$\frac{\partial \rho'}{\partial t} + \rho_0 \frac{\partial u'}{\partial x} = 0, \tag{5.1}$$

$$\rho_0 \frac{\partial u'}{\partial t} + \frac{\partial p'}{\partial x} = 0, \tag{5.2}$$

$$\rho_0 \frac{\partial h'}{\partial t} - \frac{\partial p'}{\partial t} = -\frac{\partial q^{R'}}{\partial x}. \tag{5.3}$$

These are identical to the one-dimensional form of equations (VIII 4.2), (VIII 4.3), and (VIII 4.4), except for the appearance of the radiative-transfer term on the right in (5.3). The linearized form of the state relations (2.4) and (2.5) is

$$h' = \frac{\gamma}{\gamma - 1}\left(\frac{1}{\rho_0} p' - \frac{p_0}{\rho_0^2} \rho'\right), \tag{5.4}$$

$$T' = \frac{1}{R}\left(\frac{1}{\rho_0} p' - \frac{p_0}{\rho_0^2} \rho'\right). \tag{5.5}$$

The linearization of the radiative-transfer term requires somewhat more attention. To begin, we note that the absorption coefficient, being a function of the local state, can also be written in the form $\alpha = \alpha_0 + \alpha'$.

Equation (4.2) for the optical thickness η then becomes

$$\eta(x) = \int_{x_w(t)}^{x} (\alpha_0 + \alpha')\, d\hat{x} = \alpha_0[x - x_w(t)] + \int_{x_w(t)}^{x} \alpha'\, d\hat{x},$$

where explicit recognition has now been taken of the fact that the displacement x_w of the wall from the origin is a function of time. If this displacement is assumed to be small and of the same order as the primed quantities, then η can also be written in the form

$$\eta = \alpha_0 x + \eta', \tag{5.6}$$

where

$$\eta' = -\alpha_0 x_w(t) + \int_{x_w(t)}^{x} \alpha'\, d\hat{x}.$$

The linearization is then most readily carried out on the basis of equation (4.8b). Beginning with the integral from 0 to η, we first expand $E_2(\eta - \tilde{\eta})$ as a Taylor's series with the aid of equation (5.6) as follows:

$$\begin{aligned} E_2(\eta - \tilde{\eta}) &= E_2[\alpha_0(x - \tilde{x}) + (\eta' - \tilde{\eta}')] \\ &= E_2[\alpha_0(x - \tilde{x})] + \left[\frac{dE_2(z)}{dz}\right]_{z=\alpha_0(x-\tilde{x})} \times (\eta' - \tilde{\eta}') + \cdots. \end{aligned} \tag{5.7}$$

Since $\partial T/\partial\eta|_{\tilde{\eta}} = \partial T'/\partial\eta|_{\tilde{\eta}}$, the integral can therefore be approximated to the first order in small quantities as

$$\begin{aligned} \int_0^{\eta} T^3 \left.\frac{\partial T}{\partial \eta}\right|_{\tilde{\eta}} E_2(\eta - \tilde{\eta})\, d\tilde{\eta} &\simeq T_0^3 \int_0^{\eta} \left.\frac{\partial T'}{\partial \eta}\right|_{\tilde{\eta}} E_2[\alpha_0(x - \tilde{x})]\, d\tilde{\eta} \\ &= T_0^3 \int_{x_w(t)}^{x} \left.\frac{\partial T'}{\partial x}\right|_{\tilde{x}} E_2[\alpha_0(x - \tilde{x})]\, d\tilde{x}, \end{aligned}$$

where in the last form we have used the equality

$$\left.\frac{\partial T'}{\partial \eta}\right|_{\tilde{\eta}} d\tilde{\eta} = \left.\frac{\partial T'}{\partial x}\right|_{\tilde{x}} d\tilde{x}.$$

Writing this last form as an integral from 0 to x minus an integral from 0 to $x_w(t)$, and discarding the latter integral as being of second order in small quantities, we obtain finally

$$\int_0^{\eta} T^3 \left.\frac{\partial T}{\partial \eta}\right|_{\tilde{\eta}} E_2(\eta - \tilde{\eta})\, d\tilde{\eta} \simeq T_0^3 \int_0^{x} \left.\frac{\partial T'}{\partial x}\right|_{\tilde{x}} E_2[\alpha_0(x - \tilde{x})]\, d\tilde{x}.$$

The integral from η to ∞ in equation (4.8b) can be treated in similar fashion. To linearize the remaining term we replace $(T_w^4 - T_{x=x_w}^4)$ by its linear approximation $4T_0^3(T_w' - T_{x=0}')$, where conditions at the wall have been transferred to $x = 0$ as is usual in acoustic theory (see Liepmann and Roshko, 1957, p. 207); we also replace $E_2(\eta)$ by $E_2(\alpha_0 x)$ in accord with the expansion (5.7). We thus obtain the linearized form of equation (4.8b) as follows:[2]

$$-\frac{\partial q^{R'}}{\partial x} = 8\alpha_0\sigma T_0^3\left[(T'_w - T'_{x=0})E_2(\alpha_0 x) - \int_0^x \frac{\partial T'}{\partial x}\bigg|_{\tilde{x}} E_2[\alpha_0(x-\tilde{x})]\,d\tilde{x} + \int_x^\infty \frac{\partial T'}{\partial x}\bigg|_{\tilde{x}} E_2[\alpha_0(\tilde{x}-x)]\,d\tilde{x}\right]. \quad (5.8)$$

One difficulty in the foregoing development needs to be mentioned. This arises from the fact that the first and higher derivatives of $E_2(z)$ are infinite at $z = 0$ [see Kourganoff, 1952, and equations (4.7)]. As a result, a series expansion such as that of equation (5.7) is not valid throughout the entire interval of integration, for example, at the limit $\tilde{x} = x$ in the integral considered above. Furthermore, through the effect of this singular behavior in the function $E_2(\eta)$ in equation (4.8b), derivatives of various physical quantities in a more exact treatment became infinite at $\eta = 0$, that is, at the wall (see Baldwin, 1962, p. 35). It follows that difficulties are encountered in justifying the transfer of boundary conditions from the wall to the origin, as was done above. And finally, because of this entire situation, any attempt to extend the foregoing treatment beyond the linear approximation fails completely. A detailed examination of the whole problem (see Baldwin, 1962) shows that a valid systematic expansion procedure can be obtained by transforming from the geometrical co-coordinate x to the optical thickness η in the left-hand side of the conservation equations (2.1) through (2.3). The use of such a noninertial coordinate system greatly complicates the fluid-mechanical terms in the equations but allows the expansion of the radiation term to be handled without difficulty. The details, however, are lengthy and tedious; and, fortunately for present purposes, the final result for the first approximation turns out to be the same as that obtained by the nonrigorous procedure given above. For this reason, a complete treatment will not be given.

We now have in equations (5.1) through (5.5), when supplemented by equation (5.8) for $\partial q^{R'}/\partial x$, a set of five linear equations for the five unknowns ρ', u', p', h', and T'. They can be combined into a single integro-differential equation for a potential function as follows: First the potential function φ is defined, as in equations (VIII 4.14), so that the

[2] Although this result is derived on the basis of equation (4.8b), which assumes a continuous variation of T, the linearized equation (5.8) may be applied without modification in discontinuous situations provided the derivative $\partial T'/\partial x$ at the discontinuity is interpreted as a Dirac delta function.

momentum equation (5.2) is satisfied. We thus have

$$u' = \frac{\partial \varphi}{\partial x}, \qquad p' = -\rho_0 \frac{\partial \varphi}{\partial t}. \tag{5.9}$$

To find the governing equation for φ, we first eliminate h' from the energy equation (5.3) by means of the state relation (5.4) and replace $\partial \rho'/\partial t$ in the resulting equation by means of the continuity equation (5.1). This gives

$$\frac{\partial p'}{\partial t} + \gamma p_0 \frac{\partial u'}{\partial x} = -(\gamma - 1) \frac{\partial q^{R'}}{\partial x}. \tag{5.10}$$

This is the counterpart for the present problem of equation (VIII 4.9). It will be useful to rewrite the coefficients in this relation in terms of the usual *isentropic speed of sound* $a_S \equiv \sqrt{(\partial p/\partial \rho)_s}$, given for the assumed perfect gas by the relation

$$a_S^2 = \gamma \frac{p}{\rho} = \gamma RT. \tag{5.11}$$

After substitution from (5.9), we thus obtain

$$\frac{1}{a_{S_0}^2} \frac{\partial^2 \varphi}{\partial t^2} - \frac{\partial^2 \varphi}{\partial x^2} = \frac{(\gamma - 1)}{\rho_0 a_{S_0}^2} \frac{\partial q^{R'}}{\partial x}. \tag{5.12}$$

To eliminate T', which appears in the expression for $\partial q^{R'}/\partial x$, we first differentiate the state relation (5.5) with respect to t and again replace $\partial \rho'/\partial t$ by means of the continuity equation (5.1). This gives

$$\frac{\partial T'}{\partial t} = \frac{1}{R\rho_0}\left(\frac{\partial p'}{\partial t} + p_0 \frac{\partial u'}{\partial x}\right).$$

Here it is convenient to introduce the *isothermal speed of sound* $a_T \equiv \sqrt{(\partial p/\partial \rho)_T}$, which is given for the perfect gas by

$$a_T^2 = \frac{p}{\rho} = RT. \tag{5.13}$$

Its value is accordingly smaller than that of the isentropic speed. Introducing this quantity and substituting again from (5.9), we have

$$\frac{\partial T'}{\partial t} = -\frac{a_{T_0}^2}{R}\left(\frac{1}{a_{T_0}^2} \frac{\partial^2 \varphi}{\partial t^2} - \frac{\partial^2 \varphi}{\partial x^2}\right). \tag{5.14}$$

The final equation is now obtained by putting expression (5.8) into equation (5.12), differentiating the result with respect to t, and then eliminating

$\partial T'/\partial t$ with the aid of equation (5.14). With the notation

$$A_S \equiv \frac{1}{a_{S_0}^2}\frac{\partial^2 \varphi}{\partial t^2} - \frac{\partial^2 \varphi}{\partial x^2}, \tag{5.15a}$$

$$A_T \equiv \frac{1}{a_{T_0}^2}\frac{\partial^2 \varphi}{\partial t^2} - \frac{\partial^2 \varphi}{\partial x^2}, \tag{5.15b}$$

we obtain the following integro-differential equation for the potential φ:

$$\boxed{\begin{aligned}\frac{\partial A_S}{\partial t} &= \frac{8}{\text{Bo}}\alpha_0 a_{S_0}\Bigg[-\left(\frac{1}{T_0}\frac{dT'_w}{dt} + A_T\big|_{x=0}\right)E_2(\alpha_0 x) \\ &\quad -\int_0^x \frac{\partial A_T}{\partial x}\bigg|_{\tilde{x}} E_2[\alpha_0(x-\tilde{x})]\,d\tilde{x} + \int_x^\infty \frac{\partial A_T}{\partial x}\bigg|_{\tilde{x}} E_2[\alpha_0(\tilde{x}-x)]\,d\tilde{x}\Bigg]\end{aligned}}, \tag{5.16}$$

where the symbol Bo has the meaning

$$\text{Bo} \equiv \frac{\gamma R \rho_0 a_{S_0}}{(\gamma - 1)\sigma T_0^3}. \tag{5.17}$$

The isothermal speed of sound, which plays a natural role in the foregoing analysis, is the speed of sound first proposed by Newton when he initiated acoustic theory. It gives the speed at which an acoustic disturbance must propagate when the changes in the medium are isothermal. Since under ordinary conditions the changes are in fact very close to isentropic (more specifically, reversible adiabatic), it has been superseded in general use by the isentropic speed of sound, first introduced by Laplace. The final acceptance of the latter speed took place about one hundred years ago, but only after a vigorous and bitter controversy in England in the pages of the *Philosophical Magazine* (see, e.g., Stokes, 1851). Now, because of the nonadiabatic effects introduced by our concern with radiation, the isothermal speed again plays a role.

Equation (5.16) has the same place here as does the one-dimensional wave equation in classical acoustic theory. The combination A_S is, in fact, the wave operator from the classical (i.e., isentropic) theory [see equation (VIII 4.16)]; A_T is the classical wave operator except that the isentropic sound speed is replaced by the isothermal sound speed. By containing two wave operators, the equation is reminiscent of equation (VIII 4.15), which appeared in our treatment of flow with reactive nonequilibrium. As discussed at the end of Chapter VIII, Sec. 6, equations containing several wave operators are characteristic of acoustic problems involving dissipative effects.

The dimensionless combination Bo that appears in equation (5.16) is the form, appropriate to a perfect gas, of one of the two dimensionless parameters that govern the combination of inviscid fluid motion and radiative transfer. It is called the *Boltzmann number* and for a nonperfect gas has the general form $\mathrm{Bo} \equiv h_0\rho_0 V_0/\sigma T_0^4$, where h_0 is a characteristic enthalpy and V_0 is a characteristic speed [here equal respectively to $\gamma R T_0/(\gamma - 1)$ and a_{S_0}]. It provides a measure of the relative importance of convective and radiative processes in the flow of energy. The combination $\alpha_0 a_{S_0}$ that also appears in (5.16) has the dimensions of $(\text{time})^{-1}$ and will eventually be compared with a characteristic frequency of motion of the wall. The resulting dimensionless parameter, called the *Bouguer number*, has the general form $\mathrm{Bu} \equiv \alpha_0 L$, where L is a characteristic length (equal to a_{S_0} divided by the characteristic frequency in the present case).[3] Certain properties of acoustic propagation in a radiating gas could be deduced at once from equation (5.16), much as was done from equation (VIII 4.15) for reactive flow. Such considerations, however, will be deferred until later.

Equation (5.16) can be considerably simplified through use of the exponential approximation for the E_2-function. Introducing the approximation from equation (4.9), we first write the equation as

$$\frac{\partial A_S}{\partial t} = \frac{8a}{\mathrm{Bo}}\,\alpha_0 a_{S_0}\Bigg[-\left(\frac{1}{T_0}\frac{dT'_w}{dt} + A_T\Big|_{x=0}\right)e^{-b\alpha_0 x} - \int_0^x \frac{\partial A_T}{\partial x}\bigg|_{\tilde{x}} e^{-b\alpha_0(x-\tilde{x})}\,d\tilde{x} + \int_x^\infty \frac{\partial A_T}{\partial x}\bigg|_{\tilde{x}} e^{-b\alpha_0(\tilde{x}-x)}\,d\tilde{x}\Bigg]. \tag{5.18}$$

We can now reduce this equation to a purely differential form by differentiating twice with respect to x. This gives successively

$$\frac{\partial^2 A_S}{\partial x\,\partial t} = \frac{8a}{\mathrm{Bo}}\,\alpha_0 a_{S_0}\Bigg[b\alpha_0\left(\frac{1}{T_0}\frac{dT'_w}{dt} + A_T\Big|_{x=0}\right)e^{-b\alpha_0 x} + b\alpha_0\int_0^x \frac{\partial A_T}{\partial x}\bigg|_{\tilde{x}} e^{-b\alpha_0(x-\tilde{x})}\,d\tilde{x} + b\alpha_0\int_x^\infty \frac{\partial A_T}{\partial x}\bigg|_{\tilde{x}} e^{-b\alpha_0(\tilde{x}-x)}\,d\tilde{x} - 2\frac{\partial A_T}{\partial x}\Bigg],$$

and

$$\frac{\partial^3 A_S}{\partial x^2\,\partial t} = \frac{8a}{\mathrm{Bo}}\,\alpha_0 a_{S_0}\Bigg[-b^2\alpha_0^2\left(\frac{1}{T_0}\frac{dT'_w}{dt} + A_T\Big|_{x=0}\right)e^{-b\alpha_0 x} - b^2\alpha_0^2\int_0^x \frac{\partial A_T}{\partial x}\bigg|_{\tilde{x}} e^{-b\alpha_0(x-\tilde{x})}\,d\tilde{x} + b^2\alpha_0^2\int_x^\infty \frac{\partial A_T}{\partial x}\bigg|_{\tilde{x}} e^{-b\alpha_0(\tilde{x}-x)}\,d\tilde{x} - 2\frac{\partial^2 A_T}{\partial x^2}\Bigg].$$

[3] For a general discussion of similarity parameters in radiative gas dynamics, see Goulard (1964).

The first three terms on the right in the last equation can be eliminated by substitution from the original equation (5.18). This gives finally

$$\frac{\partial^3 A_S}{\partial t\, \partial x^2} + \frac{16a}{\text{Bo}} \alpha_0 a_{S_0} \frac{\partial^2 A_T}{\partial x^2} - b^2 \alpha_0^2 \frac{\partial A_S}{\partial t} = 0. \tag{5.19}$$

With A_S and A_T as defined by (5.15), this is a linear fifth-order differential equation for the potential φ. Since it is a purely differential equation, it is easier to solve than the more exact integro-differential equation (5.16). The discussion of the equation and its solution will be taken up in Secs. 7 and 8, after we have rederived it by a different approach.

6 DIFFERENTIAL APPROXIMATION

The exponential approximation, although it does lead to a purely differential equation, is restricted to the one-dimensional case. We would like, if possible, to have a method, valid for all values of the absorption coefficient, that will reduce the complete, three-dimensional equations for a grey gas to a purely differential set. Such a method is developed in this section. Like the exponential approximation as originally given, it is not limited to linearized problems but proceeds from the nonlinear equations. As will be seen, the results, when linearized and specialized to the one-dimensional case, reproduce the differential equation (5.19).

The method proceeds by abandoning any attempt to solve the equation of radiative transfer itself. Instead we are content to satisfy only certain moments of the equation. Since the equation of radiative transfer is like the Boltzmann equation, the method is akin to those in which a number of moments of the Boltzmann equation are satisfied in kinetic theory (cf. Chapter X, Secs. 9 and 10). We begin by defining the following moments of the intensity I, where the second equality follows from the equations of Chapter XI, Sec. 2:

$$I_A \equiv \int_0^{4\pi} I \, d\Omega = cu^R, \tag{6.1a}$$

$$I_{B_i} \equiv \int_0^{4\pi} I \, l_i \, d\Omega = q_i^R, \tag{6.1b}$$

$$I_{C_{ij}} \equiv \int_0^{4\pi} I \, l_i l_j \, d\Omega = cp_{ij}^R. \tag{6.1c}$$

The first of our moment equations is obtained by integrating the transfer equation (3.7a) over the entire range of solid angle. Since T is independent of direction, this gives the scalar equation

$$\frac{\partial I_{B_j}}{\partial x_j} = -\alpha(I_A - 4\sigma T^4). \tag{6.2}$$

Except for differences in notation, this merely reproduces equation (3.6b). The second equation is obtained by multiplying the transfer equation through by l_i and then integrating. This gives the new vector equation

$$\frac{\partial I_{C_{ij}}}{\partial x_j} = -\alpha I_{B_i}. \tag{6.3}$$

Even if T were known, these two equations would not be a determinate set, since they provide only four scalar equations for the ten quantities I_A, I_{B_i}, and $I_{C_{ij}}$.[4]

To make the set determinate, we must introduce some assumption. We recall from equation (XI 2.14) that when the radiation at a point is isotropic the radiative pressure and energy density are related by

$$p_{ij}^{R^0} = \tfrac{1}{3}u^{R^0}\,\delta_{ij}.$$

We shall assume that the same relation supplies a satisfactory approximation in a nonisotropic situation, that is, that the radiative pressure and energy density are related *as if* the radiation were isotropic. This quasi-isotropic assumption was first introduced in one-dimensional problems in astrophysics, where it is known as the *Milne-Eddington approximation*. In view of relations (6.1a) and (6.1c), this assumption can be written as

$$I_{C_{ij}} = \tfrac{1}{3}I_A\,\delta_{ij}. \tag{6.4}$$

This relation can be used to eliminate $I_{C_{ij}}$ from equation (6.3). If we also set $I_{B_i} = q_i^R$ in accord with (6.1b), equations (6.2) and (6.3) then become

$$\boxed{\begin{aligned} \frac{\partial q_j^R}{\partial x_j} &= -\alpha(I_A - 4\sigma T^4) \qquad (6.5)\\ \frac{\partial I_A}{\partial x_i} &= -3\alpha q_i^R \qquad (6.6)\end{aligned}}\ .$$

These equations will be used in place of (3.6) and (3.7). Together with the conservation and state equations (2.1) through (2.5), they provide eleven scalar equations for the eleven unknowns ρ, u_i, p, h, q_i^R, T, and I_A. In

[4] In arriving at ten quantities, we have made use of the fact that $I_{C_{ij}} = I_{C_{ji}}$ for $i \neq j$, so that there are only six independent components of $I_{C_{ij}}$.

contrast to the original set of equations, the present, more approximate set is of purely differential form. We note that although u^R and p_{ij}^R were not included in the conservation equations, they are in fact not zero and do play an important part in the foregoing treatment.

The results just obtained can be found in a more systematic manner as the first approximation in the so-called *spherical-harmonic method.* This method was introduced for the solution of the transfer equation in astrophysics and has since been highly developed in neutron-transport theory (see, for example, Davison, 1957, and Weinberg and Wigner, 1958).[5] It proceeds by expanding the intensity $I = I(x_i, l_i)$ in an infinite series of spherical harmonics as follows:

$$I(x_i, l_i) = \sum_{n=0}^{\infty} \sum_{m=-n}^{n} A_n^m(x_i) Y_n^m(l_i). \tag{6.7}$$

Here the Y_n^m's are the spherical harmonics (see, for example, Irving and Mullineux, 1959, p. 194), and the A_n^m's are complex functions that are to be determined. Differential equations for these functions are found by substituting the series (6.7) into the transfer equation (3.7a), multiplying the result by $\bar{Y}_n^m(l_i)$, the complex conjugate of Y_n^m, and then integrating over the complete range of solid angle with due regard for the orthogonality properties of the spherical harmonics (for details, see Cheng, 1965). The resulting infinite set of equations is completely equivalent to the original transfer equation (3.7a). An approximate, finite set, known as the Nth approximation, is obtained by truncating the series (6.7) with the terms $n = N$. It is known from neutron-transport theory that an approximation with an odd value of N is more accurate than the one with the succeeding even value and is therefore to be preferred (see Davison, 1957, p. 127). In the first approximation ($N = 1$ and hence $n = 0$ and 1), there are four functions A_n^m and hence four corresponding differential equations. With the aid of certain relations between the first four A_n^m's and the quantities I_A and q_i^R, these four equations can be rewritten to reproduce equations (6.5) and (6.6). The correspondingly truncated series (6.7) can also be rewritten as

$$I = \frac{1}{4\pi}(I_A + 3q_j^R l_j). \tag{6.8}$$

[5] The transport of neutrons, as in a nuclear reactor, is governed by a transfer equation similar to that used here. In such problems, however, scattering cannot be neglected. The same is also true in many of the problems of stellar structure considered in astrophysics. With the inclusion of scattering, the transfer equation itself is of integro-differential form. In the present context the integro-differential character arises from the coupling of a differential transfer equation with the flow equations.

The reader can satisfy himself that when this approximation is substituted into the definition (6.1c) for $I_{C_{ij}}$, the result reproduces equation (6.4). Our previous results, including the Milne-Eddington approximation itself, are thus contained as the first approximation in the spherical-harmonic method. As can be seen from equation (6.8), they correspond to a certain simplification regarding the directional dependence of I. We may conjecture, therefore, that they will give integrated quantities such as I_A and q_i^R with better accuracy than they do the directional dependence of I itself.[6]

Equation (6.8) was not required in our original derivation of equations (6.5) and (6.6). It is needed, however, for the solution of those equations when the boundary conditions are given in terms of variables other than those appearing in the equations themselves (see Cheng, 1965; also Exercise 6.1).

It is of interest to examine the asymptotic form of equations (6.5) and (6.6) for the emission-dominated and thick-gas situations. To do this we introduce dimensionless variables as follows:

$$\bar{x}_i = \frac{x_i}{L}, \qquad \bar{\alpha} = \frac{\alpha}{\alpha_r}, \qquad \bar{T} = \frac{T}{T_r}, \qquad \bar{I}_A = \frac{I_A}{\sigma T_r^4}, \qquad \bar{q}_i^R = \frac{q_i^R}{\sigma T_r^4},$$

where L is a characteristic length in the flow field and α_r and T_r are characteristic reference values of α and T. In terms of these variables, equations (6.5) and (6.6) become

$$\frac{\partial \bar{q}_j^R}{\partial \bar{x}_j} = -(\alpha_r L)\bar{\alpha}(\bar{I}_A - 4\bar{T}^4), \tag{6.9}$$

$$\frac{\partial \bar{I}_A}{\partial \bar{x}_i} = -(\alpha_r L)3\bar{\alpha}\bar{q}_i^R. \tag{6.10}$$

We assume that $\bar{x}_i$ is of order one, so that $\partial \bar{q}_j^R/\partial \bar{x}_j$ is of the same order as $\bar{q}_i^R$ and similarly for $\partial \bar{I}_A/\partial \bar{x}_i$ and $\bar{I}_A$.

The emission-dominated form of the approximation is found by requiring that $\alpha_r L \ll 1$ in accord with equation (3.2). It follows from (6.10) that $\bar{I}_A$ is then of an order smaller than $\bar{q}_i^R$. It is therefore negligible in equation (6.9) in comparison with the other terms, and that equation becomes

$$\frac{\partial \bar{q}_j^R}{\partial \bar{x}_j} = 4(\alpha_r L)\bar{\alpha}\bar{T}^4.$$

[6] For an essentially similar procedure applied directly to the one-dimensional case and carried to the approximation $N = 3$, see Traugott (1963).

Converting back to dimensional variables, we obtain

$$-\frac{\partial q_j^R}{\partial x_j} = -4\alpha\sigma T^4. \tag{6.11}$$

This is identical to the emission-dominated equation (3.3) if α is taken as the Planck mean.

The thick-gas case is found by taking $\alpha_r L \gg 1$ in conformity with equation (3.4). Now it follows from (6.10) that $\bar{I}_A$ is of an order larger than $\bar{q}_i^R$. The term on the left in equation (6.9) is therefore negligible in comparison with the other terms, and that equation gives

$$\bar{I}_A = 4\bar{T}^4.$$

Substituting this result into (6.10) we obtain

$$16\bar{T}^3\frac{\partial \bar{T}}{\partial \bar{x}_i} = -(\alpha_r L)3\bar{\alpha}\bar{q}_i^R,$$

or, in terms of dimensional variables,

$$q_i^R = -\frac{16\sigma T^3}{3\alpha}\frac{\partial T}{\partial x_i}. \tag{6.12}$$

This is identical to the correct thick-gas equation (3.5) if α is taken as the Rosseland mean.

Thus, with the appropriate identification of α, the present differential approximation reproduces the asymptotic formulas known to be correct from the exact equations. Solutions based on the approximation will therefore behave correctly for small and large values of $\alpha_r L$ (provided in the former case that the boundary values of I are sufficiently low that the situation is in fact emission-dominated). These results increase our confidence in the possible accuracy of the method in the intermediate range where $\alpha_r L \cong 1$.

Exercise 6.1. As mentioned in the text, the truncated series (6.8) is useful for establishing the boundary conditions needed for use with the differential approximation of this section. As an example consider a solid planar wall coincident with the x_2,x_3-plane and bounding a semi-infinite expanse of gas lying in the positive x_1-direction. Assume that each point on the wall radiates as a black body with a temperature $T_w(x_2, x_3, t)$, which is a known function of position and time. By matching the frequency-integrated energy flux Q_+^R from the wall as calculated from (a) the true black-body intensity and (b) the truncated series (6.8), obtain the following relation:

$$\sigma T_w^4(x_2, x_3, t) = \tfrac{1}{4}[I_A(0, x_2, x_3, t) + 2q_1^R(0, x_2, x_3, t)].$$

For an appropriate problem, this supplies a boundary condition on the quantities that define the radiative field in the differential approximation (see Exercise 7.2).

7 ACOUSTIC EQUATION

Since they are nonlinear, the three-dimensional equations of radiative flow, even with the differential approximation of the preceding section, are still difficult to solve. We therefore examine again the corresponding linearized equations.

As in Sec. 5, the linearization is made relative to a uniform (and hence equilibrium) gas at rest. The linearized equivalent of the conservation equations (2.1), (2.2), and (2.3), now taken in their full three-dimensional form, is as follows:

$$\frac{\partial \rho'}{\partial t} + \rho_0 \frac{\partial u_j'}{\partial x_j} = 0, \tag{7.1}$$

$$\rho_0 \frac{\partial u_i'}{\partial t} + \frac{\partial p'}{\partial x_i} = 0, \tag{7.2}$$

$$\rho_0 \frac{\partial h'}{\partial t} - \frac{\partial p'}{\partial t} = -\frac{\partial q_j^{R'}}{\partial x_j}. \tag{7.3}$$

The linearized state equations are again given by (5.4) and (5.5). To linearize the differential radiation equations (6.5) and (6.6), we write $q_i^R = q_{0i}^R + q_i^{R'}$ and $I_A = I_{A_0} + I_A'$, where, since the basic flow is in equilibrium, we must have $q_{0i}^R = 0$ and $I_{A_0} \equiv \int I_0 \, d\Omega = \int (\sigma T_0^4/\pi) \, d\Omega = 4\sigma T_0^4$. To the linearized approximation we can also write in equation (6.5) that $T^4 = T_0^4 + 4T_0^3 T'$. With these expressions, the linearized equivalent of equations (6.5) and (6.6) is

$$\frac{\partial q_j^{R'}}{\partial x_j} = -\alpha_0 (I_A' - 16\sigma T_0^3 T'), \tag{7.4}$$

$$\frac{\partial I_A'}{\partial x_i} = -3\alpha_0 q_i^{R'}. \tag{7.5}$$

After elimination of I_A' these equations yield the following second-order equation for $q_i^{R'}$:

$$\frac{\partial}{\partial x_i}\left(\frac{\partial q_j^{R'}}{\partial x_j}\right) - 16\sigma\alpha_0 T_0^3 \frac{\partial T'}{\partial x_i} - 3\alpha_0^2 q_i^{R'} = 0. \tag{7.6}$$

The conservation equations (7.1), (7.2), and (7.3), plus the state equations (5.4) and (5.5) and the radiation equation (7.6), give ten scalar equations for the ten unknowns ρ', u_i', p', h', $q_i^{R'}$, and T'.

As before, the equations can be combined into a single equation for the potential φ, defined as follows to satisfy the momentum equation (7.2):

$$u_i' = \frac{\partial \varphi}{\partial x_i}, \qquad p' = -\rho_0 \frac{\partial \varphi}{\partial t}. \tag{7.7}$$

Proceeding as in Sec. 5, and generalizing the definitions (5.15) to

$$A_S \equiv \frac{1}{a_{S_0}^2}\frac{\partial^2 \varphi}{\partial t^2} - \frac{\partial^2 \varphi}{\partial x_j\, \partial x_j}, \tag{7.8a}$$

$$A_T \equiv \frac{1}{a_{T_0}^2}\frac{\partial^2 \varphi}{\partial t^2} - \frac{\partial^2 \varphi}{\partial x_j\, \partial x_j}, \tag{7.8b}$$

we can obtain the following three-dimensional generalization of equations (5.12) and (5.14):

$$\frac{\partial q_j^{R'}}{\partial x_j} = \frac{\rho_0 a_{S_0}^2}{\gamma - 1} A_S, \tag{7.9}$$

$$\frac{\partial T'}{\partial t} = -\frac{a_{T_0}^2}{R} A_T. \tag{7.10}$$

Equation (7.6) can be put in terms of these derivatives by differentiating it successively with respect to x_i and t. This gives

$$\frac{\partial^3}{\partial t\, \partial x_k\, \partial x_k}\left(\frac{\partial q_j^{R'}}{\partial x_j}\right) - 16\sigma\alpha_0 T_0^3 \frac{\partial^2}{\partial x_k\, \partial x_k}\left(\frac{\partial T'}{\partial t}\right) - 3\alpha_0^2 \frac{\partial}{\partial t}\left(\frac{\partial q_j^{R'}}{\partial x_j}\right) = 0.$$

With substitution from (7.9) and (7.10) we obtain finally

$$\boxed{\frac{\partial^3 A_S}{\partial t\, \partial x_k\, \partial x_k} + \frac{16}{\mathrm{Bo}}\alpha_0 a_{S_0} \frac{\partial^2 A_T}{\partial x_k\, \partial x_k} - 3\alpha_0^2 \frac{\partial A_S}{\partial t} = 0}, \tag{7.11}$$

where Bo is again given by equation (5.17) as

$$\mathrm{Bo} \equiv \frac{\gamma R \rho_0 a_{S_0}}{(\gamma - 1)\sigma T_0^3}. \tag{7.12}$$

When specialized to one dimension, equation (7.11) reproduces equation (5.19) with $a = 1$ and $b = \sqrt{3}$. The exponential approximation for the one-dimensional case, with these values of a and b, is thus contained within the present differential approximation. This fact can also be demonstrated on more general grounds that do not depend on the linearization of equations (6.5) and (6.6) (see Cheng, 1965). A comparison of the exponential approximation with $a = 1$ and $b = \sqrt{3}$ with the E_2-function that appears in the exact treatment of the one-dimensional case was given in Fig. 2.

Equation (7.11) is again analogous to equation (VIII 4.15) for reactive nonequilibrium. As before, certain properties of acoustic propagation with radiative transfer emerge at once from the differential equation. If we set $1/\text{Bo} = 0$ (corresponding formally to $T_0 = 0$ and hence to a completely cold gas), the second term disappears, and a solution can be obtained by taking $A_S = 0$. Similarly, with $\alpha_0 = 0$ (corresponding to a transparent gas), the second and third terms disappear, and a solution is again given by $A_S = 0$. In both of these cases, therefore, a solution is given by classical acoustic theory with the signal speed a_{S_0}. This is as it should be since, when the gas is either cold or transparent, no nonequilibrium radiative effects occur, and all changes in the medium are isentropic. If we set $\alpha_0 = \infty$ (opaque gas) the first two terms in effect disappear relative to the last term. Thus we can again obtain a solution by taking $A_S = 0$, corresponding once more to classical theory with the isentropic signal speed a_{S_0}. This is reasonable because, even though there may be intense radiation, it is immediately reabsorbed at its origin by the opaque gas and no radiative transfer takes place. In the remaining extreme case of $1/\text{Bo} = \infty$ (infinitely hot gas), the second term predominates relative to the first and third terms, and the solution is found by taking $A_T = 0$. We thus obtain in this case classical acoustic theory but with the isothermal signal speed a_{T_0}. This comes about because infinite temperatures lead to an infinite rate of heat transfer due to radiation, and this prevents any temperature differences from occurring in the gas.[7] The foregoing conclusions can also be deduced from the more exact integro-differential equation (5.16) (see Vincenti and Baldwin, 1962). They will also appear as limits in the solution of equation (7.11), with which we concern ourselves in the following section.

Exercise 7.1. Obtain the following equations relating the potential φ to the quantities I'_A and $q_i^{R'}$ that describe the disturbed radiative field in the present approximation:

$$\frac{\partial I'_A}{\partial t} = -\frac{\rho_0 a_{S_0}^2}{(\gamma - 1)\alpha_0}\frac{\partial A_S}{\partial t} - \frac{16\sigma T_0^3 a_{T_0}^2}{R} A_T,$$

$$\frac{\partial q_i^{R'}}{\partial t} = \frac{\rho_0 a_{S_0}^2}{3(\gamma - 1)\alpha_0^2}\frac{\partial^2 A_S}{\partial x_i \partial t} + \frac{16\sigma T_0^3 a_{T_0}^2}{3R\alpha_0}\frac{\partial A_T}{\partial x_i},$$

where A_S and A_T are given in terms of φ by equations (7.8).

[7] These limits are to be understood within the restrictions imposed by our original neglect of radiative pressure and energy density. At sufficiently high temperatures these quantities will no longer be negligible, and the values of $a_S \equiv (\partial p/\partial \rho)_s^{1/2}$ and $a_T \equiv (\partial p/\partial \rho)_T^{1/2}$ will be altered accordingly, even for the assumed perfect gas (see Sachs, 1946).

Exercise 7.2. By linearization of the relation in Exercise 6.1 and application of the expressions obtained in the foregoing exercise, derive the following equation:

$$\frac{1}{T_0}\frac{\partial T'_w}{\partial t}\bigg|_{x_2,x_3,t} = \frac{\text{Bo}}{16\alpha_0 a_{S_0}}\frac{\partial}{\partial t}\left(\frac{2}{3\alpha_0}\frac{\partial A_S}{\partial x_1} - A_S\right)\bigg|_{0,x_2,x_3,t} + \left(\frac{2}{3\alpha_0}\frac{\partial A_T}{\partial x_1} - A_T\right)\bigg|_{0,x_2,x_3,t}.$$

In the linearized differential approximation, this provides the radiative boundary condition imposed on φ by a planar black wall with given temperature distribution located at $x_1 = 0$.

8 PROPAGATION OF PLANE ACOUSTIC WAVES

To study the solution of equation (7.11) we shall examine the propagation of plane waves along a single space variable x. With the subscript notation for partial differentiation, equation (7.11) for this case can be written

$$\left(\frac{1}{a_{S_0}^2}\varphi_{tt} - \varphi_{xx}\right)_{txx} + \frac{16}{\text{Bo}}\alpha_0 a_{S_0}\left(\frac{1}{a_{T_0}^2}\varphi_{tt} - \varphi_{xx}\right)_{xx} - 3\alpha_0^2\left(\frac{1}{a_{S_0}^2}\varphi_{tt} - \varphi_{xx}\right)_t = 0. \tag{8.1}$$

As in Chapter VIII, Sec. 6, we restrict ourselves to harmonic waves; the solution then proceeds along much the same lines as before. To simplify matters we do not here associate the waves with any given boundary conditions. We therefore work directly with equation (8.1) rather than with the equation in terms of u' as was done previously.

Proceeding as in equation (VIII 6.8), we assume a solution of the form

$$\varphi(x, t) = f\left(\frac{\nu x}{a_{S_0}}\right)e^{i\nu t}, \tag{8.2}$$

where ν is the radian frequency of the waves and $f(\nu x/a_{S_0})$ is a complex function to be determined. Since we do not concern ourselves with any given boundary conditions, we do not for the present prefer either the real or the imaginary part of the solution. We also do not define dimensionless variables here, since little would be gained for present purposes. Substitution of the assumed solution into equation (8.1) leads to the following fourth-order ordinary differential equation for f:

$$\left[i + 16\frac{\text{Bu}}{\text{Bo}}\right]f^{iv} + \left[i(1 - 3\text{Bu}^2) + 16\gamma\frac{\text{Bu}}{\text{Bo}}\right]f'' - i3\text{Bu}^2 f = 0, \tag{8.3}$$

where the superscript denotes differentiation of f with respect to its argument and Bu is the Bouguer number already mentioned in Sec. 5:

$$\mathrm{Bu} \equiv \frac{\alpha_0 a_{S_0}}{\nu}. \tag{8.4}$$

As in equation (VIII 6.10), a solution of (8.3) is given by

$$f = C \exp\left(c \frac{\nu x}{a_{S_0}}\right), \tag{8.5}$$

where C and c are complex constants. Substitution of (8.5) into (8.3) and multiplication of the result by $-i$ gives the following fourth-order algebraic equation for c:

$$\left(1 - i16 \frac{\mathrm{Bu}}{\mathrm{Bo}}\right) c^4 + \left(1 - 3\mathrm{Bu}^2 - i16\gamma \frac{\mathrm{Bu}}{\mathrm{Bo}}\right) c^2 - 3\mathrm{Bu}^2 = 0. \tag{8.6}$$

Equation (8.6) will have four complex roots for c. Since c appears in the equation only in the combination c^2, these will occur in pairs, each consisting of equal positive and negative values. Denoting these pairs by c_1 and c_2, we can therefore represent the roots in terms of real quantities δ and λ by

$$c_1 = \pm(\delta_1 + i\lambda_1), \qquad c_2 = \pm(\delta_2 + i\lambda_2), \tag{8.7}$$

where the δ's are by definition positive. The solution of equation (8.6) for the δ's and λ's is lengthy and tedious, and we shall be content to exhibit numerical results below. As in Chapter VIII, Sec. 6, the λ's also turn out to be positive.

In terms of the four roots (8.7), the solution (8.2) and (8.5) can be written

$$\begin{aligned}\varphi = {} & C_1^+ \exp\left[-\delta_1 \frac{\nu x}{a_{S_0}} + i\nu\left(t - \lambda_1 \frac{x}{a_{S_0}}\right)\right] \\ & + C_2^+ \exp\left[-\delta_2 \frac{\nu x}{a_{S_0}} + i\nu\left(t - \lambda_2 \frac{x}{a_{S_0}}\right)\right] \\ & + C_1^- \exp\left[\delta_1 \frac{\nu x}{a_{S_0}} + i\nu\left(t + \lambda_1 \frac{x}{a_{S_0}}\right)\right] \\ & + C_2^- \exp\left[\delta_2 \frac{\nu x}{a_{S_0}} + i\nu\left(t + \lambda_2 \frac{x}{a_{S_0}}\right)\right].\end{aligned} \tag{8.8}$$

As in Chapter VIII, Sec. 6, the C^+- and C^--terms represent damped harmonic waves traveling in the positive and negative directions respectively. Here, however, there are *two* types of waves possible in each direction, corresponding to subscripts 1 and 2. The damping of each type of wave increases with the corresponding value of δ. The speed of the wave is determined by the value of λ and is given by

$$V = \frac{a_{S_0}}{\lambda}. \tag{8.9}$$

The value of the constants C and the final choice of the real or imaginary part of the solution are determined in each case by the boundary conditions. As can be seen from equation (8.6), the values of δ and λ are

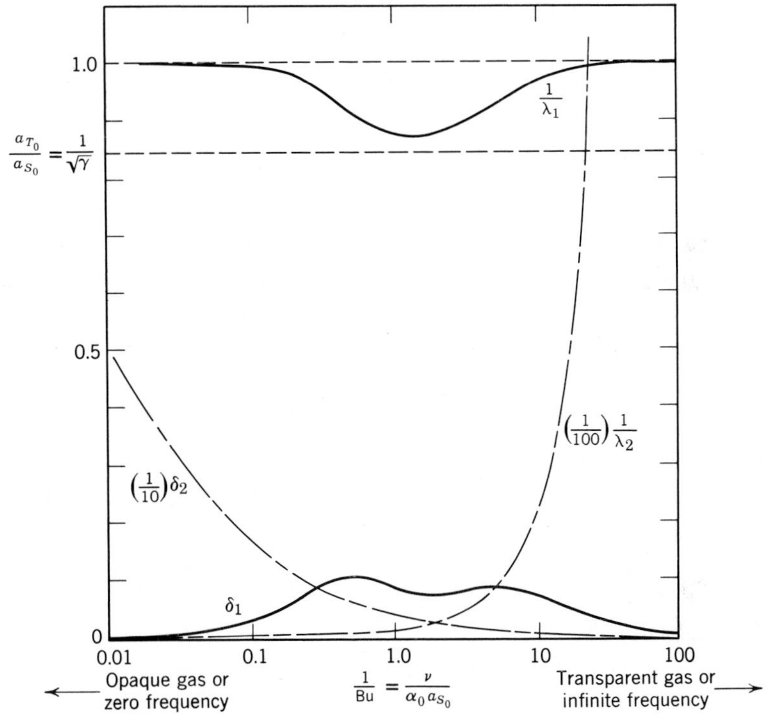

Fig. 3. Variation of δ and $1/\lambda$ with 1/Bu for Bo = 3.23, $\gamma = \frac{7}{5}$.

functions of the three dimensionless quantities Bu = $\alpha_0 a_{S_0}/\nu$, Bo = $\gamma R \rho_0 a_{S_0}/(\gamma - 1)\sigma T_0^3$, and γ. Typical values of δ and $1/\lambda$, obtained by electronic-computer calculations based on equations (8.6) and (8.7), are shown in Fig. 3 as functions of 1/Bu for $\gamma = \frac{7}{5}$ and Bo = 3.23.[8]

We look first at the results for the C_1-wave in equation (8.8). From the values of δ_1 and $1/\lambda_1$ in Fig. 3 it appears that this wave is much like the acoustic wave encountered in the study of chemical and vibrational

[8] The authors are indebted to Dr. Barrett S. Baldwin, Jr. of the Ames Research Center, NASA, for supplying these values.

nonequilibrium (cf. Chapter VIII, Fig. 5). As before, the wave speed and damping are weakly varying functions of the independent variable. For $\alpha_0 = 0$ (transparent gas) and $\alpha_0 = \infty$ (opaque gas), the speed of the wave is identically equal to a_{S_0} and the damping is zero. This is as was anticipated in the discussion of equation (7.11), where the physical reasons were explained. For intermediate values of α_0, the wave speed is somewhat smaller than a_{S_0}, approaching more or less toward the isothermal sound speed $a_{T_0} = a_{S_0}/\sqrt{\gamma}$. This decrease is accompanied by a small amount of damping. As in Chapter VIII, Sec. 6, these results can also be regarded as depending on the frequency ν. From this point of view it follows that C_1-waves of both vanishingly small and infinitely large frequency travel at the isentropic speed a_{S_0}; for intermediate frequencies the speed is somewhat less. The situation at the lower limit is in contrast to that observed in Chapter VIII, where the wave speed did *not* return to the higher of the two fundamental speeds as the frequency vanished. The damping at the low-frequency limit is obviously zero, but as in the earlier case it must be carefully considered in the high-frequency limit. Again it turns out to be non-zero. Since the speed of the C_1-wave never differs appreciably from that given by classical acoustic theory, it can be called the *modified classical wave.*

The C_2-wave has no counterpart either in classical theory or in the acoustic theory for chemical and vibrational nonequilibrium. It arises out of the fact that the governing differential equation is here of higher order than in the other theories.[9] In contrast to the situation for the modified classical wave, the speed and damping of this *radiation-induced wave* are shown in Fig. 3 to be strongly varying functions of 1/Bu. The speed varies between zero and the speed of light (infinite) as 1/Bu goes from 0 to ∞. The damping factor varies at the same time from ∞ to 0.

Although not shown in Fig. 3, the variation of δ and $1/\lambda$ with Bo, like that with Bu, is small for the C_1-wave and large for the C_2-wave. For the details of these matters, see Vincenti and Baldwin (1962). It suffices here to mention the wave speed for the C_1-wave. At $1/\text{Bo} = 0$ (completely cold gas) the speed of this wave is a_{S_0} (i.e., $1/\lambda_1 = 1$) for all values of 1/Bu. As $1/\text{Bo} \to \infty$ (infinitely hot gas), the speed approaches the lower value of a_{T_0} (i.e., $1/\lambda_1 = 1/\sqrt{\gamma}$). This is true uniformly for all values of 1/Bu except 0 and ∞, where the speed remains fixed at a_{S_0} for all values of Bo. The reader may find it helpful to picture this behavior in terms of an imagined variation of the curve for $1/\lambda_1$ in Fig. 3. A singular behavior in the limits is characteristic of other quantities in addition to $1/\lambda_1$ (see the previous reference).

[9] A secondary, nonequilibrium-induced wave, analogous to the C_2-wave discussed here, is found in acoustic theory for a viscous, heat-conducting gas (see Truesdell, 1953).

As mentioned earlier, the intensity of the C_1- and C_2-waves in any given case—that is, the values of the constants $C_1^\pm$ and $C_2^\pm$—is determined by the boundary conditions. These must now include the temperature variation as well as the motion of the walls. The details are given again by Vincenti and Baldwin (1962) or Cheng (1965); we limit ourselves here to a statement of certain of the results. For the harmonic motion of a constant-temperature wall (cf. Fig. 1), the C_1-wave is found to predominate at all conditions, but the C_2-wave is present to some extent. For a harmonic temperature variation of a motionless wall, the C_2-wave predominates at very high temperatures, but at lower temperatures the two waves are of comparable intensity. We must therefore resist the temptation to say that the C_1-waves are due to wall motion and the C_2-waves to wall-temperature variation. Actually, the two waves are coupled through the radiation of the gas. Any disturbance, even though it may tend primarily to excite one type of wave, must also in general give rise to the other. In the separate limiting cases of $1/\text{Bo} = 0$, $1/\text{Bu} = \infty$, and $1/\text{Bu} = 0$, the C_2-wave turns out to disappear entirely (i.e., $C_2^\pm = 0$). At these three limits we therefore obtain the classical results, in which wall motion produces only a C_1-wave with wave speed a_{S_0} and wall-temperature variation has no effect at all. At the limit $1/\text{Bo} = \infty$, wall motion produces only a C_1-wave with wave speed a_{T_0}. Wall-temperature variation now retains an effect, however, producing a pure C_2-wave with zero damping and infinite speed. This corresponds to a spatially uniform gas whose temperature varies with time. Thus, for an infinitely hot gas the two types of waves remain, but there is an uncoupling of the effects of wall motion and temperature variation.

The response of a semi-infinite expanse of radiating gas to the impulsive motion of a constant-temperature wall has been studied by Baldwin (1962) and Lick (1964). The results are similar to those observed for reactive nonequilibrium (see Chapter VIII, Fig. 6) but modified in a manner consistent with the findings for harmonic waves with radiation. At small time the disturbance consists essentially of a step traveling at the isentropic speed of sound a_{S_0} (see Fig. 4). A Fourier analysis of such a rapid variation for pure wall motion would be expected to yield predominantly C_1-waves of high frequency; the indicated signal is thus consistent with the results of Fig. 3. As time increases, the step decays exponentially, and the entire wave disperses. At large time the step has become insignificant, and the center of the disturbance is found to travel essentially at the lower isothermal speed a_{T_0}. Since Fourier analysis of this dispersed wave would be expected to show predominantly intermediate frequencies, this is again in keeping with Fig. 3. Up to this point the results are much like those shown in Chapter VIII, Fig. 6 for reactive nonequilibrium. In this instance, however, the center of the disturbance does not remain permanently at the lower speed. Instead it shifts back toward the higher isentropic speed as time becomes very large. This is a reflection of the return of the speed of the C_1-waves to the higher isentropic value at very low frequencies, as must predominate in the now highly dispersed wave. As discussed by Lick (1964), these results are consistent with the general findings of Whitham (1959), mentioned at the end of Chapter VIII, Sec. 6.

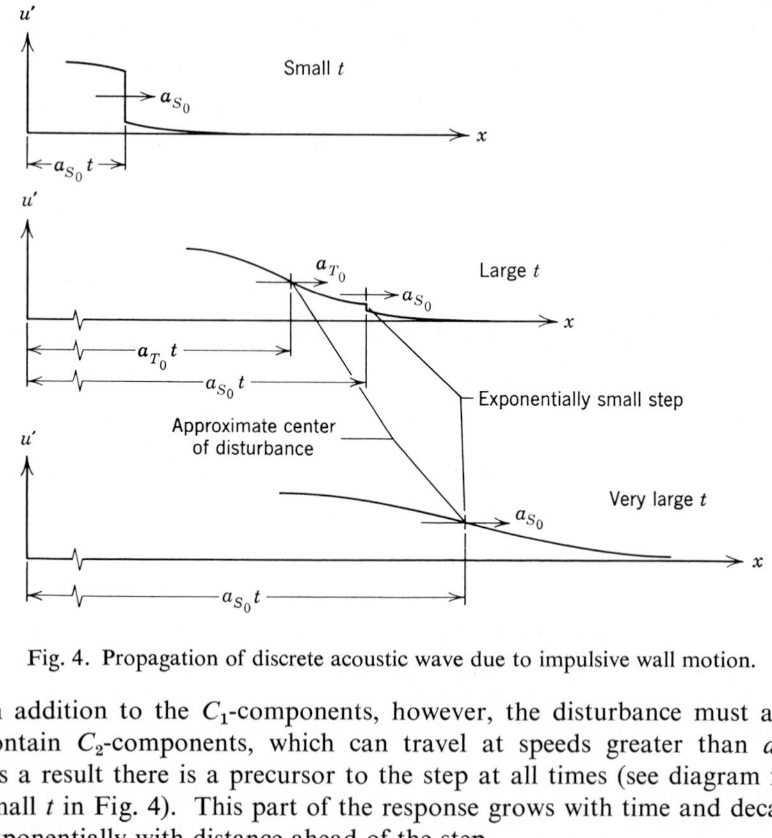

Fig. 4. Propagation of discrete acoustic wave due to impulsive wall motion.

In addition to the C_1-components, however, the disturbance must also contain C_2-components, which can travel at speeds greater than a_{S_0}. As a result there is a precursor to the step at all times (see diagram for small t in Fig. 4). This part of the response grows with time and decays exponentially with distance ahead of the step.

Exercise 8.1. The boundary condition imposed by the temperature of the wall, which is necessary for the determination of the constants C, can be obtained from the result of Exercise 7.2, which was derived on the basis of the condition of Exercise 6.1. In the one-dimensional case, an alternative boundary condition that has been proposed in place of that of Exercise 6.1 is obtained by matching the frequency-integrated intensity from the wall as given by (a) the true black-body value $I_w = \sigma T_w^4/\pi$ and (b) the truncated series (6.8) for a specific direction $l = \cos\phi$. For reasons that we shall not go into here, this specific direction is taken as given by $\cos\phi = \sqrt{3}/3$. The foregoing matching then gives in place of the condition of Exercise 6.1

$$\sigma T_w^4(t) = \tfrac{1}{4}[I_A(0, t) + \sqrt{3}\, q^R(0, t)].$$

This differs from the one-dimensional form of the condition of Exercise 6.1 only by a relatively insignificant difference in the numerical coefficient of the second term on the right. (The foregoing condition, although applicable in the one-dimensional case, is not readily generalized to more than one dimension.)

(a) Show that the present condition leads to the following equation in place of that of Exercise 7.2:

$$\frac{1}{T_0}\frac{dT_w'}{dt} = \frac{\text{Bo}}{16\alpha_0 a_{S_0}}\frac{\partial}{\partial t}\left(\frac{1}{\sqrt{3}\alpha_0}\frac{\partial A_S}{\partial x} - A_S\right) + \left(\frac{1}{\sqrt{3}\alpha_0}\frac{\partial A_T}{\partial x} - A_T\right),$$

where it is understood that the variables are now evaluated as functions of t at $x = 0$. This can be used to provide the radiative boundary condition at a black wall oscillating about $x = 0$.

(b) Show that the same condition as in (a) can be derived from the linearized integro-differential equation (5.18) and subsequent equations. Assume that the constants a and b in the exponential approximation in (5.18) are given the values consistent with the present differential approximation.

This exercise shows that in the formulation based on an integro-differential equation, the radiative boundary condition at the wall is, in effect, included in the governing equation. When we proceed to the differential approximation, however, this boundary condition is lost from the governing equation and must be independently imposed.

9 EQUATION FOR SMALL DEPARTURES FROM A UNIFORM FREE STREAM

We now turn to radiating flows that are a small departure from a uniform stream in radiative equilibrium. As in Chapter VIII, Sec. 7, the necessary governing equation can be obtained by a Galilean transformation of the acoustic equation (7.11).

As before we define a new coordinate system $\tilde{x}_1, \tilde{x}_2, \tilde{x}_3$ in which the undisturbed gas appears as a uniform stream moving with velocity U_∞ in the positive $\tilde{x}_1$-direction. All other quantities in the undisturbed stream are denoted again by subscript $(\ )_\infty$. If we suppose at once that the flow in the new coordinate system is steady (i.e., independent of time), the transformation from the old system fixed in the undisturbed gas is given by equations (VIII 7.4) as

$$\partial(\ )/\partial x_i = \partial(\ )/\partial \tilde{x}_i \quad \text{and} \quad \partial(\ )/\partial t = U_\infty \partial(\ )/\partial \tilde{x}_1.$$

We first transform the differential operators A_S and A_T of equations (7.8). Denoting the transformed operators by L_S and L_T and dropping the tilde notation as superfluous after the transformation, we obtain

$$L_S \equiv (M_{S_\infty}^2 - 1)\frac{\partial^2\varphi}{\partial x_1^2} - \frac{\partial^2\varphi}{\partial x_2^2} - \frac{\partial^2\varphi}{\partial x_3^2}, \tag{9.1a}$$

$$L_T \equiv (M_{T_\infty}^2 - 1)\frac{\partial^2\varphi}{\partial x_1^2} - \frac{\partial^2\varphi}{\partial x_2^2} - \frac{\partial^2\varphi}{\partial x_3^2}, \tag{9.1b}$$

where the isentropic and isothermal Mach numbers are defined by

$$M_{S_\infty} \equiv \frac{U_\infty}{a_{S_\infty}}, \qquad M_{T_\infty} \equiv \frac{U_\infty}{a_{T_\infty}}. \tag{9.2}$$

Applying the transformation to the differential equation (7.11), we then obtain for the governing equation

$$\boxed{\frac{\partial^3 L_S}{\partial x_1\,\partial x_k\,\partial x_k} + \frac{16}{\text{Bo}}\alpha_\infty \frac{\partial^2 L_T}{\partial x_k\,\partial x_k} - 3\alpha_\infty^2 \frac{\partial L_S}{\partial x_1} = 0}\,, \tag{9.3}$$

where the Boltzmann number Bo is now given by

$$\text{Bo} \equiv \frac{\gamma R \rho_\infty U_\infty}{(\gamma - 1)\sigma T_\infty^3}. \tag{9.4}$$

As before, the disturbance velocities u_i' are given in terms of the transformed potential φ by $u_i' = \partial\varphi/\partial x_i$. For later use we shall also need the steady-flow equivalent of equations (7.9) and (7.10). Applying the Galilean transformation as above, we obtain at once

$$\frac{\partial q_j^{R'}}{\partial x_j} = \frac{\rho_\infty a_{S_\infty}^2}{\gamma - 1} L_S, \tag{9.5}$$

$$\frac{\partial T'}{\partial x_1} = -\frac{a_{T_\infty}^2}{R U_\infty} L_T. \tag{9.6}$$

We note an essential difference between equations (7.11) and (9.3) as contrasted with the corresponding equations (VIII 4.15) and (VIII 7.7). If equation (VIII 4.15) is written out for unsteady flow in two space variables x_2 and x_3, we have

$$\overset{+}{\tau}_0 \frac{\partial}{\partial t}\left(\frac{1}{a_{f_0}^2}\frac{\partial^2 \varphi}{\partial t^2} - \frac{\partial^2 \varphi}{\partial x_2^2} - \frac{\partial^2 \varphi}{\partial x_3^2}\right) + \left(\frac{1}{a_{e_0}^2}\frac{\partial^2 \varphi}{\partial t^2} - \frac{\partial^2 \varphi}{\partial x_2^2} - \frac{\partial^2 \varphi}{\partial x_3^2}\right) = 0. \quad (9.7)$$

This may be compared with equation (VIII 7.7), which is, in three space variables,

$$\overset{+}{\tau}_\infty U_\infty \frac{\partial}{\partial x_1}\left[(M_{f_\infty}^2 - 1)\frac{\partial^2 \varphi}{\partial x_1^2} - \frac{\partial^2 \varphi}{\partial x_2^2} - \frac{\partial^2 \varphi}{\partial x_3^2}\right] + (M_{e_\infty}^2 - 1)\frac{\partial^2 \varphi}{\partial x_1^2} - \frac{\partial^2 \varphi}{\partial x_2^2} - \frac{\partial^2 \varphi}{\partial x_3^2} = 0. \quad (9.8)$$

We see that, for positive values of $(M_{f_\infty}^2 - 1)$ and $(M_{e_\infty}^2 - 1)$, the two equations are formally similar, with x_1 in the latter equation taking the place of t in the former. Thus, in the linearized theory of reactive flow, steady three-dimensional supersonic flow is formally equivalent to unsteady two-dimensional flow, with the streamwise coordinate x_1 playing the role of a "time-like" variable. The same result is obviously true for the classical Prandtl-Glauert equation as compared with the classical acoustic equation [equations (VIII 7.10) and (VIII 4.16)]. It is a consequence of the fact that in both the nonequilibrium and classical theories there exists a maximum signal speed at which a disturbance can propagate relative to the fluid. This leads in the well-known way (cf. Chapter VIII, Fig. 8) to limited regions of influence in both space (for supersonic flow) and time.

Now let us try the same comparison for radiative nonequilibrium. Obviously, the same conclusion as above follows for the operators L_S and L_T when compared with A_S and A_T. When we write the equations (7.11) and (9.3) as before, however, we have

$$\frac{\partial}{\partial t}\left(\frac{\partial^2 A_S}{\partial x_2^2} + \frac{\partial^2 A_S}{\partial x_3^2}\right) + \frac{16}{\mathrm{Bo}}\alpha_0 a_{S_0}\left(\frac{\partial^2 A_T}{\partial x_2^2} + \frac{\partial^2 A_T}{\partial x_3^2}\right) - 3\alpha_0^2\frac{\partial A_S}{\partial t} = 0 \quad (9.9)$$

and

$$\frac{\partial}{\partial x_1}\left(\frac{\partial^2 L_S}{\partial x_1^2} + \frac{\partial^2 L_S}{\partial x_2^2} + \frac{\partial^2 L_S}{\partial x_3^2}\right) + \frac{16}{\mathrm{Bo}}\alpha_\infty\left(\frac{\partial^2 L_T}{\partial x_1^2} + \frac{\partial^2 L_T}{\partial x_2^2} + \frac{\partial^2 L_T}{\partial x_3^2}\right) - 3\alpha_\infty^2\frac{\partial L_S}{\partial x_1} = 0. \quad (9.10)$$

These two equations are not formally similar by virtue of the appearance of the derivative $\partial^2/\partial x_1^2$ within the parentheses in the second equation. For radiative nonequilibrium the streamwise coordinate in steady supersonic flow is therefore *not* a time-like variable. This is a consequence of the fact that in radiative flow a signal can propagate at the speed of light, and can therefore move in the negative direction even against a supersonic stream. No signal, however, can ever propagate in a negative direction in time.[10,11]

As in Chapter VIII, Sec. 8, equation (9.3) can be solved to find the radiative flow over a wavy wall. The solution proceeds much as before, the only new, complicating element being the treatment of the radiative boundary condition at the wall. This condition is obtained for a black body by Galilean transformation of the condition of Exercise 7.2, rewritten for a wall lying in the plane $y = x_2 = 0$ (see Cheng, 1965). The final results are more or less as one would infer from our findings of Chapter VIII, Sec. 8 plus the comparison of radiative acoustic theory and the acoustic theory of that chapter. The wave system is now made up of two distinct waves propagating along lines inclined rearward from the vertical. One is a modified classical wave similar to that found for the wavy wall in the reactive flow of Chapter VIII; the other is a radiation-induced wave peculiar to the present problem. The damping and direction of propagation of the two waves are different, depending on the values of the radiative parameters and the Mach number. As in reactive flow, a pressure drag exists on the wall even at subsonic speeds. This is a reflection of the nonequilibrium character of the radiative transfer. In view of the similarity to our previous analyses, we omit the details and proceed instead to the study of the radiating flow through a normal shock wave.

10 LINEARIZED FLOW THROUGH A NORMAL SHOCK WAVE

When the temperature downstream of a shock wave is sufficiently high, the compressed gas at the rear of the wave will radiate. The radiative signal traveling at the speed of light, which is much greater than the wave

[10] For x_1 to be a time-like variable in steady radiative flow, the speed of flow would have to be greater than the speed of light. This relativistically impossible situation has been referred to by Harvard Lomax as "superscenic" flow.

[11] This difference is also reflected in the characteristic directions of equations (9.9) and (9.10). Consideration of the fifth-order terms shows that for equation (9.9) in the one-dimensional case (x_2 only) the five characteristic directions are given by

$$\frac{dt}{dx_2} = \pm \frac{1}{a_{S_0}}, \quad x_2 = \text{constant}, \quad t = \text{constant},$$

where the last direction is doubly covered. The characteristics are thus all real. For equation (9.10) in the two-dimensional case (x_1 and x_2) the characteristic directions are

$$\frac{dx_2}{dx_1} = \pm \frac{1}{\sqrt{M_{S\infty}^2 - 1}}, \quad x_2 = \text{constant}, \quad \frac{dx_2}{dx_1} = \pm i.$$

Two of the characteristic directions are thus imaginary even when $M_{S\infty} > 1$.

speed, then propagates into the region at the front of the wave. The gas flowing into the wave is thus heated by radiation from the gas flowing out of it.

As we have observed, the compression in the viscous portion of a shock wave takes place in a distance of several mean free paths of the molecules. A layer of gas of this thickness is transparent to radiation under almost all conditions. For present purposes the viscous region may therefore be regarded as a discontinuity across which the translation and rotation

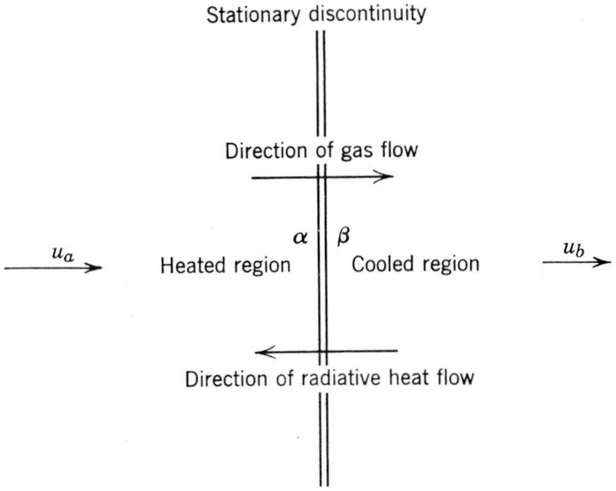

Fig. 5. Radiative flow through a stationary normal shock wave.

of the molecules change instantaneously from one equilibrium condition to another. If other types of molecular nonequilibrium are absent, as we assume for simplicity, this discontinuity is preceded and followed by regions of molecular equilibrium but radiative nonequilibrium. In the first of these the gas gains heat at the expense of radiative loss from the second. This is depicted in Fig. 5, which shows the flow from the viewpoint of an observer fixed relative to the wave. As in Chapter VIII, Sec. 9, we shall sometimes speak of the discontinuity as the shock wave. We can equally well refer to the entire region of change, including the discontinuity and the heated and cooled regions, as the shock wave; from this point of view the discontintuity is then a secondary, imbedded shock. Whatever the terminology, if the flow is steady and no solid surfaces are present, all the net heat lost by the gas in the cooled region must be absorbed by the gas in the heated region. The overall changes from infinity upstream (subscript a) to infinity downstream (subscript b) are therefore adiabatic, despite the nonadiabatic processes between. Conditions at b can therefore be calculated from given conditions at a by the usual adiabatic shock-wave relations (cf. Chapter VI, Sec. 2). We also presume that there are no sources and sinks of radiation at infinity. The conditions at infinity upstream and downstream thus correspond to radiative equilibrium, that is $I_a = (\sigma/\pi)T_a^4$, $I_b = (\sigma/\pi)T_b^4$, and $q_a^R = q_b^R = 0$.

To bring the analysis within the scope of our linearized equations, we assume that the changes in velocity, pressure, etc., in the heated and cooled regions (but not necessarily across the discontinuity) are small. This will be true if the temperature level, although high enough to cause non-negligible flux of energy by radiation, is still low enough that such flux is small compared with that by convection. Mathematically speaking, this means that the reciprocal of the Boltzmann numbers for the flow far upstream and downstream, as defined by

$$\frac{1}{\mathrm{Bo}_a} \equiv \frac{(\gamma - 1)\sigma T_a^3}{\gamma R \rho_a u_a} \quad \text{and} \quad \frac{1}{\mathrm{Bo}_b} \equiv \frac{(\gamma - 1)\sigma T_b^3}{\gamma R \rho_b u_b}, \tag{10.1}$$

must be small compared with unity [cf. equation (5.17) et seq.]. Since $\rho_b u_b = \rho_a u_a$ by conservation of mass and $T_b > T_a$ by the usual shock relations, it will be sufficient if $1/\mathrm{Bo}_b$ is small. Our linearized solution corresponds, in fact, to the first approximation in an expansion in terms of $1/\mathrm{Bo}_b$.

With the foregoing assumption, the flow in the heated and cooled regions can be analyzed on the basis of equation (9.3), which governs small departures from a uniform stream. The appropriate reference stream in each case is the flow far upstream and far downstream, respectively. If we denote the reference flow in general by subscript r, equation (9.3), applied to one-dimensional flow and written in terms of $u' = \partial\varphi/\partial x$, is

$$\frac{d^2}{dx^2}\left(\frac{d^2u'}{dx^2}\right) + \frac{16}{\mathrm{Bo}_r}\alpha_r \frac{\beta_{T_r}^2}{\beta_{S_r}^2}\frac{d}{dx}\left(\frac{d^2u'}{dx^2}\right) - 3\alpha_r^2\frac{d^2u'}{dx^2} = 0, \tag{10.2}$$

where $\beta_{S_r}^2 \equiv M_{S_r}^2 - 1$ and $\beta_{T_r}^2 \equiv M_{T_r}^2 - 1$. This equation plays the same role here as did equation (VIII 9.1) for the reacting flow behind a normal shock wave. Equation (10.2) is a second-order equation for d^2u'/dx^2. It has a solution of the form

$$\frac{d^2u'}{dx^2} = Ce^{c\alpha_r x}, \tag{10.3}$$

where C and c are real constants. Substitution of (10.3) into (10.2) gives the following quadratic equation for c:

$$c^2 + \frac{16}{\mathrm{Bo}_r} \frac{\beta_{T_r}^2}{\beta_{S_r}^2} c - 3 = 0.$$

Solving this equation and expanding the result, we find the following two solutions accurate to the first order in $1/\mathrm{Bo}_r$:

$$\left.\begin{matrix} c_1 \\ c_2 \end{matrix}\right\} = \pm\sqrt{3} - \frac{8}{\mathrm{Bo}_r} \frac{\beta_{T_r}^2}{\beta_{S_r}^2}, \tag{10.4}$$

where the upper sign goes with c_1 and the lower sign with c_2. Since c_1 is positive and c_2 negative, the two resulting solutions (10.3) represent flows with disturbances decaying to zero at $x = -\infty$ and $x = +\infty$, respectively. The first is thus suitable for representing the flow upstream of the shock wave and the second the flow downstream. Taking the shock wave to be located at $x = 0$, we thus write

$$x < 0\colon \quad \frac{d^2u'}{dx^2} = C_1 e^{c_1 \alpha_a x},$$

$$x > 0\colon \quad \frac{d^2u'}{dx^2} = C_2 e^{c_2 \alpha_b x}.$$

Integrating these equations twice and setting the additional constants of integration equal to zero to satisfy the conditions at $\mp\infty$, we obtain, in terms of new constants $D \equiv C/c^2\alpha_r^2$,

$$x < 0\colon \quad u' = D_1 e^{c_1 \alpha_a x}, \tag{10.5a}$$

$$x > 0\colon \quad u' = D_2 e^{c_2 \alpha_b x}. \tag{10.5b}$$

We remember throughout that u' is measured from u_a for $x < 0$ and from u_b for $x > 0$. The existence of two possible exponential solutions to equation (10.2) is in contrast to the situation for equation (VIII 9.1), where only a single solution existed. This is a reflection of the presence in radiative flow of a mechanism for upstream influence that is not present in reactive flow.

To evaluate the constants D_1 and D_2, we use conditions on the radiation through the discontinuity. Since the discontinuity contains no material, there is no absorption or emission within it. The intensity I in any given direction must therefore be the same on both sides, and so must all moments of the intensity. Let us denote the upstream and downstream sides of the discontinuity by α and β. For the moments (6.1a) and (6.1b) that figure in the present radiation approximation we then have in particular

$$I_{A_\alpha} = I_{A_\beta},$$

$$q_\alpha^R = q_\beta^R.$$

To put these in terms of disturbance quantities we write [see material following equation (7.3)] $I_A = I_{A_r} + I'_A = 4\sigma T_r^4 + I'_A$ and $q^R = q^{R'}$. We thus obtain

$$I'_{A_\alpha} - I'_{A_\beta} = 4\sigma(T_b^4 - T_a^4), \tag{10.6a}$$

$$q_\alpha^{R'} = q_\beta^{R'}. \tag{10.6b}$$

These equations provide two conditions on the quantities appearing in our original linearized radiation equations (7.4) and (7.5). They are sufficient for the calculation of the constants D_1 and D_2.

To carry through the calculation we must express the disturbance quantities in terms of u'. This can be done for $q^{R'}$ by writing equation (9.5) for one-dimensional flow and in terms of u'. Integrating from the reference state, where $q^{R'} = 0$ and $u' = 0$, we thus obtain

$$q^{R'} = \frac{\rho_r a_{S_r}^2 \beta_{S_r}^2}{\gamma - 1} u'. \tag{10.7}$$

For I'_A we write equation (7.5) in one-dimension, replace $q^{R'}$ by equation (10.7), and integrate from the reference state. This gives

$$I'_A = -3 \frac{\alpha_r \rho_r a_{S_r}^2 \beta_{S_r}^2}{\gamma - 1} \int_{x_r}^{x} u' \, dx. \tag{10.8}$$

Substituting these expressions into the conditions (10.6), replacing u' by means of (10.5), and performing the necessary integrations from $x_r = \mp\infty$ to $x = 0$, we find the following simultaneous equations for D_1 and D_2:

$$\rho_a a_{S_a}^2 \beta_{S_a}^2 \frac{D_1}{c_1} - \rho_b a_{S_b}^2 \beta_{S_b}^2 \frac{D_2}{c_2} = -\frac{4\sigma}{3}(\gamma - 1)(T_b^4 - T_a^4),$$

$$\rho_a a_{S_a}^2 \beta_{S_a}^2 D_1 - \rho_b a_{S_b}^2 \beta_{S_b}^2 D_2 = 0.$$

Solution of these equations gives

$$D_1 = -\frac{4c_1c_2}{3(c_2 - c_1)} \times \frac{\sigma(\gamma - 1)}{\rho_a a_{S_a}^2 \beta_{S_a}^2}(T_b^4 - T_a^4),$$

$$D_2 = -\frac{4c_1c_2}{3(c_2 - c_1)} \times \frac{\sigma(\gamma - 1)}{\rho_b a_{S_b}^2 \beta_{S_b}^2}(T_b^4 - T_a^4).$$

With the aid of equation (10.4) for c_1 and c_2, and after some algebraic manipulation, these can finally be rewritten to the first order in $1/\text{Bo}_b$ as

$$D_1 = -\frac{2}{\sqrt{3}} \times \frac{u_a(T_b/T_a)}{\text{Bo}_b \beta_{S_a}^2}\left[1 - \left(\frac{T_a}{T_b}\right)^4\right], \tag{10.9a}$$

$$D_2 = -\frac{2}{\sqrt{3}} \times \frac{u_a(u_b/u_a)}{\text{Bo}_b \beta_{S_b}^2}\left[1 - \left(\frac{T_a}{T_b}\right)^4\right]. \tag{10.9b}$$

The final solution for u' is now found by putting these expressions into (10.5) with c_1 and c_2 given by (10.4). The result can be represented as follows, where the symbols are explained below:

$$x \lessgtr 0: \quad \boxed{\frac{u'}{u_a} = \mp \frac{2}{\sqrt{3}} \times \frac{f(M_{S_a})}{\text{Bo}_b}\left[1 - \frac{1}{g(M_{S_a})}\right] e^{\pm\sqrt{3}\,\eta}} . \tag{10.10}$$

The upper signs in this expression apply for $x < 0$, where the corresponding optical variable η is given by $\eta = \alpha_a x$; the lower signs apply for $x > 0$, where the corresponding variable is $\eta = \alpha_b x$. The function f represents the two equal combinations

$$f \equiv \frac{T_b/T_a}{\beta^2_{S_a}} = -\frac{u_b/u_a}{\beta^2_{S_b}} . \tag{10.11a}$$

The equality can be established by expressing the various quantities in terms of the upstream Mach number M_{S_a} with the aid of the standard shock-wave relations for a perfect gas (see Liepmann and Roshko, 1957, p. 59). We thus obtain the common value

$$f(M_{S_a}) = \frac{[2\gamma M^2_{S_a} - (\gamma - 1)][(\gamma - 1)M^2_{S_a} + 2]}{(\gamma + 1)^2 M^2_{S_a}(M^2_{S_a} - 1)} . \tag{10.11b}$$

The function g is defined by

$$g(M_{S_a}) \equiv (M^2_{S_a} - 1)^4 f^4(M_{S_a}). \tag{10.12}$$

In writing equation (10.10), the terms in $1/\text{Bo}_r$ that appear in the exponential when (10.4) is put into (10.5) have been discarded as contributing only second-order quantities to u'/u_a.

The solution for the heat flux $q^{R'}$ can be obtained by substituting (10.10) into the relation (10.7). We thus find, with the aid of the equality (10.11a),

$$x \lessgtr 0: \quad \frac{q^{R'}}{\rho_a u_a^3} = -\frac{2}{\sqrt{3}} \times \frac{(M^2_{S_a} - 1)f(M_{S_a})}{(\gamma - 1)M^2_{S_a}\text{Bo}_b}\left[1 - \frac{1}{g(M_{S_a})}\right] e^{\pm\sqrt{3}\,\eta}. \tag{10.13}$$

The temperature variation is most readily found by writing equation (9.6) for one-dimensional flow in terms of u' and integrating from the reference state. The result can be written

$$T' = -\beta^2_{T_r}\frac{T_r}{u_r}u'.$$

The final expressions are found by substituting for u' from equation (10.10) and using equation (10.11a) plus the shock relations for a perfect gas and the equality $M_T^2 = \gamma M_S^2$. We obtain finally

$$x < 0: \quad \frac{T'}{T_b} = \frac{2}{\sqrt{3}}\left(\frac{\gamma M_{S_a}^2 - 1}{M_{S_a}^2 - 1}\right)\frac{1}{\text{Bo}_b}\left[1 - \frac{1}{g(M_{S_a})}\right]e^{\sqrt{3}\,\eta}, \tag{10.14a}$$

$$x > 0: \quad \frac{T'}{T_b} = \frac{2}{\sqrt{3}}\left(\frac{\gamma(3-\gamma)M_{S_a}^2 - (3\gamma - 1)}{(\gamma+1)(M_{S_a}^2 - 1)}\right)\frac{1}{\text{Bo}_b}\left[1 - \frac{1}{g(M_{S_a})}\right]e^{-\sqrt{3}\,\eta}. \tag{10.14b}$$

Equation (10.10) shows that the velocity disturbance due to radiation is antisymmetric in the optical variable η (although not in the physical variable x, because of the difference between α_a and α_b). To assess the sign of the disturbance, we observe that M_{S_a} is always greater than one and hence from equation (10.11b) that $f(M_{S_a})$ is always positive. It can also be shown that $g(M_{S_a})$ is greater than one, so that the combination $(1 - 1/g)$ is likewise positive. The velocity disturbance is therefore negative ahead of the discontinuity and positive behind it. The heat flux according to equation (10.13) is symmetric in η and is always negative. The temperature disturbance given by equations (10.14) is neither symmetric nor antisymmetric, and its sign depends on the Mach number of the wave as well as the region being considered. Ahead of the discontinuity equation (10.14a) shows it to be positive for all values of M_{S_a}. Behind the discontinuity, however, the coefficient that depends on M_{S_a} in equation (10.14b) changes from negative to positive as M_{S_a} increases through the value $M_{S_a} = [(3\gamma - 1)/\gamma(3-\gamma)]^{1/2}$ (equal to 1.20 for $\gamma = \frac{7}{5}$). The temperature disturbance behind the discontinuity may thus be negative for a sufficiently weak wave but is positive otherwise. The results also show that at a distance $\eta = 1$ from the discontinuity all of the disturbances due to radiation die off to $1/\exp(\sqrt{3}) = 0.177$ of the value at the discontinuity. Where the present approximation is valid, the effects of radiation are thus appreciable over a physical distance x of roughly $1/\alpha_a$ upstream of the discontinuity and $1/\alpha_b$ downstream. All these conclusions will be apparent in the numerical results of the next section.

The dividing value of M_{S_a}, deduced from equation (10.14b) with regard to the downstream temperature effect, corresponds to $M_{T_b} = 1$ and hence to $\beta_{T_b}^2 = 0$ in the earlier relation between T' and u'. For values of M_{S_a} very close to this value, nonlinear effects would be expected to be important. Such effects would also be important for M_{S_a} very close to unity, and hence $\beta_{S_a}^2$ and $\beta_{S_b}^2$ very close to zero. In these situations, the linearized analysis is not valid.

The results of equations (10.10), (10.13), and (10.14) take on a particularly simple form when the static pressure and enthalpy of the flow ahead of the shock wave are negligible compared with the dynamic quantities.

We can attain this limiting situation formally by letting $T_a \to 0$ for fixed values of ρ_a and u_a, which is equivalent to letting $M_{Sa} \equiv u_a/(\gamma R T_a)^{1/2} \to \infty$ for a fixed value of $1/\text{Bo}_b$ (see Exercise 10.1). With the aid of equations (10.11b) and (10.12), we thus obtain

$$x \lessgtr 0: \quad \frac{u'}{u_a} = \mp \frac{4\gamma(\gamma - 1)}{\sqrt{3}(\gamma + 1)^2 \text{Bo}_b} e^{\pm\sqrt{3}\,\eta}, \tag{10.15}$$

$$x \lessgtr 0: \quad \frac{q^{R\prime}}{\rho_a u_a^3} = - \frac{4\gamma}{\sqrt{3}(\gamma + 1)^2 \text{Bo}_b} e^{\pm\sqrt{3}\,\eta}, \tag{10.16}$$

$$x < 0: \quad \frac{T'}{T_b} = \frac{2\gamma}{\sqrt{3}\,\text{Bo}_b} e^{\sqrt{3}\,\eta}, \tag{10.17a}$$

$$x > 0: \quad \frac{T'}{T_b} = \frac{2\gamma(3 - \gamma)}{\sqrt{3}(\gamma + 1)\text{Bo}_b} e^{-\sqrt{3}\,\eta}. \tag{10.17b}$$

Since the ratio $(3 - \gamma)/(\gamma + 1)$ is less than one, the last two equations show that the temperature disturbance upstream of the discontinuity is larger than that downstream.

Exercise 10.1. (a) Show that for $M_{S_a} \gg 1$, expression (10.1) for $1/\text{Bo}_b$ can be approximated by

$$\frac{1}{\text{Bo}_b} \simeq \frac{8\gamma^3(\gamma - 1)^4 \sigma T_a^{5/2}}{(\gamma + 1)^6(\gamma R)^{3/2} \rho_a} M_{S_a}^5 = \frac{8\gamma^3(\gamma - 1)^4 \sigma u_a^5}{(\gamma + 1)^6 (\gamma R)^4 \rho_a}.$$

(b) The first of these expressions shows that for fixed values of T_a and ρ_a, there is a limit to how large M_{S_a} can be without violating our basic assumption that $1/\text{Bo}_b$ is small. Consider air with $T_a = 273°\text{K}$ and ρ_a equal to one-tenth of the standard value of 1.29×10^{-3} gm/cm³. Making the admittedly unrealistic assumption that air behaves at all temperatures as a perfect gas with $\gamma = \frac{7}{5}$, what is the maximum value that M_{S_a} can have without $1/\text{Bo}_b$ exceeding 0.1?

(For a fixed value of T_a as just assumed, $M_{S_a} \equiv u_a/(\gamma R T_a)^{1/2}$ can be made large only by increasing u_a. If, on the other hand, u_a is held fixed, the increase in M_{S_a} must be accomplished by decreasing T_a. If ρ_a is held fixed as before, the second of the above expressions shows that in this latter situation $1/\text{Bo}_b$ can remain at some fixed small value no matter how large the value of M_{S_a}.)

11 NONLINEAR FLOW THROUGH A NORMAL SHOCK WAVE

We conclude with a brief discussion of the more general, nonlinear treatment of radiating flow through a normal shock wave. As already pointed out, the overall changes across the wave from a position of zero radiative flux upstream to a similar position downstream are governed by the usual adiabatic shock-wave relations. These relations, as obtained by the usual considerations of the overall changes with no regard as to how they are accomplished (see Liepmann and Roshko, 1957, p. 56), were listed in Chapter VI, Sec. 2 as follows:

$$\rho_b u_b = \rho_a u_a, \tag{11.1a}$$

$$p_b + \rho_b u_b^2 = p_a + \rho_a u_a^2, \tag{11.1b}$$

$$h_b + \tfrac{1}{2}u_b^2 = h_a + \tfrac{1}{2}u_a^2. \tag{11.1c}$$

The nonlinear conservation equations for the detailed changes within the wave are obtained by specializing equations (2.1), (2.2), and (2.3) to one-dimensional steady flow. This gives the differential equations

$$\frac{d}{dx}(\rho u) = 0, \tag{11.2a}$$

$$\frac{dp}{dx} + \rho u \frac{du}{dx} = 0, \tag{11.2b}$$

$$\rho u \frac{dh}{dx} + \rho u^2 \frac{du}{dx} + \frac{dq^R}{dx} = 0, \tag{11.2c}$$

where the one-dimensional forms of (2.2) and (2.3) have been combined to obtain (11.2c). These equations can be integrated at once in regions of continuous flow to obtain

$$\rho u = \Gamma, \tag{11.3a}$$

$$p + \Gamma u = \Gamma \delta_1, \tag{11.3b}$$

$$\Gamma(h + \tfrac{1}{2}u^2) + q^R = \Gamma \delta_2, \tag{11.3c}$$

where Γ, δ_1, and δ_2 are constants of integration. The constants for the heated and cooled regions of the wave can be evaluated at infinity upstream and downstream respectively, at both of which locations $q^R = 0$. When this is done, it follows from equations (11.1) that the constants are equal for the two regions. Equations (11.3) thus apply at all points in the wave with the same value of the constants, irrespective of the presence of an imbedded discontinuity.

If equations (11.3) are applied from one side to the other of the discontinuity itself (Fig. 5), we obtain in particular

$$\rho_\beta u_\beta = \rho_\alpha u_\alpha, \tag{11.4a}$$

$$p_\beta + \rho_\beta u_\beta^2 = p_\alpha + \rho_\alpha u_\alpha^2, \tag{11.4b}$$

$$\rho_\beta u_\beta(h_\beta + \tfrac{1}{2}u_\beta^2) + q_\beta^R = \rho_\alpha u_\alpha(h_\alpha + \tfrac{1}{2}u_\alpha^2) + q_\alpha^R. \tag{11.4c}$$

Since there can be no absorption or emission within the discontinuity, we have $q_\beta^R = q_\alpha^R$ as in the preceding section, and the last of these equations becomes

$$h_\beta + \tfrac{1}{2}u_\beta^2 = h_\alpha + \tfrac{1}{2}u_\alpha^2. \tag{11.4d}$$

Equations (11.4a, b, and d) that apply across the discontinuity are thus the same as the adiabatic equations (11.1) that apply across the entire wave. This is a result of the fact that the flow across the discontinuity is adiabatic despite the heat flux through it. It is important to note, however, that the sum of the two terms on each side of equation (11.4d) in any given case is *not* equal to the corresponding sum in (11.1c).

The quantitative solution of equations (11.3) must proceed by numerical methods. For this purpose these equations must be supplemented by equations of state $h = h(p, \rho)$ and $T = T(p, \rho)$ and by an equation giving q^R in terms of the distribution of T. The latter equation is provided by the integral expression for the one-dimensional heat flux obtained in Exercise 4.1, but with the wall eliminated and the first integration beginning at $-\infty$ instead of 0. The solution is somewhat involved and is complicated by the fact that we do not really know at the outset whether discontinuities do or do not exist within the wave. This question was first discussed by Zel'dovich (1957) using the exponential approximation (4.9) to reduce the system of equations to purely differential form. The key to the problem is to take as the primary variable of interest the heat flux q^R, which is a continuous function at all times [see material preceding equations (10.6)]. Other variables such as T and u, in which discontinuities may exist, are then regarded as secondary variables expressible in terms of q^R through the conservation and state equations. Using this approach and employing "phase-plane" techniques that take advantage of certain properties of the various quantities, Zel'dovich was able to establish the existence of imbedded discontinuities in strong shock waves and to deduce the qualitative features of the solution. Later the problem was studied independently along similar lines by Heaslet and Baldwin (1963), who give quantitative results for the wave profiles for the complete range of parameters.

For brevity we limit ourselves to a discussion of the results of Heaslet and Baldwin, reproduced in Fig. 6. These results, which were obtained by numerical calculations on an electronic computer, are shown for nine combinations of the following dimensionless parameters used in the original paper:

$$\theta_\infty \equiv \frac{1}{4}\left(\frac{u_a - u_b}{\delta_1}\right)^2,$$

$$\frac{K}{K'} \equiv \frac{16a\gamma^3(\gamma-1)^2\sigma\,\delta_1^6}{\sqrt{2}\,b(\gamma+1)^8\sqrt{\theta_\infty}\,R^4\Gamma}.$$

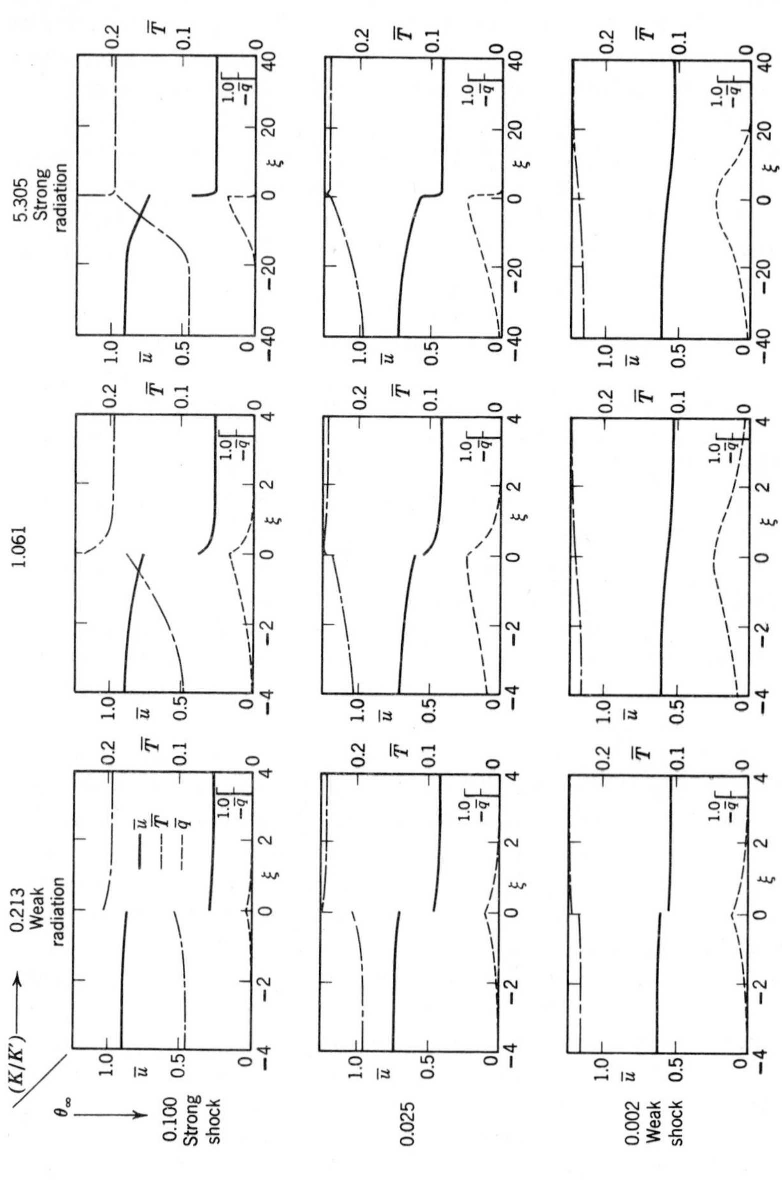

Fig. 6. Shock-wave profiles in a radiating gas for $\gamma = \frac{7}{5}$ (after Heaslet and Baldwin, 1963); ——— $\bar{u}$, — — — — $\bar{T}$, - - - - $\bar{q}$.

In the second of these, a and b are the constants in the exponential approximation. The value of θ_∞ is a measure of the strength of the shock wave. For a given value of θ_∞, the value of K/K' is a measure of the level of radiation. It thus plays a role here akin to that of $1/\text{Bo}_b$ in the preceding section. The dependent variables in each of the plots are dimensionless measures of the velocity, temperature, and heat flux as defined by

$$\bar{u} \equiv \frac{1}{\delta_1} u, \qquad \bar{T} \equiv \frac{R}{\delta_1^2} T, \qquad \bar{q}^R \equiv \frac{2(\gamma - 1)}{(\gamma + 1)\theta_\infty \Gamma\, \delta_1^2} q^R.$$

The dependent variable in each case is an optical length ξ defined by

$$\xi \equiv b \int_0^x \alpha \, d\hat{x}.$$

If the temperature varies appreciably through the wave, the value of α varies and the profile in terms of actual physical distance differs from that shown in the figure. All results shown are for $\gamma = \frac{7}{5}$.

For weak radiation ($K/K' = 0.213$) the results agree with the linearized solution of the preceding section. The velocity disturbance from conditions far upstream and downstream is antisymmetric in the optical variables. The heat flux is symmetric. The temperature disturbance has no symmetry properties; it is usually positive and is generally larger upstream of the discontinuity. The negative temperature disturbance that the linearized theory predicts downstream of a weak shock wave is confirmed by the calculations for $\theta_\infty = 0.002$.

For a strong shock wave ($\theta_\infty = 0.100$), the various disturbances increase in magnitude as the level of radiation increases. The symmetry properties also disappear. As a result, the disturbances for strong radiation ($K/K' = 5.305$) extend over many radiative mean free paths upstream of the discontinuity but are confined to a fraction of a mean free path downstream. If this very narrow downstream region is considered as part of the imbedded shock, that shock is seen to be essentially isothermal. An analytical solution for this situation has been given by Raizer (1957); analytical solutions for strong radiation for all strengths of shock wave are also discussed in the paper by Heaslet and Baldwin.

For the intermediate shock strength ($\theta_\infty = 0.025$) the trends are much the same as for a strong shock up to intermediate levels of radiation ($K/K' = 1.061$). A minor difference occurs in the appearance of a small temperature peak removed a short distance downstream of the discontinuity. For strong radiation, however, a new result is evident in that the discontinuity disappears entirely. It is replaced instead by a narrow region of rapid but continuous variation. Thus, when the radiation is sufficiently intense, a continuous profile can exist purely as a result of the dissipative effects of the radiative nonequilibrium.

For the weak wave ($\theta_\infty = 0.002$) the trends of the forward part of the profile with increasing radiation level are as before. The rear part behaves differently, however, in that it becomes dispersed at the higher radiation levels rather than compressed as was the case for the stronger waves. The change from a discontinuous to a continuous profile occurs at a smaller value of K/K' than it did for the wave of intermediate strength.

The continuous profiles observed here are analogous to the fully dispersed shock wave that was observed to exist for reactive flow in Chapter VIII, Sec. 12. The complicated conditions for the existence of a continuous profile with radiation have been examined in detail by Mitchner and Vinokur (1963). In general, it is found that a continuous profile can exist provided that (a) the upstream Mach number is less than $[(2\gamma - 1)/\gamma(2 - \gamma)]^{1/2}$, and (b) the radiation level is above some value that decreases with decreasing Mach number. If these conditions are satisfied, a viscous shock wave need not be introduced to achieve a transition from supersonic to subsonic speed.

Finally, we may mention that the shock-wave equations have been solved without the exponential approximation by Pearson (1964). The solution proceeds by numerical iteration on the governing integral equation obtained by combining equations (11.3) and the appropriate expression for q^R. The results agree very closely with those obtained with the aid of the exponential approximation by Heaslet and Baldwin. This supports the belief that this approximation—or the completely equivalent differential approximation in three dimensions—is a reliable working tool.

The foregoing discussion has dealt entirely with steady shock waves and with radiative nonequilibrium only. For recent work on the effects of radiation on unsteady shock waves, see Olfe (1964) and Wang (1964). For consideration of the coupled effects of ionizational nonequilibrium see Ferrari and Clarke (1964) and of translational nonequilibrium see Traugott (1965).

Exercise 11.1. Using equations (11.4) and (11.1), show that to the first order in the disturbance quantities used in Sec. 10, the velocity changes across the imbedded shock in a perfect gas must satisfy the linear relation

$$u_b u'_\alpha + u_a u'_\beta = -2\frac{\gamma - 1}{\gamma + 1} \times \frac{q_\beta^{R'}}{\rho_a u_a}.$$

Show that the linearized solution obtained in Sec. 10 does in fact satisfy this jump condition.

References

Baldwin, B. S., Jr, 1962, The Propagation of Plane Acoustic Waves in a Radiating Gas. *NASA*, TR R-138.

Cheng, P., 1965, *Study of the Flow of a Radiating Gas by a Differential Approximation*, Ph.D. Dissertation, Stanford University (available from University Microfilms, Ann Arbor, Mich).

Davison, B., 1957, *Neutron Transport Theory*, Oxford.

Ferrari, C., and J. H. Clarke, 1964, On Photoionization ahead of a Strong Shock Wave; in *Supersonic Flow, Chemical Processes and Radiative Transfer*, ed. by D. B. Olfe and V. Zakkay, Macmillan.

Goulard, R., 1964, Similarity Parameters in Radiation Gas-Dynamics; in *High Temperatures in Aeronautics*, ed. by C. Ferrari, Pergamon.

Heaslet, M. A., and B. S. Baldwin, 1963, Predictions of the Structure of Radiation-Resisted Shock Waves. *Phys. Fluids*, vol. 6, no. 6, p. 781.

Irving, J., and N. Mullineaux, 1959, *Mathematics in Physics and Engineering*, Academic.

Kourganoff, V., 1952, *Basic Methods in Transfer Problems*, Oxford.

Lick, W. J., 1964, The Propagation of Small Disturbances in a Radiating Gas. *J. Fluid Mech.*, vol. 18, pt. 2, p. 274.

Liepmann, H. W., and A. Roshko, 1957, *Elements of Gasdynamics*, Wiley.

Lighthill, M. J., 1960, Dynamics of a Dissociating Gas. Part 2. Quasi-equilibrium Transfer Theory. *J. Fluid Mech.*, vol. 8, pt. 2, p. 161.

Mitchner, M., and M. Vinokur, 1963, Radiation Smoothing of Shocks with and without a Magnetic Field. *Phys. Fluids*, vol. 6, no. 12, p. 1682.

Olfe, D. B., 1964, The Influence of Radiant-Energy Transfer on One-Dimensional Shock Wave Propagation; in *Supersonic Flow, Chemical Processes and Radiative Transfer*, ed. by D. B. Olfe and V. Zakkay, Macmillan.

Pearson, W. E., 1964, On the Direct Solution of the Governing Equations for Radiation-Resisted Shock Waves. *NASA*, TN D-2128.

Prokof'ev, V. A., 1962, The Equations of Relativistic Radiation Hydrodynamics. *Sov. Phys.-Doklady*, vol. 6, no. 10, p. 861.

Raizer, Iu. P., 1957, On the Structure of the Front of Strong Shock Waves in Gases. *Sov. Phys. JETP*, vol. 5, no. 6, p. 1242.

Sachs, R. G., 1946, Some Properties of Very Intense Shock Waves. *Phys. Rev.*, vol. 69, nos. 9 and 10, p. 514.

Sen, H. K., and A. W. Guess, 1957, Radiation Effects in Shock-Wave Structure. *Phys. Rev.*, vol. 108, no. 3, p. 560.

Simon, R., 1963, The Conservation Equations of a Classical Plasma in the Presence of Radiation. *J. Quant. Spectr. and Radiative Transfer*, vol. 3, no. 1, p. 1.

Stokes, G. G., 1851, An Examination of the Possible Effect of the Radiation of Heat on the Propagation of Sound. *Phil. Mag.*, vol. 1, p. 305.

Traugott, S. C., 1963, A Differential Approximation for Radiative Transfer with Application to Normal Shock Structure; in *Proceedings of the 1963 Heat Transfer and Fluid Mechanics Institute*, ed. by A. Roshko et al., Stanford University Press.

Traugott, S. C., 1965, Shock Structure in a Radiating, Heat Conducting, and Viscous Gas. *Phys. Fluids*, vol. 8, no. 5, p. 834.

Truesdell, C., 1953, Precise Theory of the Absorption and Dispersion of Forced Plane Infinitesimal Waves according to the Navier-Stokes Equations. *J. Rat. Mech. Anal.*, vol. 2, p. 643.

Vincenti, W. G., and B. S. Baldwin, Jr., 1962, Effect of Thermal Radiation on the Propagation of Plane Acoustic Waves. *J. Fluid. Mech.*, vol. 12, pt. 3, p. 449.

Wang, K. C., 1964, The 'Piston Problem' with Thermal Radiation. *J. Fluid Mech.*, vol. 20, pt. 3, p. 447.

Weinberg, A. M., and E. P. Wigner, 1958, *The Physical Theory of Neutron Chain Reactors*, University of Chicago Press.

Whitham, G. B., 1959, Some Comments on Wave Propagation and Shock Wave Structure with Application to Magnetohydrodynamics. *Comm. Pure Appl. Math.*, vol. XII, no. 1, p. 113.

Zel'dovich, Ia. B., 1957, Shock Waves of Large Amplitude in Air. *Sov. Phys. JETP*, vol. 5, no. 5, p. 919.

Zhigulev, V. N., Ye. A. Romishevskii, and V. K. Vertushkin, 1963, Role of Radiation in Modern Gasdynamics. *AIAA J.*, vol. 1, no. 6, p. 1473.

Appendix

1 DEFINITE INTEGRALS

A definite integral that appears frequently in kinetic theory and statistical mechanics is the following:

$$I_n(a) \equiv \int_0^\infty x^n e^{-ax^2}\, dx,$$

where $a > 0$ and n is a nonnegative integer. Values of this integral for specific values of n are

$$I_0(a) = \frac{1}{2}\left(\frac{\pi}{a}\right)^{1/2}, \qquad I_1(a) = \frac{1}{2a},$$

$$I_2(a) = \frac{1}{4}\left(\frac{\pi}{a^3}\right)^{1/2}, \qquad I_3(a) = \frac{1}{2a^2},$$

$$I_4(a) = \frac{3}{8}\left(\frac{\pi}{a^5}\right)^{1/2}, \qquad I_5(a) = \frac{1}{a^3}, \text{ etc.}$$

Starting with I_0 and I_1, which will be established below, each integral can be found from the one above it by means of the relation

$$I_{n+2}(a) = -\frac{dI_n}{da},$$

which follows obviously from the defining formula. The integral from $-\infty$ to $+\infty$ is twice the value given above if n is even and zero if n is odd.

To establish $I_1(a)$, we make the substitution $z = ax^2$, $dz = 2ax\, dx$ and obtain

$$I_1(a) \equiv \int_0^\infty xe^{-ax^2}\, dx = \frac{1}{2a}\int_0^\infty e^{-z}\, dz = -\frac{1}{2a}\, e^{-z}\Big|_0^\infty = \frac{1}{2a}.$$

To find $I_0(a) \equiv \int_0^\infty e^{-ax^2}\, dx$, we form

$$I_0^2(a) = \int_0^\infty e^{-ax^2}\, dx \int_0^\infty e^{-ay^2}\, dy = \int_0^\infty \int_0^\infty e^{-a(x^2+y^2)}\, dx\, dy.$$

Transforming to plane polar coordinates, where $x^2 + y^2 = r^2$ and $dx\,dy = r\,dr\,d\theta$, we find

$$I_0^2(a) = \int_0^{\pi/2} d\theta \int_0^{\infty} re^{-ar^2}\,dr = \frac{\pi}{2} I_1(a) = \frac{\pi}{4a}$$

or finally

$$I_0(a) = \frac{1}{2}\left(\frac{\pi}{a}\right)^{1/2}.$$

2 FUNDAMENTAL PHYSICAL CONSTANTS

Values for the fundamental physical constants that appear in the text are listed below in CGS units. The values are taken from a consistent set of values recommended by the Committee on Fundamental Constants of the National Academy of Sciences—National Research Council (see *Physics Today*, vol. 17, no. 2, 1964, p. 48).

Speed of light in vacuum, c	2.997925×10^{10} cm sec^{-1}
Avogadro's number, $\hat{N}$	6.02252×10^{23} mole^{-1}
Planck constant, h	6.6256×10^{-27} erg sec
Universal gas constant, $\hat{R}$	8.3143×10^{7} erg °K^{-1} mole^{-1}
Standard number density for perfect gas, n_0	2.68699×10^{19} cm^{-3}
Boltzmann constant, k	1.38054×10^{-16} erg °K^{-1}
Stefan-Boltzmann constant, σ	5.6697×10^{-5} erg cm^{-2} sec^{-1} °K^{-4}

3 PHYSICAL CONSTANTS FOR CONSTITUENTS OF AIR

Values of the physical constants for certain of the important constituents of high-temperature air, some of which are given in various places in the text, are summarized in the following table:

	O_2	N_2	NO	O	N
Molecular weight, $\hat{M}$, gm mole^{-1}	32.0	28.0	30.0	16.0	14.0
Characteristic temperature for rotation, Θ_r, °K	2.1	2.9	2.5	—	—
Characteristic temperature for vibration, Θ_v, °K	2,270	3,390	2,740	—	—
Characteristic temperature for dissociation, Θ_d, °K	59,500	113,000	75,500	—	—
Characteristic temperature for single ionization, Θ_i, °K	142,000	181,000	108,000	158,000	169,000

Subject Index

Symbol Index

Mathematical notation is an obvious problem in a book that combines a number of often separate disciplines. We have attempted to follow a reasonably consistent notation throughout, while still retaining symbols that are widely accepted as standard. The result is a compromise and contains a number of inconsistencies that are logical only from the point of view of general usage. Many symbols also appear with different meanings in different places, but the differences in both meaning and position are sufficient that this should cause no problem.

As usual, extensive properties of a system are denoted by capital letters. The corresponding intensive (or specific) properties measured per unit mass are then denoted by the corresponding lower-case letter. Quantities measured per mole are indicated by the addition of a caret (ˆ) over the symbol and quantities measured per molecule by the addition of a tilde (˜). These rules have been violated in specific instances, however, where following them would have led to inconsistencies with accepted practice.

Because of its utility in nonequilibrium kinetic theory, we have throughout adopted the Cartesian subscript notation for vectors and tensors. A quantity such as u_i thus denotes a vector with Cartesian components u_1, u_2, u_3 and magnitude $u = (u_1^2 + u_2^2 + u_3^2)^{1/2}$. The repeated-subscript notation for summation (e.g., $u_ju_j = u_1^2 + u_2^2 + u_3^2$) is explained where first used on page 246. The meaning of tensor quantities such as τ_{ij} is indicated on page 326.

An index of the main symbols and the pages on which they are introduced follows. Page numbers referring to the same meaning for a given symbol are separated by commas; numbers referring to different meanings are separated by semicolons. Symbols that appear only briefly in a single section are not included.

SUBSCRIPTS

SUPERSCRIPTS

OVERLINES